Nanotechnology in Electronics

Nanotechnology in Electronics

Materials, Properties, Devices

Edited by Visakh P.M., Artem Semkin, Raneesh Balakrishnan, and Saša Lazović

WILEY-VCH

Editors

Dr. Visakh P.M.
Department of Physical Electronics
TUSUR University
Lenin Avenue 30
634050 Tomsk
Russia

Dr. Artem Semkin
TUSUR University
Department of Microwave and Quantum
Radio Engineering, TUSUR University
Lenin Avenue 30
634050 Tomsk
Russia

Dr. Raneesh Balakrishnan
Department of Physics
Catholicate College
Pathanamthitta
689645, Kerala
India

Dr. Saša Lazović
Biomimetics Laboratory
Innovation Center, Institute of Physics
Belgrade
Pregrevica 118
11080 Belgrade
Serbia

Cover Image: © MirageC/Getty Images

Library of Congress Card No.:
applied for

British Library Cataloguing-in-Publication Data
A catalogue record for this book is available from the British Library.

Bibliographic information published by the Deutsche Nationalbibliothek
The Deutsche Nationalbibliothek lists this publication in the Deutsche Nationalbibliografie; detailed bibliographic data are available on the Internet at <http://dnb.d-nb.de>.

Print ISBN: 978-3-527-34673-8
ePDF ISBN: 978-3-527-82421-2
ePub ISBN: 978-3-527-82423-6
oBook ISBN: 978-3-527-82422-9

Typesetting Straive, Chennai, India
Printing and Binding CPI Group (UK) Ltd, Croydon, CR0 4YY

C096947_260922

Contents

Preface *xiii*

1 Nanotechnology in Electronics, Materials Properties, and Devices: State of the Art and Future Challenges *1*
P. M. Visakh and Raneesh Balakrishnan
1.1 Graphene-based Nanoelectronic Biosensors *1*
1.2 Zinc Oxide Piezoelectric Nanogenerators for Low-frequency Applications *2*
1.3 Investigation of the Hot Carrier–induced Degradation in Nanoscale Junctionless MOSFETs: A Reliability-based Analysis *3*
1.4 Study of Electrostatic and Dispersion Forces in Nanoelectromechanical Systems (NEMS) *4*
1.5 Nanomaterials for Wearable, Flexible, and Stretchable Strain/Pressure Sensors *5*
1.6 Conductive Nanomaterials for Printed and Flexible Electronics Application *6*
1.7 Metal-oxide Semiconductors for Noninvasive Diagnosis of Breast Cancer *8*
1.8 Down-conversion Photoluminescence Properties of ZrO_2: Ln^{3+} (Ln = Eu, Sm, Er, Tb, Ho, Tm, Pr, Gd, Dy) Films Formed by Plasma Electrolytic Oxidation *10*
1.9 Multiferroics for Spintronic Applications *10*
1.10 Quartz Tuning Fork Based Nanosensors *12*
References *13*

2 Graphene-based Nanoelectronic Biosensors *25*
Muhammad Aamir, Ahmed Shuja, Saba Ashraf, and Javeed Akhtar
2.1 Introduction on Graphene *25*
2.2 History of Graphene *26*
2.3 Properties of Graphene *27*
2.4 Fundamentals of G-Derivatives *27*
2.5 Synthesis of Graphene *28*
2.5.1 Graphene-based Nanoelectronics *30*

2.5.2 Graphene-based Biosensors *32*
2.5.3 Graphene in Electrochemical Biosensing Platforms *33*
2.5.4 Field-effect Transistors *34*
2.5.5 Optical Platform for Biosensing *34*
2.6 Applications of Graphene-based Biosensors *35*
2.6.1 Graphene-based Electrochemical Biosensors *35*
2.6.2 Detection of Acute Myocardial Infarction *36*
2.6.3 Detection of Lung Cancer *37*
2.6.4 Detection of Asthma *37*
2.6.5 Detection of Diabetes *38*
2.6.5.1 Electrochemical Enzymatic Glucose Biosensor *38*
2.6.5.2 Nonenzymatic Glucose Biosensors *40*
2.6.6 Detection of Cholesterol *41*
2.6.6.1 Enzymatic Detection of Cholesterol *41*
2.6.6.2 Enzyme-free Biocatalytic Oxidation of Cholesterol *41*
2.6.7 Detection of Hydrogen Peroxide *42*
2.6.8 Hydrogen Peroxide Detection in Living Cells *43*
2.6.9 Nucleic Acid Biosensors *43*
2.6.10 Detection of Enzymes *44*
2.6.11 Food Toxin Sensing *45*
2.6.12 Heavy Metal Detection *45*
2.6.13 Detection of Pesticides *45*
2.6.14 Graphene-based Fluorescent Biosensors *46*
2.6.15 Detection of Small Molecules *46*
2.6.16 Detection of Nucleic Acids *47*
2.6.17 Detection of Pathogens and Food Toxins *47*
2.6.18 Detection of Toxic Heavy Metal Ions *47*
References *48*

3 Zinc Oxide Piezoelectric Nanogenerators for Low-frequency Applications *63*
Zohreh Moarref, Negar Honaramiz Fahim, and Majid Montazer
3.1 Introduction of Zinc Oxide *63*
3.1.1 Structure of ZnO NPs *63*
3.1.2 Crystal Structure of ZnO NPs *63*
3.1.3 Methods for Synthesis of ZnO NPs *66*
3.1.3.1 Mechanochemical Methods *66*
3.1.3.2 Sol–gel Synthesis *66*
3.1.3.3 Hydrothermal Method *67*
3.1.3.4 Liquid-phase Synthesis *67*
3.1.3.5 Controlled Precipitation *67*
3.1.3.6 Vapor Transport Synthesis *68*
3.1.4 Piezoelectric Effect of ZnO *69*
3.2 Zinc Oxide Piezoelectric Nanogenerators *70*
3.2.1 Nanogenerators *71*

3.2.1.1 Piezoelectric Nanogenerators (PENGs) *71*
3.2.1.2 Triboelectric Nanogenerators (TENGs) *73*
3.2.1.3 Pyroelectric Nanogenerators *74*
3.2.1.4 Hybrid Nanogenerators *75*
3.2.2 Zinc Oxide Piezoelectric Nanogenerators *76*
3.3 Zinc Oxide Piezoelectric Nanogenerators for Low-frequency Applications *79*
3.3.1 Approaches for Scavenging Low-frequency Vibrations *79*
3.3.2 Device Structures *79*
3.3.2.1 Arc Shaped *79*
3.3.2.2 Cantilever *80*
Conclusion *85*
References *86*

4 Investigation of Hot Carrier–induced Degradation in Nanoscale Junctionless MOSFETs: A Reliability-based Analysis *97*
Nidhal Abdelmalek, Fayçal Djeffal, and Toufik Bentrcia
4.1 Introduction *97*
4.2 Overview of the Junctionless Paradigm *98*
4.3 Simulation Framework of Hot Carrier Degradation *102*
4.4 Creation of Interface Traps *104*
4.5 Performance Degradation Due to Hot Carrier Effect *108*
4.6 Hot Carrier Degradation in Digital Applications *118*
4.6.1 Static Analysis *118*
4.6.2 Transient Analysis *119*
4.7 Concluding Remarks *122*
References *122*

5 Study of Electrostatic and Dispersion Forces in Nanoelectromechanical Systems (NEMS) *127*
Masoud SoltanRezaee and Amin Farrokhabadi
5.1 Introduction *127*
5.2 Electrostatic Forces *131*
5.2.1 Rectangular Beam–Plate *131*
5.2.2 Wire–Plate *131*
5.2.3 Carbon Nanotube (CNT) Sheets *132*
5.2.4 Rectangular Tweezers *132*
5.2.5 Carbon Nanotube (CNT) Tweezers *134*
5.3 Fringing Field Effects *135*
5.3.1 Palmer's Model *136*
5.3.2 Mejis–Fokkema Model *136*
5.3.3 Other Models *136*
5.4 Van der Waals Force *137*
5.5 Rectangular Beam–Sheets *138*

5.5.1 Wire–Plate *138*
5.5.2 Carbon Nanotube (CNT) Sheets *138*
5.5.3 Rectangular Tweezers *139*
5.5.4 Circular Tweezers *140*
5.5.5 Carbon Nanotube (CNT) Tweezers *140*
5.6 Casimir Force *141*
5.6.1 Rectangular Beam–Plate *142*
5.6.2 Wire–Plate *142*
5.6.3 Carbon Nanotube (CNT) Sheets *143*
5.6.4 Rectangular Tweezers *143*
5.6.5 Circular Tweezers *143*
5.6.6 Carbon Nanotube (CNT) Tweezers *144*
5.7 Other Theories Related to the Casimir Force *144*
5.7.1 Proximity Force Approximation (PFA) *144*
5.7.2 Dirichlet and Neumann Modes *148*
5.8 Freestanding Phenomenon *149*
5.8.1 Detachment Length *149*
5.8.2 Surface Layer and Size-dependent Effects *150*
5.9 Summary *151*
References *151*

6 Nanomaterials for Wearable, Flexible, and Stretchable Strain/Pressure Sensors *155*
Ali Sedighi and Majid Montazer
6.1 Introduction *155*
6.2 Wearable Strain/Pressure Sensors *156*
6.2.1 Piezoresistive Sensors *157*
6.2.2 Capacitive Sensor *167*
6.2.3 Piezoelectric Sensors *171*
6.2.4 Triboelectric Sensors *173*
6.3 Applications *174*
6.3.1 Movement Monitoring and Daily Performance Tracking *174*
6.3.2 Health Monitoring *176*
6.3.3 Human–machine Interface, Soft Robotics, and Artificial Skin *178*
6.4 Conclusion and Outlook *184*
References *185*

7 Conductive Nanomaterials for Printed and Flexible Electronics Application *207*
K. Thiyagarajan, S.G. Rahul, G.K. Rajini, Debashis Maji, and Arunkumar Chandrasekhar
7.1 Introduction *207*
7.2 Printing Technology and Challenges with Fabrication of Electronics *208*
7.2.1 Inkjet Printing *209*

7.2.2 Screen Printing *210*
7.2.3 Slot Die Coating *210*
7.2.4 Electrohydrodynamic (EHD) Printing *211*
7.2.5 Gravure Printing *211*
7.2.6 Flexographic Printing *212*
7.2.7 Roll-to-roll (R2R) Printing *212*
7.2.8 Printable Nanomaterials Requirements *212*
7.3 Synthesis and Preparation of Nanomaterial-based Inks *213*
7.3.1 Metallic-based Inks *214*
7.3.1.1 Silver Nanoparticles *215*
7.3.1.2 Silver Nanowires *222*
7.3.2 1D and 2D Material-based Inks *225*
7.3.2.1 CNT Ink *225*
7.3.2.2 Graphene Ink *228*
7.4 Outlooks and Perspectives *235*
References *235*

8 Metal-oxide Semiconductors for Noninvasive Diagnosis of Breast Cancer *241*
Mohan Velumani, Ivneet Banga, Anirban Paul, Asokan Prasanth, Samir Ranjan Meher, Elizabeth Rufus, Sriram Muthukumar, Shalini Prasad, and Zachariah C. Alex
8.1 Introduction *241*
8.2 Sensing Material and Techniques *244*
8.3 Biomarkers for Noninvasive Diagnosis of Breast Cancer *246*
8.3.1 Body Metabolism for VOC Generation *247*
8.3.2 Various Components of Human Breath and Its Related Diseases *249*
8.3.3 Breast Cancer–related VOCs *250*
8.4 Sensing Elements *251*
8.4.1 Various Materials for VOC Sensing *251*
8.4.2 Metal Oxides for VOC Sensing *252*
8.4.3 Significance of Composite Metal Oxides *252*
8.5 Fabrication Methods *252*
8.5.1 Thin Film Deposition *252*
8.5.1.1 Vacuum-based Techniques *254*
8.5.1.2 Chemical Routes *254*
8.5.2 Thick Film Deposition *255*
8.5.3 Growth of Nanomaterials *255*
8.5.3.1 Physical Methods *256*
8.5.3.2 Chemical Methods *257*
8.6 Noninvasive Techniques for Breast Cancer Diagnosis *257*
8.6.1 Selected Ion Flow Tube Mass Spectrometry (SIFT-MS) *257*
8.6.2 Proton Transfer Reaction Mass Spectrometry (PTR-MS) *258*
8.6.3 Gas Chromatography–Mass Spectrometry (GC-MS) *258*
8.6.4 Differential Mobility Spectrometer (DMS) *261*

8.6.5 Chemiresistive Sensing Mechanism *261*
8.6.6 Fiber-optic Sensors *262*
8.6.6.1 Evanescent Wave Fiber-optic VOC Sensors *263*
8.6.6.2 Lossy Mode Resonance Fiber-optic VOC Sensors *263*
8.6.7 Surface Plasmon Resonance Sensor *264*
8.6.8 Calorimetric Sensors *265*
8.7 Conclusion *266*
Acknowledgements *266*
References *267*

9 Down-conversion Photoluminescence Properties of $ZrO_2 : Ln^{3+}$ (Ln = Eu, Sm, Er, Tb, Ho, Tm, Pr, Gd, Dy) Films Formed by Plasma Electrolytic Oxidation *279*
Aleksandar Ćirić and Stevan Stojadinović
9.1 Introduction *279*
9.2 Experiment *280*
9.3 Results and Discussion *281*
9.3.1 Morphology, Chemical, and Phase Composition *281*
9.3.2 PL of ZrO_2 Films *283*
9.3.2.1 PL of Undoped ZrO_2 *283*
9.3.2.2 PL of $ZrO_2 : Eu^{3+}$ *283*
9.3.2.3 PL of $ZrO_2 : Sm^{3+}$ *286*
9.3.2.4 PL of $ZrO_2 : Er^{3+}$ *286*
9.3.2.5 PL of $ZrO_2 : Tb^{3+}$ *287*
9.3.2.6 PL of $ZrO_2 : Ho^{3+}$ *288*
9.3.2.7 PL of $ZrO_2 : Tm^{3+}$ *288*
9.3.2.8 PL of $ZrO_2 : Pr^{3+}$ *290*
9.3.2.9 PL of $ZrO_2 : Gd^{3+}$ *292*
9.3.2.10 PL of $ZrO_2 : Dy^{3+}$ *292*
9.4 CIE Chromaticity of $ZrO_2 : Ln^{3+}$ *295*
9.5 Conclusion *295*
Acknowledgments *296*
References *296*

10 Multiferroics for Spintronic Applications *301*
Jini K. Jose and Raneesh Balakrishnan
10.1 Magnetoelectric Multiferroic Materials *301*
10.2 Spintronics *302*
10.2.1 Fundamental Aspects of Spintronics *302*
10.2.2 Giant Magnetoresistance *303*
10.2.3 Tunneling Magnetoresistance *304*
10.3 Spintronic Devices *305*
10.3.1 Spin Valve *306*
10.3.2 Multiferroic Tunnel Junctions *307*
10.3.3 Spin FET *309*

10.3.4 Spin LED *311*
10.4 Summary *311*
References *312*

11 Quartz Tuning Fork–Based Nanosensors *317*
S. Abraham Sampson and Suwarna S. Datar
11.1 Introduction *317*
11.2 Chemical Sensors *318*
11.3 Quartz Tuning Forks (QTFs) *319*
11.4 Early QTF Development *324*
11.5 QTF as a Sensor *325*
11.5.1 Mass-loaded QTFs as Sensors *328*
11.5.2 Polymer Wire/Film Modified QTF Sensor *331*
11.5.2.1 QTFs Modified with Polymer Wire Bridges *331*
11.5.2.2 QTFs Modified with Polymer Film Bridges *338*
11.5.2.3 Improving Selectivity of Polymer-modified QTF Sensors and Classification of VOCs *342*
11.6 Conclusions *347*
References *351*

Index *357*

Preface

Nanotechnology is one of the most talked about technologies today. The application of nanotechnology in modern electronic devices, especially memory devices, nanoscale transistors, nanosensors, etc., improves the devices' functionalities. Not only is the power consumption in nanoelectronic devices reduced drastically but also the weight and thickness of the devices are decreased. This book summarizes recent research accomplishments in the area of nanotechnology-based electronics, its properties, and applications.

A brief description of the book content and an introduction to nanotechnology in electronics and current challenges are given in the first chapter. The second chapter offers a review of graphene-based nanoelectronic biosensors. This chapter covers the introduction to graphene, its properties, and some novel potential applications as biosensors. Zinc oxide piezoelectric nanogenerators for low-frequency applications are described in Chapter 3. Different types of ZnO nanostructure-based nanogenerators are discussed. Chapter 4 deals with the investigation of hot carrier-induced degradation in nanoscale junctionless MOSFETs. The fifth chapter discusses the electrostatic and dispersion forces in nano-electromechanical systems. In addition, the effects of different electric fringing field corrections and other effects related to the Casimir force are separately reported. The sixth chapter deals with nanomaterials that can be used as wearable, flexible, and stretchable strain/pressure sensors. A detailed review of conductive nanomaterials for printed and flexible electronics applications is covered in the seventh chapter, with the fundamentals, features, and progress in various printing technologies.

The application of different metal oxide semiconductors for the non-invasive diagnosis of breast cancer is discussed in Chapter 8. This review outlines breast cancer-related volatile organic compounds and the detection mechanisms currently used for diagnosis. The ninth chapter presents the down-conversion photoluminescence properties of $ZrO_2:Ln^{3+}$ (Ln = Eu, Sm, Er, Tb, Ho, Tm, Pr, Gd, Dy) films formed by plasma-electrolytic oxidation. Chapter 10 provides an overview of multiferroics for spintronics applications. This chapter reviews the fundamental aspects and applications of multiferroic-based spintronic devices. The eleventh chapter covers many of the latest research accomplishments in quartz tuning fork-based nanosensors.

In this book, chapter authors have reviewed different aspects of nanotechnology in the field of electronics and related advancements. This book is a valuable reference source for faculties, professionals, research fellows, senior graduate students, and researchers working in the field of nanotechnology/electronics. The use of nanomaterial in electronics is considered as a promising area of research, a lot of research activities are going on, and some knowledge of nanoelectronics is helpful to bring a change in the current techniques and applications. Finally, we would like to express our sincere gratitude to all the contributors of this book, who made excellent support to the successful completion of this venture. We also thank the publisher Wiley for recognizing the demand and importance of nanotechnology in the field of electronics.

Visakh P.M.
Artem Semkin
Raneesh Balakrishnan
Saša Lazović

1

Nanotechnology in Electronics, Materials Properties, and Devices: State of the Art and Future Challenges

P. M. Visakh[1] and Raneesh Balakrishnan[2]

[1] *Faculty of Electronic Engineering, Department of Physical Electronics, TUSUR University, Tomsk 634050, Russia*
[2] *Department of Physics, Catholicate College, Pathanamthitta, Kerala, 689645, India*

1.1 Graphene-based Nanoelectronic Biosensors

Graphene is the thinnest, strongest material, with one-particle thickness, and is 200 times stronger than steel. Graphene exhibits outstanding electrical and thermal conductivity with remarkable light absorption capability. Graphene is a material that could change the world in the true sense, with its infinite potential for integration practically in any industry. Graphene has diverse forms and can be combined with other materials to form various materials of extraordinary characteristics. Researchers are very keen to explore and patent graphene and gain expertise about its various properties and potential applications, mainly in cells, transistors, computer chips, supercapacitors, energy production, etc.

The common graphene derivatives are pristine graphene, reduced graphene oxide, graphene quantum dots (GQDs), and polycrystalline graphene, which are extensively used in biosensing applications. Pristine graphene consists of single crystalline grains that have flawless lattices made up of defect-free hexagonal ring networks of sp^2 hybridized carbons atoms [1]. Graphene oxide also mainly consists of covalently bonded sp^2-hybridized carbon atoms but has some defects due to the following reasons: disrupted sp^3-hybridized carbon atoms and the oxygen functional groups like carboxyl group located at the edges and the basal plane are saturated with hydroxyl and epoxy groups [2]. The presence of these groups makes graphene oxide hydrophilic depending on their oxidation state [3].

Nanoelectronic devices are made up of nanomaterials such as nanowires, graphene, carbon nanotubes (CNTs), and transition metal dichalcogenides. These nanomaterials must possess extraordinary characteristics such as high surface-area-to-volume ratio, low power consumption, high charge mobility, and excellent compatibility to fit in modern electronic devices. As nanomaterials display unusual properties that are not obtainable at the micro scale, nanoelectronics holds

Nanotechnology in Electronics: Materials, Properties, Devices, First Edition.
Edited by Visakh P.M., Artem Semkin, Raneesh Balakrishnan, and Saša Lazović.

significant importance. Graphene is a recently discovered nanomaterial known for its outstanding mechanical, electrical, and optical properties [4–6]. To develop an efficient graphene-based electrochemical biosensor, the graphene oxide nanosheets were decorated with Au NPs by aryldiazonium bridges fabricated on glassy carbon electrode (GCE). Then, the glucose oxidase enzyme was immobilized on the graphene oxide–Au NP nanocomposites. The biosensor showed a high sensitivity of about 42 μA mM^{-1} cm^{-2} with a wide detection range of 0.3–20 mM. This glucose biosensor showed excellent selectivity for glucose detection [7].

N-doped graphene–chitosan–glucose oxidase biosensors showed an excellent sensitivity of 226.24 μA mM^{-1} M^{-2} 120 due to high porosity of graphene and high conductivity of graphene sheets, which facilitate enhanced electron transfer. The CuO nanoparticle (NP) composites with graphene oxide have shown a high sensitivity of 262.52 μA mM^{-1} cm^{-2} with an LOD of 0.69 μM. This biosensor was also used for human serum detection and showed a sensitivity of 285.38 μA mM^{-1} cm^{-2} [8]. Similarly, biosensors based on Pt-NiO–reduced graphene oxide nanocomposites showed a sensitivity of 832.95 μA mM^{-1} cm^{-2} and an LOD of 2.67 μM. This high performance was attributed to porosity of Pt-NiO–reduced graphene oxide nanocomposites [9]. Moreover, the Cu-Co–reduced graphene oxide nanocomposites were electrochemically deposited over the pencil graphite electrode and showed a high sensitivity of 240 μA mM^{-1} cm^{-2} and an LOD of 0.15 μM [10].

1.2 Zinc Oxide Piezoelectric Nanogenerators for Low-frequency Applications

Nanoparticles can be fabricated through solid-phase, liquid-phase, and gas-phase methods. Chemical bath deposition [11], green chemistry synthesis [12], and wet chemical methods [13] are other recently used techniques for the synthesis of ZnO NPs. Vertical ZnO nanorods were grown hydrothermally on a gold-coated PTFE (polytetrafluoroethylene). Spin coating was used to deposit 0.005 M zinc acetate dehydrated in ethanol on the surface of gold-coated PTFE film. Then, the substrate was heated at 90 °C for 30 minutes. The seed layer was uniformly deposited in a two-step process of deposition and decomposition to cover the substrate. ZnO nanorods were hydrothermally grown on the seed layer. Zinc nitrate hexahydrate and hexamethylenetetramine (HMT) were used to hold the seed layer at 90 °C for three hours [14].

Masuda et al. [15] demonstrated a simple liquid-phase synthesis method to control the morphology and crystal size of ZnO. ZnO nanowires of 100 nm long and about 50 nm wide were successfully synthesized at 50 °C. They also found that nanowires had no branches without aggregations. Several techniques have been explored to synthesize and develop both direct and alternating current zinc oxide piezoelectric nanogenerators [16]. Controlled precipitation is typically used to produce ZnO NPs in large quantities with reproducible properties for use in industrial products. In this process, zinc salt solution is spontaneously reduced by a reducing agent.

Breaking down of agglomerations is an important process in this method. Therefore, temperature, pH, type of aqueous media, time of precipitation, and the raw materials are most important parameters in the precipitation of ZnO NPs, which have impact on the size and final characteristics of the nanoparticles [17]. Aladpoosh et al. [18] used the ZnO-controlled precipitation method, a green synthesis method using natural plant source, namely, Keliab and zinc acetate. Wurtzite structure and nanorod shapes result from in situ synthesis of ZnO on the cellulosic chains of cotton using the green synthesis method. The principle of nanogenerators was first used by Wang's research group in 2006 [19]. They used a conductive tip of an atomic force microscope (AFM) to deflect the aligned semiconductor ZnO nanowires and successfully obtained a measurable piezoelectric voltage output of around 10 mV. They grew ZnO nanorods on a sapphire substrate using a vapor–liquid solid method, which caused the gold particles to remain on the top of each rod after their growth.

A Schottky contact was formed between the conductive AFM tip and the semiconductor ZnO nanowires, which is a key factor in current generation by piezoelectric nanogenerators. The stretched side of the rod with a positive potential produced a reverse bias with the Schottky junction, leading to no current to screen the polarization. When the contact between the tip and the compressed, negatively polarized side of the rod was disconnected, the junction was forward biased, and the current was able to flow and screen the polarization. Monitoring the open-circuit voltage reflects an uncertainty in the measurement of this type of device [20]. Zinc oxide is an environmentally friendly multifunctional semiconductor possessing outstanding piezoelectric effect, piezoelectric coupling coefficient, and thermal and mechanical stability at room temperature. Owing to its non-centrosymmetric hexagonal wurtzite crystal structure and polar Zn—O bonds, ZnO has unique piezoelectric properties as the core of studies in the ceramic industry [21, 22]. For the first time in 2009, Xi and coworkers [23] fabricated a ZnO-nanotube-based piezoelectric nanogenerator. They used a solution chemical method to synthesize hexagonal ZnO nanotube arrays at temperatures below 100 °C.

ZnO-based nanogenerators can be incorporated into various substrates. In 2012, Khan et al. [24] grew ZnO nanorods on a silver-coated cotton fabric using a low-temperature, aqueous chemical growth method. An external force was applied on ZnO nanorods via an AFM tip, where the flow of charges was taken into account and a Schottky barrier formed between the electrode and the nanorods.

1.3 Investigation of the Hot Carrier–induced Degradation in Nanoscale Junctionless MOSFETs: A Reliability-based Analysis

In the past decades, reliability of MOSFET devices was a major concern in the electronics industry: the aging phenomenon degraded the device operation and lifetime severely. Extensive research has been carried out to identify the causes of aging and to seek viable solutions. Channel and substrate hot carrier injection

(HCI and SCI), channel hot electron/hole (CHE/H), drain avalanche injection, and radiation were found to be the main degradation mechanisms[25–31]. Hot carrier–induced degradation is of prime interest for designers as the process aggravates with downscaling. Degradation is essentially due to the enhancement of the electric field with dimension reduction, in the lucky electron model representation, providing the carriers enough energy to be injected through the oxide layer. These carriers are most likely to be captured by the existing oxide traps or to break the silicon bounds and thus giving rise to a localized and nonuniform pile-up of interface states near the drain–channel junction [32]. A energy-driven paradigm proposed by Rauch et al. suggests that the electron–electron scattering mechanism, governed by energy rather than the electric field, is responsible for aggravation of hot carrier degradation in downscaled devices [33].

The concept of junctionless transistor was proposed by Lee et al. [34] as a simple way to overcome process limitations. Indeed, as downscaling of channel lengths trends toward less than 10 nm, controlling process-induced variability becomes imperative to conserve the properties of transistors. At these scales, inherent scattering and diffusion of shallow-extension impurities into ultra-short channels pose a serious problem. As high doping concentration gradient is primordial, the extension formation with the actual process techniques implies additional complex and costly operations [35]. Contrary to IM MOSFETs, measures highlighted an enhancement in the effective mobility with the increasing gate bias [36]. The enhancement is so significant that the effective mobility exceeds the corresponding bulk mobility especially in the accumulation regime [37]. Surface mobility also significantly improved in the accumulation layer of heavily doped devices, which indicates a higher current in the accumulation regime of the junctionless field-effect transistor (JLFET). This effect is attributed to the impact of the accumulation layer on screening of ionized impurities, yielding a reduction of Coulomb scattering [38]. In heavily doped devices, measurements highlight a correlation between the carrier density and the Coulomb scattering limitation, where a significant improvement in the effective mobility with respect to the carrier density is observed [39, 40]. Also, it was demonstrated that enhanced electrostatic controllability provided by multigates and thinner equivalent oxide thickness (EOT) yielded higher carrier density and improved mobility.

1.4 Study of Electrostatic and Dispersion Forces in Nanoelectromechanical Systems (NEMS)

Due to the outstanding electrical and mechanical properties of nanostructures, researchers are interested in novel generations of ultrasmall devices such as micro- and nanoelectromechanical systems (MEMS and NEMS, respectively), which are used in switches, actuators, sensors, AFMs, mass and force detectors, etc. New effects appearing due to the scale alteration in geometry introduce several additional terms into the theory of nanoscale structures, while their influence on other scales is negligible. One of these effects is dispersion forces. Dispersion forces or quantum interactions, including the Casimir and van der Waals (vdW) regimes, play a key role in the behavior of nanodevices [41–43], related to the structure dimensions/configurations and the gap between electrically actuated electrodes [44]. The total electromagnetic Casimir energy is the sum of energies of the Dirichlet and Neumann modes. However, in large separations, the Dirichlet mode is dominant, while the Neumann mode is negligible [45]. It should be noted that the large separation assumption is valid for most real nano-tweezers since the gap between the arms is relatively larger than the radius of the arms ($D \gg R$). Typically, for ($D < 100R$), the contribution of the Neumann mode is negligible compared to the Dirichlet mode, and the asymptotic approximation provides an excellent estimation of the Casimir force [46]. In this section, practical mathematical models by assuming the large separation approximation (Dirichlet mode) are developed.

1.5 Nanomaterials for Wearable, Flexible, and Stretchable Strain/Pressure Sensors

Hardness and rigidity of micro and smart electronics cause mechanical deformation because of brittleness and inflexibility of their material components. This has restricted the application of non-flexible electronics that were replaced with electronic devices containing polymer and plastic substrates as flexible electronics in the past decades. In electronic devices, flexibility implies lightweight, elasticity, non-brittle, ductility, and easy fabrication [47]. Flexible electronics aroused great interest among researchers, industries, organizations, and military sectors to design and manufacture hybrid composite structures with lightweight flexible electronic devices [48]. On the other hand, the rapid growth of technologies, especially Internet in our daily life, has resulted in the advent of wearable technology in recent years [49]. Wearable and flexible sensors and biosensors have been widely explored in research and industry [50–52]. Sensors convert one form of energy into another form for processing [53]. Sensors detect environmental changes using conductive materials and send data to the computer for processing [54]. Conventional detection methods have been replaced by wearable and stretchable sensing electronics with nanomaterials [55]. Structures and designs of wearable sensors as well as nanomaterials determine sensor applications and their performance. Nanomaterials and nanocomposites depending on their properties, shape, size, structure, porosity, and fabrication method impart different sensor properties to wearable substrates. Strain/pressure sensors measure deformations, pressure stimuli, and mechanical properties [56, 57]. Metallic and semiconducting films were conventionally used as strain sensors and exhibited brittleness and rigidity with low stretchability [58]. These bulk materials could not attach to wearable substrates.

Nanomaterials including metal and metal-oxide nanoparticles and nanowires, conducting polymers (CPs), graphene, and CNTs have been explored due to their larger surface area compared to bulk materials. Superior mechanical and electrical properties and use of lower active materials are some of the advantages of the integration of nanostructures and hybrid nanomaterials with flexible supports. The supports can also maintain their stretchability, wearability, and flexibility [59–64]. Flexible wearable piezoresistive strain/pressure sensors are composed of conductive sensing elements in addition to elastic polymers or other flexible/stretchable substrates. The contact resistance of the conductive part changes with strain/pressure stimuli, detected by electrical measuring systems. Different conductive nanomaterials have been explored as conductive fillers of elastomer composites including metal nanoparticles, nanowires and thin films, CPs, and carbon nanomaterials due to their desirable electrical and mechanical properties [65–71]. PDMS, polyurethane (PU), silicon rubber, Ecoflex, and Lycra are also explored as elastic substrates or components for conductive nanomaterials [72, 73]. On the other hand, incorporation of metallic nanomaterials, CNTs, and graphene with flexible wearable substrates propose desirable gauge factor ranges, introducing different structures and resulting in strains inside the matrix during stretching [74, 75]. Disconnection between nanomaterials decrease the contact area between them, leading to higher electrical resistance.

Graphene is another impressive carbon-based nanomaterial for resistance-based strain sensors due to its high electrical conductivity and mechanical properties [76–78]. Applied strain can change and partially destroy the hexagonal structure of the edges of graphene films [79]. Graphene can be combined with elastic polymers, which are generally grown through chemical vapor deposition or derived from exfoliated graphite. The shape and structure of graphene determine the flexibility and stretchability of sensors. To increase the stretchability and sensitivity of graphene in strain/pressure sensors, certain structures and engineering strategies were designed [80, 81]. In another work, macroscopic, flat, uniform graphene ribbons were fabricated through a modified wet spinning method [82]. Carbon black (CB) and graphite were also integrated with textiles to show piezoresistivity [83–85]. Natural biomaterials including natural textiles such as silk and cotton can be converted into conductive carbon materials via hydrothermal and pyrolysis processes [86, 87]. Carbonized silk nanofiber membranes integrated with PDMS film were also reported as skin-like pressure sensors.

1.6 Conductive Nanomaterials for Printed and Flexible Electronics Application

Printing is a process whereby formulated inks are deposited on the top of the substrate material for data transfer just as graphics, images, text, and the forms reappear as an image-carrying medium called substrate. Accordingly, printed and flexible electronics follow a set of printing methodologies to build electronic devices on flexible substrates. In the bubble-jet printing model, there is a print head with a heating unit. This heating unit is responsible for the preparation of ink bubbles and also forces the ink drop out of the nozzle. In the piezoelectric printing model, the nozzle has a ceramic piezoelectric tile that gets expanded into the ink reservoir inside the print head and forces the ink drop out of the nozzle. The technology that enables to print inks on a variety of substrates with flexible or rigid structure materials is called screen printing. The entire process is relatively simple, flexible, and economical. The ink passes through a stencil made of a porous material during printing and gets deposited onto the substrate material. A delicate mesh made of nylon is used as a whole screen and stretched over a wooden or aluminum frame with a fitness of 150–400 fiber in.$^{-1}$

A high-resolution printing technique that uses an electric field to produce an essential fluid flow for ink delivery onto a substrate is called an electrohydrodynamic printing. It produces two to five smaller droplets compared to the nozzle size. The positive impact on the shape regulation of nanoparticles in the droplets developed for printing is due to the electric field existing between the substrate and the nozzle. It is also preferred in 3D structure additive manufacturing. Magdassi et al. prepared concentrated citrate-reduced silver nanocolloids as ink for inkjet printing on surfaces such as glass and paper. The material was prepared by redispersion following lyophilization of newly prepared colloids. Carboxymethyl cellulose, a polymer stabilizer, was used as a binder and stabilizer. The material was characterized by its high electrical conductivity, reactivity, and optical properties [88]. Shankar et al. [89] synthesized Ag NPs by a reduction method using tin acetate as a reducing agent with a passivating shell of dodecylamine (DDA). This process was carried out in the toluene medium. The aim was to formulate nonaqueous conductive ink to print circuits for antennas and radio frequency identification (RFID) tags on surfaces such as Epson photo paper through aerosol jet printing. Lai et al. presented a review article on synthesis of silver NP and inks. Also, control of chemical concentration is significant to maintain the size and morphology of nanoparticles. The most suggested method to synthesize nanoparticles is chemical reduction, which requires a reducing agent to decrease the ion solution to the physical nanoparticle form. The authors concluded that most printing metal inks in the market are silver based owing to its remarkable electrical properties [90]. Zhou et al. performed a study on DDA-protected Ag NPs by modifying a small amount of dodecanethiol as the co-protective agent to formulate conductive silver ink for inkjet printing. The study revealed that electrical properties of silver can be enhanced quickly by heat treating at 90 °C, thus proving its suitability to print electronic circuits [91]. A similar team also synthesized water-dispersible Ag NPs using polyacrylic acid as a capping agent to inkjet print on paper substrates. Following the treatment of the Ag NPs with the poly(diallyldimethylammonium chloride) [PDAC] solution at room temperature, the nanoparticles coalesced together after the detachment of polyacrylic acid (PAA). The films showed changes in the morphology and electrical properties. Which is significant identification for printed electronic devices [92]. Zhang et al. synthesized Ag NPs for paper-based electronics through inkjet printing. Conductivity of the printed silver tracks was enhanced by a multiple-step room temperature (RT) post-processing method. Electrical properties, mechanical flexibility, and strain sensitivity of the material are proven effective to serve the 3D electronics in future [93]. Kastner et al. and Shahariar developed Ag NPs for inkjet printing of conductive lines on textiles. The reactive silver ink contained silver in the complexed form, and no dispersions were found. The inkjet printer electronic circuit paths on the fabric were found to withstand washing tests, moisture, and heat [94, 95]. Liang et al. developed water-based Ag nanowires to fabricate wearable thin-film transistors and stretchable conductors through screen printing on polyethylene terephthalate (PET) sheets. A new water-based Ag nanowire ink was prepared using a mixture of (hydroxypropyl) methylcellulose, fluorosurfactant, distilled water (DI) as a solvent, and an antifoaming agent as an organic binder. Investigations were carried out to connect the rheological properties of the ink with its composition [96].

Patil et al. performed a study on inkjet printing of silver nanowires synthesized by the polyol method. The study focused on conductivity improvement printed pattern stability on flexible surfaces and techniques followed. The process was optimized by varying the concentration of isopropyl alcohol and the Ag nanowire mixture suspended in ethylene glycol, which was synthesized. The printed patterns obtained using the developed ink demonstrated high conductivity and strength [97]. Zhao et al. performed their study on selective dopamine determination in human serum using inkjet-printed multi-walled CNT chips. The CNT ink was examined with homogeneous double layers to increase the efficiency of dopamine detection. This method successfully resulted in the demonstration of the samples. The low-cost approach for the ink formulation paves a way for futuristic applications in printed sensors [98].

1.7 Metal-oxide Semiconductors for Noninvasive Diagnosis of Breast Cancer

In recent years, research in biomedical engineering has enabled clinicians to achieve early medical diagnosis with help of advanced technologies. Although these conventional diagnostic techniques help in diagnosis of cancer, they have certain drawbacks: high cost, painful to patients, and may not always give accurate results, especially in the early stage of the disease.

Recent research has revealed that the overall cost of cancer diagnosis and treatment is indeed decreasing for patients treated in the early stages; hence, the prognostic management of cancer remains a topic of key interest in the scientific community, from fabrication of devices to even machine learning [99]. Over the past few decades, advances in molecular biology and biotechnologies have achieved significant milestones to identify cellular changes at the molecular level, which may be a promising approach for early disease detection. Moreover, present technologies can quantify cellular changes at an early phase of disease development. This may help in identifying particular pathophysiological processes and underlying conditions to build more accurate devices [100–103]. The early diagnosis of cancer is crucial for patient survival; therefore, sensitive and specific methods are required for early cancer diagnosis. These methods aid clinicians in successfully diagnosing and treating patients, thus, increasing their chances of survival [104].

The conventional diagnostic methods for cancer detection involve histological characterization such as staining based on cell morphology and analyzing the slides using microscopic techniques. These are invasive techniques that require human sample collection by taking a biopsy sample followed by sample processing techniques that involve cell fixation and further analysis using a microscope to detect cancer cells [105–107]. Breast cancer is a result of interplay of various factors that lead to genetic mutations. Hereditary and environmental factors play a major role in causing these mutations, which inactivate the DNA repair genes in the breast cells [108]. The main biomarkers associated with targeted therapy of breast cancer are estrogen receptor and progesterone receptor. Growth factors such as epidermal growth factors, mainly ErbB-1 (EGFR) and ErbB-2 (ERBB2, HER-2/neu, HER2), are also regulated in breast cancer tumorigenesis [109, 110]. EGFR overexpression is correlated to overexpression of HER-2, levels of which are used for targeted therapy by using the drug trastuzumab [111]. However, breast cancer type is classified based on gene expression profiling, which requires a large amount of tumor tissue for analysis and is highly expensive and also a difficult procedure. Moreover, gene expression in an individual is subject to changes due to post-transcriptional modifications and may lead to upregulation or downregulation of certain genes [112]. The diagnosis of breast-cell-related diseases requires an extensive biopsy and sampling for confirmation. These invasive diagnostic techniques cause distress until the diagnosis is confirmed through biopsy, and sometimes even after treatment there is a high chance of relapse; therefore, the patient has to get a biopsy done every few months. Some benign tumors, such as premalignant hyperplasia, when left untreated can also lead to invasive breast cancer, which is one of the most common malignancies diagnosed in women.

Breath analysis is widely being explored as a noninvasive diagnostic test for detection of breast cancer [113]. Increased oxidative stress is a key factor regulated in breast cancer, which causes damage to the cellular DNA, lipids, and proteins. Also, cytochrome P450 enzyme, expressed in breast cancer, causes lipid peroxidation of polyunsaturated fatty acids and leads to overexpression of volatile organic compounds such as alkanes and alkane derivatives in breath [114, 115]. Apart from lipid peroxidation, free radical–induced oxidation of certain amino acids, proteins, and peptides has also produced hydrocarbons and monoaldehydes [116, 117]. Recently, implementation of optical fibers in sensor application is rapidly increasing due to their ease of fabrication, low cost, and immunity to electromagnetic interferences. Especially in volatile organic compound (VOC) gas sensor applications, optical fibers are widely used because of their fast response and sensitivity toward gas molecules. Different types of optical fibers have been reported in sensor applications such as plastic optical fibers, silica optical fibers, fiber Bragg grating optical fibers [118], photonic crystal optical fibers [119]. In addition to the geometry and types of optical fibers, various sensing mechanisms have been reported, including absorption-based sensors, evanescent mode [120], plasmonic mode optical sensors such as surface plasmon resonance [121], localized surface plasmon resonance [122], and lossy mode resonance [123].

1.8 Down-conversion Photoluminescence Properties of ZrO_2: Ln^{3+} (Ln = Eu, Sm, Er, Tb, Ho, Tm, Pr, Gd, Dy) Films Formed by Plasma Electrolytic Oxidation

In recent years, plasma electrolytic oxidation (PEO) is a very convenient process for formation of Ln ion–doped metal-oxide semiconductor films such as ZrO_2 [124–132], Al_2O_3 [133–139], TiO_2 [140, 141], Zn [142], Nb_2O_5 [143], HfO_2 [144], Y_2O_3 [145], and Gd_2O_3 [146] by PEO of pure Zr, Al, Ti, Zn, Nb, Hf, Y, and Gd metals, respectively, in an electrolyte containing suitable lanthanide oxide (Ln_2O_3) particles. The formation mechanism of the PEO film is complex, as it involves electro-, thermo-, and plasma-chemical reactions in the electrolyte and metal–electrolyte interface, due to locally high temperature (10^3–10^4 K) and pressure (~10^2 MPa) at the micro-discharge sites [147].

During the PEO process of zirconium in an alkaline electrolyte containing Ln_2O_3 particles, ZrO_2 films are formed by PEO as a result of the oxidation of zirconium substrate via a sequence of localized micro-discharging events. Zr^{4+} ions formed at the zirconium/oxide interface ($Zr \rightarrow Zr^{4+} + 4e^-$) enter the micro-discharge channels and react with O^{2-}/OH^- ions from the electrolyte. Metal oxides are eminent host matrices for Ln^{3+} due to their outstanding physicochemical characteristics such as good thermal, chemical, and physical stability, wide bandgap, broad optical transparency, low phonon frequency. Among them, zirconium oxide (zirconia, ZrO_2) is an excellent host material because its low phonon energy of about 470 cm^{-1} can increase the probability of radiative transition occurrence and reduce non-radiative multiphonon relaxation of excited Ln^{3+} throughout the vibrational bands of the ZrO_2 host lattice, thus foreseeing a high yield of intra-4f transitions from the Ln^{3+} [148–150].

The Ln_2O_3 particles take part in PEO process through the electrophoretic and micro-discharging mechanisms [151]. Ln_2O_3 particles have a negative zeta potential in alkaline media; thus, its particles are negatively charged. Due to the electrophoretic effect, the Ln_2O_3 particles move toward the zirconium anode. Melting point of Ln_2O_3 particles (~2400 °C) is much lower than locally high temperatures (~7000 °C) which induced at the micro-discharging sites [152], allowing the molten Ln_2O_3 particles to react with ZrO_2 and form Ln^{3+} ion–doped ZrO_2 films.

1.9 Multiferroics for Spintronic Applications

Multiferroics are a special class of materials that simultaneously possess two or more primary ferroic order parameters in the same phase such as ferroelectricity, ferromagnetism, ferroelasticity, and ferrotoroidicity [153]. In the last few decades, a significant number of research works have been done on multiferroics. A material that simultaneously possesses multiferroic orders is not sufficient. There should be a strong coupling interaction between different orders to achieve potential applications. Magnetoelectric coupling refers to the linear magnetoelectric effect [154]. One can manipulate electric polarization by a magnetic field and magnetization by an electric field [155], which makes them potential candidates in device fabrication for different applications – memory devices [156–158], spintronics, sensors [159, 160], etc. Spintronics, spin-based electronics, is a multidisciplinary and fast-developing research area. Spintronic research is closely related to physics, chemistry, materials science, electrical engineering, and computer science. In spintronic devices, electron spin is employed in addition to electron charge, whereas in conventional electronic devices electron charge is used to transfer and store data. The properties such as intrinsic non – volatility, smaller size, high speed, ultralow-power consumption make spin-based devices a potential candidate in the microelectronics industry.

Spintronics is the study of the role played by electron spin in solid-state physics and in the fabrication of possible devices that explicitly exploit spin properties in addition to charge degrees of freedom. Numerous experiments have been carried out using polarized neutron reflectometry (PNR), which clearly illustrates Fe/Cr superlattices that lead to the giant magnetoresistance (GMR) effect [161], but PNR has some limitations too [162]. In 1994, Daughton et al. reported the design of a magnetic sensor to detect very weak magnetic fields, which was considered the first GMR-based device [163].

The major achievement of GMR in read heads for magnetic hard-disk drives was made by IBM in 1997 [164]. The innovation of GMR-based devices has dramatically increased the storage capacity of hard disks from 1 to 20 gigabytes. Nowadays all conventional electronic devices are based on the electron flow, which is considered a binary system (on/off state). If electrons flow in a circuit, it is the on state, and if there is no electron flow, it is the off state. Researchers have been focusing on the fabrication of new spintronic devices that offer an increase in the power of device up to 10 000 times compared to conventional electronic devices. The magnetic materials required for the fabrication of spintronic devices should have a stable structure, good electrical properties, and the ability to be applied in thin film form. A novel material $(Fe_{0.66}Dy_{0.24}Tb_{0.1})_3O_{7-x}$ (FDTO) with a semiconducting behavior, suitable electrical conductivity, and a high magnetic moment was reported, all of which can be controlled by altering the oxygen concentration during deposition. Recently, double-perovskite Pr_2FeCrO_6 (PFCO) was reported. Studies revealed that PFCO could be a promising multiferroic material for spintronic applications. The field dependence of magnetoresistance (MR) for PFCO showed MR% of 44% with a clear distinction of resistance states in parallel and antiparallel states between 0 and ±5 KOe observed [165].

A considerable change in MR with field revealed that PFCO was a potential candidate for the magnetic tunnel junction (MTJ) in spintronics using multiferroic materials. Spin valves, fabricated in 1991 [166],led to a drastic change in magnetic sensors [167, 168], read heads of magnetic sensors [169, 170], and magnetic random-access memory (MRAM). The spin valve structure in either current-in-plane (CIP) or current-perpendicular-to-plane (CPP) form can be used to analyze the GMR effect.

1.10 Quartz Tuning Fork Based Nanosensors

Though the commercially available quartz tuning fork (QTF) was first designed as a timing crystal for wrist watches in the early 1970s, it held, and still holds, much promise as a scientific tool. However, it would be more than a decade before the QTF's potential as a sensor was explored.

QTFs are single-crystal quartz mechanical oscillators, which are relatively inexpensive and have been used in wrist watches and other high precision timing circuits for decades. Piezoelectricity of quartz, which enables the conversion of electrical signals into mechanical vibrations and vice versa, allows the QTFs to be excited electrically and their resonant frequencies to be read out electrically. These QTFs can be considered mechanical transducers since changes in the mechanical properties of a system can be monitored by measuring the changes in the resonant frequency of the system, which is measured in the form of an electrical signal. Owing to their high precision resonance frequency, high quality (Q) factor, high mechanical stability, near-zero thermal deflection, and low power consumption, they are excellent candidates for use in timing circuits, probe microscopy techniques, gyroscopes, micro balances, and sensors [171, 172]. The QTF was modified by surface coating the prongs with a thin layer of polystyrene (PS) by dipping the fork in a solution of PS and aniline and allowing aniline to evaporate off the fork. For human IgG detection, the PS-coated QTFs were again dipped in anti-human IgG and allowed to dry. For self-excited QTFs, the prongs were coated with varying concentrations of PS and their responses were measured.

In 2003, S. Boussaad and N. J. Tao published a landmark study "Polymer wire chemical sensor using a microfabricated tuning fork" in *Nano Letters*. This was a milestone in the development of QTFs as a nanosensor since the study explored the possibility of using more than simple mass loading to detect analytes. They successfully formed a bridge by attaching a nitrocellulose/toluene sulfonamide formaldehyde resin wire between the two tines of a QTF.

The QTFs were attached with polyvinylcinnamate (PVCN) wires, and the wires were crosslinked by exposing them to UV radiation for one to three minutes, thereby the wires could shrink and their stiffness increased. A comparison of the frequency responses of QTFs before and after crosslinking the wires showed that crosslinking increased the sensitivity and Q factor of the sensor by amplifying the change in the modulus of the wire due to the presence or absence of ethanol vapors [173].

In another study, published in 2016, the electrospun wires of polymethyl methacrylate (PMMA) onto the QTF and formed polymer bridges. Then, they exposed the sensors to water vapor in order to use them as humidity sensors. The resonance frequency of the sensors reduced with an increase in the humidity level due to adsorption of the water molecules onto the PMMA wires, leading to a decrease in the modulus of the polymer bridges. The amount of decrease, for a given humidity level, was found to be a function of the number of wires present [174].

References

1 Rodriguez-Perez, L., Herranz, M.Á., and Martin, N. (2013). The chemistry of pristine graphene. *Chem. Commun.* 49: 3721–3735.

2 Dreyer, D.R., Park, S., Bielawski, C.W., and Ruoff, R.S. (2010). The chemistry of graphene oxide. *Chem. Soc. Rev.* 39: 228–240.

3 Eigler, S. and Hirsch, A. (2014). Chemistry with graphene and graphene oxide – challenges for synthetic chemists. *Angew. Chem. Int. Ed.* 53: 7720–7738.

4 Bhoria, R.S. (2015). Graphene: potential material for nanoelectronics applications. *Indian J. Pure Appl. Phys.* 53: 501–513.

5 Li, Z., Ma, K., Cheng, Z. et al. (2017). Fabrication of electrochemical immunosensor for cardiac biomarker troponin I determination and its potential for acute myocardial infarction diagnosis. *Int. J. Electrochem. Sci.* 12: 2389–2399.

6 Bhatnagar, D., Kaur, I., and Kumar, A. (2017). Ultrasensitive cardiac troponin I antibody based nanohybrid sensor for rapid detection of human heart attack. *Int. J. Biol. Macromol.* 95: 505–510.

7 Qi, M., Zhang, Y., Cao, C. et al. (2016). Increased sensitivity of extracellular glucose monitoring based on AuNP decorated GO nanocomposites. *RSC Adv.* 6: 39180–39187.

8 Song, J., Xu, L., Zhou, C. et al. (2013). Synthesis of graphene oxide based CuO nanoparticles composite electrode for highly enhanced nonenzymatic glucose detection. *ACS Appl. Mater. Interfaces* 5: 12928–12934.

9 Wang, L., Lu, X., Wen, C. et al. (2015). One-step synthesis of Pt–NiO nanoplate array/reduced graphene oxide nanocomposites for nonenzymatic glucose sensing. *J. Mater. Chem. A* 3: 608–616.

10 Justice Babu, K., Sheet, S., Lee, Y.S., and Gnana Kumar, G. (2018). Three-dimensional dendrite Cu–Co/reduced graphene oxide architectures on a disposable pencil graphite electrode as an electrochemical sensor for nonenzymatic glucose detection. *ACS Sustain. Chem. Eng.* 6: 1909–1918.

11 Manthina, V. and Agrios, A.G. (2016). Single-pot Zno nanostructure synthesis by chemical bath deposition and their applications. *Nano-Struct. Nano-Objects* 7: 1–11.

12 Ghayempour, S. and Montazer, M. (2017). Ultrasound irradiation based in-Situ synthesis of star-like tragacanth gum/zinc oxide nanoparticles on cotton fabric. *Ultrason. Sonochem.* 34: 458–465.

13 Gayen, R. and Paul, R. (2018). Phosphorous doping in vertically aligned Zno nanorods grown by wet-chemical method. *Nano-Struct. Nano-Objects* 13: 163–169.

14 Fakhri, P., Amini, B., Bagherzadeh, R. et al. (2019). Flexible hybrid structure piezoelectric nanogenerator based on ZnO nanorod/Pvdf nanofibers with improved output. *RSC Adv.* 9: 10117–10123.

15 Masuda, Y., Kinoshita, N., and Koumoto, K. (2012). Liquid phase morphology control of ZnO nanowires, ellipse particles, hexagonal rods, and particle in aqueous solutions. *ISRN Nanotechnol.* 2012: 1–6.

16 Choi, M.Y., Choi, D., Jin, M.J. et al. (2009). Mechanically powered transparent flexible charge-generating nanodevices with piezoelectric ZnO nanorods. *Adv. Mater.* 21: 2185–2189.

17 Rodrıguez-Paéz, J., Caballero, A., Villegas, M. et al. (2001). Controlled precipitation methods: formation mechanism of ZnO nanoparticles. *J. Eur. Ceram. Soc.* 21: 925–930.

18 Aladpoosh, R. and Montazer, M.J. (2015). The role of cellulosic chains of cotton in biosynthesis of ZnO nanorods producing multifunctional properties: mechanism, characterizations and features. *Carbohydr. Polym.* 126: 122–129.

19 Wang, Z.L. and Song, J. (2006). Piezoelectric nanogenerators based on zinc oxide nanowire arrays. *Science* 312: 242–246.

20 Wang, Z.L. (2009). Energy harvesting using piezoelectric nanowires – a correspondence on "energy harvesting using nanowires?" by Alexe et al. *Adv. Mater.* 21: 1311–1315.

21 Özgür, Ü., Alivov, Y.I., Liu, C. et al. (2005). A comprehensive review of Zno materials and devices. *J. Appl. Phys.* 98: 11.

22 Moezzi, A., McDonagh, A.M., and Cortie, M.B. (2012). Zinc oxide particles: synthesis, properties and applications. *Chem. Eng. J.* 185: 1–22.

23 Xi, Y., Song, J., Xu, S. et al. (2009). Growth of ZnO nanotube arrays and nanotube based piezoelectric nanogenerators. *J. Mater. Chem.* 19: 9260–9264.

24 Savarimuthu, K., Rajamanickam, G., Shankararajan, R. et al. (2017). Experimental study on flexible ZnO based nanogenerator using Schottky contact for energy harvesting applications. *IEEE Trans. Nanotechnol.* 16: 469–476.

25 Takeda, E. (1993). Hot-carrier effects in scaled MOS devices, microelectronics reliability. *Microelectron. Reliab.* 33: 1687–1711.

26 Roblin, P., Samman, A., and Bibyk, S. (1988). Simulation of hot-electron trapping and aging of nMOSFETs. *IEEE Trans. Electron Devices* 35: 2229–2237.

27 Tam, S., Ko, P.K., and Hu, C. (1984). Lucky-electron model of channel hot-electron injection in MOSFET's. *IEEE Trans. Electron Devices* 31: 1116–1125.

28 Hänsch, W., Schwerin, A.V., and Hofmann, F. (1990). A new self-consistent modeling approach to investigating MOSFET degradation. *IEEE Electron Device Lett.* 11: 362–364.

29 Doyle, B., Bourcerie, M., Marchetaux, J.C., and Boudou, A. (1990). Interface state creation and charge trapping in the medium-to-high gate voltage range ($V_d/2 \geq V_g \geq V_d$) during hot-carrier stressing of n-MOS transistors. *IEEE Trans. Electron Devices* 37: 744–754.

30 Chen, Z., Hess, K., Lee, J. et al. (2000). On the mechanism for interface trap generation in MOS transistors due to channel hot carrier stressing. *IEEE Electron Device Lett.* 21: 24–26.

31 A. Bravaix, C. Guerin, V. Huard et al. (2009). Hot-carrier acceleration factors for low power management in DC-AC stressed 40 nm NMOS node at high temperature. Reliability Physics Symposium, 2009 IEEE International, Montreal, QC, Canada, 26–30 April 2009.

32 Djeffal, F., Ghoggali, Z., Dibi, Z., and Lakhdar, N. (2009). Analytical analysis of nanoscale multiple gate MOSFETs including effects of hot-carrier induced interface charges. *Microelectron. Reliab.* 49: 377–381.

33 Rauch, S., La Rosa, G., and Guarin, F. (2001). Role of e-e scattering in the enhancement of channel hot carrier degradation of deep-submicron nMOSFETs at high Vgs conditions. *IEEE Trans. Device Mater. Reliab.* 1: 113–119.

34 Lee, C.W., Afzalian, A., Dehdashti Akhavan, N. et al. (2009). Junctionless multigate field-effect transistor. *Appl. Phys. Lett.* 94 (1–2): 053511.

35 Colinge, J.P., Lee, C.W., Afzalian, A. et al. (2010). Nanowire transistors without junctions. *Nat. Nanotechnol.* 5: 225–229.

36 Rios, R., Cappellani, A., Armstrong, M. et al. (Sep. 2011). Comparison of junctionless and conventional trigate transistors with Lg down to 26 nm. *IEEE Electron Device Lett.* 32: 1170–1172.

37 Jeon, D.Y. (Dec. 2013). Revisited parameter extraction methodology for electrical characterization of junctionless transistors. *Solid State Electron.* 90: 86–93.

38 Colinge, J.P., Lee, C.W., Dehdashti Akhavan, N. et al. (2012). Junctionless transistors: physics and properties. In: *Semiconductor-On-Insulator Materials for Nanoelectronics Applications* (ed. A. Nazarov, J.-P. Colinge, F. Balestra, et al.), 187–200. Berlin Heidelberg: Springer.

39 Goto, K.I., Yu, T.H., Wu, J. et al. (2012). Mobility and screening effect in heavily doped accumulation-mode metal-oxide-semiconductor field-effect transistors. *Appl. Phys. Lett.* 101: 073503.

40 Rudenko, T., Nazarov, A., Yu, R. et al. (Sep. 2013). Electron mobility in heavily doped junctionless nanowire SOI MOSFETs. *Microelectron. Eng.* 109: 326–329.

41 Tavakolian, F., Farrokhabadi, A., SoltanRezaee, M., and Rahmanian, S. (2019). Dynamic pull-in of thermal cantilever nanoswitches subjected to dispersion and axial forces using nonlocal elasticity theory. *Microsyst. Technol.* 25 (1): 19–30.

42 Esfahani, S., Khadem, S.E., and Mamaghani, A.E. (2019). Size-dependent nonlinear vibration of an electrostatic nanobeam actuator considering surface effects and inter-molecular interactions. *Int. J. Mech. Mater. Des.* 15 (3): 489–505.

43 SoltanRezaee, M., Bodaghi, M., Farrokhabadi, A., and Hedayati, R. (2019). Nonlinear stability analysis of piecewise actuated piezoelectric microstructures. *Int. J. Mech. Sci.* 160: 200–208.

44 Lamoreaux, S.K. (2005). The Casimir force: background, experiments, and applications. *Rep. Prog. Phys.* 68 (1): 201.

45 Chan, H., Bao, Y., Zou, J. et al. (2008). Measurement of the Casimir force between a gold sphere and a silicon surface with nanoscale trench arrays. *Phys. Rev. Lett.* 101 (3): 030401.

46 Rahi, S.J., Emig, T., Jaffe, R.L. et al. (2008). Casimir forces between cylinders and plates. *Phys. Rev. A* 78 (1): 012104.

47 Wong, W.S. and Salleo, A. (2009). *Flexible Electronics: Materials and Applications*, vol. 11. Springer Science & Business Media.

48 Zuber, K., Fabretto, M., Hall, C., and Murphy, P. (2008). Improved PEDOT conductivity via suppression of crystallite formation in Fe (III) tosylate during vapor phase polymerization. Macromol. *Rapid Commun.* 29 (18): 1503–1508.

49 McCann, J. and Bryson, D. (2009). *Smart Clothes and Wearable Technology*. Elsevier.

50 Patel, S., Park, H., Bonato, P. et al. (2012). A review of wearable sensors and systems with application in rehabilitation. *J. Neuroeng. Rehabil* 9 (1): 21.

51 Sedighi, A., Montazer, M., and Mazinani, S. (2019). Synthesis of wearable and flexible NiP0. 1-SnOx/PANI/CuO/cotton towards a non-enzymatic glucose sensor. *Biosens. Bioelectron.* 135: 192–199.

52 Sedighi, A., Montazer, M., and Samadi, N. (2014). Synthesis of nano Cu_2O on cotton: morphological, physical, biological and optical sensing characterizations. *Carbohydr. Polym.* 110: 489–498.

53 Jayathilaka, W.A.D.M., Qi, K., Qin, Y. et al. (2019). Significance of nanomaterials in wearables: a review on wearable actuators and sensors. *Adv. Mater.* 31 (7): 1805921.

54 Wu, W. (2017). Inorganic nanomaterials for printed electronics: a review. *Nanoscale* 9 (22): 7342–7372.

55 Wang, X., Liu, Z., and Zhang, T. (2017). Flexible sensing electronics for wearable/attachable health monitoring. *Small* 13 (25): 1602790.

56 Lou, Z., Li, L., Wang, L., and Shen, G. (2017). Recent progress of self-powered sensing systems for wearable electronics. *Small* 13 (45): 1701791.

57 Wang, C., Xia, K., Wang, H. et al. (2019). Advanced carbon for flexible and wearable electronics. *Adv. Mater.* 31 (9): 1801072.

58 Wagner, S. and Bauer, S. (2012). Materials for stretchable electronics. *MRS Bull.* 37 (3): 207–213.

59 Yao, S., Swetha, P., and Zhu, Y. (2018). Nanomaterial-enabled wearable sensors for healthcare. *Adv. Healthc. Mater* 7 (1): 1700889.

60 Wang, Z., Zhou, H., Chen, W. et al. (2018). Dually synergetic network hydrogels with integrated mechanical stretchability, thermal responsiveness, and electrical conductivity for strain sensors and temperature alertors. *ACS Appl. Mater. Interfaces* 10 (16): 14045–14054.

61 Wang, Y., Hao, J., Huang, Z. et al. (2018). Flexible electrically resistive-type strain sensors based on reduced graphene oxide-decorated electrospun polymer fibrous mats for human motion monitoring. *Carbon* 126: 360–371.

62 Heo, Y., Hwang, Y., Jung, H.S. et al. (2017). Secondary sensitivity control of silver-nanowire-based resistive-type strain sensors by geometric modulation of the elastomer substrate. *Small* 13 (23): 1700070.

63 Zhong, J., Zhong, Q., Hu, Q. et al. (2015). Stretchable self-powered fiber-based strain sensor. *Adv. Funct. Mater.* 25 (12): 1798–1803.

64 Lee, J., Shin, S., Lee, S. et al. (2018). Highly sensitive multifilament fiber strain sensors with ultrabroad sensing range for textile electronics. *ACS Nano* 12 (5): 4259–4268.

65 Hou, C., Wang, H., Zhang, Q. et al. (2014). Highly conductive, flexible, and compressible all-graphene passive electronic skin for sensing human touch. *Adv. Mater.* 26 (29): 5018–5024.

66 Yao, H.B., Ge, J., Wang, C.F. et al. (2013). A flexible and highly pressure-sensitive graphene–polyurethane sponge based on fractured microstructure design. *Adv. Mater.* 25 (46): 6692–6698.

67 Zhu, B., Niu, Z., Wang, H. et al. (2014). Microstructured graphene arrays for highly sensitive flexible tactile sensors. *Small* 10 (18): 3625–3631.

68 Jung, S., Kim, J.H., Kim, J. et al. (2014). Reverse-micelle-induced porous pressure-sensitive rubber for wearable human–machine interfaces. *Adv. Mater.* 26 (28): 4825–4830.

69 Park, J., Lee, Y., Hong, J. et al. (2014). Tactile-direction-sensitive and stretchable electronic skins based on human-skin-inspired interlocked microstructures. *ACS Nano* 8 (12): 12020–12029.

70 Choong, C.L., Shim, M.B., Lee, B.S. et al. (2014). Highly stretchable resistive pressure sensors using a conductive elastomeric composite on a micropyramid array. *Adv. Mater.* 26 (21): 3451–3458.

71 Gong, S., Schwalb, W., Wang, Y. et al. (2014). A wearable and highly sensitive pressure sensor with ultrathin gold nanowires. *Nat. Commun.* 5 (1): 1–8.

72 Bae, S.-H., Lee, Y., Sharma, B.K. et al. (2013). Graphene-based transparent strain sensor. *Carbon* 51: 236–242.

73 Zhang, F., Zang, Y., Huang, D. et al. (2015). Flexible and self-powered temperature–pressure dual-parameter sensors using microstructure-frame-supported organic thermoelectric materials. *Nat. Commun.* 6 (1): 1–10.

74 Gong, S., Lai, D.T., Su, B. et al. (2015). Highly stretchy black gold E-skin nanopatches as highly sensitive wearable biomedical sensors. *Adv. Electron.* 1 (4): 1400063.

75 Gong, S., Lai, D.T., Wang, Y. et al. (2015). Tattoolike polyaniline microparticle-doped gold nanowire patches as highly durable wearable sensors. *ACS Appl. Mater. Interfaces* 7 (35): 19700–19708.

76 Bae, G.Y., Pak, S.W., Kim, D. et al. (2016). Linearly and highly pressure-sensitive electronic skin based on a bioinspired hierarchical structural array. *Adv. Mater.* 28 (26): 5300–5306.

77 Boland, C.S., Khan, U., Backes, C. et al. (2014). Sensitive, high-strain, high-rate bodily motion sensors based on graphene–rubber composites. *ACS Nano* 8 (9): 8819–8830.

78 Yang, T., Jiang, X., Zhong, Y. et al. (2017). A wearable and highly sensitive graphene strain sensor for precise home-based pulse wave monitoring. *ACS Sens.* 2 (7): 967–974.

79 Li, Y., Jiang, X., Liu, Z., and Liu, Z. (2010). Strain effects in graphene and graphene nanoribbons: the underlying mechanism. *Nano Res.* 3 (8): 545–556.

80 Tian, H., Shu, Y., Cui, Y.-L. et al. (2014). Scalable fabrication of high-performance and flexible graphene strain sensors. *Nanoscale* 6 (2): 699–705.

81 Hempel, M., Nezich, D., Kong, J., and Hofmann, M. (2012). A novel class of strain gauges based on layered percolative films of 2D materials. *Nano Lett.* 12 (11): 5714–5718.

82 Sun, J., Li, Y., Peng, Q. et al. (2013). Macroscopic, flexible, high-performance graphene ribbons. *ACS Nano* 7 (11): 10225–10232.

83 Wang, F., Zhu, B., Shu, L., and Tao, X. (2013). Flexible pressure sensors for smart protective clothing against impact loading. *Smart Mater. Struct.* 23 (1): 015001.

84 Shyr, T.-W., Shie, J.-W., and Jhuang, Y.-E. (2011). The effect of tensile hysteresis and contact resistance on the performance of strain-resistant elastic-conductive webbing. *Sensors* 11 (2): 1693–1705.

85 Cochrane, C., Koncar, V., Lewandowski, M., and Dufour, C. (2007). Design and development of a flexible strain sensor for textile structures based on a conductive polymer composite. *Sensors* 7 (4): 473–492.

86 Li, Y., Samad, Y.A., and Liao, K. (2015). From cotton to wearable pressure sensor. *J. Mater. Chem. A* 3 (5): 2181–2187.

87 Wang, C., Li, X., Gao, E. et al. (2016). Carbonized silk fabric for ultrastretchable, highly sensitive, and wearable strain sensors. *Adv. Mater.* 28 (31): 6640–6648.

88 S. Magdassi, A. Bassa, Y. Vinetsky, and A. Kamyshny, "Silver nanoparticles as pigments for water-based ink-jet inks," *Chem. Mater.*, vol. 15, no. 11, pp. 2208–2217, 2003, doi:https://doi.org/10.1021/cm021804b.

89 Shankar, R., Groven, L., Amert, A. et al. (2011). Non-aqueous synthesis of silver nanoparticles using tin acetate as a reducing agent for the conductive ink formulation in printed electronics. *J. Mater. Chem.* 21 (29): 10871–10877.

90 Lai, C.Y., Cheong, C.F., Mandeep, J.S. et al. (2014). Synthesis and characterization of silver nanoparticles and silver inks: review on the past and recent technology roadmaps. *J. Mater. Eng. Perform.* 23 (10): 3541–3550.

91 Zhou, X., Li, W., Wu, M. et al. (2014). Enhanced dispersibility and dispersion stability of dodecylamine-protected silver nanoparticles by dodecanethiol for ink-jet conductive inks. *Appl. Surf. Sci.* 292: 537–543.

92 Huang, Q., Shen, W., Xu, Q. et al. (2014). Room-temperature sintering of conductive Ag films on paper. *Mater. Lett.* 123: 124–127.

93 Zhang, X.Y., Xu, J.J., Wu, J.Y. et al. (2017). Seeds triggered massive synthesis and multi-step room temperature post-processing of silver nanoink – application for paper electronics. *RSC Adv.* 7 (1): 8–19.

94 Kastner, J., Faury, T., Außerhuber, H.M. et al. (2017). Silver-based reactive ink for inkjet-printing of conductive lines on textiles. *Microelectron. Eng.* 176: 84–88.

95 Shahariar, H., Kim, I., Soewardiman, H., and Jur, J.S. (2019). Inkjet printing of reactive silver ink on textiles. *ACS Appl. Mater. Interfaces* 11 (6): 6208–6216.

96 Liang, J., Tong, K., and Pei, Q. (2016). A water-based silver-nanowire screen-print ink for the fabrication of stretchable conductors and wearable thin-film transistors. *Adv. Mater.* 28 (28): 5986–5996.

97 P. Patil, S. Patil, P. Kate, and A. A. Kulkarni, "Inkjet printing of silver nanowires on flexible surfaces and methodologies to improve the conductivity and stability of the printed patterns," *Nanoscale Adv.*, 2021, 31(1), 240–248.

98 Zhao, J., Yu, Y., Weng, B. et al. (2013). Sensitive and selective dopamine determination in human serum with inkjet printed Nafion/MWCNT chips. *Electrochem. Commun.* 37: 32–35.

99 Kourou, K., Exarchos, T.P., Exarchos, K.P. et al. (2015). Machine learning applications in cancer prognosis and prediction. *Comput. Struct. Biotechnol. J.* 13: 8–17.

100 Solassol, J., Marin, P., Demettre, E. et al. (2005). Proteomic detection of prostate-specific antigen using a serum fractionation procedure: potential implication for new low-abundance cancer biomarkers detection. *Anal. Biochem.* 338 (1): 26–31.

101 Singh, S., Gill, A.A.S., Nlooto, M., and Karpoormath, R. (2019). Prostate cancer biomarkers detection using nanoparticles based electrochemical biosensors. *Biosens. Bioelectron.* 137: 213–221.

102 Singh, P., Pandey, S.K., Singh, J. et al. (2016). Biomedical perspective of electrochemical nanobiosensor. *Nano-Micro Lett.* 8 (3): 193–203.

103 Wei, F., Patel, P., Liao, W. et al. (2009). Electrochemical sensor for multiplex biomarkers detection. *Clin. Cancer Res. An Off. J. Am. Assoc. Cancer Res.* 15 (13): 4446–4452.

104 Tripathi, P.P., Arami, H., Banga, I. et al. (2018). Cell penetrating peptides in preclinical and clinical cancer diagnosis and therapy. *Oncotarget* 9 (98): 37252–37267.

105 Alturkistani, H.A., Tashkandi, F.M., and Mohammedsaleh, Z.M. (2015). Histological stains: a literature review and case study. *Glob. J. Health Sci.* 8 (3): 72–79.

106 Zeng, C., Vangveravong, S., Xu, J. et al. (2007). Subcellular localization of sigma-2 receptors in breast cancer cells using two-photon and confocal microscopy. *Cancer Res.* 67 (14): 6708–6716.

107 Kaufmann, R., Müller, P., Hildenbrand, G. et al. (2011). Analysis of Her2/neu membrane protein clusters in different types of breast cancer cells using localization microscopy. *J. Microsc.* 242 (1): 46–54.

108 DeVita, V.T., Lawrence, T.S., and Rosenberg, S.A. (2008). *DeVita, Hellman, and Rosenberg's Cancer : Principles & Practice of Oncology*. Philadelphia, PA: Wolters Kluwer/Lippincott Williams & Wilkins.

109 Cleator, S., Heller, W., and Coombes, R.C. (2007). Triple-negative breast cancer: therapeutic options. *Lancet. Oncol.* 8 (3): 235–244.

110 Phillips, M., Cataneo, R.N., Ditkoff, B.A. et al. (2003). Volatile markers of breast cancer in the breath. *Breast J.* 9 (3): 184–191.

111 Dean-Colomb, W. and Esteva, F.J. (2008). Her2-positive breast cancer: herceptin and beyond. *Eur. J. Cancer* 44 (18): 2806–2812.

112 Kodadek, T. (2001). Protein microarrays: prospects and problems. *Chem. Biol.* 8 (2): 105–115.

113 Shuster, G., Gallimidi, Z., Reiss, A.H. et al. (2011). Classification of breast cancer precursors through exhaled breath. *Breast Cancer Res. Treat.* 126 (3): 791–796.

114 Kasapović, J., Pejić, S., Stojiljković, V. et al. (2010). Antioxidant status and lipid peroxidation in the blood of breast cancer patients of different ages after chemotherapy with 5-fluorouracil, doxorubicin and cyclophosphamide. *Clin. Biochem.* 43 (16–17): 1287–1293.

115 Mense, S.M., Remotti, F., Bhan, A. et al. (2008). Estrogen-induced breast cancer: alterations in breast morphology and oxidative stress as a function of estrogen exposure. *Toxicol. Appl. Pharmacol.* 232 (1): 78–85.

116 Kessler, W. and Remmer, H. (1990). Generation of volatile hydrocarbons from amino acids and proteins by an iron/ascorbate/GSH system. *Biochem. Pharmacol.* 39 (8): 1347–1351.

117 Clemens, M.R., Remmer, H., and Waller, H.D. (1984). Phenylhydrazine-induced lipid peroxidation of red blood cells in vitro and in vivo: monitoring by the production of volatile hydrocarbons. *Biochem. Pharmacol.* 33 (11): 1715–1718.

118 Jiang, T., Ren, L., Jia, Z. et al. (2017). Application of FBG based sensor in pipeline safety monitoring. *Appl. Sci.* 7 (6): 540.

119 Lopez-Torres, D., Elosua, C., and Arregui, F.J. (2020). Optical fiber sensors based on microstructured optical fibers to detect gases and volatile organic compounds – a review. *Sensors* 20 (9): 2555.

120 Narasimman, S., Balakrishnan, L., Meher, S.R. et al. (2017). Influence of surface functionalization on the gas sensing characteristics of ZnO nanorhombuses. *J. Alloys Compd.* 706: 186–197.

121 Lee, K.H., Zhang, Y.-Z., Jiang, Q. et al. (2020). Ultrasound-driven two-dimensional $Ti_3C_2T_x$ MXene hydrogel generator. *ACS Nano* 14 (3): 3199–3207.

122 Baldini, F., Brenci, M., Chiavaioli, F. et al. (2012). Optical fibre gratings as tools for chemical and biochemical sensing. *Anal. Bioanal. Chem.* 402 (1): 109–116.

123 Wang, Q. and Zhao, W.-M. (2018). A comprehensive review of lossy mode resonance-based fiber optic sensors. *Opt. Lasers Eng.* 100: 47–60.

124 Ćirić, A. and Stojadinović, S. (2019). Photoluminescence properties of Pr^{3+} doped ZrO_2 formed by plasma electrolytic oxidation. *J. Alloys Compd.* 803: 126–134.

125 Stojadinović, S. and Vasilić, R. (2016). Formation and photoluminescence of Eu^{3+} doped zirconia coatings formed by plasma electrolytic oxidation. *J. Lumin.* 176: 25–31.

126 Stojadinović, S., Tadić, N., and Vasilić, R. (2016). Photoluminescence of Sm^{3+} doped ZrO_2 coatings formed by plasma electrolytic oxidation of zirconium. *Mater. Lett.* 164: 329–332.

127 Stojadinović, S., Tadić, N., and Vasilić, R. (2019). Down- and up-conversion photoluminescence of $ZrO_2 : Ho^{3+}$ and $ZrO_2 : Ho^{3+}/Yb^{3+}$ coatings formed by plasma electrolytic oxidation. *J. Alloys Compd.* 785: 1222–1232.

128 Ćirić, A. and Stojadinović, S. (2019). Photoluminescence studies of ZrO_2 : Tm^{3+}/Yb^{3+} coatings formed by plasma electrolytic oxidation. *J. Lumin.* 214: 116568.

129 Stojadinović, S., Tadić, N., and Vasilić, R. (2018). Down-conversion photoluminescence of ZrO_2 : Er^{3+} coatings formed by plasma electrolytic oxidation. *Mater. Lett.* 219: 251–255.

130 Stojadinović, S., Tadić, N., and Vasilić, R. (2018). Structural and photoluminescent properties of ZrO_2 : Tb^{3+} coatings formed by plasma electrolytic oxidation. *J. Lumin.* 197: 83–89.

131 Ćirić, A. and Stojadinović, S. (2020). Photoluminescence of ZrO_2 : Gd^{3+} and ZrO_2 : Dy^{3+} coatings formed by the plasma electrolytic oxidation. *J. Alloys Compd.* 832: 154907.

132 Stojadinović, S., Tadić, N., and Vasilić, R. (2019). Photoluminescence properties of Er^{3+}/Yb^{3+} doped ZrO_2 coatings formed by plasma electrolytic oxidation. *J. Lumin.* 208: 296–301.

133 Stojadinović, S. and Vasilić, R. (2018). Eu^{2+} photoluminescence in Al_2O_3 coatings obtained by plasma electrolytic oxidation. *J. Lumin.* 199: 240–244.

134 Stojadinović, S., Tadić, N., and Vasilić, R. (2017). Photoluminescence of Sm^{2+}/Sm^{3+} doped Al_2O_3 coatings formed by plasma electrolytic oxidation of aluminum. *J. Lumin.* 192: 110–116.

135 Stojadinović, S. and Vasilić, R. (2018). Photoluminescence of Ce^{3+} and Ce^{3+}/Tb^{3+} ions in Al_2O_3 host formed by plasma electrolytic oxidation. *J. Lumin.* 203: 576–581.

136 Stojadinović, S. and Ćirić, A. (2019). Ce^{3+}/Eu^{2+} doped Al_2O_3 coatings formed by plasma electrolytic oxidation of aluminum: photoluminescence enhancement by $Ce^{3+} \rightarrow Eu^{2+}$ energy transfer. *Coatings* 9: 819.

137 Stojadinović, S. and Ćirić, A. (2020). Dy^{3+} and Dy^{3+}/Ce^{3+} doped Al_2O_3 coatings obtained by plasma electrolytic oxidation: Photoluminescence and energy transfer from Ce^{3+} to Dy^{3+}. *J. Lumin.* 226: 117403.

138 Stojadinović, S. and Ćirić, A. (2020). Highly-increased photoluminescence of Pr^{3+} due to the $Eu^{2+} \rightarrow Pr^{3+}$ energy transfer in Al_2O_3 coatings formed by plasma electrolytic oxidation. *J. Lumin.* 226: 117407.

139 Stojadinović, S. and Ćirić, A. (2021). Sensitizing effect of Ce^{3+} on Tm^{3+} blue emission in Al_2O_3:Tm^{3+}/Ce^{3+} coatings formed by plasma electrolytic oxidation. *J. Mater. Sci. Mater. Electron.* 32: 7841–7847.

140 Stojadinović, S., Radić, N., Grbić, B. et al. (2016). Structural, photoluminescent and photocatalytic properties of TiO_2:Eu^{3+} coatings formed by plasma electrolytic oxidation. *Appl. Surf. Sci.* 370: 218–228.

141 Stojadinović, S., Tadić, N., Radić, N. et al. (2018). Effect of Tb^{3+} doping on the photocatalytic activity of TiO_2 coatings formed by plasma electrolytic oxidation of titanium. *Surf. Coating. Technol.* 337: 279–289.

142 Stojadinović, S. and Ćirić, A. (2021). Photoluminescence of ZnO:Eu^{3+} and ZnO:Tb^{3+} coatings formed by plasma electrolytic oxidation of pure zinc substrate. *J. Lumin.* 235: 118022.

143 Stojadinović, S. and Vasilić, R. (2016). Orange–red photoluminescence of $Nb_2O_5:Eu^{3+}$, Sm^{3+} coatings formed by plasma electrolytic oxidation of niobium. *J. Alloys Compd.* 685: 881–889.

144 Stojadinović, S., Tadić, N., Ćirić, A., and Vasilić, R. (2018). Photoluminescence properties of Eu^{3+} doped HfO_2 coatings formed by plasma electrolytic oxidation of hafnium. *Opt. Mater.* 77: 19–24.

145 Ćirić, A. and Stojadinović, S. (2020). Structural and photoluminescence properties of Y_2O_3 and $Y_2O_3:Ln^{3+}$ (Ln = Eu, Er, Ho) films synthesized by plasma electrolytic oxidation of yttrium substrate. *J. Lumin.* 217: 116762.

146 Ćirić, A. and Stojadinović, S. (2019). Photoluminescence of Gd_2O_3 and $Gd_2O_3:Ln^{3+}$ (Ln = Eu, Er, Ho) formed by plasma electrolytic oxidation of pure gadolinium substrate. *Opt. Mater.* 99: 109546.

147 Kaseem, M., Fatimah, S., Nashrah, N., and Ko, Y.G. (2021). Recent progress in surface modification of metals coated by plasma electrolytic oxidation: principle, structure, and performance. *Prog. Mater. Sci.* 117: 100735.

148 Li, J., Meng, S., Niu, J., and Lu, H. (2017). Electronic structures and optical properties of monoclinic ZrO_2 studied by first-principles local density approximation + U approach. *J. Adv. Ceram.* 6: 43–49.

149 Cabello, G., Lillo, L., Caro, C. et al. (2008). Structure and optical characterization of photochemically prepared ZrO_2 thin films doped with erbium and europium. *J. Non-Cryst. Solids* 354: 3919–3928.

150 Martínez-Olmos, R.C., Guzmán-Mendoza, J., Báez-Rodríguez, A. et al. (2015). Synthesis, characterization and luminescence studies in $ZrO_2:Dy^{3+}$ and $ZrO_2:Dy^{3+}$, Gd^{3+} films deposited by the Pyrosol method. *Opt. Mater.* 46: 168–174.

151 Lu, X., Mohedano, M., Blawert, C. et al. (2016). Plasma electrolytic oxidation coatings with particle additions – a review. *Surf. Coat. Technol.* 307: 1165–1182.

152 Stojadinović, S., Radić-Perić, J., Vasilić, R., and Perić, M. (2014). Spectroscopic investigation of direct current (DC) plasma electrolytic oxidation of zirconium in citric acid. *Appl. Spectrosc.* 68: 101–112.

153 Schmid, H. (1994). Multi-ferroicmagnetoelectrics. *Ferroelectrics* 162 (1): 317–338.

154 Fiebig, M. (2005). Revival of the magnetoelectric effect. *J. Phys. D. Appl. Phys.* 38 (8): R123–R152.

155 Ramesh, R. and Spaldin, N.A. (2009). Multiferroics: progress and prospects in thin films. *Nanosci. Technol.* 3: 20–28.

156 Hill, N.A. (2000). Why are there so few magnetic ferroelectrics? *J. Phys. Chem. B* 104 (29): 6694–6709.

157 Israel, C., Mathur, N.D., and Scott, J.F. (2008). A one-cent room-temperature magnetoelectric sensor. *Nat. Mater.* 7 (2): 93–94.

158 Veev, J.P., Duan, C.-G., Burton, J.D. et al. (2009). Magnetic tunnel junctions with ferroelectric barriers: prediction of four resistance states from first principles. *Nano Lett.* 9 (1): 427–432.

159 Jahns, R., Greve, H., Woltermann, E. et al. (2012). Sensitivity enhancement of magnetoelectric sensors through frequency-conversion. *Sens. Actuators A Phys.* 183: 16–21.

160 Jahns, R., Piorra, A., Lage, E. et al. (2013). Giant magnetoelectric effect in thin-film composites. *J. Am. Ceram. Soc.* 96 (6): 1673–1681.

161 Moyerman, S., Eckert, J.C., Borchers, J.A. et al. (2006). Magnetic structure variations during giant magnetoresistance training in spin valves with picoscale antiferromagnetic layers. *J. Appl. Phys.* 99 (8): 8–11.

162 Li, M., Cui, W., Yu, J. et al. (2015). Magnetic proximity effect and interlayer exchange coupling of ferromagnetic/topological insulator/ferromagnetic trilayer. *Phys. Rev. B Condens. Matter Mater. Phys.* 91 (1): 1–7.

163 Daughton, J., Brown, J., Chen, E. et al. (1994). Magnetic field sensors using GMR multilayer. *IEEE Trans. Magn.* 30 (6): 4608–4610.

164 Prinz, G.A. (1998). Magnetoelectronics. *Science* 282 (5394): 1660–1663.

165 Ravi, S. (2020). High Curie temperature and room temperature magnetoresistance in Pr_2FeCrO_6 material for spintronics applications. *Mater. Lett.* 278: 128448.

166 Dieny, B., Speriosu, V.S., Parkin, S.S.P. et al. (1991). Giant magnetoresistive in soft ferromagnetic multilayers. *Phys. Rev. B* 43 (1): 1297–1300.

167 Heim, D.E., Fontana, R.E., Tsang, C. et al. (1994). Design and operation of spin valve sensors. *IEEE Trans. Magn.* 30 (2): 316–321.

168 Freitas, P.P., Leal, J.L., Melo, L.V. et al. (1994). Spin-valve sensors exchange-biased by ultrathin TbCo films. *Appl. Phys. Lett.* 65 (4): 493–495.

169 Leal, L., Oliveira, N., Rodrigues, L. et al. (1994). Unshielded spin-valve sensors exchange-biased by thin TbCo layers. *IEEE Trans. Magn.* 30 (6): 3831–3833.

170 Tsang, C., Fontana, R.E., Lin, T. et al. (1994). Design, fabrication & testing of spin-valve read heads for high density recording. *IEEE Trans. Magn.* 30 (6): 3801–3806.

171 Kosterev, A.A., Tittel, F.K., Serebryakov, D.V. et al. (2005). Applications of quartz tuning forks in spectroscopic gas sensing. *Rev. Sci. Instrum* 76 (4): 043 105–1–043 105–9.

172 Valkering, A., Mares, A., Untiedt, C. et al. (2005). A force sensor for atomic point contacts. *Rev. Sci. Instrum.* 76 (10): 103 903–1–103 903–5.

173 Lee, S., Yun, M., and Jeon, S. (2014). A tunable microresonator sensor based on a photocrosslinking polymer microwire. *App. Phy. Lett* 104: 053 506–1–053 506–4.

174 Kim, W., Yun, M., Lee, S., and Jeon, S. (2016). Enhanced sensitivity of quartz tuning fork sensors using electrospun polymer wires. *RSC Adv.* 6: 31 131–31 134.

2

Graphene-based Nanoelectronic Biosensors

Muhammad Aamir[1], Saba Ashraf[2,3], Ahmed Shuja[3], and Javeed Akhtar[*1]*

[1] *Mirpur University of Science and Technology (MUST), Functional Nanomaterials Lab (FNL), Department of Chemistry, Allama Iqbal Road, Mirpur, AJK 10250, Pakistan*
[2] *Department of Chemistry, Rawalpindi Women University 6th Road, Satellite Town, Rawalpindi, Pakistan,*
[3] *International Islamic University, Centre for Advanced Electronics & Photovoltaic Engineering, Al-Farabi Research Complex, Islamabad, 44000, Pakistan*

2.1 Introduction on Graphene

The word "carbon" is derived from the Latin word "carbo," meaning charcoal. Of four allotropic forms of carbon, graphite is the most common one. Graphene mostly consists of a single layer of carbon atoms closely packed into a two-dimensional (2D) honeycomb-like structure, which was named by Boehm, Setton, and Stump in 1994. Graphene possesses remarkable thermal and electrical properties [1]. In graphene, each carbon is sp^2 hybridized in a hexagonal arrangement. Graphene, the fundamental unit of graphite, used in pencil tips, batteries, and numerous other modern applications, is a single-atom-thick sheet of graphite, called the "magic material" that have properties of both metals and semiconductors [2]. The characteristic features of graphene are as follows:

- Greater mobility of electrons ($\sim$15 000 $cm^2\ V^{-1}\ s^{-1}$) at normal temperature compared to silicon ($\sim$1400 $cm^2\ V^{-1}\ s^{-1}$)
- Low resistivity, which is approximately 35% less than that of copper
- Large breaking strength ($\sim$40 $N\ m^{-1}$) and hence stronger than diamond
- Ballistic transport

Although graphene was the most discussed topic for more than six decades, free-state existence of graphene remained a mystery. In 2004, graphene was separated from graphite by Dr. Andre Geim and his coworkers, at The University of Manchester [3]. Graphene is the slenderest material with one-particle thickness and is 200 times stronger than steel. Graphene exhibits outstanding electrical and thermal properties with remarkable light absorption capability. It is a material that could change the world in the true sense, with its infinite potential for integration practically in any industry. Graphene is an extremely diverse material and can be combined with other materials (gases and metals) to form various materials

Nanotechnology in Electronics: Materials, Properties, Devices, First Edition.
Edited by Visakh P.M., Artem Semkin, Raneesh Balakrishnan, and Saša Lazović.

of remarkable characteristics. Researchers are very much keen to explore and patent graphene and to gain expertise about its various properties and potential applications mostly in cells, transistors, computer chips, energy production, supercapacitors, DNA sequencing, water filtration, aerials, touchscreens (LCD or OLED), solar cells, and spintronic products. Andre Geim and Konstantin Novoselov won the Nobel Prize for physics in 2004 for isolation of graphene. Following that achievement, graphene has received significant attention. This chapter describes fundamental properties of graphene and utilization in novel high-end technological devices as electronic and biosensors applications. A number of real case studies from the literature have been highlighted and discussed.

2.2 History of Graphene

It is significantly important to consider a lesser number of graphite layers to understand the development of graphene research. For about 500 years, graphite has been regarded as one of the most plentiful and naturally occurring materials. It has been employed in various instruments in the Middle Ages too as we are using it in pencils [4]. The history of graphene is almost 1.5 century old. In 1859, highly layered forms of graphene oxide were identified by an English chemist, Benjamin Collins Brodie. He assumed that it was a new form of carbon (molecular weight = 33) [5]. In 1916, the structure of graphite was determined through the powder diffraction method. Kohlschütter and Haenni reported the details of its structure in 1918 [6, 7]. In 1947, Wallace investigated the theory of graphene, which paved a clear route for a better understanding of improved electrical properties of graphite. Ruess and Vogt published the transmission electron microscope (TEM) images of graphite in 1948 [8]. Boehm and coworkers identified single layers of graphene using TEM and X-ray diffraction techniques in 1962. In 1975, van Bommel et al. heated hexagonal silicon carbide (SiC) in vacuum to get graphene layers. In 1986, a single layer of graphite was described by the term "graphene," which was also initially used to denote carbon nanotubes [9].

Many efforts were made to fabricate graphitic thin films using mechanical exfoliation until 1990. Only 50–100 layers were able to achieve before 2004 [10]. Graphene sheets were first observed by Boehm and Hofmann, which was made possible by their assumption that single layers must be available among byproducts. Moreover, Boehm et al. were the first to use the term "graphene" in 1986 [11]. Initially, thin films of graphene were fabricated by the exfoliation process that has same similarities to the drawing technique. By this method, multilayered structures of about 10 nm thickness were obtained. One of the primary patents related to the formation of graphene, entitled "Nano-scaled Graphene Plates," was filed in October 2002 and approved in 2006. Two years later, in 2004, Andre Geim and Kostya Novoselov at The University of Manchester extracted single-layer crystallites from bulk graphite and transferred them onto thin SiO_2 grown on a silicon wafer through micromechanical cleavage. SiO_2 electrically isolated graphene and formed weak interactions with it, resulting in the formation of charge-free layers of graphene.

The charge density of graphene can be varied widely using Si under SiO_2 as it acts as a "back gate" electrode in this regard [12]. In order to complete the history of graphene, some earlier ideas were also discussed. Thomas Ebbesen and Hidefumi Hiura visualized a probability of graphene-based nanoelectronics in 1995 [13]. Even before, in 1970, Hans-Joachim Teuschler verified a patent that described field-effect transistors utilizing "pyrolytic graphite" instead of Si. Graphene had actually been the main topic of interest for many centuries; however, its reality was not clear. The key to success was high-throughput visual recognition of graphene on an appropriately selected substrate, which gives a little but observable optical difference [14]. Graphene has outstanding structural, electrical, and mechanical properties, and after six years of assumption, Novoselov and Geim were awarded with the Nobel Prize in physics "for innovative experiments about the two-dimensional material graphene." In 2006, Ruoff's group demonstrated the dispersion of single-layered graphene for the first time. They reduced graphene oxide (GO) prepared by Hummers' method using hydrazine hydrate. Exfoliation method using Ruoff's pillar was used to deposit graphene flakes on oxidized degenerately doped silicon wafers to make field-effect transistors.

The second method involved the graphitization of SiO_2. The third method involved chemical vapor deposition growth on metals [15]. Two-dimensional (2D) graphite or "graphene" was first separated through mechanical exfoliation. The development of chemical vapor deposition as a manufacturing strategy concedes the fabrication of graphene with higher area coverage than was previously possible with exfoliation-based methods.

2.3 Properties of Graphene

The properties of G-derivatives are controlled by stacked layers. Single- or few-layered graphene exhibit distinct properties. In general, thin graphene films exhibit interesting properties compared to thick graphene (bulk graphite). For example, the properties of graphene having over 100 layers are similar to bulk graphite. G-derivatives are known for their electronic properties. The electron mobility of graphene is 15 000 $cm^2\,V^{-1}\,s^{-1}$ at room temperature, whereas it shows low resistance at low temperatures. Moreover, G-derivatives have shown a wide electrochemical potential window, low charge resistance, and a well-defined redox peak under controlled diffusion. The electron transfer in G-derivatives depends on the surface functional groups. The greater the number of oxygenated functional groups on the graphene surface, the higher the rate of electron transfer.

2.4 Fundamentals of G-Derivatives

The common G-derivatives are pristine graphene, reduced graphene oxide, graphene quantum dots (GQDs), and polycrystalline graphene. These derivatives have been extensively used in biosensing applications. Pristine graphene is a single-crystalline grain that has flawless lattices made up of defect-free hexagonal ring networks of sp^2 hybridized carbon atoms [16]. GO also mainly consists of covalently bonded sp^2-hybridizd carbon atoms but have some defects because of the disrupted sp^3-hybridized carbon atoms and because oxygen functional groups like the carboxyl group located at the edges and basal plane are saturated with hydroxyl and epoxy groups [17]. The presence of these groups make GO hydrophilic depending on the oxidation state of the groups [14]. The reduction of GO results in the formation of reduced graphene oxide (RGO). The RGO obtained by thermal reduction has more defects and thus facilitates the heterogeneous electron transfer in the electrochemical process of sp^2-hybridized carbon atoms at the edges and edges.

The chemical and electrochemical reduction of GO to obtain RGO has some oxygen-containing functional groups. The water dispersibility properties of GO and RGO can be controlled by the presence of hydrophilicity-facilitating groups. Thus, GO and RGO water suspensions can be used to fabricate biosensors. Moreover, the binding ability of G-derivatives with various biomolecules depends on their C/O ratio. Obviously, a higher C/O ratio in G-derivatives provides strong binding between graphene and biomolecules [18].

Another derivative polycrystalline graphene consists of single-crystalline grains of graphene oriented in distinct directions with mosaic-like assemblies having grain boundaries, forming the topological defects like dislocations [19]. GQDs are nanocrystalline graphene units facilitated with exciton confinements and quantum size effects, which impart photoluminescence properties to the GQDs tunable with size and surface chemistry [20]. These G-derivatives have interesting intrinsic features such as purity in structure, defects, and functional groups. These features can be tuned by manipulating the synthetic routes. For example, pristine graphene has some outstanding properties due to its defect-free crystalline structures, but high production cost of pristine graphene makes it hard to scalable.

Besides pure graphene materials, nanocomposites of graphene are also found useful in biosensing applications. These nanocomposites are formed by hybridizing graphene with polymers, metal nanoparticles, and surface modifications via functionalization with proteins or enzymes. In this context, GO and RGO are important as they have reactive oxygen-containing functional groups and surface defects that allow control over nucleation and growth of metal/metal-oxide nanoparticles [21–23].

2.5 Synthesis of Graphene

Several methods have been reported to obtain single-layer graphene [24], of which mechanical exfoliation is the simplest method [3]. Single-layer graphene can be obtained by repeated application of the scotch tape. Due to the relatively weak Vander Waals forces between graphene sheets, splitting of these sheets becomes easy [25]. It is one of the simplest methods and provides flawless single-layer graphene, but the overall percentage yield is very low. Therefore, this method is not

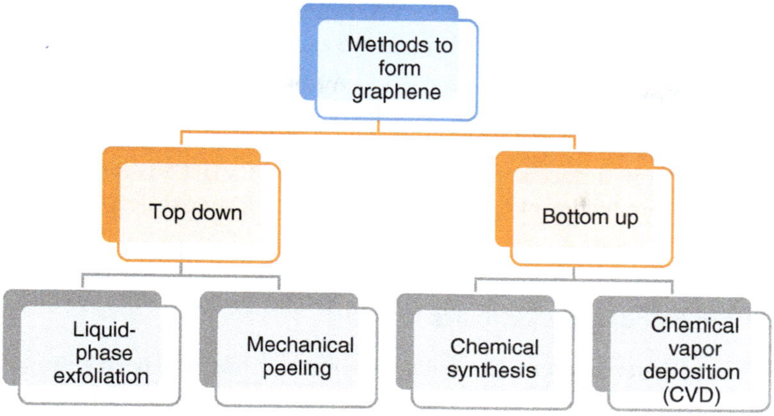

Figure 2.1 Schematic diagram of graphene synthesis methods.

ideal for the industrial production of graphene to use in flexible electronic devices (Figure 2.1).

Hummers' method is one of the most common chemical synthesis methods [26]. In this method, strong acids and oxidizing agents are used to oxidize graphite. Due to the hydrophilic nature of graphene oxide, water molecules can interpolate between the graphene layers, which ultimately increase the distance between the graphene sheets, and ultrasonication provides exfoliation of graphene oxide sheets [27]. Hummers' method provides GO sheets that contain various functional groups such as —OH, —COOH, which alter the unique attributes of GO sheets. Thus, graphite should be reduced chemically with no functional groups on its surface.

The deposition based on solution-processed reduced graphene oxide was reported by G. Eda et al. [23, 28]. This method is well known for uniform film deposition and to get controlled thickness of graphene. However, deposition parameters are the key factors controlling electronic and optical properties of reduced graphene oxide films ranging from a single layer to few layers. Transistor characteristics were reported in reduced graphene oxide films prepared by mechanical exfoliation with a shunt resistance of 43 $k\Omega\,sq^{-1}$. One of the major advantages of chemical synthesis is easy fabrication of graphene films and at the same time, an important obstacle is the loss of electrical properties of reduced graphene films for the application as transparent electrodes in flexible electronics.

High-quality uniform graphene films can be produced by chemical vapor deposition (CVD). In the CVD technique, transition metals are used as catalysts such as nickel (Ni), platinum (Pt), copper (Cu). CVD involves two mechanisms for uniform graphene production: one is related to segregation and precipitation and other one is surface growth according to the solubility of carbon-to-metal catalysts. Nickel has highest solubility for carbon. Carbon released from hydrocarbon decomposition is absorbed at high temperatures (~1000 °C), it appears on the metal surface, and upon cooling of the substrate produces graphene. Metal catalysts such as copper (Cu), which has low solubility, will facilitate the surface growth of graphene. Single-layer graphene with greater surface area can be formed by nucleation and with the help of continuous growth. Once the catalytic metals are etched away, graphene can be easily shifted to the given substrate.

Bae et al. reported a large-scale production of graphene with chemical doping, which resulted in low shunt resistance (R_{sh} ~125 Ω sq^{-1}). It was achieved on a copper foil as a source of a large surface area metal catalyst using CVD. CVD is the most useful method for the production of single-layer graphene on a large scale for flexible electronics.

2.5.1 Graphene-based Nanoelectronics

In spite of advancements in microelectronics, the industry has now been shifting to nanoelectronics. The materials that we used in nanoelectronics are termed as nanomaterials, and they have at least one dimension less than 100 nm. In nanoelectronics, the methodology is to design such systems in which nanomaterials are used as basic building blocks. Therefore, nanoelectronic devices are based on nanomaterials such as nanowires, graphene, carbon nanotubes, and transition metal dichalcogenides. These materials must possess extraordinary characteristics such as high surface-area-to-volume ratio, low power consumption, high charge mobility, and excellent compatibility to fit in modern electronic devices. As nanomaterials display unusual properties that are not obtainable at the micro scale, nanoelectronics holds significant importance. Graphene is one of the most recently discovered nanomaterials known for its outstanding mechanical, electrical, and optical properties, which are not present in any other material on the earth. As graphene is the most thinnest, strongest, stiffest and highly conductive material, it has claimed various applications [29].

Flexible, nonrigid electronic devices are in high demand compared to rigid devices. To make these electronic devices more practical, it is recommended that fluorine-doped tin oxide (FTO) and indium-tin-oxide (ITO) should be replaced due to their high cost, fragile nature, and instability. Graphene has emerged as an excellent alternative material with its unique mechanical and opto-electronic properties. Graphene is being used as a flexible substrate in electronic devices such as transistors, solar cells, touch displays, fuel cells, gas sensors, and batteries, as shown in Figure 2.2.

As graphene is a highly flexible transparent conductor, it is used in LED, OLED, and LCD screens. Graphene is easily stretchable and highly transparent. By doping and using different chemical methods, sheet resistance of graphene can be adjusted so that it can be made suitable for LEDs, OLEDs, etc. In flexible, organic field-effect transistors, graphene is used as an ideal electrode material because of its low contact resistance [30]. Another most important application of graphene is supercapacitors. Materials used in supercapacitors should have high specific surface area and high conductivity. Other carbon materials can be used in supercapacitors. They have high surface area, but they possess low conductivity. Chemically reduced graphene oxide is considered best in making supercapacitors because of its high conductance, large specific surface area, and chemical stability [31]. Graphene

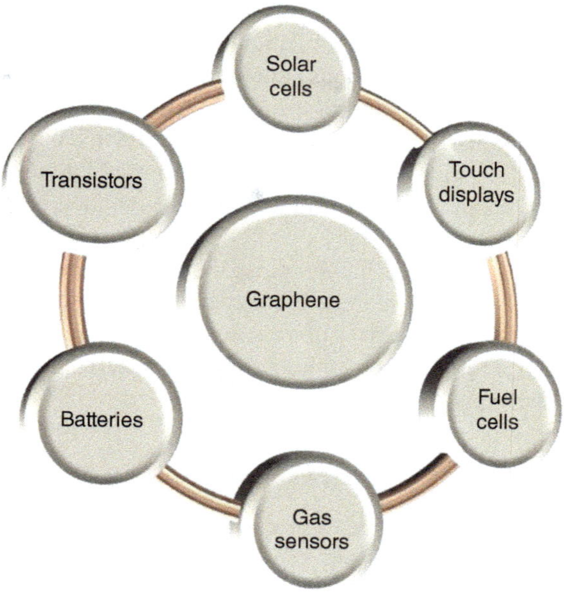

Figure 2.2 Applications of graphene to flexible electronics such as solar cells, transistors, batteries, gas sensors, fuel cells, and touch displays.

has paved a new path for electronic devices. Its application in FETs has opened a new gate of electronics devices. As its zero bandgap restricts its use in transistors, one-dimensional structures of graphene with narrow widths and smooth edges called graphene nano-ribbons (GNRs) have been investigated for transistor applications. These GNR-based FETs offer excellent switching speed and high carrier mobility [32].

Li-ion battery is another area of application in which graphene has been used. Because of reversibility and good specific capacity, graphite was used as the anode material in Li-ion batteries initially. Graphene-based anode has been itemized for Li-ion batteries due to its better electrical conductivity than graphite. Graphene electrodes offer high surface area and chemical tolerance. Li-ion batteries made up of graphene electrodes possess ten times more charging capacity than conventional batteries [33].

Graphene has gained considerable attention because of its large specific surface area and low Johnson noise. Transmission of charge occurring at the surface of graphene causes a change in the Fermi level and hence in its conductivity. The variation in conductivity with the increase in surface absorption makes it a suitable material for making sensors to detect gases and biomolecules [34].

Graphene is used in optical modulators by tuning its Fermi level to match optical absorption. The bandwidth of optical modulators made up of graphene ranges from 1.3 to 1.6 μm [35].

Graphene possesses the piezoelectric property. Piezoelectric substances produce electric current due to physical pressure and vice versa. Graphene itself is not piezoelectric by nature; it can be made a piezoelectric material by punching holes into it and applying pressure to it.

Due to this exceptional property of graphene, it is used in applications where precise control of mechanical motion is required [36]. To achieve best results, graphene can be used with various nanoparticles and act as a good photocatalyst. For example, ZnO/graphene nanocomposites perform much better than pure ZnO nanoparticles in the degradation of methyl blue dye under UV light. Other nanoparticles such as Ag, TiO_2 nanoparticles have also been used for making graphene nanocomposites, which can play its role in the photodegradation of methyl orange and methyl blue dyes [37].

As graphene is highly transparent and a good conductor, it can be used in solar cells. Efficiency can be increased by using graphene as the top electrode in solar cells. Conductivity of titanium oxide can be increased by adding graphene, which brings 52.4% more current into the solar cell circuit and in turn increase in solar cell efficiency [38].

2.5.2 Graphene-based Biosensors

Devices that can measure a biological or biochemical reaction in the form of physiochemical signals are known as biosensors [39, 40]. They can be classified as in vivo and in vitro. In vivo biosensing applications require the biocompatibility of transducers [41]. Biosensors can be used to sense heart rate, pulse, body motion, blood pressure, glucose, blood oxygen level, cholesterol, DNA, RNA, proteins, cancer, heavy metals, toxin in food and blood, and antibodies [42–44]. Biosensors can be of various types such as electrochemical [45], colorimetric [46], fluorescent [39, 47], and potentiometric [48].

Biosensors consist of two major components [39]:

1. Surface-related states that interact selectively with analytes in a biological medium such as blood and serum
2. Transducers that can measure the analyte binding with the surface

The key advantages of biosensors include specificity to detect the required analyte and fast response time.

Clark is considered the father of biosensors. He reported the first electrochemical biosensor by entrapping glucose oxidase enzyme onto the surface of an oxygen electrode using dialysis membrane [49, 50]. This amperometric sensor was commercialized in 1975. Later, in the 1980s, antibodies were immobilized onto the transducer surface to make SPR biosensors. SPR biosensors have gained an enormous success due to real-time signaling of antibody–antigen reactions [51] and many challenges like efficiency, high sensitivity, fast response time, high shelf life, selectivity, and easy to handle for recognition of biological events. Moreover, like other sensors, in biosensing systems, the pM LOD is desirable [39, 52].

G-derivatives and their nanocomposites with various semiconductors, metals, metal oxides, etc. have received considerable attention for biosensing applications [39]. Graphene-based biosensors operate through various sensing mechanisms including electrochemical, electrical, and optical [39]. In general, for biosensors operating through these sensing mechanisms, the following properties are highly desirable:

- High electron transport rate, high charge carrier mobility, and lowest electrical noise are required for electrochemical biosensing and electrical biosensing [53, 54].
- For fluorescent biosensors, the FRET process should be efficient.

G-derivatives fulfill these requirements. Moreover, G-derivatives provide high defects on the edges, and thus, are an ideal system for electron transfer to biological and chemical moieties. Monolayer G-derivatives show high optical transparency, which is ideal for optical sensing systems.

2.5.3 Graphene in Electrochemical Biosensing Platforms

The oxygen content, defects, impurities, and thickness of G-derivatives control the heterogeneous electron transfer (HET) process. GO has various functional groups and are highly disordered compared to single and polycrystalline graphene. After the reduction of GO to produce RGO, which still has oxygen-containing functional groups depending on the synthetic routes. The oxygen content in graphene decreases the HET, whereas the HET can be enhanced by increasing the defects. Thus, approaches to achieve effective reduction are important. In this context, the laser-scribed reduction approach to develop graphene thin films is an effective method with an increased number of edge-like structures with high surface area [55]. Thin films fabricated by vacuum filtration have shown well-oriented GO sheets. The drop casting or inkjet printing produces increased edge-like structures. Moreover, the structure of graphene is affected by the fabrication of graphene thin films on rough substrates [56, 57].

High electrical conductivity, high electrochemical stability, and large surface area are the major features that make graphene an excellent electrode material. As mentioned before, the structural defects and edge defects provide a platform for fast electron transfer kinetics.

Therefore, graphene-based electrodes are suitable for biosensing at very low electrochemical potentials [21, 58]. In electrochemical reactions, charge transfer takes place at the electrode surface. The electrochemical reactions involve adsorption, electrolyte resistance, and mass transfer to the electrode surface from the solution [59]. Based on the electrochemical reactions, electrochemical biosensing can be measured by using voltammetric, impedometric, or potentiometric methods. In voltammetric methods, the electron transfer reaction is derived by controlling the electrode potential. The current associated with this electron transfer reaction is measured by $[Fe(CN)_6]^{3-}$ or metals. The shifting in the current value is caused by quantitative perturbation. The voltammetric technique is the most widely used strategy due to its simplicity, affectability, and cost-effectiveness. In the case of impedometric measurements, modifications in the interfacial features of the electrode surface are measured. This method helps to determine the electron transfer kinetics and electron diffusion properties of the sensing reaction. Thus, it helps to quantitatively estimate electrochemical reactions [60–62]. For example, immunosensors and aptasensors are based on this technique [63]. In potentiometers, the potential difference between two electrodes is measured. The electrochemical activities of the system are measured by using this method [62].

2.5.4 Field-effect Transistors

These devices provide a platform that is fully integrated into electronic chips and enable complete electronic detection of biomolecules. In FET biosensing, the charge carrier density at the biorecognition layer is changed by electric charge distribution. For FET biosensors, graphene is one of the ideal materials due to the semimetallic bandgap that can be easily adjusted through modifying the surface chemistry of graphene. FET biosensing devices are ideal for charge molecules. Thus, biomolecules like DNA (due to charged phosphate backbones) and proteins can be detected by using graphene-based FET devices. For example, single or multi-layered graphene films deposited by CVD are found to be highly sensitive to DNA hybridization. 10 nM of ssDNA could even generate a large shift in gate voltage. The attachment of DNA or proteins with graphene shifts the Dirac point and thus results in the quantitative detection of protein molecules. The shift in the Dirac point is caused by the change in surface charges. FETs were fabricated using pristine or polycrystalline graphene and RGO [64–66].

2.5.5 Optical Platform for Biosensing

G-derivatives are optically active and have the ability to quench fluorescence by energy transfer. The florescence quenching ability of graphene depends on two parameters: the number of graphene layers stacked over one another and the degree of oxidation. For example, the more oxidized GO has less efficiency of fluorescence quenching [18, 67]. Moreover, the long-range energy transfer process can occur in the graphene layers up to 30 nm [68]. This distance-dependent fluorescence quenching process is called fluorescence resonance energy transfer (FRET). The electron–hole pairs are found in sp^2-hybridized carbon domains embedded in the sp^3 matrix. The recombination of these electron–hole pairs results in photoluminescence properties of graphene.

The tuning of emission wavelengths in G-derivatives can be achieved by modulating the oxygen-containing functional groups, lateral size of graphene sheets, and degree of oxidation [69–71].

The utilization of G-derivatives in biosensing devices is mainly based on the interaction of biomolecules and G-derivatives. These interactions include hydrophobic interactions and strong π–π stacking interactions. These interactions between G-derivatives and biomolecules result in a simple, sensitive, and versatile platform for biosensing even in the absence of an analyte. Fluorescence sensing is based on the energy transfer process that takes place between GO and photoexcited quantum dots (QDs) [52, 72].

2.6 Applications of Graphene-based Biosensors

2.6.1 Graphene-based Electrochemical Biosensors

Electrochemical biosensors are integrated devices consisting of two basic units: a bio-receptor to recognize the biological content and an electrode to read the response of biomarkers with high selectivity and sensitivity. The nature of bio-receptors is very crucial for selectivity and sensitivity of the device. Similarly, the electrode transfers electrical signals through its interaction with the biomolecules [73, 74].

G-derivative-based electrochemical electrodes have been used in various biosensing devices such as glassy carbon electrodes (GCEs) [74–76], screen printed electrodes (SPEs) [77, 78] integrated electrodes (IDEs) [79, 80], field-effect transistors (FETs) [81, 82], and paper electrodes (PEs) [83, 84]. These device structures modified with G-derivatives possess some interesting advantages.

a) GCE:
- Excellent conductivity
- Good sensitivity
- Enhanced electrochemical active sites
- Anti-interference ability
- Enhanced stability
- Reproducibility
- Good electrochemical activity

These enhanced properties of the GCE modified with G-derivatives are due to the good binding ability of the G-derivatives with the surface of carbon electrodes with some additional linkers [74, 85–87].

b) SPE
- Disposability
- Portability
- Low fabrication cost
- Reproducibility
- Easy to scale up

In SPE devices, the composition of G-derivative paste is very crucial to determine the properties of the electrode [88, 89].

c) IDE
- Facile fabrication
- Two-electrode system
- Inexpensive
- Easy scale-up

G-derivative-modified IDEs show a rapid detection of biomolecules and provide a platform for fast diagnosis of various bio-analytes [90].

d) FET
- Ultrasensitive transducers
- High specificity
- Rapid response
- Good stability
- Real-time detection capability

G-derivative-based FETs show versatility in structure compared to other transducers [91–96].

e) PE
- Flexible
- Disposable
- Lightweight
- Simple fabrication
- Low cost
- Good sensitivity [97]

2.6.2 Detection of Acute Myocardial Infarction

Cardiovascular diseases cause sudden death without showing any major symptoms. The major cause of developing cardiovascular diseases is plaques in the vessels that produce abnormalities in the blood flow. This blockage condition is called acute myocardial infarction [98, 99]. Therefore, early-stage detection of this problem is very important.

Cardiac injury results in the release of cardiac troponin I (cTnl) by myocardial cells. This biomarker has shown highest specificity compared to creatine kinase–MB isoenzyme and myoglobin. In healthy humans, the concentration of cTnl in blood is 0.4 ng ml^{-1}. In minor cardiac injury, the level of cTnl rises up to 0.2–1.4 ng ml^{-1} within three to six hours, whereas in major cardiac injury, this level goes up to 100 ng ml^{-1}, which remains in the bloodstream for five to nine days [79, 100, 101]. The electrochemical biosensing of cTnl biomarker is sensitive and selective. The sensitivity and selectivity can further be increased by adjusting the electrode surface with G-derivatives. G-derivative-based electrochemical biosensors can detect the cTnl concentration in blood in pg ml^{-1} and fg ml^{-1}.

For example, an immunosensor with the architecture of IDE-graphene-2-aminobenzyl amine-anti-cTnl showed a wide range of antigen detection (0.01–1 ng ml^{-1}) with a detection limit of 10 pg ml^{-1}. This system has shown good detection time of about 10 minutes. [102]. Likewise, electroactive graphene multi-walled carbon nanotube–based films were used as the bio-electrode for cTnl detection. This GCE-G-MWCNT-1-pyrene butyric acid *N*-hydroxysuccinimide ester–anti-cTnl system showed cTnl detection limit of 0.001–10 ng ml^{-1} in human blood serum. The lowest detection limit observed in this case was 0.94 pg ml^{-1} [103]. The fabrication of this transducer involved some extensive washing steps. The immobilization of anti-cTnl was improved by fabricating the porous graphene oxide nanomaterial on the surface of GCE. This partially reduced graphene oxide provided a large surface-to-volume ratio. This porous graphene-oxide-based GCE system has shown a detection limit of 70 pg ml^{-1} [103]. Graphene oxide and poly-(diallyldimethylammonium chloride) nanocomposite was immobilized with anti-cTnl, which shows a low detection limit of 24 pg ml^{-1} [104]. In another electrochemical biosensing platform, the ultrasensitive detection of cTnl was performed by embedding polyamidoamine dendrimers on graphene quantum dots. This composite deposited on SPE followed by anti-cTnl immobilization resulted in a highly sensitive system for the detection of cTnl. This Au–graphene quantum dots–polyamidoamine electrode showed a detection limit of 20 fg ml^{-1} with a very high electrode sensitivity of 109.23 μA cm^{-2} μg^{-1}. This electrode also showed a wide range response of about 10–6 ng 6 μl^{-1} of cTnl in human serum. The electrode was also found highly specific for the detection of antigen cTnl [105].

2.6.3 Detection of Lung Cancer

Cancer is one of the most fatal diseases causing deaths at a very high rate. Among the various types of cancers, lung cancer is ranked first, accounting for 13% of the total cancer cases reported [106]. In lung cancer, uncontrolled growth of cells takes place in one or both lungs. Two most common types of lung cancer are small-cell lung cancer and non-small-cell lung cancer. The biomarker for cancer is carcinoembryonic antigen (CEA), and it is associated with lung cancer, colon cancer, breast cancer, ovarian cancer, and pancreatic cancer [107].

The PE-based electrochemical biosensor was fabricated by using PEDOT : PSS and reduced graphene oxide composite for CEA detection. The PE was fabricated by dipping the Whatman paper into an aqueous suspension of reduced graphene oxide, PEDOT : PSS, and ethylene glycol followed by immobilization of anti-CEA by physical adsorption. This electrochemical biosensor displayed high sensitivity of 25.8 μA ng^{-1} ml cm^{-2} with a detection range of 2–8 ng ml^{-1} [107]. In another approach, the water-soluble graphene functionalized with PEG and pyrenebutyric acid modified with gold nanoparticle nanocomposites for sensitive detection of CEA exhibited excellent performance with a wide response range of 0.1–1000 ng ml^{-1} with a detection limit of 0.06 ng ml^{-1} [103].

GCE modified with reduced graphene oxide and bimetallic nano-rings of Ag-Pt nanocomposites immobilized with anti-CEA for efficient detection of CEA displayed good performance with a wide detection range of 5–50 fg ml^{-1} and a low detection limit of 1.43 fg ml^{-1}. This electrode also exhibited enhanced stability, reproducibility, and selectivity [108].

2.6.4 Detection of Asthma

Asthma is a respiratory disease causing inflammation, swelling, and muscle tightness in the airways. A noninvasive detection method has been used to detect this disease [109]. Recently, a nitrite biomarker was detected through an exhaled breath condensate using a graphene-based electrochemical sensor. The SPE was deposited by reduced graphene oxide. The biosensor showed good sensitivity [110, 111].

2.6.5 Detection of Diabetes

Diabetes is a life-threatening metabolic disorder in which the body is not able to produce or react with the insulin hormone, thereby resulting in a spike in the blood glucose level. Diabetes is classified into type 1, 2, and gestational diabetes. Diabetes is diagnosed by the glucose level in blood. The normal glucose level in healthy persons is between 3.9 and 6.2 mM under fasting and after two hours of food, this level should be in the range 3.9–7.8 mM [74].

Among all the biosensors used for the detection of various diseases, 85% of them belong to the glucose level sensing. These are classified into three types:

1. In the first type of glucose biosensor, detection is based on the measurement of the oxygen level after reaction of the glucose oxidase–based enzyme electrode with glucose. This reaction produces hydrogen peroxide (H_2O_2). The glucose level is estimated by the amount of H_2O_2 produced in the enzymatic reaction [112].
2. The second type of glucose biosensor is based on electron transfer facilitated by mediators such as flavin adenine dinucleotide (FAD) active sites of G-derivatives and electrode surfaces. These biosensors exhibit disadvantages such as lower total energy and low stability [112].
3. The third type of biosensor is based on direct electron transfer (DET) between FAD-G-derivatives and electrode surfaces. Like mediated electron transfer, the DET faces various challenges such as reliable glucose detection and operational stability [112].

2.6.5.1 Electrochemical Enzymatic Glucose Biosensor

Electrochemical enzymatic biosensors exhibit various advantages such as simplicity, high selectivity and sensitivity. An enzymatic sensor was developed by immobilizing the glucose oxidase enzyme onto the electrode surface [113]. To develop an efficient graphene-based electrochemical biosensor, graphene oxide nanosheets were decorated with Au NPs by arylazonium bridging fabricated on GCE. Finally, the glucose oxidase enzyme was immobilized on the graphene oxide–Au NP nanocomposites. These biosensors showed a high sensitivity of about 42 $\mu A\ mM^{-1}\ cm^{-2}$ with a wide range of 0.3–20 mM. This system displayed fast electron transfer.

Moreover, this glucose biosensor showed excellent selectivity for glucose detection. The real-time sensing ability of this biosensor was also determined by choosing two cell lines [114].

The glucose oxidase enzyme was also immobilized onto G-derivatives for the fabrication of the electrode for glucose biosensing using the DET process. However, the DET process between the biomarker and electrode surface is a major challenge due to the complex nature of the glucose oxidase enzyme. To overcome issues such as the long electron tunneling distance and the DET process between electrode surface and immobilized enzyme, various efforts have been made. The short electron tunneling distance enhances the sensitivity of biosensors [115, 116]. For example, single-layer reduced graphene oxide was functionalized with (3-anibopropyl)triethoxysilane (APTES) and the glucose oxidase enzyme was covalently immobilized to fabricate the GCE–APTES–RGO–glucose oxidase electrode [116].

The simplest electrode device was fabricated by directly immobilizing the glucose oxidase enzyme onto the reduced graphene oxide to fabricate a GCE–glucose oxidase–reduced graphene oxide–based system and it exhibited a sensitivity of 1.85 μA mM^{-1} cm^{-2} [117]. On the other hand, an electrochemical biosensor fabricated by CVD showed a sensitivity of 0.087 μA mM^{-1} cm^{-2} and good LOD of 0.12 μM. A high sensitivity of 42 mA M^{-1} cm^{-2} was observed in a DET-based electrode fabricated by using the reduced graphene oxide adsorbed on phenothiazine [118].

Graphene-bridged glucose oxidase enzymes were fabricated by layer-by-layer assembly. This system have shown LOD as low as 0.154 mM for glucose sensing. The LOD of glucose was further decreased by using N-doped graphene, which shows an LOD of 0.01 mM. This high performance of N-doped graphene oxide was attributed to the high electronic state density and availability of free electrons. The nitrogen doping modifies the density states near the Fermi level, which increases the electron transfer rate. The N-doped graphene oxide facilitates the electrochemical reduction of H_2O_2 by the breakdown of the O—O bond [119].

The direct electrochemistry of glucose oxidase immobilized on the porous graphene–modified electrodes has also been studied. In this context, an enzymatic biosensor based on the N-doped graphene oxide was fabricated using the CVD method. To immobilize the N-doped glucose oxidase, the chitosan biopolymer was used as a binder. The fabricated N-doped graphene–chitosan–glucose oxidase biosensor exhibited a sensitivity of 226.24 μA mM^{-1} M^{-2} [120]. This excellent response was attributed to the high porosity of graphene and high conductivity of graphene sheets, which enable enhanced electron transfer. Compared to that, graphene quantum dots immobilized onto glucose oxidase on carbon ceramics electrode (CCE) displayed a sensitivity of 0.085 μA μM^{-1} cm^{-2} and an LOD of 1.73 μM [121].

To increase the stability and catalytic function of G-derivatives in glucose biosensing, the G-derivatives were grafted with different polymers such as polyethylene glycol [122], chitosan [123], and polyvinylpyrrolidone [124]. The polymer G-derivative-based composites exhibit interesting advantages: the polymer enables effective immobilization of redox enzyme over the surface of G-derivatives, maintain good stability and biological functionality, suppress leaching effect, and allow efficient electron transfer [125, 126]. The chitosan-loaded graphene nanocomposite electrodes have shown a high loading capacity of the glucose oxidase enzyme. This nanocomposite showed DET sensitivity of 37.9 μA mM^{-1} cm^{-2} and an LOD of 0.02 mM. The chitosan biopolymers have amino and hydroxyl groups that provide better immobilization of the glucose oxidase enzyme [127]. Besides effective glucose oxidase enzyme immobilization based on amino and hydroxyl groups, carboxylic acid groups of graphene oxide allow covalent immobilization of the glucose oxidase enzyme. In this context, EDC and NHS provided this functionality, and the resultant graphene oxide and glucose oxidase composite exhibited glucose sensitivity of 8.045 mA M^{-1} cm^{-2} [21].

G-derivatives and inorganic nanoparticle nanocomposites were also explored for glucose sensing. The G-derivatives obtained by chemical methods consist of increased defects or vacancies and functional groups such as carbonyl, hydroxyl, and epoxides that act as a suitable platform for immobilization of inorganic nanoparticles and glucose oxidase enzymes. The G-derivatives and inorganic nanocomposites were involving metallic nanoparticles or metal-oxide nanoparticles or bimetallic nanoparticles [128]. The functionalization of these nanoparticles is an effective approach for specific-site conjugation of biomolecules. For example, the noble metal–based G-derivative nanocomposites provide effective site-specific enzyme conjugation. The Pt nanoparticle composites with G-derivatives have been used for biosensors that have shown an LOD of 0.3 μM for glucose detection and 75% sensitivity [129]. The reduced graphene oxide nanocomposites with tannic acid–mediated Au nanoparticles show glucose biosensing with a sensitivity of 18.73 mA mM^{-1} cm^{-2} [130]. Compared to monometallic nanocomposites, bimetallic nanocomposites have exhibited better electrochemical activity. The electrochemical deposition of Pt—Pd nanoparticles on reduced graphene oxide was immobilized by glucose oxidase enzymes. The fabricated biosensor showed a sensitivity of 24 μA mM^{-1} cm^{-2} [131]. Moreover, the glucose biosensor was fabricated by the composite of MnO_2 nanoparticles on the surface of graphene. The graphene–MnO_2–glucose oxidase–Nafion showed a high sensitivity of 3.30 μA mM^{-1} cm^{-2} with an LOD of 10 μM [131].

2.6.5.2 Nonenzymatic Glucose Biosensors

Enzymatic biosensors exhibit high sensitivity and selectivity with highest catalytic functionalities. A majority of enzymatic biosensors are associated with various disadvantages such as multistep fabrication process, stability, expensive enzymes, and influence of external environment. These disadvantages with enzymatic sensors have diverted researchers' attention toward the development of nonenzymatic glucose sensors. Nonenzymatic sensors have the following advantages: low cost in fabrication, fast response, very low level detection, and good stability [132, 133]. Glucose oxidase enzymes are affected by various parameters such as pH, temperature, humidity, and pressure. Glucose oxidase enzymes are comparatively stable between pH 2 and 8.

The structures of G-derivatives have shown unique electrochemical properties and therefore can be used for the fabrication of nonenzymatic glucose biosensors. A nonenzymatic biosensor was fabricated using inorganic nanoparticles and G-derivative-based nanocomposites. Although the biosensor was found less selective and operated in alkaline media, it displayed good stability, enhanced sensitivity, and low detection limit in the micro to nanomolar range. For example, nitrogen-doped graphene oxide and Cu nanoparticle composites showed good sensitivity [134]. $NiCo_2O_4$-nitrogen-doped reduced graphene oxide and ionic liquid nanocomposites were deposited onto the GCE for glucose sensing. These nanocomposites were displayed remarkable electrocatalytic activity for glucose sensing. The sensitivity of this electrode was found to be 3.76 mA mM^{-1} cm^{-2} with a wide response between 0.001 and 4.555 mM [135]. Later, nickel hydroxide nanorods and nitrogen-doped graphene oxide nanocomposites were fabricated on the GCE. This biosensor exhibited a high sensitivity of 3.214 mA mM^{-1} cm^{-2} and the lowest detection limit of 0.12 μM.

It also exhibited good reproducibility, long-term stability, and selectivity for glucose, fructose, and lactose [136]. On the other hand, Cu@Ni core–shell nanoparticles and reduced graphene oxide nanocomposites deposited on GCE showed a lower sensitivity of 0.78 mA mM^{-1} cm^{-2} [137].

The CuO nanoparticles and graphene oxide composite showed a high sensitivity of 262.52 μA mM^{-1} cm^{-2} with an LOD of 0.69 μM. This biosensor was also used to detect human serum and showed a sensitivity of 285.38 μA mM^{-1} cm^{-2} [138]. Likewise, a biosensor based on Pt—NiO–reduced graphene oxide nanocomposites showed a sensitivity of 832.95 μA mM^{-1} cm^{-2} and an LOD of 2.67 μM. This high performance was attributed to the porosity of Pt—NiO–reduced graphene oxide nanocomposites [139]. Moreover, the Cu—Co–reduced graphene oxide nanocomposites were electrochemically deposited over the pencil graphite electrode and showed a high sensitivity of 240 μA mM^{-1} cm^{-2} and an LOD of 0.15 μM [140].

2.6.6 Detection of Cholesterol

The mammalian cell membranes, steroid hormones, and bile acids are made up of cholesterol and its esters with fatty acid esters. The concentration of cholesterol in the bloodstream is the gauge for the diagnosis of hypertension, myocardial infraction, atherosclerosis, and cardiopathy [141, 142]. The normal cholesterol level in the bloodstream is 5.17 mM, which can vary based on gender, age, and body weight [142].

Electrochemical cholesterol biosensing can be determined by two approaches.

(i) *Enzymatic detection of cholesterol*

The cholesterol oxidase is used to oxidize cholesterol into the cholest-4,3-one and hydrogen peroxide. The estimation of cholesterol is monitored by the concentration of hydrogen peroxide [143].

(ii) *Enzyme-free biocatalytic oxidation of cholesterol* [144]

2.6.6.1 Enzymatic Detection of Cholesterol

Cholesterol can be detected by using graphene nanocomposites. For example, redox enzyme– functionalized graphene oxide was used for biosensing of cholesterol. The prepared biosensor showed a sensitivity of around 5.71 μA mM^{-1} cm^{-2} with an LOD of 0.1 μM [143]. Pd nanoparticles and reduced graphene oxide nanocomposites showed enhanced cholesterol sensing sensitivity of 5.12 μA μM^{-1} cm^{-2} and an LOD of 0.05 μM [145]. The bimetallic nanoparticle- based nanocomposites were also used for the detection of cholesterol. Similarly, the titanium oxide nanostructures fabricated on graphene showed a sensitivity of 3.82 μA μM^{-1} cm^{-2} with an LOD of 6 μM [146]. Compared to metal nanoparticle–based nanocomposites, this biosensor exhibited lower performance due to lower electrocatalytic activity and conductivity.

2.6.6.2 Enzyme-free Biocatalytic Oxidation of Cholesterol

Enzyme-free detection of cholesterol exhibits many advantages such as simple fabrication process, reproducibility, stability, and cost-effectiveness [144]. Cholesterol biosensing was performed by using β-cyclodextrin-modified graphene and methylene blue redox indicator. This system displayed a sensitivity of 0.101 μA mg^{-1} dl^{-2}. Molecularly imprinted polymer and graphene composites also showed high sensitivity with an LOD of 0.1 nM with an advantage of fast response time [144]. Besides, polymer, graphene nanocomposites, the NiO-graphene composite. This biosensor displayed a sensitivity of 40.6 mA μM^{-1} cm^{-2} [147]. It also featured long-term stability, reproducibility and facilitated cholesterol detection in milk samples.

2.6.7 Detection of Hydrogen Peroxide

Hydrogen peroxide is a byproduct of various enzymatic reactions and acts a mediator in food, medical, and environmental applications. Hydrogen peroxide possesses oxidizing and reducing properties. Oxidase enzymes in living organisms produce hydrogen peroxide. Graphene-based nanocomposites were used for hydrogen peroxide sensing [148]. In this context, multilayer composites of (phthalocyaninato)(-porphyrinato) europium(III) triple decker compound (Pc)Eu(Pc) and graphene oxide were used for hydrogen oxide detection. This fabricated biosensor possessed high electrical conductivity and large surface area of graphene oxide. The triple decker and graphene oxide sensor showed a sensitivity of 7.4 μA mM^{-1} with an LOD of 0.017 μM [149].

Noble metal nanoparticles such as Pt, Ag, Pd, Au, Cu fabricated on graphene oxide were also used for the development of electrochemical biosensors for hydrogen peroxide. Ag nanoparticles and porous graphene composites for hydrogen peroxide sensing displayed an LOD of 14.9 μM [150]. Similarly, bimetallic systems were also used to develop hydrogen peroxide biosensors. Pt-Ni nanowires were fabricated on reduced graphene oxide to develop composites that showed a wide sensing range. Another bimetallic system based on Au and Pt alloy microspheres was fabricated on reduced graphene oxide. This biosensing system exhibited a sensitivity of 1117.0 μA mM^{-1} cm^{-2}, a wide sensing range of 0.005–4.0 mM, and an LOD of 0.008 μM [151].

Compared to nonporous G-derivative-based nanocomposites, porous G-derivative-based nanocomposites exhibit better biosensing of hydrogen peroxide. For example, an electrochemical biosensor based on Au nanoparticles and porous graphene oxide nanocomposites showed a sensitivity of 75.9 μA mM^{-1} cm^{-2} and an LOD of 0.1 μM [152]. Moreover, it shows a wide linear range of 0.5 μM to 4.9 mM. This biosensor displayed enhanced electrochemical detection of H_2O_2 compared to nonporous graphene oxide–based biosensors [152].

The flexible biosensors were fabricated for the detection of hydrogen peroxide. The flexible electrode was designed and fabricated by using Au nanoparticles and freestanding graphene oxide paper by oil–water interface that produces highly dense monolayer assemblies on the surface of graphene oxide derivatives. This flexible biosensor showed improved sensitivity of 236.8 μA mM^{-1} cm^{-2} with an LOD of 2.0 μM [153]. Moreover, the Cu_2O microcrystals were fabricated on the surface of porous graphene aerogel to produce a biosensor. This fabricated biosensor showed reduction of hydrogen peroxide and thus improved the sensing performance with a wide range of 5.0 μM to 10.56 mM and an LOD of 3.78 μM [154].

2.6.8 Hydrogen Peroxide Detection in Living Cells

Hydrogen peroxide detection is also used for sensing cancer cells. Hydrogen peroxide was detected in human serum and also in the secretions of Raw 264.7 cells by using a biosensor based on nitrogen and sulfur co-doped graphene quantum dots and graphene nanosheets. This sensor exhibited a wide detection range of 0.4–33 μM with an LOD of 26 nM. This biosensor possessed high electrical conductivity, more available doping sites, and edges with high surface area [155].

Pt nanoparticles were decorated over the graphene carbon nanotube–based paper electrode to develop a real-time biosensor for sensing H_2O_2 in living cells that showed a sensitivity of 1.41 μA $μM^{-1}$ cm^{-2} with an LOD of 10 nM [156]. Later, Pt nanoparticles were fabricated on porous graphene for detecting H_2O_2 secreted from living cells. The biosensor exhibited a sensitivity of 341.14 μA mM^{-1} cm^{-2} with a wide detection range of 1.0–1477 μM with an LOD of 0.5 μM.

The improved performance was attributed to uniform distribution of Pt nanoparticles on the enhanced surface area and active sites on porous graphene nanostructure [157]. Similarly, a biosensor based on Au nanoparticles decorated on the surface of nitrogen-doped graphene quantum dots showed improved sensitivity and a selectivity of 186.22 μA mM^{-1} cm^{-2} with an LOD of 0.12 μM for hydrogen peroxide detection in human cervical cancer cells [158]. Likewise, various bimetallic nanoparticle-decorated G-derivative based-biosensors were used for detecting hydrogen peroxide in the living cells [159].

2.6.9 Nucleic Acid Biosensors

Nucleic acids have specific sequences and are diverse. Single-stranded DNA can be detected using nucleic acid biosensors. The selective and sensitive detection of nucleic acid is very important to diagnose genetic diseases [160]. Electrochemical biosensor-based detection of nucleic acids is based on a probe immobilized onto the surface of the sensor and changes in the signals generated. G-derivative-based biosensors have displayed high selectivity and sensitivity for DNA/RNA detection. G-derivatives can easily form π—π bonds with DNA/RNA due to their conjugated system. Theoretical studies have shown that the molecular interaction between DNA/RNA and G-derivatives are predominantly weak noncovalent π–π interactions [161, 162].

Au nanoparticles fabricated on the reduced graphene oxide have shown high selectivity for DNA with a detection range of 0.1 μM to 0.1 fM with an LOD of 35 aM [163]. Another approach involves the integration of Au nanoparticles with G-derivatives for biosensors. The Au nanorods were fabricated onto the surface of graphene oxide nanosheets to develop DNA sensors. The biosensor exhibited a wide detection range from 1.0 nM to 10 fM and an LOD of 3.5 fM [164].

A label-free electrochemical aptasensor was developed. The self-assembly of DNA aptamers on VS_2 and Au nanoparticles was fabricated on graphene for biosensing of DNA. This VS_2–Au–graphene composite-based biosensor exhibited a detection range of 5.0×10^{-13}–5.0×10^{-10} M with an LOD of 52 fM for the deletion of a specific gene sequence in H5N1 [164].

MicroRNAs are involved in various diseases such as cancer, central nervous system diseases, diabetes, viral infection, and kidney diseases [165]. Therefore, detection of microRNA is very important in diagnosis and prevention of such diseases. In this context, thio-functionalized Au nanoparticles were decorated on graphene oxide followed by deposition on GCE surface to detect plasma microRNA-155. This biosensor showed a wide range of detection from 2.0 fM to 8.0 pM with an LOD of 0.6 fM and it was used to detect breast cancer cells [166].

2.6.10 Detection of Enzymes

Nicotinamide adenine dinucleotide (NADH) and NAD+/NADH redox couple are generally involved in various enzymatic reactions [167]. To detect NADH, a graphene oxide–based electrode exhibited high electrocatalytic oxidation of NADH at a reduced potential of 1.35 V [168]. A biosensor based on Au nanoparticles and reduced graphene oxide nanocomposites displayed a sensitivity of 0.916 $\mu A\,mM^{-1}\ cm^{-2}$ with an LOD of 1.13 nM. Detection of NADH by this biosensor was not interfered by ascorbic acids, glucose, glutathione, and quinine bioanalytes [169]. Enhanced sensitivity was observed for NADH by nitrogen-doped graphene-based electrodes with a stability of over 15 days with small contamination. Besides, graphene-based electrodes, iron nitride, and nitrogen-doped graphene nanocomposites were used for NADH detection [170]. Among the dopamine, ascorbic acid, and uric acid biosensing. Dopamine is a neurotransmitter found in the brain, and it controls various physiological functions in the human body. Also, it is used to treat disorders such as Parkinson's disease [171]. Electrochemical biosensing can be used to detect these biochemicals. Interestingly, dopamine, ascorbic acid, and uric acid are found to have similar electrooxidation peak potentials. Therefore, a majority of studies focused on to separate the electrooxidation peaks of these biomolecules by enhancing selectivity and sensitivity [172].

Ascorbic acid is also found in the mammalian brain. While uric acid is the end product of purine metabolism in humans, its high level can result in diseases such as hyperuricemia, gout [128, 173]. G-derivative-based biosensors have been used for the detection of dopamine, uric acid, and ascorbic acid. For example, graphene-oxide-treated GCE was used for detection of dopamine along with ascorbic acid as a bioanalyte [173]. Oxidation of ascorbic acid was found difficult on this electrode surface due to the electrostatic interaction between graphene oxide and ascorbic acid. On the other hand, the dopamine aromatic ring and graphene oxide lead to the π–π stacking, which promotes oxidation. This system exhibited an LOD of 0.27 μM [173].

The electrocatalytic activity and sensitivity of G-derivative-based biosensors can be increased by doping with nitrogen or modification with polymers or decorating with various inorganic metals or metal-oxide nanoparticles [125, 174, 175]. For example, an enzyme-free urea biosensor was developed using SnO_2 quantum dots and reduced graphene oxide. This biosensor showed high electrochemical activity with a wide detection range of 1.6×10^{-14}–3.9×10^{-12} M and a very low LOD of 11.7 fM [176].

2.6.11 Food Toxin Sensing

Many environmental conditions such as temperature, insect damage, humidity lead to increase in infection and toxin production. For example, mycotoxins are produced by fungi such as Aspergillus flavus, Aspergillus ochraceus, Aspergillus parasiticus, Penicillium verrucosum. Mycotoxins can readily grow on various crops such as dried fruits, beans, nuts, cereals [177]. Efficient detection of these toxins is of paramount importance. Water and food contamination by pathogenic microorganisms causes 1.8 million deaths per year worldwide [178].

Electrochemical biosensing of aflatoxin-B1 was performed by covalently binding monoclonal antibodies onto the reduced graphene oxide surface. This biosensor based on reduced graphene oxide decorated by anti-aflatoxin-B1 showed a sensitivity of 68 $\mu A\, ng^{-1}\, ml^{-1} cm^{-2}$ with an LOD of 0.12 $ng\, ml^{-1}$ [179]. Likewise, immunosensors based on antibody-immobilized Au nanoparticles and reduced graphene oxide nanocomposites have shown good stability of 56 days and the detection of aflatoxin-B1 with a sensitivity of 182.4 $\mu A\, ng^{-1}\, ml^{-1}\, cm^{-2}$ and an LOD of 0.1 $ng\, ml^{-1}$ [180]. The same electrode architecture except the use of Ni nanoparticles displayed improved electron transfer rate and a high sensitivity of 129.6 $\mu A\, ng\, ml^{-1}\, cm^{-2}$ with an LOD of 0.16 $ng\, ml^{-1}$ for aflatoxin-B1 [181].

2.6.12 Heavy Metal Detection

Human body can accumulate various heavy toxic metal ions through food chains. These highly toxic pollutants are resistant to biodegrade and can have adverse effects on human health [182]. Therefore, it is important to develop new methods for detection of various toxic heavy metal ions in food, water, and environment [183]. Among various methods, the electrochemical approach is a simple, low-cost, and real-time monitoring facility for pollutants. G-derivatives and their nanocomposites can be used to remove toxic heavy metals in aqueous solutions.

For example, graphene oxide was functionalized with L-cysteine for the detection of Cu(II), Cd(II), Hg(II), and Pb(II). This biosensor showed a wide detection range of 0.4–2.0 mM for Cd^{2+}, Cu^{2+}, and Hg^{2+} and 0.4–1.2 mM for Pb^{2+} [184]. Likewise, graphene oxide and ruthenium(II) bipyridine complex ($[Ru(bpy)_3]^{2+}$) nanocomposites showed detection limits of 2.8 nM for Cd(II), 1.41 nM for Pb(II), 2.3 nM for As(III), and 1.6 nM for Hg(II) ions [185]. A more effective electrochemical biosensor was also designed by the cysteine-functionalized graphene oxide electrode for the detection of Pb^{2+} ions in water. This biosensor showed a wide detection range of 1.4–28 ppb with an LOD of 0.07 ppb. This detection value was found to be twofold lower than 10 ppb [186].

2.6.13 Detection of Pesticides

The term "pesticides" broadly covers all chemicals that are used to control weeds, insects, pests, etc. and to increase food production. Among various pesticides, herbicides and insecticides are widely used to protect plants; they are found to be toxic to humans. These pesticides readily dissolve in water and are resistant to decompose, producing adverse health effects [187]. Various biosensors have been developed for sensitive and selective detection of pesticides. There are four major classes of pesticides on the basis of their structure [188]:

1. Chlorinated hydrocarbons
2. Organophosphates
3. Carbamates
4. Chlorophenoxy acids

Organophosphate-based pesticides are potentially neurotoxic [188]. Exposure to these pesticides significantly inhibits the acetylcholinesterase enzyme, which is present in the brain and hydrolyzes neurotransmitters, acetylcholine, and choline esters. Acetylcholinesterase enzyme–based biosensors can be used to detect organophosphates. Moreover, acetylcholinesterase can be efficiently immobilized onto the surface of nanomaterials for pesticide detection [189]. Electrochemical biosensors of graphene nanocomposites deposited on GCEs can be used for the detection of pesticides such as methyl parathion [40, 190], carbofuran [40], chlorpyrifos [40, 191], imidacloprid [192], methyl parathion [190], phoxim [193]. Other G-derivative-based nanocomposites have also been used as biosensors for various pesticides.

2.6.14 Graphene-based Fluorescent Biosensors

Fluorescence-based sensing platforms show many advantages including high sensitivity, fast detection, and easy processing. These platforms are important and ideal for detecting various chemicals in biological and biochemical fields. Fluorescent biosensors work based on the Forster resonance energy transfer (FRET) mechanism and can be explained by using nanomaterials with improved selectivity and sensitivity [194].

Graphene is one among such nanomaterials with superior fluorescence quenching ability and a strong acceptor in the FRET process. Moreover, functionalities like carboxylic acids can be used to achieve high quenching behavior. Therefore, G-derivative nanocomposites have been used for fluorescence sensing of biomolecules [194].

2.6.15 Detection of Small Molecules

G-derivative-based nanocomposites can be used to detect hydrogen peroxide, glucose, dopamine, and glucose [194]. For example, graphene quantum dots functionalized with tyramine were used for fluorescent sensing of hydrogen peroxide, cholesterol, glucose, xanthine, and L-lactate in blood samples [195]. These fluorescent biosensors can be used for the diagnosis of various metabolic diseases. Moreover, graphene oxide was also used to detect dopamine. Graphene oxide shows good affinity for dopamine and effective near-infrared fluoresce quenching [196].

2.6.16 Detection of Nucleic Acids

Graphene oxide–based fluorescent biosensors have been widely used for the detection of nucleic acids such as DNA, RNA, and proteins [197]. For example, a fluorescent biosensor based on nitrogen- and sulfur-doped reduced graphene oxide was used to detect hepatitis B virus (HBV) DNA and HIV DNA. This biosensor showed an LOD of 2.4 nM with a wide linear range of 5–100 nM. Likewise, graphene quantum dots functionalized with pyrene-MB for the detection of micro RNAs showed a linear range from 0.1 to 200 nM with an LOD of 100 pM. These biosensors have shown low-intensity signals [198]. To increase the signal intensity that helps to increase the sensitivity of biosensors, various strategies such as ligation chain reaction (LCR), polymerase chain reaction (PCR), rolling circle amplification (RCA) are used [199].

2.6.17 Detection of Pathogens and Food Toxins

Pathogens and food toxins can be detected by G-derivatives through strong fluorescence quenching [200]. For example, a graphene quantum dot and Au-nanoparticle-based FRET biosensor was used to detect Staphylococcus aureus by monitoring the fluorescence quenching efficiency. Around 87% fluorescence was quenched with an LOD of 1 nM [201]. Likewise, graphene oxide was used to detect S. Paratyphi A [202].

2.6.18 Detection of Toxic Heavy Metal Ions

G-derivative-based fluorescent probes have been explored for the detection of various metal ions like Cu^{2+}, Hg^{2+}, Ag^{+}, Pb^{2+}, and Fe^{3+}. G-derivative-based fluorescent probes have shown enhanced selectivity and sensitivity [197]. For example, graphene oxide was immobilized onto a specific oligonucleotide to detect Ag^{+} ions. This fluorescent sensor showed an LOD of 20 nM [203]. Likewise, a graphene-oxide-based self-assembled fluorescent chemodosimeter was fabricated to detect the presence of F^{-} ions. This graphene-based chemodosimeter showed fivefold high sensing performance, short detection time, and twofold high sensitivity compared to a chemodosimeter probe without graphene oxide [204]. Moreover, graphene quantum dots have emerged as efficient sensors for the detection of various metal ions compared to graphene-oxide probes. The enhanced fluorescent sensing performance of graphene quantum dots was attributed to excellent photostability, high solubility, high chemical stability, biocompatibility, and small size [205]. The hydroxyl group (—OH) on the surface of graphene quantum dots results in a strong coordination between the graphene quantum dots and Fe^{3+} ions. This strong coordination leads to fluorescence quenching of graphene quantum dots with an LOD of 0.45 mM and high selectivity for Fe^{3+} ions [206]. Likewise, dopamine-functionalized graphene quantum dots were used to detect Fe^{3+} ions. This sensor showed an LOD of 7.6 nM [207].

References

1 Geim, A. and Novoselov, K. (2010). The rise of graphene. In: *Nanoscience and Technology: A Collection of Reviews from Nature Journals* (ed. P. Rodgers), 11–19. Singapore: World Scientific.

2 Lee, Y.L. and Lo, Y.S. (2009). Highly efficient quantum-dot-sensitized solar cell based on co-sensitization of CdS/CdSe. *Adv. Funct. Mater.* 19: 604–609.

3 Huang, J.Y., Ding, F., Yakobson, B.I. et al. (2009). In situ observation of graphene sublimation and multi-layer edge reconstructions. *Proc. Natl. Acad. Sci.* 106: 10103–10108.

4 Allen, M.J., Tung, V.C., and Kaner, R.B. (2009). Honeycomb carbon: a review of graphene. *Chem. Rev.* 110: 132–145.

5 Dreyer, D.R., Park, S., Bielawski, C.W., and Ruoff, R.S. (2010). The chemistry of graphene oxide. *Chem. Soc. Rev.* 39: 228–240.

6 Lee, B., Park, S.-Y., Kim, H.-C. et al. (2008). Conformal Al_2O_3 dielectric layer deposited by atomic layer deposition for graphene-based nanoelectronics. *Appl. Phys. Lett.* 92: 203102.

7 Novoselov, K., Fal'ko, V.I., Colombo, L. et al. (2012). A roadmap for graphene. *Nature* 490: 192–200.

8 Xu, Z., Zheng, Q.-S., and Chen, G. (2007). Elementary building blocks of graphene-nanoribbon-based electronic devices. *Appl. Phys. Lett.* 90: 223115.

9 Oshima, C. and Nagashima, A. (1997). Ultra-thin epitaxial films of graphite and hexagonal boron nitride on solid surfaces. *J. Phys. Condens. Matter* 9: 1.

10 Kitko, K.E. and Zhang, Q. (2019). Graphene-based nanomaterials: from production to integration with modern tools in neuroscience. *Front. Syst. Neurosci.* 13: 26.

11 Boehm, H.P., Setton, R., and Stumpp, E. (1994). Nomenclature and terminology of graphite intercalation compounds (IUPAC Recommendations 1994). *Pure Appl. Chem.* 66: 1893–1901.

12 Novoselov, K.S., Geim, A.K., Morozov, S.V. et al. (2004). Electric field effect in atomically thin carbon films. *Science* 306: 666–669.

13 Ebbesen, T.W. and Hiura, H. (1995). Graphene in 3-dimensions: towards graphite origami. *Adv. Mater.* 7: 582–586.

14 Hong, B.J., An, Z., Compton, O.C., and Nguyen, S.T. (2012). Tunable biomolecular interaction and fluorescence quenching ability of graphene oxide: application to "turn-on" DNA sensing in biological media. *Small* 8: 2469–2476.

15 Geim, A. (2012). Graphene prehistory. *Phys. Scr.* 2012: 014003.

16 Rodriguez-Perez, L., Herranz, M.Á., and Martin, N. (2013). The chemistry of pristine graphene. *Chem. Commun.* 49: 3721–3735.

17 Eigler, S. and Hirsch, A. (2014). Chemistry with graphene and graphene oxide – challenges for synthetic chemists. *Angew. Chem. Int. Ed.* 53: 7720–7738.

18 Yazyev, O.V. and Chen, Y.P. (2014). Polycrystalline graphene and other two-dimensional materials. *Nat. Nanotechnol.* 9: 755.

19 Ponomarenko, L., Schedin, F., Katsnelson, M. et al. (2008). Chaotic dirac billiard in graphene quantum dots. *Science* 320: 356–358.

20 Liu, Y., Yu, D., Zeng, C. et al. (2010). Biocompatible graphene oxide-based glucose biosensors. *Langmuir* 26: 6158–6160.

21 Yin, P.T., Kim, T.-H., Choi, J.-W., and Lee, K.-B. (2013). Prospects for graphene – nanoparticle-based hybrid sensors. *Phys. Chem. Chem. Phys.* 15: 12785–12799.

22 Liu, J., Fu, S., Yuan, B. et al. (2010). Toward a universal "adhesive nanosheet" for the assembly of multiple nanoparticles based on a protein-induced reduction/decoration of graphene oxide. *J. Am. Chem. Soc.* 132: 7279–7281.

23 Adetayo, A. and Runsewe, D. (2019). Synthesis and fabrication of graphene and graphene oxide: a review. *Open J. of Comp. Mater.* 9: 207.

24 Van Noorden, R. (2012). Beyond sticky tape. *Nature* 483: S32.

25 William, S., Hummers, J., and Offeman, R.E. (1958). Preparation of graphitic oxide. *J. Am. Chem. Soc.* 80: 1339–1339.

26 Li, D., Müller, M.B., Gilje, S. et al. (2008). Processable aqueous dispersions of graphene nanosheets. *Nat. Nanotechnol.* 3: 101.

27 Becerril, H., Mao, J., Liu, Z. et al. (2008). Evaluation of solution-processed reduced graphene oxide films as transparent conductors. *ACS Nano* 2: 463–470.

28 Eda, G., Fanchini, G., and Chhowalla, M. (2008). Large-area ultrathin films of reduced graphene oxide as a transparent and flexible electronic material. *Nat. Nanotechnol.* 3: 270–274.

29 Bhoria, R.S. (2015). Graphene: potential material for nanoelectronics applications. *Indian J. Pure Appl. Phys.* 53: 501–513.

30 Wang, X., Zhi, L., and Müllen, K. (2008). Transparent, conductive graphene electrodes for dye-sensitized solar cells. *Nano Lett.* 8: 323–327.

31 Wang, Y., Shi, Z., Huang, Y. et al. (2009). Supercapacitor devices based on graphene materials. *J. Phys. Chem. C* 113: 13103–13107.

32 Son, Y.-W., Cohen, M.L., and Louie, S.G. (2006). Energy gaps in graphene nanoribbons. *Phys. Rev. Lett.* 97: 216803.

33 Meyer, J.C., Geim, A.K., Katsnelson, M.I. et al. (2007). The structure of suspended graphene sheets. *Nature* 446: 60.

34 Dan, Y., Lu, Y., Kybert, N.J. et al. (2009). Intrinsic response of graphene vapor sensors. *Nano Lett.* 9: 1472–1475.

35 Lu, Y., Lerner, M.B., John Qi, Z. et al. (2012). Graphene-protein bioelectronic devices with wavelength-dependent photoresponse. *Appl. Phys. Lett.* 100: 033110.

36 Madden, J.D., Vandesteeg, N.A., Anquetil, P.A. et al. (2004). Artificial muscle technology: physical principles and naval prospects. *IEEE J. Ocean. Eng.* 29: 706–728.

37 Zhang, L., Du, L., Cai, X. et al. (2013). Role of graphene in great enhancement of photocatalytic activity of ZnO nanoparticle – graphene hybrids. *Phys. E Low-Dimens. Syst. Nanostruct.* 47: 279–284.

38 Wang, Y., Tong, S.W., Xu, X.F. et al. (2011). Interface engineering of layer-by-layer stacked graphene anodes for high-performance organic solar cells. *Adv. Mater.* 23: 1514–1518.

39 Szunerits, S. and Boukherroub, R. (2018). Graphene-based biosensors. *Interface Focus* 8: 20160132.

40 Yang, L., Wang, G., and Liu, Y. (2013). An acetylcholinesterase biosensor based on platinum nanoparticles–carboxylic graphene–Nafion-modified electrode for detection of pesticides. *Anal. Biochem.* 437: 144–149.

41 Liu, Q., Wu, C., Cai, H. et al. (2014). Cell-based biosensors and their application in biomedicine. *Chem. Rev.* 114: 6423–6461.

42 Du, Y. and Dong, S. (2016). Nucleic acid biosensors: recent advances and perspectives. *Anal. Chem.* 89: 189–215.

43 Huang, X., Liu, Y., Yung, B. et al. (2017). Nanotechnology-enhanced no-wash biosensors for in vitro diagnostics of cancer. *ACS Nano* 11: 5238–5292.

44 Bereza-Malcolm, L.T., Mann, G.L., and Franks, A.E. (2014). Environmental sensing of heavy metals through whole cell microbial biosensors: a synthetic biology approach. *ACS Synth. Biol.* 4: 535–546.

45 Heller, A. and Feldman, B. (2008). Electrochemical glucose sensors and their applications in diabetes management. *Chem. Rev.* 108: 2482–2505.

46 Gaddes, D., Reeves, W.B., and Tadigadapa, S. (2017). Calorimetric biosensing system for quantification of urinary creatinine. *ACS Sens.* 2: 796–802.

47 Tamura, T. and Hamachi, I. (2014). Recent progress in design of protein-based fluorescent biosensors and their cellular applications. *ACS Chem. Biol.* 9: 2708–2717.

48 De Wael, K., Daems, D., Van Camp, G., and Nagels, L.J. (2012). Use of potentiometric sensors to study (bio) molecular interactions. *Anal. Chem.* 84: 4921–4927.

49 Qlark, L.C.J. (1956). Monitor and control of blood and tissue oxygen tensions. *ASAIO J.* 2: 41–48.

50 Clark, L.C. Jr., and Lyons, C. (1962). Electrode systems for continuous monitoring in cardiovascular surgery. *Ann. N. Y. Acad. Sci.* 102: 29–45.

51 Liedberg, B., Nylander, C., and Lunström, I. (1983). Surface plasmon resonance for gas detection and biosensing. *Sens. Actuators* 4: 299–304.

52 Morales-Narváez, E., Baptista-Pires, L., Zamora-Gálvez, A., and Merkoçi, A. (2017). Graphene-based biosensors: going simple. *Adv. Mater.* 29: 1604905.

53 Ambrosi, A., Chua, C.K., Bonanni, A., and Pumera, M. (2014). Electrochemistry of graphene and related materials. *Chem. Rev.* 114: 7150–7188.

54 Viswanathan, S., Narayanan, T.N., Aran, K. et al. (2015). Graphene–protein field effect biosensors: glucose sensing. *Mater. Today* 18: 513–522.

55 El-Kady, M.F. and Kaner, R.B. (2013). Scalable fabrication of high-power graphene micro-supercapacitors for flexible and on-chip energy storage. *Nat. Commun.* 4: 1475.

56 Hyun, W.J., Park, O.O., and Chin, B.D. (2013). Foldable graphene electronic circuits based on paper substrates. *Adv. Mater.* 25: 4729–4734.

57 Baptista-Pires, L., Mayorga-Martínez, C.C., Medina-Sánchez, M. et al. (2016). Water activated graphene oxide transfer using wax printed membranes for fast patterning of a touch sensitive device. *ACS Nano* 10: 853–860.

58 Zhou, M., Zhai, Y., and Dong, S. (2009). Electrochemical sensing and biosensing platform based on chemically reduced graphene oxide. *Anal. Chem.* 81: 5603–5613.

59 Ershler, B. (1947). Investigation of electrode reactions by the method of charging-curves and with the aid of alternating currents. *Discuss. Faraday Soc.* 1: 269–277.

60 Bakker, E. and Qin, Y. (2006). Electrochemical sensors. *Anal. Chem.* 78: 3965–3984.

61 Hayat, A., Barthelmebs, L., and Marty, J.-L. (2012). Electrochemical impedimetric immunosensor for the detection of okadaic acid in mussel sample. *Sens. Actuators B Chem.* 171–172: 810–815.

62 Bakker, E. and Pretsch, E. (2005). Potentiometric sensors for trace-level analysis. *TrAC Trends Anal. Chem.* 24: 199–207.

63 Ambrosi, A., Chua, C.K., Latiff, N.M. et al. (2016). Graphene and its electrochemistry – an update. *Chem. Soc. Rev.* 45: 2458–2493.

64 He, R.X., Lin, P., Liu, Z.K. et al. (2012). Solution-gated graphene field effect transistors integrated in microfluidic systems and used for flow velocity detection. *Nano Lett.* 12: 1404–1409.

65 He, Q., Sudibya, H.G., Yin, Z. et al. (2010). Centimeter-long and large-scale micropatterns of reduced graphene oxide films: fabrication and sensing applications. *ACS Nano* 4: 3201–3208.

66 Liu, S. and Guo, X. (2012). Carbon nanomaterials field-effect-transistor-based biosensors. *NPG Asia Mater.* 4: e23.

67 Chen, Z., Berciaud, S., Nuckolls, C. et al. (2010). Energy transfer from individual semiconductor nanocrystals to graphene. *ACS Nano* 4: 2964–2968.

68 Morales-Narváez, E., Hassan, A.R., and Merkoçi, A. (2013). Graphene oxide as a pathogen-revealing agent: sensing with a digital-like response. *Angew. Chem. Int. Ed.* 52: 13779–13783.

69 Maiti, R., Midya, A., Narayana, C., and Ray, S.K. (2014). Tunable optical properties of graphene oxide by tailoring the oxygen functionalities using infrared irradiation. *Nanotechnology* 25: 495704.

70 Jang, M.H., Ha, H.D., Lee, E.S. et al. (2015). Is the chain of oxidation and reduction process reversible in luminescent graphene quantum dots? *Small* 11: 3773–3781.

71 Mei, Q., Chen, J., Zhao, J. et al. (2016). Atomic oxygen tailored graphene oxide nanosheets emissions for multicolor cellular imaging. *ACS Appl. Mater. Interfaces* 8: 7390–7395.

72 Korhonen, J.T., Hiekkataipale, P., Malm, J. et al. (2011). Inorganic hollow nanotube aerogels by atomic layer deposition onto native nanocellulose templates. *ACS Nano* 5: 1967–1974.

73 Gerard, M., Chaubey, A., and Malhotra, B.D. (2002). Application of conducting polymers to biosensors. *Biosens. Bioelectron.* 17: 345–359.

74 Taniselass, S., Arshad, M.M., and Gopinath, S.C. (2019). Graphene-based electrochemical biosensors for monitoring noncommunicable disease biomarkers. *Biosens. Bioelectron.* 130: 276–292.

75 Wang, C., Wang, Q., Sun, D. et al. (2014). Immunogenic alteration in laparoscopic common bile duct exploration. *J. Surg. Res.* 187: 302–309.

76 Pan, H.-Z., Yu, H.-W., Wang, N. et al. (2015). Electrochemical DNA biosensor based on a glassy carbon electrode modified with gold nanoparticles and graphene for sensitive determination of Klebsiella pneumoniae carbapenemase. *J. Biotechnol.* 214: 133–138.

77 Martín, A., Hernández-Ferrer, J., Martínez, M.T., and Escarpa, A. (2015). Graphene nanoribbon-based electrochemical sensors on screen-printed platforms. *Electrochim. Acta* 172: 2–6.

78 Yang, J., Yu, J.-H., Rudi Strickler, J. et al. (2013). Nickel nanoparticle–chitosan-reduced graphene oxide-modified screen-printed electrodes for enzyme-free glucose sensing in portable microfluidic devices. *Biosens. Bioelectron.* 47: 530–538.

79 Tuteja, S.K., Priyanka, Bhalla, V. et al. (2014). Graphene-gated biochip for the detection of cardiac marker Troponin I. *Anal. Chim. Acta* 809: 148–154.

80 Tuteja, S.K., Kukkar, M., Suri, C.R. et al. (2015). One step in-situ synthesis of amine functionalized graphene for immunosensing of cardiac marker cTnI. *Biosens. Bioelectron.* 66: 129–135.

81 Farid, S., Meshik, X., Choi, M. et al. (2015). Detection of Interferon gamma using graphene and aptamer based FET-like electrochemical biosensor. *Biosens. Bioelectron.* 71: 294–299.

82 Mukherjee, S., Meshik, X., Choi, M. et al. (2015). A graphene and aptamer based liquid gated FET-like electrochemical biosensor to detect adenosine triphosphate. *IEEE Trans. Nanobiosci.* 14: 967–972.

83 Sun, G., Zhang, L., Zhang, Y. et al. (2015). Multiplexed enzyme-free electrochemical immunosensor based on ZnO nanorods modified reduced graphene oxide-paper electrode and silver deposition-induced signal amplification strategy. *Biosens. Bioelectron.* 71: 30–36.

84 Ruecha, N., Rangkupan, R., Rodthongkum, N., and Chailapakul, O. (2014). Novel paper-based cholesterol biosensor using graphene/polyvinylpyrrolidone/polyaniline nanocomposite. *Biosens. Bioelectron.* 52: 13–19.

85 Wang, C., Du, J., Wang, H. et al. (2014). A facile electrochemical sensor based on reduced graphene oxide and Au nanoplates modified glassy carbon electrode for simultaneous detection of ascorbic acid, dopamine and uric acid. *Sens. Actuators B Chem.* 204: 302–309.

86 Raj, M.A. and John, S.A. (2013). Fabrication of electrochemically reduced graphene oxide films on glassy carbon electrode by self-assembly method and their electrocatalytic application. *J. Phys. Chem. C* 117: 4326–4335.

87 Chen, L., Tang, Y., Wang, K. et al. (2011). Direct electrodeposition of reduced graphene oxide on glassy carbon electrode and its electrochemical application. *Electrochem. Commun.* 13: 133–137.

88 Randviir, E.P., Brownson, D.A.C., Metters, J.P. et al. (2014). The fabrication, characterisation and electrochemical investigation of screen-printed graphene electrodes. *Phys. Chem. Chem. Phys.* 16: 4598–4611.

89 Karuwan, C., Wisitsoraat, A., Chaisuwan, P. et al. (2017). Screen-printed graphene-based electrochemical sensors for a microfluidic device. *Anal. Methods* 9: 3689–3695.

90 Ohno, R., Ohnuki, H., Wang, H. et al. (2013). Electrochemical impedance spectroscopy biosensor with interdigitated electrode for detection of human immunoglobulin A. *Biosens. Bioelectron.* 40: 422–426.

91 Kakatkar, A., Abhilash, T., De Alba, R. et al. (2015). Detection of DNA and poly-L-lysine using CVD graphene-channel FET biosensors. *Nanotechnology* 26: 125502.

92 Lee, G., Yang, G., Cho, A. et al. (2016). Defect-engineered graphene chemical sensors with ultrahigh sensitivity. *Phys. Chem. Chem. Phys.* 18: 14198–14204.

93 Pearce, R., Iakimov, T., Andersson, M. et al. (2011). Epitaxially grown graphene based gas sensors for ultra sensitive NO_2 detection. *Sens. Actuators B Chem.* 155: 451–455.

94 Trung, T.Q., Tien, N.T., Kim, D. et al. (2014). A flexible reduced graphene oxide field-effect transistor for ultrasensitive strain sensing. *Adv. Funct. Mater.* 24: 117–124.

95 Cai, B., Wang, S., Huang, L. et al. (2014). Ultrasensitive label-free detection of PNA–DNA hybridization by reduced graphene oxide field-effect transistor biosensor. *ACS Nano* 8: 2632–2638.

96 Park, J.W., Park, S.J., Kwon, O.S. et al. (2014). Polypyrrole nanotube embedded reduced graphene oxide transducer for field-effect transistor-type H_2O_2 biosensor. *Anal. Chem.* 86: 1822–1828.

97 Ghosale, A., Shrivas, K., Shankar, R., and Ganesan, V. (2017). Low-cost paper electrode fabricated by direct writing with silver nanoparticle-based ink for detection of hydrogen peroxide in wastewater. *Anal. Chem.* 89: 776–782.

98 Negahdary, M., Behjati-Ardakani, M., Sattarahmady, N. et al. (2017). Electrochemical aptasensing of human cardiac troponin I based on an array of gold nanodumbbells – applied to early detection of myocardial infarction. *Sens. Actuators B Chem.* 252: 62–71.

99 Fathil, M.F.M., Md Arshad, M.K., Gopinath, S.C.B. et al. (2015). Diagnostics on acute myocardial infarction: cardiac troponin biomarkers. *Biosens. Bioelectron.* 70: 209–220.

100 Singal, S., Srivastava, A.K., Biradar, A.M. et al. (2014). Pt nanoparticles-chemical vapor deposited graphene composite based immunosensor for the detection of human cardiac troponin I. *Sens. Actuators B Chem.* 205: 363–370.

101 Wu, Q., Li, S., Sun, Y., and Wang, J. (2017). Hollow gold nanoparticle-enhanced SPR based sandwich immunoassay for human cardiac troponin I. *Microchim. Acta* 184: 2395–2402.

102 Singal, S., Srivastava, A.K., Dhakate, S. et al. (2015). Electroactive graphene-multi-walled carbon nanotube hybrid supported impedimetric immunosensor for the detection of human cardiac troponin-I. *RSC Adv.* 5: 74994–75003.

103 Kazemi, S.H., Ghodsi, E., Abdollahi, S., and Nadri, S. (2016). Porous graphene oxide nanostructure as an excellent scaffold for label-free electrochemical biosensor: detection of cardiac troponin I. *Mater. Sci. Eng. C* 69: 447–452.

104 Li, Z., Ma, K., Cheng, Z. et al. (2017). Fabrication of electrochemical immunosensor for cardiac biomarker troponin I determination and its potential for acute myocardial infarction diagnosis. *Int. J. Electrochem. Sci.* 12: 2389–2399.

105 Bhatnagar, D., Kaur, I., and Kumar, A. (2017). Ultrasensitive cardiac troponin I antibody based nanohybrid sensor for rapid detection of human heart attack. *Int. J. Biol. Macromol.* 95: 505–510.

106 Stewart, B. and Wild, C. (2014). *World Cancer Report 2014*. France: International Agency for Research on Cancer *Back to cited text* **2015**.

107 Kumar, S., Kumar, S., Srivastava, S. et al. (2015). Reduced graphene oxide modified smart conducting paper for cancer biosensor. *Biosens. Bioelectron.* 73: 114–122.

108 Wang, R., Feng, J.-J., Xue, Y. et al. (2018). A label-free electrochemical immunosensor based on AgPt nanorings supported on reduced graphene oxide for ultrasensitive analysis of tumor marker. *Sens. Actuators B Chem.* 254: 1174–1181.

109 Carter, S.R., Davis, C.S., and Kovacs, E.J. (2012). Exhaled breath condensate collection in the mechanically ventilated patient. *Respir. Med.* 106: 601–613.

110 Tang, K., Shao, X., Liu, F. et al. (2017). Correlation between nitric oxide content in exhaled breath condensate and the severity of acute respiratory distress syndrome. *Int. J. Clin. Exp. Pathol.* 10: 7350–7355.

111 Bajaj, P. and Ishmael, F.T. (2013). Exhaled breath condensates as a source for biomarkers for characterization of inflammatory lung diseases. *J. Anal. Sci. Methods Instrum.* 3: 17.

112 Chen, C., Xie, Q., Yang, D. et al. (2013). Recent advances in electrochemical glucose biosensors: a review. *RSC Adv.* 3: 4473–4491.

113 Zhu, Z., Garcia-Gancedo, L., Flewitt, A.J. et al. (2012). A critical review of glucose biosensors based on carbon nanomaterials: carbon nanotubes and graphene. *Sensors* 12: 5996–6022.

114 Qi, M., Zhang, Y., Cao, C. et al. (2016). Increased sensitivity of extracellular glucose monitoring based on AuNP decorated GO nanocomposites. *RSC Adv.* 6: 39180–39187.

115 Ianniello, R.M., Lindsay, T.J., and Yacynych, A.M. (1982). Differential pulse voltammetric study of direct electron transfer in glucose oxidase chemically modified graphite electrodes. *Anal. Chem.* 54: 1098–1101.

116 Wang, Z., Zhou, X., Zhang, J. et al. (2009). Direct electrochemical reduction of single-layer graphene oxide and subsequent functionalization with glucose oxidase. *J. Phys. Chem. C* 113: 14071–14075.

117 Unnikrishnan, B., Palanisamy, S., and Chen, S.-M. (2013). A simple electrochemical approach to fabricate a glucose biosensor based on graphene–glucose oxidase biocomposite. *Biosens. Bioelectron.* 39: 70–75.

118 Osikoya, A.O., Parlak, O., Murugan, N.A. et al. (2017). Acetylene-sourced CVD-synthesised catalytically active graphene for electrochemical biosensing. *Biosens. Bioelectron.* 89: 496–504.
119 Liu, J., Kong, N., Li, A. et al. (2013). Graphene bridged enzyme electrodes for glucose biosensing application. *Analyst* 138: 2567–2575.
120 Guo, J., Zhang, T., Hu, C., and Fu, L. (2015). A three-dimensional nitrogen-doped graphene structure: a highly efficient carrier of enzymes for biosensors. *Nanoscale* 7: 1290–1295.
121 Razmi, H. and Mohammad-Rezaei, R. (2013). Graphene quantum dots as a new substrate for immobilization and direct electrochemistry of glucose oxidase: application to sensitive glucose determination. *Biosens. Bioelectron.* 41: 498–504.
122 Mao, H.Y., Laurent, S., Chen, W. et al. (2013). Graphene: promises, facts, opportunities, and challenges in nanomedicine. *Chem. Rev.* 113: 3407–3424.
123 Qian, L. and Lu, L. (2014). Three dimensional porous graphene–chitosan composites from ice-induced assembly for direct electron transfer and electrocatalysis of glucose oxidase. *RSC Adv.* 4: 38273–38280.
124 Shan, C., Yang, H., Song, J. et al. (2009). Direct electrochemistry of glucose oxidase and biosensing for glucose based on graphene. *Anal. Chem.* 81: 2378–2382.
125 Tiwari, J.N., Vij, V., Kemp, K.C., and Kim, K.S. (2015). Engineered carbon-nanomaterial-based electrochemical sensors for biomolecules. *ACS Nano* 10: 46–80.
126 Georgakilas, V., Tiwari, J.N., Kemp, K.C. et al. (2016). Noncovalent functionalization of graphene and graphene oxide for energy materials, biosensing, catalytic, and biomedical applications. *Chem. Rev.* 116: 5464–5519.
127 Kang, X., Wang, J., Wu, H. et al. (2009). Glucose oxidase–graphene–chitosan modified electrode for direct electrochemistry and glucose sensing. *Biosens. Bioelectron.* 25: 901–905.
128 Krishnan, S.K., Singh, E., Singh, P. et al. (2019). A review on graphene-based nanocomposites for electrochemical and fluorescent biosensors. *RSC Adv.* 9: 8778–8881.
129 Claussen, J.C., Kumar, A., Jaroch, D.B. et al. (2012). Nanostructuring platinum nanoparticles on multilayered graphene petal nanosheets for electrochemical biosensing. *Adv. Funct. Mater.* 22: 3399–3405.
130 Çakıroğlu, B. and Özacar, M. (2017). Tannic acid modified electrochemical biosensor for glucose sensing based on direct electrochemistry. *Electroanalysis* 29: 2719–2726.
131 Hossain, M. and Park, J.Y. (2014). Amperometric glucose biosensor based on Pt-Pd nanoparticles supported by reduced graphene oxide and integrated with glucose oxidase. *Electroanalysis* 26: 940–951.
132 Zheng, D., Hu, H., Liu, X., and Hu, S. (2015). Application of graphene in electrochemical sensing. *Curr. Opin. Colloid Interface Sci.* 20: 383–405.
133 Labib, M., Sargent, E.H., and Kelley, S.O. (2016). Electrochemical methods for the analysis of clinically relevant biomolecules. *Chem. Rev.* 116: 9001–9090.

134 Maaoui, H., Singh, S.K., Teodorescu, F. et al. (2017). Copper oxide supported on three-dimensional ammonia-doped porous reduced graphene oxide prepared through electrophoretic deposition for non-enzymatic glucose sensing. *Electrochim. Acta* 224: 346–354.

135 Rao, H., Zhang, Z., Ge, H. et al. (2017). Enhanced amperometric sensing using a $NiCo_2O_4$/nitrogen-doped reduced graphene oxide/ionic liquid ternary composite for enzyme-free detection of glucose. *New J. Chem.* 41: 3667–3676.

136 Zhang, Y., Lei, W., Wu, Q. et al. (2017). Amperometric nonenzymatic determination of glucose via a glassy carbon electrode modified with nickel hydroxide and N-doped reduced graphene oxide. *Microchim. Acta* 184: 3103–3111.

137 Wu, K.-L., Cai, Y.-M., Jiang, B.-B. et al. (2017). Cu@Ni core–shell nanoparticles/reduced graphene oxide nanocomposites for nonenzymatic glucose sensor. *RSC Adv.* 7: 21128–21135.

138 Song, J., Xu, L., Zhou, C. et al. (2013). Synthesis of graphene oxide based CuO nanoparticles composite electrode for highly enhanced nonenzymatic glucose detection. *ACS Appl. Mater. Interfaces* 5: 12928–12934.

139 Wang, L., Lu, X., Wen, C. et al. (2015). One-step synthesis of Pt–NiO nanoplate array/reduced graphene oxide nanocomposites for nonenzymatic glucose sensing. *J. Mater. Chem. A* 3: 608–616.

140 Justice Babu, K., Sheet, S., Lee, Y.S., and Gnana Kumar, G. (2018). Three-dimensional dendrite Cu–Co/reduced graphene oxide architectures on a disposable pencil graphite electrode as an electrochemical sensor for nonenzymatic glucose detection. *ACS Sustain. Chem. Eng.* 6: 1909–1918.

141 Ahmadalinezhad, A. and Chen, A. (2011). High-performance electrochemical biosensor for the detection of total cholesterol. *Biosens. Bioelectron.* 26: 4508–4513.

142 Sekretaryova, A.N., Beni, V., Eriksson, M. et al. (2014). Cholesterol self-powered biosensor. *Anal. Chem.* 86: 9540–9547.

143 Dey, R.S. and Raj, C.R. (2013). Redox-functionalized graphene oxide architecture for the development of amperometric biosensing platform. *ACS Appl. Mater. Interfaces* 5: 4791–4798.

144 Lee, Y.-J. and Park, J.-Y. (2010). Nonenzymatic free-cholesterol detection via a modified highly sensitive macroporous gold electrode with platinum nanoparticles. *Biosens. Bioelectron.* 26: 1353–1358.

145 Dey, R.S. and Raj, C.R. (2014). Enzyme-integrated cholesterol biosensing scaffold based on in situ synthesized reduced graphene oxide and dendritic Pd nanostructure. *Biosens. Bioelectron.* 62: 357–364.

146 Komathi, S., Muthuchamy, N., Lee, K., and Gopalan, A. (2016). Fabrication of a novel dual mode cholesterol biosensor using titanium dioxide nanowire bridged 3D graphene nanostacks. *Biosens. Bioelectron.* 84: 64–71.

147 Rengaraj, A., Haldorai, Y., Kwak, C.H. et al. (2015). Electrodeposition of flower-like nickel oxide on CVD-grown graphene to develop an electrochemical non-enzymatic biosensor. *J. Mater. Chem. B* 3: 6301–6309.

148 Zhang, R. and Chen, W. (2017). Recent advances in graphene-based nanomaterials for fabricating electrochemical hydrogen peroxide sensors. *Biosens. Bioelectron.* 89: 249–268.

149 Yu, Z., Zou, L., Chen, Y., and Jiang, J. (2016). (Pc) Eu (Pc) Eu [trans-T ($COOCH_3$) 2PP]/GO hybrid film-based nonenzymatic H_2O_2 electrochemical sensor with excellent performance. *ACS Appl. Mater. Interfaces* 8: 30398–30406.

150 Zhan, B., Liu, C., Shi, H. et al. (2014). A hydrogen peroxide electrochemical sensor based on silver nanoparticles decorated three-dimensional graphene. *Appl. Phys. Lett.* 104: 243704.

151 Bai, Z., Dong, W., Ren, Y. et al. (2018). Preparation of nano Au and Pt alloy microspheres decorated with reduced graphene oxide for nonenzymatic hydrogen peroxide sensing. *Langmuir* 34: 2235–2244.

152 Xi, Q., Chen, X., Evans, D.G., and Yang, W. (2012). Gold nanoparticle-embedded porous graphene thin films fabricated via layer-by-layer self-assembly and subsequent thermal annealing for electrochemical sensing. *Langmuir* 28: 9885–9892.

153 Tee, S.Y., Teng, C.P., and Ye, E. (2017). Metal nanostructures for non-enzymatic glucose sensing. *Mater. Sci. Eng. C* 70: 1018–1030.

154 Wang, C., Sun, Y., Yu, X. et al. (2016). Ag–Pt hollow nanoparticles anchored reduced graphene oxide composites for non-enzymatic glucose biosensor. *J. Mater. Sci. Mater. Electron.* 27: 9370–9378.

155 Zhang, T., Gu, Y., Li, C. et al. (2017). Fabrication of novel electrochemical biosensor based on graphene nanohybrid to detect H_2O_2 released from living cells with ultrahigh performance. *ACS Appl. Mater. Interfaces* 9: 37991–37999.

156 Sun, Y., He, K., Zhang, Z. et al. (2015). Real-time electrochemical detection of hydrogen peroxide secretion in live cells by Pt nanoparticles decorated graphene–carbon nanotube hybrid paper electrode. *Biosens. Bioelectron.* 68: 358–364.

157 Liu, J., Bo, X., Zhao, Z., and Guo, L. (2015). Highly exposed Pt nanoparticles supported on porous graphene for electrochemical detection of hydrogen peroxide in living cells. *Biosens. Bioelectron.* 74: 71–77.

158 Ke, W., Fang, G., Liu, Q. et al. (2015). Low-temperature solution-processed tin oxide as an alternative electron transporting layer for efficient perovskite solar cells. *J. Am. Chem. Soc.* 137: 6730–6733.

159 Sun, Y., Luo, M., Meng, X. et al. (2017). Graphene/intermetallic PtPb nanoplates composites for boosting electrochemical detection of H_2O_2 released from cells. *Anal. Chem.* 89: 3761–3767.

160 He, Y., Huang, G., and Cui, H. (2013). Quenching the chemiluminescence of acridinium ester by graphene oxide for label-free and homogeneous DNA detection. *ACS Appl. Mater. Interfaces* 5: 11336–11340.

161 Zhu, X., Sun, L., Chen, Y. et al. (2013). Combination of cascade chemical reactions with graphene–DNA interaction to develop new strategy for biosensor fabrication. *Biosens. Bioelectron.* 47: 32–37.

162 Dontschuk, N., Stacey, A., Tadich, A. et al. (2015). A graphene field-effect transistor as a molecule-specific probe of DNA nucleobases. *Nat. Commun.* 6: 6563.

163 Wang, J., Shi, A., Fang, X. et al. (2015). An ultrasensitive supersandwich electrochemical DNA biosensor based on gold nanoparticles decorated reduced graphene oxide. *Anal. Biochem.* 469: 71–75.

164 Han, X., Fang, X., Shi, A. et al. (2013). An electrochemical DNA biosensor based on gold nanorods decorated graphene oxide sheets for sensing platform. *Anal. Biochem.* 443: 117–123.

165 Chen, Y.-X., Huang, K.-J., and Niu, K.-X. (2018). Recent advances in signal amplification strategy based on oligonucleotide and nanomaterials for microRNA detection-a review. *Biosens. Bioelectron.* 99: 612–624.

166 Azimzadeh, M., Rahaie, M., Nasirizadeh, N. et al. (2016). An electrochemical nanobiosensor for plasma miRNA-155, based on graphene oxide and gold nanorod, for early detection of breast cancer. *Biosens. Bioelectron.* 77: 99–106.

167 Li, Z., Huang, Y., Chen, L. et al. (2013). Amperometric biosensor for NADH and ethanol based on electroreduced graphene oxide–polythionine nanocomposite film. *Sens. Actuators B Chem.* 181: 280–287.

168 Tabrizi, M.A. and Zand, Z. (2014). A facile one-step method for the synthesis of reduced graphene oxide nanocomposites by NADH as reducing agent and its application in NADH sensing. *Electroanalysis* 26: 171–177.

169 Govindhan, M., Amiri, M., and Chen, A. (2015). Au nanoparticle/graphene nanocomposite as a platform for the sensitive detection of NADH in human urine. *Biosens. Bioelectron.* 66: 474–480.

170 Mutyala, S. and Mathiyarasu, J. (2016). A highly sensitive NADH biosensor using nitrogen doped graphene modified electrodes. *J. Electroanal. Chem.* 775: 329–336.

171 Savitt, J.M., Dawson, V.L., and Dawson, T.M. (2006). Diagnosis and treatment of Parkinson disease: molecules to medicine. *J. Clin. Invest.* 116: 1744–1754.

172 Falat, L. and Cheng, H. (1982). Voltammetric differentiation of ascorbic acid and dopamine at an electrochemically treated graphite/epoxy electrode. *Anal. Chem.* 54: 2108–2111.

173 Gao, F., Cai, X., Wang, X. et al. (2013). Highly sensitive and selective detection of dopamine in the presence of ascorbic acid at graphene oxide modified electrode. *Sens. Actuators B Chem.* 186: 380–387.

174 Zhang, H., Gai, P., Cheng, R. et al. (2013). Self-assembly synthesis of a hierarchical structure using hollow nitrogen-doped carbon spheres as spacers to separate the reduced graphene oxide for simultaneous electrochemical determination of ascorbic acid, dopamine and uric acid. *Anal. Methods* 5: 3591–3600.

175 Weng, X., Cao, Q., Liang, L. et al. (2013). Simultaneous determination of dopamine and uric acid using layer-by-layer graphene and chitosan assembled multilayer films. *Talanta* 117: 359–365.

176 Dutta, D., Chandra, S., Swain, A.K., and Bahadur, D. (2014). SnO_2 quantum dots-reduced graphene oxide composite for enzyme-free ultrasensitive electrochemical detection of urea. *Anal. Chem.* 86: 5914–5921.

177 Rotariu, L., Lagarde, F., Jaffrezic-Renault, N., and Bala, C. (2016). Electrochemical biosensors for fast detection of food contaminants–trends and perspective. *Trends Anal. Chem.* 79: 80–87.

178 Silva, N.F., Magalhães, J.M., Freire, C., and Delerue-Matos, C. (2018). Electrochemical biosensors for Salmonella: state of the art and challenges in food safety assessment. *Biosens. Bioelectron.* 99: 667–682.

179 Srivastava, S., Kumar, V., Ali, M.A. et al. (2013). Electrophoretically deposited reduced graphene oxide platform for food toxin detection. *Nanoscale* 5: 3043–3051.

180 Srivastava, S., Abraham, S., Singh, C. et al. (2015). Protein conjugated carboxylated gold@ reduced graphene oxide for aflatoxin B1 detection. *RSC Adv.* 5: 5406–5414.

181 Srivastava, S., Kumar, V., Arora, K. et al. (2016). Antibody conjugated metal nanoparticle decorated graphene sheets for a mycotoxin sensor. *RSC Adv.* 6: 56518–56526.

182 Aragay, G., Pons, J., and Merkoçi, A. (2011). Recent trends in macro-, micro-, and nanomaterial-based tools and strategies for heavy-metal detection. *Chem. Rev.* 111: 3433–3458.

183 Vilela, D., Parmar, J., Zeng, Y. et al. (2016). Graphene-based microbots for toxic heavy metal removal and recovery from water. *Nano Lett.* 16: 2860–2866.

184 Muralikrishna, S., Sureshkumar, K., Varley, T.S. et al. (2014). In situ reduction and functionalization of graphene oxide with L-cysteine for simultaneous electrochemical determination of cadmium (II), lead (II), copper (II), and mercury (II) ions. *Anal. Methods* 6: 8698–8705.

185 Gumpu, M.B., Veerapandian, M., Krishnan, U.M., and Rayappan, J.B.B. (2017). Simultaneous electrochemical detection of Cd (II), Pb (II), As (III) and Hg (II) ions using ruthenium (II)-textured graphene oxide nanocomposite. *Talanta* 162: 574–582.

186 Seenivasan, R., Chang, W.-J., and Gunasekaran, S. (2015). Highly sensitive detection and removal of lead ions in water using cysteine-functionalized graphene oxide/polypyrrole nanocomposite film electrode. *ACS Appl. Mater. Interfaces* 7: 15935–15943.

187 Aktar, W., Sengupta, D., and Chowdhury, A. (2009). Impact of pesticides use in agriculture: their benefits and hazards. *Interdiscip. Toxicol.* 2: 1–12.

188 Gunnell, D., Eddleston, M., Phillips, M.R., and Konradsen, F. (2007). The global distribution of fatal pesticide self-poisoning: systematic review. *BMC Public Health* 7: 357.

189 Grutzendler, J. and Morris, J.C. (2001). Cholinesterase inhibitors for Alzheimer's disease. *Drugs* 61: 41–52.

190 Jeyapragasam, T., Saraswathi, R., Chen, S.-M., and Chen, T.-W. (2017). Acetylcholinesterase biosensor for the detection of methyl parathion at an electrochemically reduced graphene oxide-Nafion modified glassy carbon electrode. *Int. J. Electrochem. Sci.* 12: 4768–4781.

191 Yang, L., Wang, G., Liu, Y., and Wang, M. (2013). Development of a biosensor based on immobilization of acetylcholinesterase on NiO nanoparticles–carboxylic graphene–Nafion modified electrode for detection of pesticides. *Talanta* 113: 135–141.

192 Zhang, M., Zhao, H., Xie, T. et al. (2017). Molecularly imprinted polymer on graphene surface for selective and sensitive electrochemical sensing imidacloprid. *Sens. Actuators B Chem.* 252: 991–1002.

193 Chao, M. and Chen, M. (2014). Electrochemical determination of phoxim in food samples employing a graphene-modified glassy carbon electrode. *Food Anal. Methods* 7: 1729–1736.

194 Gao, M. and Tang, B.Z. (2017). Fluorescent sensors based on aggregation-induced emission: recent advances and perspectives. *ACS Sens.* 2: 1382–1399.

195 Li, N., Than, A., Wang, X. et al. (2016). Ultrasensitive profiling of metabolites using tyramine-functionalized graphene quantum dots. *ACS Nano* 10: 3622–3629.

196 Chen, J.-L., Yan, X.-P., Meng, K., and Wang, S.-F. (2011). Graphene oxide based photoinduced charge transfer label-free near-infrared fluorescent biosensor for dopamine. *Anal. Chem.* 83: 8787–8793.

197 Zhang, H., Zhang, H., Aldalbahi, A. et al. (2017). Fluorescent biosensors enabled by graphene and graphene oxide. *Biosens. Bioelectron.* 89: 96–106.

198 Chen, L., Song, L., Zhang, Y. et al. (2016). Nitrogen and sulfur codoped reduced graphene oxide as a general platform for rapid and sensitive fluorescent detection of biological species. *ACS Appl. Mater. Interfaces* 8: 11255–11261.

199 Kim, J., Park, S.-J., and Min, D.-H. (2016). Emerging approaches for graphene oxide biosensor. *Anal. Chem.* 89: 232–248.

200 Mitra, R. and Saha, A. (2016). Reduced graphene oxide based "turn-on" fluorescence sensor for highly reproducible and sensitive detection of small organic pollutants. *ACS Sustain. Chem. Eng.* 5: 604–615.

201 Shi, J., Chan, C., Pang, Y. et al. (2015). A fluorescence resonance energy transfer (FRET) biosensor based on graphene quantum dots (GQDs) and gold nanoparticles (AuNPs) for the detection of mecA gene sequence of Staphylococcus aureus. *Biosens. Bioelectron.* 67: 595–600.

202 Yan, X., Li, W., Liu, K., and Deng, L. (2015). Highly sensitive fluorescent aptasensor for Salmonella paratyphi A via DNase I-mediated cyclic signal amplification. *Anal. Methods* 7: 10243–10250.

203 Wen, Y., Xing, F., He, S. et al. (2010). A graphene-based fluorescent nanoprobe for silver (I) ions detection by using graphene oxide and a silver-specific oligonucleotide. *Chem. Commun.* 46: 2596–2598.

204 Wang, C., Yang, S., Yi, M. et al. (2014). Graphene oxide assisted fluorescent chemodosimeter for high-performance sensing and bioimaging of fluoride ions. *ACS Appl. Mater. Interfaces* 6: 9768–9775.

205 Zhou, S., Xu, H., Gan, W., and Yuan, Q. (2016). Graphene quantum dots: recent progress in preparation and fluorescence sensing applications. *RSC Adv.* 6: 110775–110788.

206 Zheng, Y., Liu, Z., Jing, Y. et al. (2015). An acetylcholinesterase biosensor based on ionic liquid functionalized graphene–gelatin-modified electrode for sensitive detection of pesticides. *Sens. Actuators B Chem.* 210: 389–397.

207 Dutta Chowdhury, A. and Doong, R.-A. (2016). Highly sensitive and selective detection of nanomolar ferric ions using dopamine functionalized graphene quantum dots. *ACS Appl. Mater. Interfaces* 8: 21002–21010.

3

Zinc Oxide Piezoelectric Nanogenerators for Low-frequency Applications

Zohreh Moarref, Negar Honaramiz Fahim, and Majid Montazer

Amirkabir University of Technology, Textile Department, Functional Fibrous Structures & Environmental Enhancement (FFSEE), No. 350, Hafez Ave, Valiasr Square, Tehran 1591634311, Iran

3.1 Introduction of Zinc Oxide

3.1.1 Structure of ZnO NPs

Zinc oxide (ZnO) is a multifunctional, environmentally friendly, and highly stable material, possessing high chemical inactivity, piezoelectric coupling coefficient, thermal and mechanical stability at room temperature [1–3]. ZnO is a potential semiconductor material largely used in electronic industries because of its wide direct energy band (~3.4 eV) and high bond energy (~60 meV). It can be significantly shaped into different dimensions (one, two, and three dimensions) and structures. Various shapes of ZnO such as nanotubes, star-like, wires, comb-like, springs, belts, ribbons, rings, spheres, nanorods are synthesized through different methods [4–10].

3.1.2 Crystal Structure of ZnO NPs

Most of the II–VI binary compound semiconductors crystallize in cubic or hexagonal wurtzite (Wz) structures [11, 12], in which each anion is surrounded by four cations at the corners of a tetrahedron and vice versa. The considerable ionic properties of ZnO have increased its bandgap compared to typical sp^3 covalent bonds in the other tetrahedral coordination structure in similar compounds. ZnO is a semiconductor with ionic characteristics residing at the borderline between the covalent and ionic bonds [13, 14]. Rocksalt (B1), zinc blende (B3), and wurtzite (B4) are three known crystal structures, which are schematically depicted in Figure 3.1.

The symmetric crystalline structure of wurtzite is stable under ambient conditions. The zinc blende structure can be stabilized only by growth on cubic substrates. The rocksalt or Rochelle salt (NaCl) structure can be achieved under relatively high-pressure conditions similar to GaN [13].

Nanotechnology in Electronics: Materials, Properties, Devices, First Edition.
Edited by Visakh P.M., Artem Semkin, Raneesh Balakrishnan, and Saša Lazović.

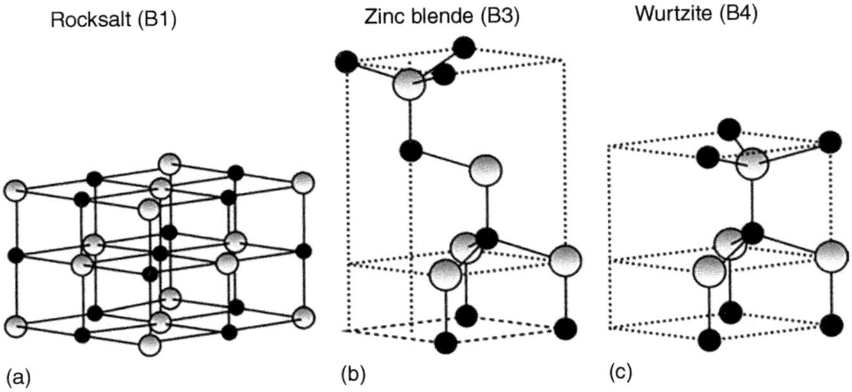

Figure 3.1 (a) Cubic rocksalt (B1), (b) cubic zinc blende (B3), and (c) hexagonal wurtzite (B4) as the ZnO crystal structures shown in stick and ball representation. Gray and black spheres indicate Zn and O atoms, respectively. Source: [15] / with permission of AIP Publishing.

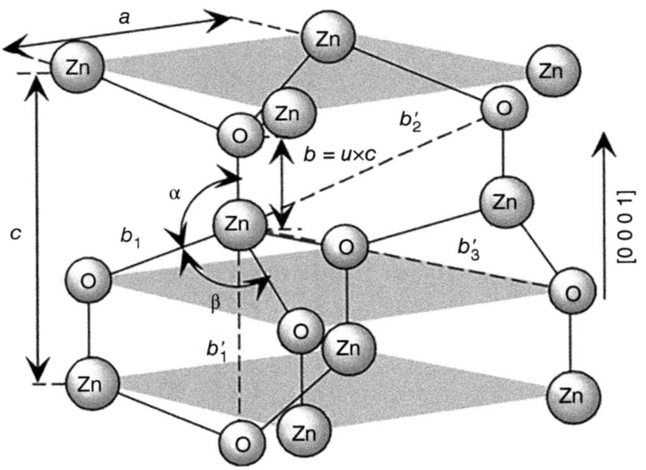

Figure 3.2 Wurtzite ZnO crystal structure. *a*, *c*, and *u* are lattice constants in the basal plane and in the basal direction, and the bond length or the nearest-neighbor distance *b* is calculated by *c*. α and β are bond angles, and b′1, b′2, and b′3 denote the type of second-nearest-neighbor distances. ($u = 0.375$ and $\alpha = 109.47°$ in an ideal crystal). Source: [16] / Hindawi/ CC BY 3.0.

Two important lattice parameters are a and c, which are in the ratio of $c/a = \sqrt{8}/3 = 1.633$ (in an ideal wurtzite structure), which indicates a hexagonal unit cell of the ZnO wurtzite structure schematically represented in **Figure** 3.2.

The structure includes two percolated two hexagonal close-packed (hcp) sublattices. Each sublattice has one type of atom that transposes regarding each other along the c-axis. In an ideal wurtzite structure, the amount of displacement would be $u = 3/8 = 0.375$, in which u is the length of the bond parallel to the c-axis, in units of c.

Each sublattice consists of four atoms per unit cell. Four atoms from group VI are surrounded by group II atoms or vice versa, which are coordinated at the edges of a tetrahedron. The fundamental parameter u is determined as the b (bond length or nearest-neighbour distance) dividing by c is defined as u value, which is the anion–cation bond length or the nearest-neighbor distance, and the basal lattice parameter. The basal plane lattice parameter is universally described by a and the unit cell height, perpendicular to the basal lattice plane, is denoted by c [17].

The main crystallographic vectors of the wurtzite crystal structure are $\vec{a} = (1/2, \sqrt{3/2}, 0)$, $\vec{b} = a(1/2, -\sqrt{3/2}, 0)$, and $\vec{c} = a(0, 0, c/a)$. Cartesian coordinates of the basis atoms in wurtzite are $(0, 0, 0)$, $(0, 0, uc)$, $a(1/2, \sqrt{3}/6, c/2a)$, and $a(1/2, \sqrt{3}/6)$, $[u + 1/2]c/a)$ [17].

In a real ZnO crystal, due to the changes in the c/a ratio or the u value, the ideal wurtzite crystal arrangement is deviated. From the experimental results, the deviation from the observed structure, in comparison to GaN, could be attributed to the absence of zinc blende phase under ambient conditions [14, 18].

There exists a profound correlation between the c/a ratio and the u basal parameter. Accordingly, when the c/a ratio declines, u parameter increases so that the four tetrahedral distances remain constant, while a wide range of polar interactions distorts tetrahedral angles [14, 17]. In the following relation, the slight difference between the two bond lengths will disappear and they will become equal Eqs. (3.1–3.3):

$$u = \left(\frac{1}{3}\right)\left(\frac{a^2}{c^2}\right) + \frac{1}{4} \tag{3.1}$$

The closest-neighbor bond length b along the c-direction and b_1 along the off c-axis are calculated as follows:

$$b = cu \text{ and } b_1 = \sqrt{\frac{1}{3}a^2 + \left(\frac{1}{2} - u\right)^2 c^2} \tag{3.2}$$

Furthermore, the closest-neighbor distances, b'_1 (along the c-direction), b'_2, and b'_3, are determined as the second-closest neighbors with the bond lengths [18].

$$b'_1 = c(1 - u), \quad b'_2 = \sqrt{a^2 + (uc)^2}, \quad \text{and} \quad b'_3 = \sqrt{\frac{4}{3}a^2 + c^2\left(\frac{1}{2} - u\right)^2} \tag{3.3}$$

The bond angles α and β are calculated from Eq. (3.4):

$$\begin{aligned} \alpha &= \pi/2 \begin{matrix} u + \frac{1}{2} \\ - \\ +\cos^{-1}\left[\left(\sqrt{1 + 3\left(\frac{c}{a}\right)^2 (2)}\right)^{-1}\right] \end{matrix}, \\ \beta &= 2\sin^{-1}\left[\left(\sqrt{\frac{4}{3} + 4\left(\frac{c}{a}\right)^2 \left(-u + \frac{1}{2}\right)^2}\right)^{-1}\right] \end{aligned} \tag{3.4}$$

Close-packed (1 1 1) planes in zinc blende and rocksalt (Rochelle salt) crystals and the resultant (0 0 0 1) basal planes in the wurtzite crystal are different from (1́ 1́ 1́) and (0 0 0 1́) planes. Because these three crystalline structures do not have inversion symmetry, the crystals display polarization, which shows the direction of the bonds.

Based on the convention, the polarity is considered Zn polarity when the [0 0 0 1] axis points from the face of the O plane to the Zn plane, known as the *z* positive direction. Therefore, the bonds along the *c*-direction are from Zn cation to O anion. By the same argument, when the bonds along the *c*-direction are from anion (O) to cation (Zn), the polarity of O is considered the polarity of the structure [14].

Polarity affects many properties of a material such as plasticity, etching, growth, surface tension, pyroelectricity, piezoelectricity. In addition to the primary polar plane (0 0 0 1) and the related direction ⟨0001⟩ in a wurtzite ZnO crystal, which is the most conventional surface and direction for growth, this crystal structure possesses many other planes and secondary directions, resulting in appropriate polarization of the structure [13, 16].

3.1.3 Methods for Synthesis of ZnO NPs

The techniques for the synthesis of zinc nanoparticles can be divided into solid-, liquid-, and gas-phase processes: solid phase – mechanical ball milling and mechanochemical methods; liquid phase – laser ablation, exploding wire, solution reduction, and decomposition method; gas phase – gas evaporation, explosion wire, and laser ablation and spark discharges. Chemical bath deposition [19], green chemistry synthesis [5, 10], and wet chemical method [20] are other recently used techniques for the synthesis of ZnO NPs.

3.1.3.1 Mechanochemical Methods

Mechanochemical synthesis includes a combination of chemical and mechanical processes in molecular dimensions involving mechanical and chemical separation of mechanically strained solids. Ball milling is typically different from the mechanochemical synthesis. In a typical ball milling method, medium size reduction under atmospheric condition produces nanosized or micron-sized powder particles. Solid-state displacement reaction during ball milling is the initial process in the mechanochemical method that can produce nanosized particles down to 5 nm embedded within larger by-product phase particles [21, 22].

Li et al. [22] used a new environmentally friendly mechanochemical method for synthesizing several metal-oxide semiconductors (MoO_3, TiO_2, and ZnO). They indicated that about 10 g of $H_{0.5}MoO_3$ and vacancy-rich TiO_2 and ZnO can be easily obtained by grinding semiconductor powders with ascorbic acid at room temperature. Uniformity, simplicity, high efficiency, and scalability are the important benefits of this method. The synthesis was confirmed by crystal structures, size and shape of metal-oxide semiconductors using XRD analysis and SEM images.

3.1.3.2 Sol–gel Synthesis

The sol–gel synthesis of nanoparticles produces metal-oxide materials from small inorganic molecules, which is appropriate for preparing different coatings on other material surfaces based on the chemical reaction of a certain solution. This process involves hydrolysis, condensation, particle growth, and gel formation, which are affected by temperature, reactants, type of reaction medium, and pH. The main advantages of sol–gel synthesis are low temperature, suitable mechanical stability, possibility of casting in different shapes, and good solution resistance [23].

Zinc oxide nanoparticles (ZnO NPs) through the sol–gel method can be prepared without coating and stabilizer agents using ultrasonic irradiation, which affects the properties of nanoparticles. The results revealed that the structure of nanoparticles was uniform with less aggregations and reduced particle sizes due to the increased ultrasonic irradiation power and time [24].

3.1.3.3 Hydrothermal Method

The hydrothermal method is a simple and environmentally friendly technique as it does not need organic solvents or additional processes such as grinding and calcination to obtain the final product. In this synthesis method, at a temperature above 100 °C and high pressure in an aqueous medium, a blend of materials is heated in an autoclave for several days, and nucleation and growth of crystals occur. Different crystal shapes and dimensions are obtained by engineering the process temperature and pressure [25].

To improve the electrical performance, novel hybrid piezoelectric structures based on electrospun polyvinylidene fluoride (PVDF) nanofibers and vertically grown ZnO nanorods were reported. Vertical ZnO nanorods were grown on a gold-coated PTFE (polytetrafluoroethylene) by the hydrothermal method. Spin coating was used to deposit 0.005 M of zinc acetate dehydrated in ethanol on the gold-coated PTFE film surface. Then, the substrate was heated at 90 °C for 30 minutes. The seed layer was uniformly deposited in a two-step process of deposition and decomposition to cover the substrate. ZnO nanorods were hydrothermally grown on the seed layer. Zinc nitrate hexahydrate and hexamethylenetetramine (HMTA) were used to hold the seed layer at 90 °C for three hours [26].

3.1.3.4 Liquid-phase Synthesis

Nanoparticles of narrow sizes and high purity are produced by pulsed laser ablation deposition method. In this technique, core–shell or narrow-sized nanoparticles are synthesized by degradation of a solid metal immersed in an aqueous solution by using a laser beam [27, 28].

Masuda et al. [28] demonstrated a simple liquid-phase synthesis method to control the morphology and crystal size of ZnO. ZnO nanowires (NWs) of 100 nm long and about 50 nm wide were successfully synthesized at 50 °C. They also indicated that nanowires had no branches without aggregations.

3.1.3.5 Controlled Precipitation

Controlled precipitation is a typical method used to obtain ZnO nanoparticles in large quantities with repeatable properties for use in industrial products. In this process, zinc salt solution is spontaneously reduced by a reducing agent. After a while, the precursor of ZnO is precipitated from the solution. Then ZnO is thermally treated and subjected to milling to remove impurities. Breaking down of agglomerations is an important step in this method. Therefore, temperature, pH, type of aqueous medium, time of precipitation, and the selected zinc salt are most vital

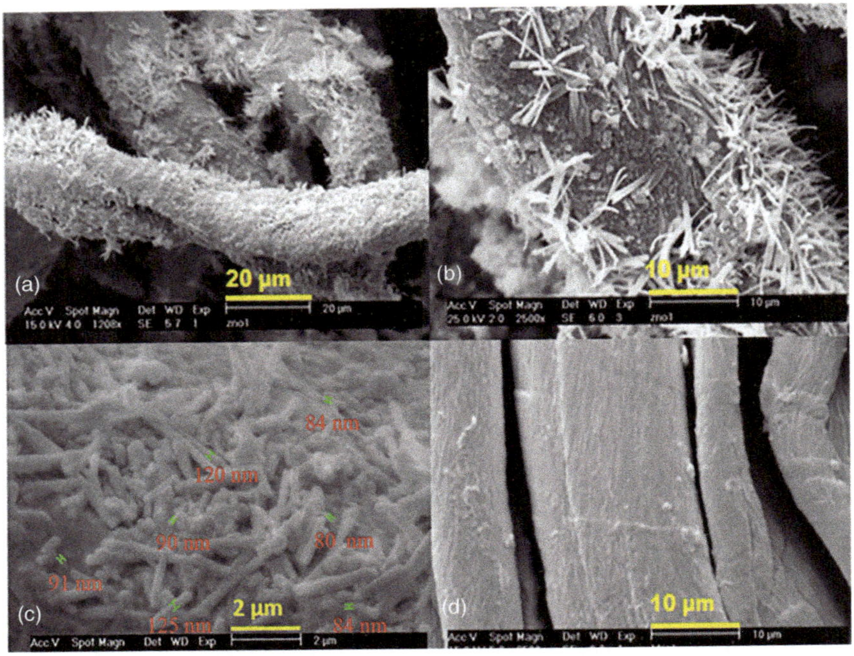

Figure 3.3 SEM images of (a–c) ZnO nanorods synthesized on cellulosic fibers and (d) untreated cellulosic fibers. Source: [10] / with permission of ELSEVIER.

parameters in precipitation of ZnO NPs, which have influence on the size and final characteristics of the nanoparticles [29].

Aladpoosh et al. [10] used the ZnO controlled precipitation method, a green synthesis process using the natural plant source, namely, Keliab and zinc acetate. Wurtzite structure and nanorod shapes are obtained from in situ synthesis of ZnO on the cellulosic chains of cotton through the green synthesis method as depicted in Figure 3.3.

Ghayempour et al. [5] obtained urchin-like ZnO nanorod arrays using Tragacanth gum, a clean and natural reducing and stabilizing agent for preparation at low temperatures.

3.1.3.6 Vapor Transport Synthesis

Vapor transport synthesis is one of the most commonly used methods to synthesize ZnO nanostructures. Depending on the nanostructure formation mechanism, vapor transport synthesis can be classified into free vapor–solid and catalyst-assisted vapor–liquid–solid (VLS) [30, 31].

Using a vapor transport method, Wu et al. [32] synthesized ZnO nanostructures by the reaction of Zn vapor with H_2O vapor and O_2 gas. The growth orientation of ZnO strongly depends on the oxygen species. ZnO nanoneedles were efficiently developed on a glass using the VLS method, which was confirmed by using a glass covered with Ag thin layer as a catalyst. Different Ag layer thicknesses and Ar flow rates were applied for the growth of ZnO nanoneedles at 450 °C.

Experimental results revealed that the growth of ZnO nanoneedles on the surface of the Ag catalyst–coated glass with 10 nm thickness at the Ar flow rate of 40 mL min^{-1} was one of the parameters influencing the nucleation and growth of nanoneedles with high crystallinity. They also discovered needle–wire connections that significantly contribute to the electrical features of ZnO devices. This approach has potentiality for various applications such as high-performance gas sensor devices [31].

3.1.4 Piezoelectric Effect of ZnO

In the piezoelectricity phenomenon, an electric field is generated when a material is subjected to mechanical strain or vice versa [30, 33–35].

The primary reason for the piezoelectric property is the nonsymmetric crystal structure of some particular materials that triggers a response by changing the polarity created by the application of mechanical strain. Piezoelectric materials can be used as sensors, actuators, and energy harvesters because a material under strain behaves as a capacitor that stores electrostatic potential energy. Therefore, piezoelectricity allows transformation of mechanical and electrical energies [33, 36, 37].

The electrical energy output is relatively low at the macroscale in comparison to the mechanical energy required to deform a material. Therefore, use of piezoelectric materials is generally limited in actuators, which are used in atomic force microscopy and pressure sensors in some cases. In contrast, by extracting relatively low mechanical energy from natural sources such as wind, human body or muscular movements, sea waves, and flowing water, an adequate amount of small forces to deform piezoelectric nanostructures in the nanoscale can be generated [36, 38, 39].

Hence, thin films and NWs are the building blocks suitable for next-generation energy harvesting devices. Moreover, the size of these structures can be reduced to the nanoscale (thin films and NWs) to enhance the conversion efficiency significantly. The size reduction is mainly due to the following reasons: (i) before failure, nanomaterials tolerate rather large critical deformations since the generated electrostatic potential (V) is proportional to the applied strain (ε) and the piezoelectric coefficient (d_{33}), namely $V \propto d_{33}\varepsilon$; (ii) because of surface effects and high surface-to-volume ratios, the properties of the material are often enhanced at the nanoscale compared to the bulk [40, 41].

Nevertheless, electromechanical coupling and potential size effects are not addressed adequately. There exist some challenges including: (i) difficulty in manipulating the samples at the nanoscale, (ii) performing suitable electrical measurements responsible for contact resistances, and (iii) current and voltage measurements with high enough resolution. These challenges seem to cause a large scatter in experimentally measured piezoelectric coefficients of ZnO nanostructures. For instance, besides the piezoelectric coefficient of ZnO nanorods ranging from 4.41 to 7.5 pm V^{-1} with diameters in the 150–500 nm range, a range of 14.3 to 26.7 pm V^{-1} was found through piezoresponse force microscopy for ZnO nanobelts of tens of nanometers in thickness. On the contrary, for a 230 nm ZnO NW, a resonance shift method resulted in rather a high value of 12 000 pm V^{-1}, compared to the bulk value of 12.4 pm V^{-1} [26, 34, 41, 42].

Briefly, quantum mechanical computations showed that both GaN and ZnO NWs have larger piezoelectric coefficients due to size effects. Diameters relatively larger than 2.5 nm for GaN are desired to be considered bulk values. The results revealed that the potential of NW systems in actuation, sensing, and energy harvesting applications will go a long way to be fully exploited. Especially, by developing self-powered energy harvesting devices, piezoelectric output voltage in proportion to the piezoelectric coefficient can be magnitude by orders of 1–2 [13, 17, 34, 43]. This can be done by decreasing the size of NWs that are used in the design of systems, suggesting new research directions. Especially, experimental confirmation for theoretical predictions is needed to acquire further understanding of the phenomenon. Apparently, measurement of piezoelectric coefficients in NWs of diameters less than 20 nm is more challenging and a very interesting topic in the nanotechnology community. Furthermore, in this regard, in order for modeling of large NWs, which cannot be investigated with density functional theory (DFT), development of new computational tools based on classical molecular dynamics has been suggested [44].

3.2 Zinc Oxide Piezoelectric Nanogenerators

One of the future generation challenges would be restriction on available energy sources, such as fossil fuels. Therefore, significant efforts must be devoted to discovering, developing, and optimizing energy conservation techniques. Harvesting energy from ambient sources can be a viable solution. In this regard, piezoelectric nanogenerators (PENGs) have attracted a great deal of attention. They are capable of harvesting mechanical or thermal energy from wasted resources such as human movements, wind, acoustic and ultrasonic waves, fluid flow and converting it into low-cost, clean, efficient, and sustainable electrical energy that is suitable to be used for powering nanodevices [45–47].

The concept of NGs was introduced by Wang's research group in 2006 [46]. They utilized a conductive tip of an atomic force microscope (AFM) to deflect the aligned semiconductor ZnO NWs and successfully produced a measurable piezoelectric output voltage around 10 mV. They grew ZnO nanorods on a sapphire substrate using the VLS method, in which the gold particles remained on the top of each rod after growth. This formed a Schottky contact between the conductive AFM tip and the semiconductor ZnO NWs, which is a key factor in the current generation of the PENGs. They showed that the stretched side of the rod with a positive potential produced a reverse bias with the Schottky junction and this led to no current to screen the polarization. Once the contact between the tip and the compressed negatively polarized side of the rod was disconnected, the junction was forward biased, and the current would be able to flow and screen the polarization. Monitoring the open-circuit voltage generally reflects uncertainty in measurements of this type of device. The current flow is used to explain voltage measurements, while there is no direct relation between current flow and voltage descriptions in such complex situations involving nonlinear and capacitive elements. Further, the source of the output voltage in this method is primarily controversial [48].

3.2.1 Nanogenerators

NGs mainly work based on four different approaches including triboelectric, piezoelectric, pyroelectric, and their combinations. NGs supplement a new space to the principles of conversion of mechanical energy into electrical energy via piezoelectric and triboelectric effects. The performance of piezoelectric systems is better than triboelectric ones [49]. Moreover, some NGs use thermal energy to produce electricity based on pyroelectric effects [26, 49–54].

3.2.1.1 Piezoelectric Nanogenerators (PENGs)

Piezoelectricity means stress-induced electricity when the central symmetry in the crystal structure breaks down under external forces, thereby forming a piezoelectric potential or piezopotential (Figure 3.4). Similarly, subjecting a piezoelectric material to an external electric field causes mechanical deformations [55–57]. Opposite poles deposited in the amorphous regions of the polymer structure help to form piezoelectric effects [58]. Besides, the difference in atomic electronegativity results in polarity of bonds in polymer chains, causing piezoelectric effects. For instance, Fukada in 1974 [59] explained the piezoelectric effect in both β- and α-helix forms of biopolymers, in which different atoms within the polymer structure form chiral carbon atoms in the polymer backbone, thereby leading to an electrical gradient.

Also, the wurtzite-structured ZnO crystal as a modular material system causes the tetrahedrally coordinated Zn^{2+} and O^{2-} to stake layer by layer along the *c*-axis (Figure 3.5a). The center of the positive and negative charges coincides under stress-free conditions in the undistributed state. Applying an external force along the c-direction deforms the structure of the ZnO crystal. Therefore, the center of the positive and negative charges will no longer coincide and an electric dipole is formed, thereby generating the piezopotential – the basic principle of piezopotential formation (Figure 3.5b). For energy conversion, in the load-deformed crystal free electrons are driven to flow through an external circuit to quantitatively screen the

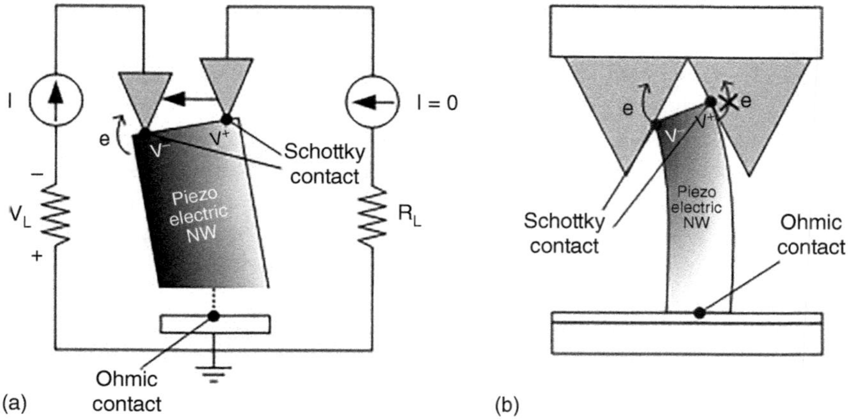

Figure 3.4 Working principle of piezoelectric nanogenerators where an individual NW is subjected to the force of an AFM tip [55].

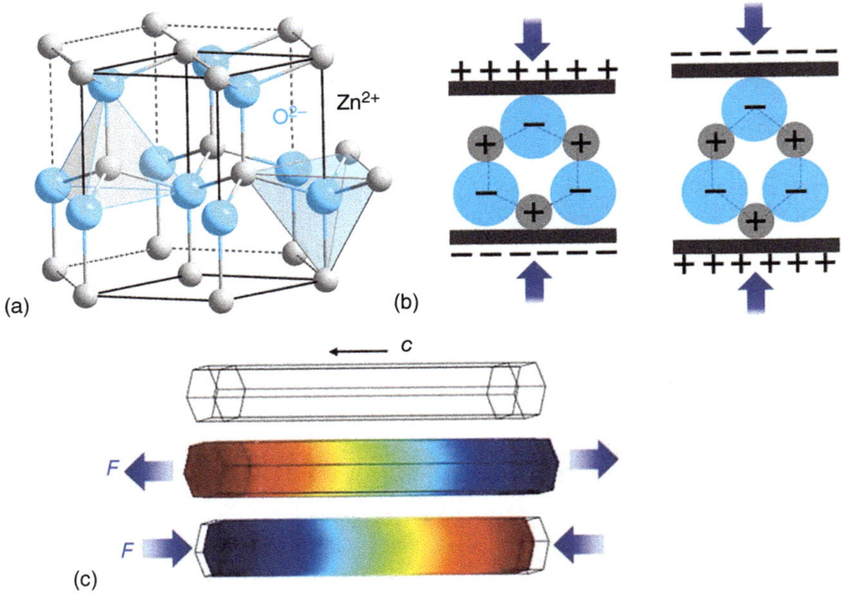

Figure 3.5 (a) Atomic model of the wurtzite structure of ZnO. (b) Piezoelectric properties and different piezopotentials in tension and compression modes of the material. (c) Distribution of the piezoelectric potential in ZnO NWs grown along the *c*-axis under axial strain. Source: [61] / with permission of AIP Publishing.

piezopotential and reach a new equilibrium state (Figure 3.5c) [61]. Overall, an NG produces a steady stream of pulse current flowing through the external circuit by changing the piezopotential continuously through a dynamic strain phenomenon [62]. The formation of piezopotential and the model of power generation can be applied to various piezoelectric materials such as PZT, $BaTiO_3$ (BTO), PVDF. PENGs are a combination of piezoelectric materials and flexible substrates. Therefore, the two key factors for manufacturing PENGs are material selection and structural design.

The possibility of production and operation of PENGs in macro-, micro-, and nanoscales is an advantage of piezoelectric products [63–65].

The polarization or surface charge density and applied stress can be calculated using the equation $P_i = d_{ij} \times T_j$, where T_j is the applied mechanical stress along the direction j, P_i is the induced polarization along the direction i, and d_{ij} is the piezoelectric coefficient [66]. For instance, cantilever beam-type PENGs can be polarized in the d_{33} and d_{31} piezoelectric production modes, where "$_{31}$" or "$_{33}$" represents the direction of polarization and applied stress [67]. The d_{31} mode electrodes are perpendicular to the polarization direction, while in the d_{33} mode, the direction of the electrodes and polarization is the same [68, 69].

The generated piezoelectric alternating current (AC) must turn to direct current (DC) using a rectifier and be stored in the operational supercapacitor [70].

3.2.1.2 Triboelectric Nanogenerators (TENGs)

Triboelectrification is a universal and ubiquitous phenomenon, which has been known since the ancient Greek era, and is considered a negative effect due to no reasonable cognition and utilization technology. Energy harvesting using triboelectric effect has attracted significant attention since the invention of the first triboelectric generator in 2012. Energy harvesting using triboelectric nanogenerators (TENGs) is based on the electrostatic induction and triboelectric effect. This novel developed TENG can convert abundant mechanical energy into electricity as a useful mechanical energy harvesting technology [71–74].

TENGs have promising applications in harvesting mechanical energy in small and large scales due to their high efficiency, lightweight, environmentally friendly nature, pervasive accessibility, and low cost. They are able to scavenge various types of energies that are pervasive otherwise wasted in our daily life such as human motion [75–77], vibration [78, 79], mechanical rigging [80–82], wind [83–85], and flowing water [83, 86, 87]. As Figure 3.6 indicates, the first TENG was made of a polyethylene terephthalate (PET) layer, Kapton film, and two electrodes on both sides [71]. Contacting the layers by friction, different electron attraction of the two polymer films generates opposite charges in equal amounts at the surface of the two films. An electron will move from the PET layer with a high electrochemical potential to the surface of the Kapton film with a low electrochemical potential, causing positive and negative charges on the surfaces. This makes an interface dipole layer called the triboelectric potential layer. As the two films contact each other and separate, the alternative potential drives the electrons in the external circuit to minimize the energy created by the triboelectric potential [71, 73, 74, 88].

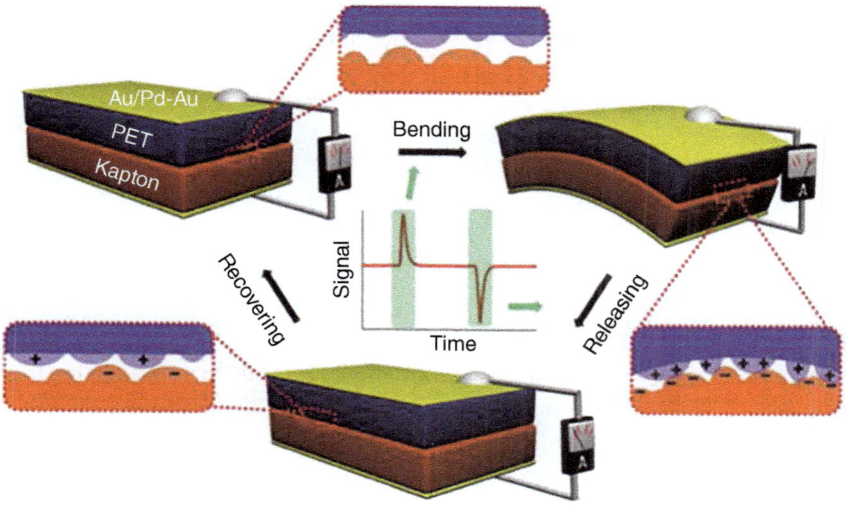

Figure 3.6 Schematic illustration of the construction and working principle of the first TENG. Source: [71] / with permission of Elsevier.

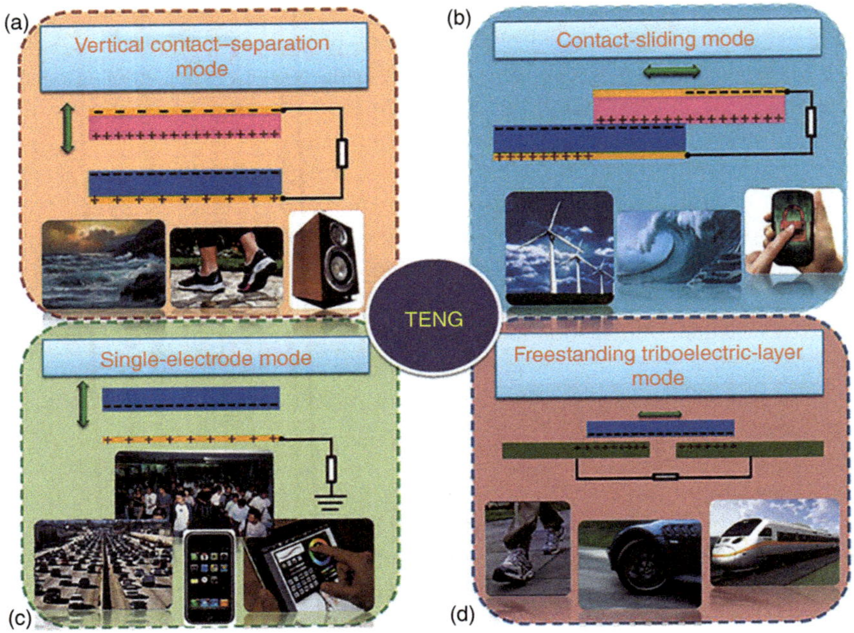

Figure 3.7 Triboelectric nanogenerators according to four different operation modes. (a) Vertical contact, (b) lateral slip, (c) single electrode, and (d) non-connected electrode. Source: [73] / with permission of Royal Society of Chemistry.

TENGs can be categorized into four different operation modes:

1. Vertical contact–separation: An electrode is located on one surface of each two-pole dielectric film. Once two dielectric films physically contact, opposite charges appear on their surface (Figure 3.7a).
2. Lateral sliding: Two dielectric films slip parallel to the surface and triboelectric charges are created on their surface. Electrons flow on the top and bottom electrodes to screen the polarization (Figure 3.7b).
3. Single-electrode: In this mode, the ground is considered the reference electrode for harvesting energy from a liberally touching object devoid of using an electric conductor (Figure 3.7c).
4. Freestanding triboelectric-layer: This involves a triboelectric layer with rough sections and two fixed electrodes with interdigital designs (Figure 3.7d) [73].

Sufficient load separation and friction area are the two main factors affecting the performance of TENGs. The load separation is affected by the electronegativity of the materials used in the device.

3.2.1.3 Pyroelectric Nanogenerators

Pyroelectricity is one of the properties of certain anisotropic layers defined as the temperature-related polarization. Net dipole moment and the temperature change (or bound charge) cause changes in pyroelectric materials. Extensive distribution of thermal energy is an advantage of pyroelectric NGs. Pyroelectric NGs are widely

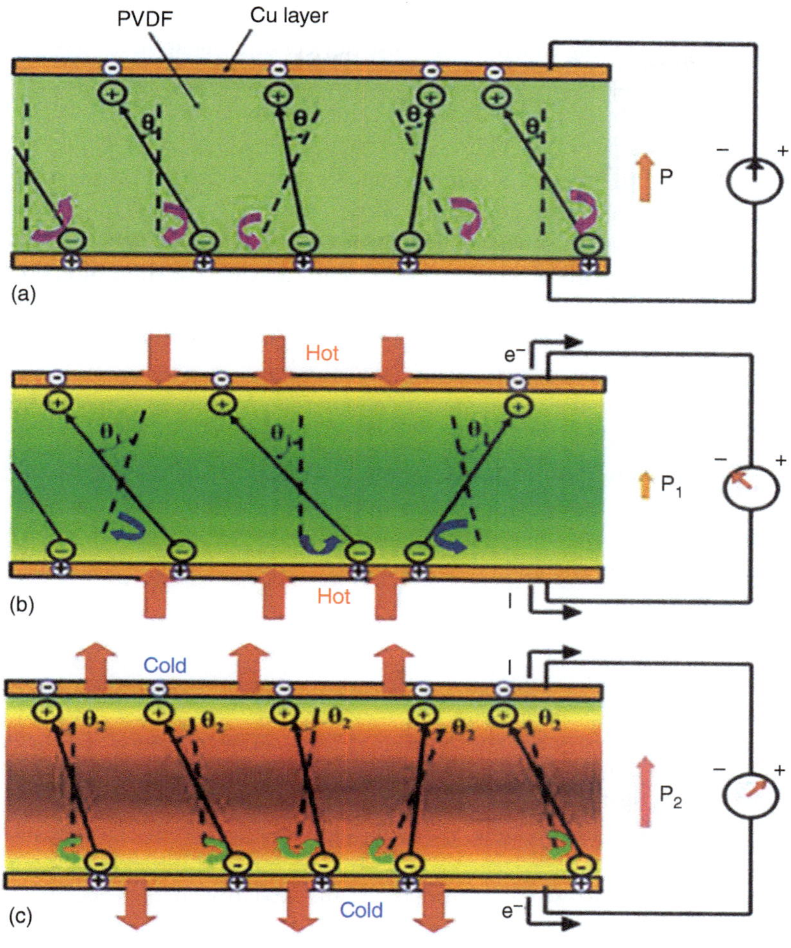

Figure 3.8 Pyroelectric effect. Source: [89] / with permission of Royal Society of Chemistry.

used to scavenge waste heat from industries such as petroleum refining and steel/glass manufacturing industries or to harvest natural geothermal heat such as volcanic heat or solar heat.

Figure 3.8 indicates a pyroelectric NG made of two copper-plated electrodes, a pyro-active PVDF film, and a contact wire from the electrodes for measuring the generated electricity. The device is pack-coated with polyvinyl chloride (PVC) and used for harvesting the heat energy from hot/cold water. The open-circuit output voltage and short-circuit current reached a maximum of 192 V and 12 μA, respectively, under a temperature gradient of 80 °C [89].

3.2.1.4 Hybrid Nanogenerators

To scavenge different kinds of energy, miniaturize the device size, and compensate for the respective shortcomings in each type, hybrid energy harvesters are needed for future wearable power sources, which have received significant attention [26, 90–92]. The energy harvesting capability of these devices and the ability to sense a variety of signals are strengthened due to the synergistic effects of individual components. PENGs and TENGs are more likely to be combined into one device to superimpose the electrical output from their distinct mechanisms due to their relevant impedance and structural features and, more importantly, the capability to convert mechanical energy into electricity. High piezoelectric output current and high triboelectric output voltage are the important advantages of PENGs and TENGs, which can be combined using appropriate structural designs and polarization direction modulations [93]. Researchers have shown that the matched direction of piezoelectric polarized and triboelectric induced charges generate more power, while their reverse direction significantly reduces electricity generation.

There are three different coupling modes to combine PENGs and TENGs based on the number of electrodes and their combination methods. In mode I, PENGs and TENGs share a pair of electrodes [94–98]; in mode II, they share one common electrode [47, 51]; and in mode III, they work independently [99–102]. Electrical generation of each mode can be divided into the following steps due to the contact/separation state of triboelectric materials and compress/release state of piezoelectric materials:

1. TENG contacted and PENG pressed
2. TENG contacted and PENG released
3. TENG separated and PENG released

An r-shaped piezo–triboelectric hybrid nanogenerator (PTHNG) with PVDF piezoactive layer, nanostructured Al electrodes, and micro-patterned polydimethylsiloxane (PDMS) layer (Figure 3.9) reached the output voltage, current density, and volume power density of the PENG up to 52.8 V, 20.75 $\mu A\,cm^{-2}$, and 10.95 $mW\,cm^{-3}$ and for the triboelectric NG, 240 V, 3.40 $\mu A\,cm^{-2}$, and 2.04 $mW\,cm^{-3}$, respectively [51]. As another example, a PTHNG was made by coupling triboelectric PTFE layer with ZnO–PVDF piezoactive nanocomposite film and depositing metal electrodes of gold and aluminum (Figure 3.10). ZnO nanorods improved both piezo- and triboelectric effects of PVDF, resulting in better performance of the hybrid NG. Thus, the piezo- and triboelectric open-circuit voltage enhanced from ~8 to ~12 V and from ~45 to ~66 V, respectively. The overall output voltage of the PTHNG also increased from ~55 to ~78 V. Moreover, in the presence of ZnO nanorods, the overall current density increased from ~0.30 to ~0.46 $\mu A\,cm^{-2}$ for neat PVDF [103].

3.2.2 Zinc Oxide Piezoelectric Nanogenerators

ZnO is an environmentally friendly and multifunctional semiconductor with outstanding piezoelectric effects, good piezoelectric coupling coefficient, and high thermal and mechanical stability at room temperature. Owing to the noncentrosymmetric hexagonal wurtzite crystal structure and polar Zn—O bonds, ZnO has unique piezoelectric properties, which make it a core element of study in the ceramic industry [15, 104].

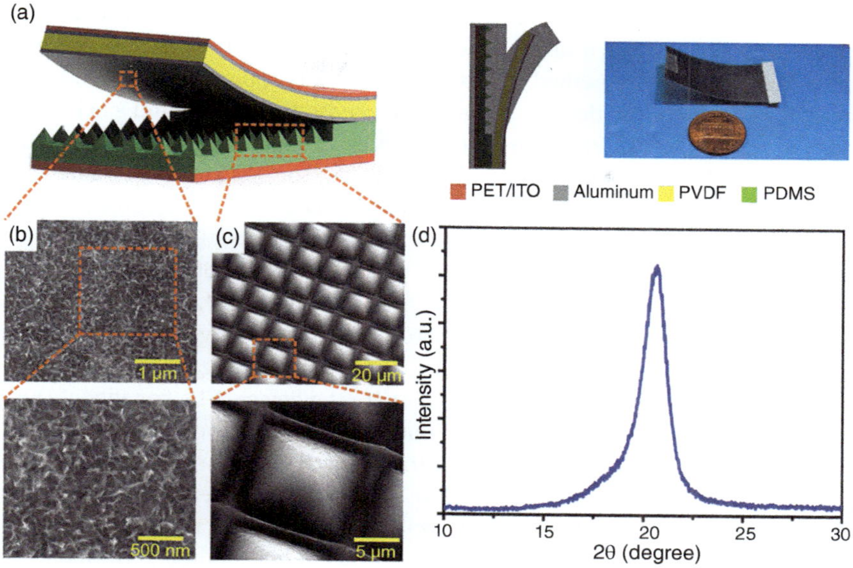

Figure 3.9 (a) Schematic picture of the r-shaped hybrid NG. (b) SEM images of the nanostructured aluminum electrode. (c) SEM images of the micro-patterned PDMS. (d) XRD analysis of the polarized PVDF. Source: [50] / with permission of American Chemical Society.

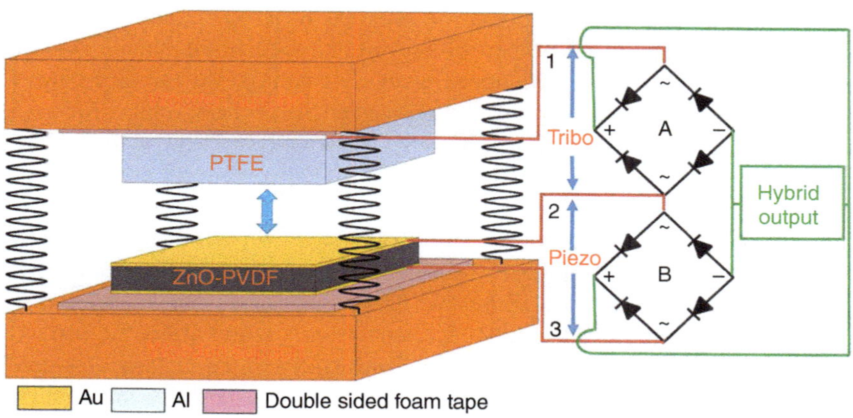

Figure 3.10 Schematic of the piezo- and triboelectric hybrid NG along with the electrical connections. The bridge rectifiers 'A' and 'B' represent the piezoelectric and triboelectric output, respectively. Source: [103] / with permission of American Chemical Society.

Several techniques have been explored to synthesize and develop both DC and AC ZnO PENGs [105]. For the first time in 2009, Xi and coworkers [106] presented a ZnO nanotube-based PENG. They utilized a solution chemical method to synthesize hexagonal ZnO nanotube arrays at temperatures below 100 °C. They studied different growth-related parameters such as time, temperature, and concentration of the solution and confirmed that ZnO nanotubes can be applied for harvesting mechanical energy.

ZnO-based NGs can be incorporated into various substrates. In 2012, Khan et al. [42] grew ZnO nanorods on a silver-coated cotton fabric using the low-temperature aqueous chemical growth method. An external force was applied to ZnO nanorods via an AFM tip, where the flow of charges was taken into account, and a Schottky barrier formed between the electrode and the nanorods. Once the AFM tip deflected, the ZnO nanorods produced an average output voltage of 9–9.5 mV. They believed that output electricity could be increased by using the cotton fabric as substrate due to more flexibility of the fabricated nanorods. Not long after in 2015, a porous polycarbonate (PC) membrane of three different pore sizes of 100, 50, and 30 nm was used as a substrate to synthesize porous ZnO nanostructures by the sol–gel method.

Nanocomposite materials are easy to process and apply without additional treatments, in particular, without removing the template. Organic track-etched polycarbonate membranes offer more flexibility to the device. The benefits of this technique are quick and high-throughput one-single step synthesis and vertically aligned one-dimensional (1D) ZnO nanostructures deposited on a flexible and insulating matrix. All three types of NGs exhibited good performance under bending stress, due to the parallel polarization axes of ZnO nanotubes to the applied strain. This model of ZnO–PC100 NG exhibited outstanding performances with a significant power density of 287.5 mW cm^{-3} [107]. In another research, Zhang et al. [108] designed a hybrid piezoelectric paper-based PENG using in situ synthesized uniformly assembled vanadium-doped ZnO (V-ZnO) mircoflowers and the bacterial cellulose (BC) matrix. To improve the piezoelectricity of neat ZnO, the *c*-axis of crystalline structure must be oriented while V-ZnO demonstrated ferroelectric (or Piezoelectric) properties as a prior requisite for enhancing the output performance by poling with applying an external high voltage. The output piezoelectric voltage, current density, and power density of the V-ZnO/BC hybrid paper-based PENG reached 1.5 V, 80 nA cm^{-2}, and 60 nW cm^{-2}, respectively. The V-ZnO/BC hybrid paper-based flexible PENG is a lightweight and cost-effective approach to develop wearable or implantable energy harvesters and self-powered sensors. Lee et al. [109] used an aqueous solution method with optimized conditions to grow two-dimensional (2D) ZnO nanosheets on an aluminum (Al) electrode with enhanced DC power generation of flexible PENGs. Synthesis and experimental conditions such as growth time and concentration of solution affect the surface morphology and crystallinity of 2D ZnO nanosheets. Increased growth time and concentration of the solution led to synthesis of more ZnO nanosheets followed by better performance of the PENG, while excessive aging and concentration caused ZnO aggregation on the surface, resulting in poor performance in generating DC power output. The open-circuit voltage, short-circuit current density, and power density reached 0.9 V, 16.5 μA cm^{-2}, and 600 nW cm^{-2}, respectively.

Overall, there are four initial aspects representing the unique merits of ZnO nanostructure-based NGs:

1. 1D nanomaterials can endure extreme elastic deformation.
2. 1D NWs are possibly free of transposition due to their small diameters and likely to extend the device lifetime because of their fatigue resistance.

3. 1D nanomaterials have a small elastic modulus and can easily deform under small external mechanical loads. They are even able to harvest energy from small mechanical disturbances.
4. Synthesis of 1D ZnO nanomaterials is substrate-independent that can be synthesized easily on diverse substrates.

The morphology of the substrates can vary among NWs, nanorods, nanoribbons, and nanobelts. The substrates can also be hard or soft with any shapes and degree of crystallinity such as graphene, silicon, polymers, or other nanomaterials. The flexibility of the synthesis method is vital for the technological use of these materials [110].

3.3 Zinc Oxide Piezoelectric Nanogenerators for Low-frequency Applications

PENGs are the most attractive and preferred devices for running various low-power electronics. Among all the low-frequency energy resources, ambient vibrations are considered the most popular phenomenon due to their ubiquitous presence, which can be utilized for energy harvesting using NGs. They can be created by mechanical disturbances from different sources.

There are many abundant low-frequency vibration-based energy resources, which are neglected and wasted such as different types of human body movements, environmental sources (i.e. wind, acoustic wave, solar, temperature gradient, tide), construction and industrial sources, and mechanical sources [111]. Mechanical movements are a considerable source to produce energy due to the availability in practical environments.

3.3.1 Approaches for Scavenging Low-frequency Vibrations

Frequency up-conversion is considered a very practical solution to enhance the output power and deal the challenges of harvesting environmental energy at low frequencies. The harvested kinetic energy from environmental vibrations is detected by a low-frequency oscillator and transferred to a high-frequency oscillator to convert into electrical energy by suitable transfer mechanisms. Mechanical impact, spring resonator, and magnetic spring are the three important and widely used mechanisms to achieve frequency up-conversion [112].

3.3.2 Device Structures

3.3.2.1 Arc Shaped

The arch-shaped configuration is one of the recently used structures for scavenging low-frequency vibrations from various mechanical sources. This configuration is mostly used for TENGs and PTHNGs [113–115].

3.3.2.2 Cantilever

Many vibration energy harvester structures consist of the shape of cantilever beams such as triangular, rectangular, cylindrical. Under excitation of vibration, oscillation is applied to the entire beam, and its magnitude gradually increases from the fixed end to unfixed end. The resonant frequency (ω_r) of the cantilever beam–based vibration energy harvesting is

$$\omega_r = \sqrt{\frac{k}{m}} \tag{3.5}$$

where k and m are the cantilever stiffness and sum of masses of the cantilever and tip, respectively [116]. Several methods regarding multiple cantilever designs were considered to extend the bandwidth of the working frequency for energy harvesters [117–119].

Fabrication of different sandwiched NGs on plastic substrates were reported by studying PENGs for low-frequency applications to maximize the scavenged piezoelectric output through different configurations [120]. ZnO NWs were hydrothermally grown on silver- and gold-coated plastic to make different sandwiched NGs to observe the output through mechanical deformation.

The three different configurations and designs are Ag—ZnO NWs–AU, Ag—ZnO NWs/ZnO NWs–Au, and Ag—ZnO NWs/PVDF/ZnO NWs–Au where the third design showed a maximum induced open-circuit output voltage of 2.4 V and short-circuit current of 152.2 μA under loading of 0.05 N force [120].

Each configuration yielded different results indicating the impact of the sandwiched configuration on the output magnitude [121].

Also, piezoelectric vertically aligned ZnO nanorods (NR)s grown on a PEDOT:PSS plastic substrate were used as a direct power source without an energy storage system. The NG was tested under different loads and fingerprint pressure as low-frequency sources. Applying different low-frequency tests showed a relatively high sensitivity to frequencies as low as 5 Hz. The stress produced by acceleration is proportional to the magnitude of the applied acceleration; therefore, the harvested piezoelectric voltage will be proportional to acceleration. An acceleration range of 0.67–5.5 ms^{-2} was obtained for a frequency range of 5–41 Hz, representing a linear correlation [122]. This sensor showed a higher sensitivity of 0.047 $Vs^2\,m^{-1}$ compared with an accelerator sensor based on ZnO NW arrays grown on Au/Si with a sensitivity of 0.025 $Vs^2\,m^{-1}$ at 500 Hz [123].

Hsu and Chen [124] designed a PENG using hydrothermally grown ZnO NWs on a flexible, transparent, and low-cost PET substrate. Figure 3.11 schematically depicts the growth and processing steps of the process. To grow ZnO NWs, the PET was immersed in an aqueous solution of 0.06 M zinc nitrate hexahydrate [$Zn(NO_3)_2$] and 0.06 M HMTA ($C_6H_{12}N_4$) at 95 °C for six hours. ZnO NWs were then coated with 20-nm-thick platinum (Pt) used as a nanostructured contact electrode through sputtering. The PENG was made of ZnO NWs, Pt/ZnO NWs, and PET flexible substrate used for low- vibration frequencies induced by a cool fan with approximately 120 Hz frequency. Without vibration, 2% bending of the sample exhibited 2.5×10^{-9} A, while the maximum piezoelectric current of the sample with

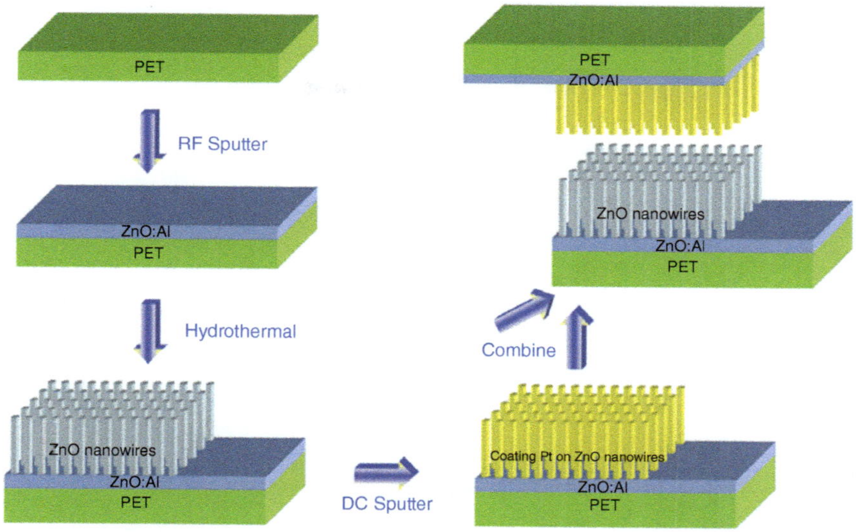

Figure 3.11 Schematic illustration of the growth and processing steps of PENGs. Source: [124] / with permission of American Chemical Society.

2% compressive bending under vibration at low frequencies was about 2.5×10^{-7} A. Bending of 2% along with vibration produced more piezoelectric current than bending of 1 or 0%, twofold or fourfold increase, respectively.

To harvest ambient low-frequency vibrations, Deng et al. [125] fabricated a tuning fork–shaped cantilever-based PENG with ZnO nanorods to provide an optimized power at lower work frequencies. Figure 3.12b shows a detailed description of the fabrication process. To fabricate the PENG, a copper layer with about 300 nm thickness was deposited on a flexible tuning fork–shaped Kapton substrate through magnetron sputtering to serve as the bottom electrode. Then, a ZnO film (about 100 nm) coated on the bottom electrode served as the seed layer through evaporation technique. ZnO, as key functional material of this design, was grown on the ZnO-seed-coated Kapton film via a low-temperature hydrothermal method [126]. Finally, a thin layer of PMMA (polymethyl methacrylate) was spin-coated on the surface of the substrate and then another copper layer was deposited as the top electrode. Usually, PENGs are designed in a cantilever configuration with piezoelectric ceramic layers and a vibrating host structure, mostly steel, for electrical power generation from vibrations under resonance excitation [127]. Both piezoelectric ceramics and steel have high Young's modulus, resulting in high natural frequency of the device. Therefore, material selection and structural design are counted as benefits. Flexible Kapton film with smaller Young's modulus results in lower resonant frequency. Piezoactive ZnO NRs possess the following advantages:, simple manufacturing process, low cost, no sintering and no polarization, and the discontinuity of ZnO NRs, indicating lower shedding and fragmentation of the functional layer [125].

Typically, cantilever-type PENGs are in a rectangular shape with optional attachment of a proof mass at the free end [128]. The Williams and Yates models are

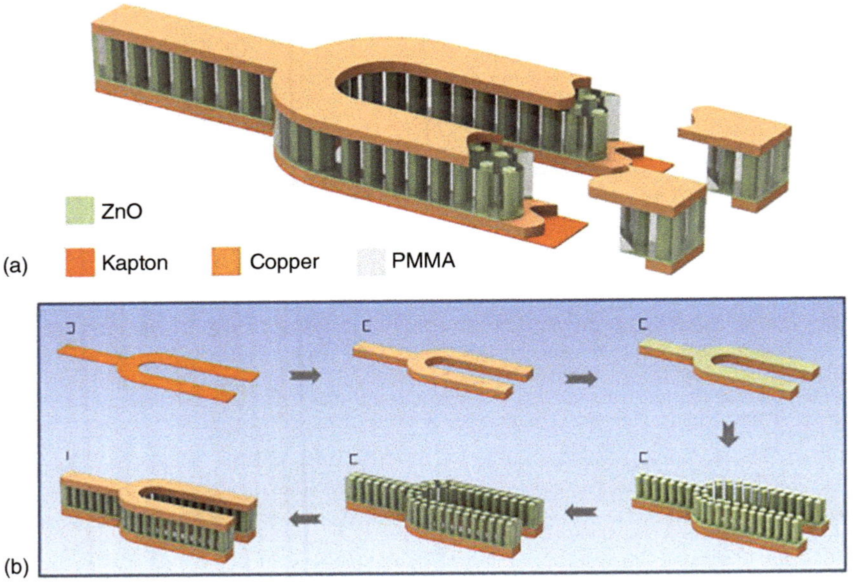

Figure 3.12 (a) Structural design of tuning fork-shaped PENG. (b) Fabrication of fork-shaped PENG (I) Kapton, (II) bottom electrode, (III) ZnO seed layer, (IV) ZnO NRs, (V) spin-coated PMMA, and (VI) top electrode. Source: [125] / with permission of Royal Society of Chemistry.

the most widely used mathematical models for this type of energy harvester [129], in which the external excitation frequency is equal to the natural frequency of the device and leads to resonance, producing the maximum power output. The natural frequency of the harvester is determined by the dimensions of the cantilever. Therefore, different shapes of the cantilever, i.e. rectangular, triangular, tuning fork, were fabricated to verify the influence of the designed structure to effectively reduce the natural frequency. The peak values of the open-circuit voltage and the short-circuit current reached 160 mV and 11 nA at 13 Hz, respectively, and a maximum instantaneous peak power of $0.92\,\mu Wcm^{-3}$ was obtained across a matched load of $9\,M\Omega$.

High piezoelectric coefficients were obtained as a result of large deformation of one-dimensional ZnO nanostructures by applying small mechanical forces [130] and revealed significant mechanical and nonsymmetric properties [131, 132].

Yin et al. [133] introduced low-frequency flexible piezoelectric DC nanogenerators based on ZnO nanorods grown on Cu wires. To produce ZnO nanorods grown on Cu wires through hydrothermal methods, first, Cu wires were modified to copper oxide by hydrochloric acid, acetone, ethanol, and deionized water in ultrasonic bath. Second, $30\,mmol\,L^{-1}$ zinc acetate dehydrate and HMTA in a ratio of 1 : 1 were utilized as solution; the solution was stirred and then transferred into a Teflon reaction chamber. Third, the Cu wires were placed vertically in the reaction chamber for formation of the uniform ZnO seed layer. After that, the ZnO nanorods were grown from the ZnO seeds by an inexpensive hydrothermal method, and finally, the ZnO seed layer of 20 nm thickness was coated on the Cu wire at room temperature by magnetron sputtering [134–136].

Uniform ZnO nanorods covered the entire surface of the Cu wires of 500 nm diameter and 5 μm length as indicated by SEM images [133]. Also, a few hundred nanometers spacing between the nanorods created suitable piezoelectric potential [35, 42, 105].

The positive side of the I–V curves showed a reduction in current with increasing strain. Output current was measured at low frequencies, e.g. less than 10 Hz indicated 50 nA cm^{-1}, which is sufficient to power some nanoscale devices [137]. Flexibility, foldability, and strong power sources in different shapes along with generation of piezoelectric effects by applying common mechanical forces such as heartbeats and footsteps are considerable benefits of NGs based on ZnO nanorods grown on Cu wires [133, 138, 139].

Song et al. [140] revealed that ZnO NWs performed as an energy converter as first, ZnO seed layer with approximately 50 nm thickness deposited on the Ti by magnetron sputtering. Then, ZnO NWs were synthesized by a wet chemical method in the solution containing $Zn(NO_3)^2$ and HMTA. Finally, the ZnO NWs were formed directly on a metal cantilever [141].

The resonance frequency of the NG cantilever could be adjusted by tuning weight of the proof mass or length of the beam. The ZnO NWs were far away from the resonant peak that a large strain could be induced at low frequencies [140]. Various frequencies were modulated for harvesting low-frequency vibrations in the natural environment. Addition of a new piece called stopper created linear vibrations, and compared to other linear vibrations, the bandwidth of the output response was broadened about 200% as the NG harvested the random frequency energy [141, 142]. New combining arrays of these NGs used as cantilevers indicated a bandwidth of 2.3 Hz that was 4.6 times broader than that of a single NG in the linear vibration mode. Generally, the entire device could be considered economical as it reduced the volume while generating the same bandwidth [140–143].

Poulin et al. [144] studied a converter to convert mechanical energy into electrical energy (and vice versa) that can recharge a lithium battery using a piezoelectric generator. The effectiveness of this mechanical harvesting structure was strongly determined by the electromechanical conversion unit, the power management unit, and the electrical load unit.

The last device in this context is schematically represented in Figure 3.13. E-beam evaporation was considered to deposit the titanium and nickel layer, respectively, on a flexible PET substrate. The ZnO NWs were grown on the nickel layer by the hydrothermal method. ZnO NWs from the solution in the presence of nickel and aluminum as galvanic cells were grown at 100 °C for four hours [145–147]. SEM images in Figure 3.13 reveal ZnO NWs of 2 μm in length and 200 nm in diameter [144].

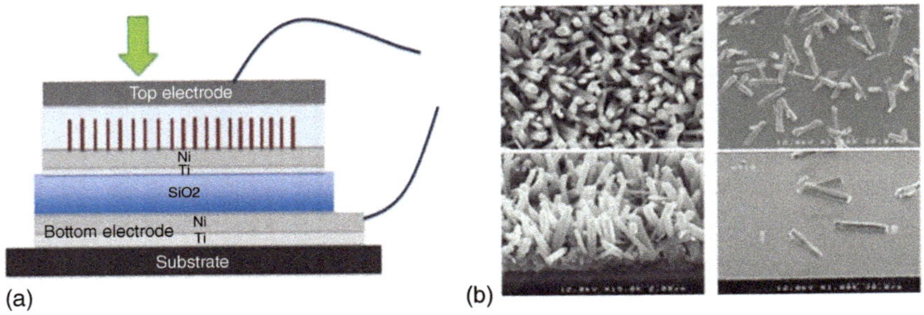

Figure 3.13 (a) Schematic of a piezoelectric generator based on ZnO NWs–polymer composite. (b) SEM images of the top and side of ZnO NWs. Source: [144] / with permission of Elsevier.

ZnO nanowires grown on the metal-coated PET substrate showed high-density piezoelectric properties when reacted with galvanic cells in their growth medium. The following are some benefits of the hydrothermal method for industrial use:

1. Large quantities of ZnO NWs were grown without using an autoclave compared to the classical hydrothermal method.
2. $0.5\,\mu m\,h^{-1}$ at 100 °C was a good growth rate for the induced potential gradient.
3. High-density ZnO NWs with suitable lattice parameter could be generated on special substrates without any seed layers and/or substrate [144, 147].

This study demonstrated that ZnO NWs could be grown on a plastic substrate, presenting a significant performance. In order that they produced a ~56 mV peak-to-peak voltage on 1 MΩ resistive load under a sinusoidal bending force of 10 Hz [144].

Fakhri et al. [26] hybridized electrospun PVDF nanofibers with hydrothermally grown ZnO nanorods to design a hybrid PENG to consider the synergistic effects of different piezoelectric materials. To fabricate the PENG, electrospinning technique was used to deposit PVDF nanofibers (NFs) on the surface of the gold-coated PTFE layer on which ZnO nanorods (NRs) were grown in advance. A conductive Al foil served as the top electrode and attached on electrospun PVDF NFs by silver paste (Figure 3.14). Pristine ZnO and PVDF NF NGs were also fabricated to explore the effect of the hybrid. Applying 4 N force at 6 Hz frequency generated an output current of 114, 212, and 456 nA for pristine ZnO, pristine PVDF, and hybrid structure, respectively. It seems that the output current of the hybrid structure was

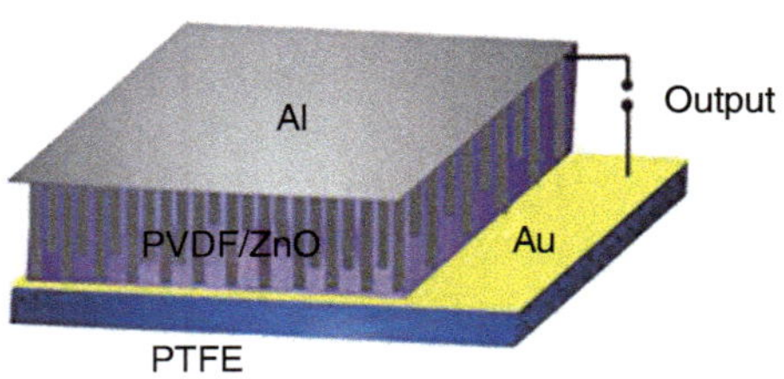

Figure 3.14 Schematic of PVDF/ZnO hybrid nanogenerator [26].

two and four times higher than the pristine PVDF and ZnO, respectively. Based on these findings, using two piezoelectric components in a hybrid NG causes a substantial improvement in the generated power more than 10 times greater than that generated by the nanogenerators based on virgin ZnO, and more than three times greater than the virgin PVDF.

Conclusion

ZnO has been considered an important industrial material for centuries due to its outstanding piezoelectric effect and coupling coefficient owing to the nonsymmetric crystalline wurtzite structure, which is currently the focus of considerable new interests. Physical properties such as high electrical and thermal conductivities, very high temperature stability, and optical absorption in the ultraviolet region and chemical properties such as mild antimicrobial effect and stability at neutral pH along with its abundance and reasonable cost ensure its use in a wide range of industries.

However, the development of smart devices requires independent, sustainable, and maintenance-free energy. Thus, energy harvesting based on nanotechnologies from the surrounding as sustainable sources suggest as an adequate solution to power the devices to ensure their operation. These novel energy harvesting structures require self-sufficient implantable devices and/or wireless sensors. Hence, there has been an increasing demand for these devices in many fields.

In energy shortage and sustainable development, ZnO is considered a promising material for energy harvesting due to its ability to convert ambient energy into electrical energy and the ease of production along with cost-effective methods. Different types of ZnO-nanostructure-based NGs that are used in low-frequency applications have been largely studied. Several techniques have been explored to synthesize and develop direct and AC ZnO PENGs. Overall, there are four initial aspects representing the unique merits of ZnO nanostructure-based NGs. First, 1D nanomaterials can endure extreme elastic deformation. Second, 1D NWs are possibly free of transposition due to their small diameters and likely to extend the device lifetime because of their fatigue resistance. Third, 1D nanomaterials have a small elastic modulus and can easily deform under small external mechanical loads. They are even able to harvest energy from small mechanical disturbances, and fourth, synthesis of 1D ZnO nanomaterial is substrate-independent type, which can be synthesized easily on diverse substrates. High piezoelectric coefficients were obtained as a result of large deformation of 1D ZnO nanostructures by applying small mechanical forces and significant mechanical and nonsymmetric properties were exhibited. The principle of sandwich NGs confirmed the opinion for a novel self-powering nanotechnology that can be used for harvesting reasonably great amounts of electrical energy from the environment for various usages. To achieve nanosystems with reliable performance and efficient energy harvesters, more investigations are needed to develop high-quality, cost-effective nanomaterial procedures and complete device fabrication steps, along with efforts devoted to emerging energy harvesting devices that perform efficiently at low frequencies.

References

1 Montazer, M., Amiri, M.M., and Mohammad Ali Malek, R. (2013). In situ synthesis and characterization of nano ZnO on wool: influence of nano photo reactor on wool properties. *Photochem. Photobiol.* 89: 1057–1063.

2 Ramezanzadeh, B., Attar, M., and Farzam, M. (2011). Effect of ZnO nanoparticles on the thermal and mechanical properties of epoxy-based nanocomposite. *J. Therm. Anal. Calorim.* 103: 731–739.

3 Noorian, S.A., Hemmatinejad, N., and Bashari, A. (2015). One-pot synthesis of Cu_2O/ZnO nanoparticles at present of folic acid to improve UV-protective effect of cotton fabrics. *Photochem. Photobiol.* 91: 510–517.

4 Pelicano, C.M. and Yanagi, H. (2019). Enhanced charge transport in Al-doped ZnO nanotubes designed via simultaneous etching and Al doping of H_2O-oxidized ZnO nanorods for solar cell applications. *J. Mater. Chem. C* 7: 4653–4661.

5 Ghayempour, S. and Montazer, M. (2017). Ultrasound irradiation based in-situ synthesis of star-like tragacanth gum/zinc oxide nanoparticles on cotton fabric. *Ultrason. Sonochem.* 34: 458–465.

6 Wang, B.-S., Li, R.-Y., Zhang, Z.-Y. et al. (2019). An overlapping ZnO nanowire photoanode for photoelectrochemical water splitting. *Catal. Today* 321: 100–106.

7 Mohammadzadeh, A., Azadbeh, M., Shokriyan, B., and Abad, S.N.K. (2020). Synthesis of ZnO nanocombs and tetrapods by catalyst-free oxidation of alpha brass powders in air atmosphere. *Ceram. Int.* 46: 2552–2557.

8 Li, G., Su, Y., Chen, X.-X. et al. (2019). Enhanced chemiresistive sensing performance of well-defined porous CuO-Doped ZnO nanobelts toward Vocs. *Nanoscale Adv.* 1: 3900–3908.

9 Haider, A.J., Sultan, F.I., Haider, M.J., and Hadi, N.M. (2019). Spectroscopic and structural properties of zinc oxide nanosphere as random laser medium. *Appl. Phys. A* 125: 260.

10 Aladpoosh, R. and Montazer, M. (2015). The role of cellulosic chains of cotton in biosynthesis of ZnO nanorods producing multifunctional properties: mechanism, characterizations and features. *Carbohydr. Polym.* 126: 122–129.

11 Datta, S., Saha-Dasgupta, T., and Sarma, D. (2008). Wannier function study of the relative stability of zinc-Blende and Wurtzite structures in the CdX (X = S, Se, Te) series. *J. Phys. Condens. Matter* 20: 445217.

12 Chen, Z. and O'Brien, S. (2008). Structure direction of Ii–Vi semiconductor quantum dot binary nanoparticle superlattices by tuning radius ratio. *ACS Nano* 2: 1219–1229.

13 Leszczynski, M., Edgar, J., Strite, S. et al. (1999). Common crystal structures of the group Iii nitrides. *Emis Datareviews Series* 23: 3–5.

14 Dulub, O., Boatner, L.A., and Diebold, U. (2002). Stm study of the geometric and electronic structure of ZnO (0001)—Zn, (0001)—O, (1010), and (1120) surfaces. *Surf. Sci.* 519: 201–217.

15 Özgür, Ü., Alivov, Y.I., Liu, C. et al. (2005). A comprehensive review of ZnO materials and devices. *J. Appl. Phys.* 98: 11.

16 Sabir, S., Arshad, M., and Chaudhari, S.K. (2014). Zinc oxide nanoparticles for revolutionizing agriculture: synthesis and applications. *Sci. World J.* 2014: 925494, 8 pages. https://doi.org/10.1155/2014/925494.

17 Leszczynski, M., Teisseyre, H., Suski, T. et al. (1996). Lattice parameters of gallium nitride. *Appl. Phys. Lett.* 69: 73–75.

18 Ambacher, O., Majewski, J., Miskys, C. et al. (2002). Pyroelectric properties of Al(In)GaN/GaN hetero- and quantum well structures. *J. Phys. Condens. Matter* 14: 3399.

19 Manthina, V. and Agrios, A.G. (2016). Single-pot ZnO nanostructure synthesis by chemical bath deposition and their applications. *Nano-Structures Nano-Objects* 7: 1–11.

20 Gayen, R. and Paul, R. (2018). Phosphorous doping in vertically aligned ZnO nanorods grown by wet-chemical method. *Nano-Structures Nano-Objects* 13: 163–169.

21 Ao, W., Li, J., Yang, H. et al. (2006). Mechanochemical synthesis of zinc oxide nanocrystalline. *Powder Technol.* 168: 148–151.

22 Li, R., Wang, J., He, Y. et al. (2019). Mechanochemical synthesis of defective molybdenum trioxide, titanium dioxide, and zinc oxide at room temperature. *ACS Sustain. Chem. Eng.* 7: 11985–11989.

23 Mahato, T., Prasad, G., Singh, B. et al. (2009). Nanocrystalline zinc oxide for the decontamination of sarin. *J. Hazard. Mater.* 165: 928–932.

24 Mahdavi, R. and Talesh, S.S.A. (2017). The effect of ultrasonic irradiation on the structure, morphology and photocatalytic performance of ZnO nanoparticles by sol-gel method. *Ultrason. Sonochem.* 39: 504–510.

25 Salahuddin, N.A., El-Kemary, M., and Ibrahim, E.M. (2015). Synthesis and characterization of ZnO nanotubes by hydrothermal method. *Int. J. Sci. Res. Publ.* 5: 3–6.

26 Fakhri, P., Amini, B., Bagherzadeh, R. et al. (2019). Flexible hybrid structure piezoelectric nanogenerator based on ZnO nanorod/PVDF nanofibers with improved output. *RSC Adv.* 9: 10117–10123.

27 Caicedo, N., Thomann, J.-S., Leturcq, R., and Lenoble, D. (2016). Aspect ratio improvement of ZnO nanowires grown in liquid phase by using step-by-step sequential growth. *CrystEngComm* 18: 5502–5511.

28 Masuda, Y., Kinoshita, N., and Koumoto, K. (2012). Liquid phase morphology control of ZnO nanowires, ellipse particles, hexagonal rods, and particle in aqueous solutions. *ISRN Nanotechnol.* 2012: 1–6.

29 Rodrıguez-Paéz, J., Caballero, A., Villegas, M. et al. (2001). Controlled precipitation methods: formation mechanism of ZnO nanoparticles. *J. Eur. Ceram. Soc.* 21: 925–930.

30 Kong, X.Y. and Wang, Z.L. (2003). Spontaneous polarization-induced nanohelixes, nanosprings, and nanorings of piezoelectric nanobelts. *Nano Lett.* 3: 1625–1631.

31 Alsultany, F.H., Majdi, H.S., Abd, H.R. et al. (2019). Catalytic growth of 1D ZnO nanoneedles on glass substrates through vapor transport. *J. Electron. Mater.* 48: 1660–1668.

32 Wu, Q.-H. (2015). ZnO nanostructures prepared using a vapour transport method. *J. Exp. Nanosci.* 10: 161–166.
33 Li, X., Sun, M., Wei, X. et al. (2018). 1D piezoelectric material based nanogenerators: methods, materials and property optimization. *Nanomaterials* 8: 188.
34 Xiang, H., Yang, J., Hou, J., and Zhu, Q. (2006). Piezoelectricity in ZnO nanowires: a first-principles study. *Appl. Phys. Lett.* 89: 223111.
35 Gao, Y. and Wang, Z.L. (2007). Electrostatic potential in a bent piezoelectric nanowire. The fundamental theory of nanogenerator and nanopiezotronics. *Nano Lett.* 7: 2499–2505.
36 Zandesh, G., Gheibi, A., Sorayani Bafqi, M. et al. (2017). Piezoelectric electrospun nanofibrous energy harvesting devices: influence of the electrodes position and finite variation of dimensions. *J. Ind. Text.* 47: 348–362.
37 Ghafarirad, H., Rezaei, S., Zareinejad, M., and Mardi, N. (2019). Charge-based hysteresis compensation in low impedance piezoelectric actuators by a modified Prandtl–Ishlinskii model. *Proc. Inst. Mech. Eng. E: J. Process Mech. Eng.* 233: 83–93.
38 Muralt, P., Polcawich, R., and Trolier-McKinstry, S. (2009). Piezoelectric thin films for sensors, actuators, and energy harvesting. *MRS Bull.* 34: 658–664.
39 Motamedi, A.S., Mirzadeh, H., Hajiesmaeilbaigi, F. et al. (2017). Piezoelectric electrospun nanocomposite comprising Au Nps/Pvdf for nerve tissue engineering. *J. Biomed. Mater. Res. A* 105: 1984–1993.
40 Korir, K., Cicero, G., and Catellani, A. (2013). Piezoelectric properties of zinc oxide nanowires: an Ab initio study. *Nanotechnology* 24: 475401.
41 Zhu, R., Wang, D., Xiang, S. et al. (2008). Piezoelectric characterization of a single zinc oxide nanowire using a nanoelectromechanical oscillator. *Nanotechnology* 19: 285712.
42 Khan, A., Ali Abbasi, M., Hussain, M. et al. (2012). Piezoelectric nanogenerator based on zinc oxide nanorods grown on textile cotton fabric. *Appl. Phys. Lett.* 101: 193506.
43 Xu, S., Qin, Y., Xu, C. et al. (2010). Self-powered nanowire devices. *Nat. Nanotechnol.* 5: 366.
44 Dai, S., Dunn, M.L., and Park, H.S. (2010). Piezoelectric constants for ZnO calculated using classical polarizable core–shell potentials. *Nanotechnology* 21: 445707.
45 Kaur, J. and Singh, H. (2019). Synthesis and fabrication of zinc oxide nanostrands based piezoelectric nanogenerator. *J. Mater. Sci.* 30: 4437–4445.
46 Wang, Z.L. and Song, J. (2006). Piezoelectric nanogenerators based on zinc oxide nanowire arrays. *Science* 312: 242–246.
47 Wang, X., Yang, B., Liu, J. et al. (2016). A flexible triboelectric-piezoelectric hybrid nanogenerator based on P (VDF-TRFE) nanofibers and PDMS/MWCNT for wearable devices. *Nat. Sci. Rep.* 6: 36409.
48 Wang, Z.L. (2009). Energy harvesting using piezoelectric nanowires–a correspondence on "energy harvesting using nanowires?" by Alexe et al. *Adv. Mater.* 21: 1311–1315.

49 Ganesh, R.S., Sharma, S.K., Abinnas, N. et al. (2017). Fabrication of the flexible nanogenerator from BTO nanopowders on graphene coated PMMA substrates by sol-gel method. *Mater. Chem. Phys.* 192: 274–281.

50 Askari, H., Khajepour, A., Khamesee, M.B. et al. (2018). Piezoelectric and triboelectric nanogenerators: trends and impacts. *Nano Today* 22: 10–13.

51 Han, M., Zhang, X.-S., Meng, B. et al. (2013). R-shaped hybrid nanogenerator with enhanced piezoelectricity. *ACS Nano* 7: 8554–8560.

52 Abbasipour, M., Khajavi, R., Yousefi, A.A. et al. (2019). Improving piezoelectric and pyroelectric properties of electrospun PVDF nanofibers using nanofillers for energy harvesting application. *Polym. Adv. Technol.* 30: 279–291.

53 Lee, J.H., Lee, K.Y., Gupta, M.K. et al. (2014). Highly stretchable piezoelectric-pyroelectric hybrid nanogenerator. *Adv. Mater.* 26: 765–769.

54 Askari, H., Hashemi, E., Khajepour, A. et al. (2019). Tire condition monitoring and intelligent tires using nanogenerators based on piezoelectric, electromagnetic, and triboelectric effects. *Adv. Mater. Technol.* 4: 1800105.

55 Self-made. https://commons.wikimedia.org/wiki/File:NG_Working_1.png, accessed 8 December 2010.

56 Liu, H., Zhong, J., Lee, C. et al. (2018). A comprehensive review on piezoelectric energy harvesting technology: materials, mechanisms, and applications. *Appl. Phys. Rev.* 5: 041306.

57 Kumar, B. and Kim, S.-W. (2011). Recent advances in power generation through piezoelectric nanogenerators. *J. Mater. Chem.* 21: 18946–18958.

58 Wang, Z.L. (2007). Nanopiezotronics. *Adv. Mater.* 19: 889–892.

59 Fukada, E. (1968). Piezoelectricity in polymers and biological materials. *Ultrasonics* 6: 229–234.

60 Fukada, E. (1974). Piezoelectric properties of organic polymers. *Ann. N.Y. Acad. Sci.* 238: 7–25.

61 Gao, Z., Zhou, J., Gu, Y. et al. (2009). Effects of piezoelectric potential on the transport characteristics of metal-ZnO nanowire-metal field effect transistor. *J. Appl. Phys.* 105: 113707.

62 Wang, Z.L., Yang, R., Zhou, J. et al. (2010). Lateral nanowire/nanobelt based nanogenerators, piezotronics and piezo-phototronics. *Mater. Sci. Eng. R Rep.* 70: 320–329.

63 He, S., Dong, W., Guo, Y. et al. (2019). Piezoelectric thin film on glass fiber fabric with structural hierarchy: an approach to high-performance, superflexible, cost-effective, and large-scale nanogenerators. *Nano Energy* 59: 745–753.

64 Gupta, K., Brahma, S., Dutta, J. et al. (2019). Recent progress in microstructure development of inorganic one-dimensional nanostructures for enhancing performance of piezotronics and piezoelectric nanogenerators. *Nano Energy* 55: 1–21.

65 Sadeghzadeh, S. and Ghazi, S.F. (2018). A size dependent dynamic model for piezoelectric nanogenerators: effects of geometry, structural and environmental parameters. *Mater. Res. Express* 5: 035508.

66 Wang, Z.L. (2010). Piezotronic and piezophototronic effects. *J. Phys. Chem. Lett.* 1: 1388–1393.

67 Kong, L.B., Li, T., Hng, H.H. et al. (2014). *Waste Energy Harvesting-Mechanical and Thermal Energies*. Springer Science & Business Media.
68 Tan, Y.K. (2011). *Sustainable Energy Harvesting Technologies: Past, Present and Future*. InTech, BoD–Books on Demand.
69 Vatansever, D., Siores, E., Hadimani, R.L., and Shah, T. (2011). Smart woven fabrics in renewable energy generation. In: *Advances in Modern Woven Fabrics Technology* (ed. S.G. Vassiliadis). InTech.
70 Ottman, G.K., Hofmann, H.F., and Lesieutre, G.A. (2003). Optimized piezoelectric energy harvesting circuit using step-down converter in discontinuous conduction mode. *IEEE Trans. Power Electron.* 18: 696–703.
71 Fan, F.-R., Tian, Z.-Q., and Wang, Z.L. (2012). Flexible triboelectric generator. *Nano Energy* 1: 328–334.
72 Wang, Z.L., Lin, L., Chen, J. et al. (2016). *Triboelectric Nanogenerators*. Springer.
73 Wang, Z.L. (2015). Triboelectric nanogenerators as new energy technology and self-powered sensors–principles, problems and perspectives. *Faraday Discuss* 176: 447–458.
74 Wang, Z.L. (2013). Triboelectric nanogenerators as new energy technology for self-powered systems and as active mechanical and chemical sensors. *ACS Nano* 7: 9533–9557.
75 Dong, K., Deng, J., Zi, Y. et al. (2017). 3D orthogonal woven triboelectric nanogenerator for effective biomechanical energy harvesting and as self-powered active motion sensors. *Adv. Mater.* 29: 1702648.
76 Wang, P., Pan, L., Wang, J. et al. (2018). An ultra-low-friction triboelectric–electromagnetic hybrid nanogenerator for rotation energy harvesting and self-powered wind speed sensor. *ACS Nano* 12: 9433–9440.
77 Xi, Y., Hua, J., and Shi, Y. (2020). Noncontact triboelectric nanogenerator for human motion monitoring and energy harvesting. *Nano Energy* 69: 104390.
78 He, J., Fan, X., Mu, J. et al. (2020). 3D full-space triboelectric-electromagnetic hybrid nanogenerator for high-efficient mechanical energy harvesting in vibration system. *Energy* 194: 116871.
79 Arora, N., Abowd, G.D., Gupta, M. et al. (2019). Thin and flexible self-powered vibration transducer employing triboelectric nanogeneration. Google Patents.
80 Wang, P., Liu, R., Ding, W. et al. (2018). Complementary electromagnetic-triboelectric active sensor for detecting multiple mechanical triggering. *Adv. Funct. Mater.* 28: 1705808.
81 Xie, Z., Zeng, Z., Wang, Y. et al. (2020). Novel sweep-type triboelectric nanogenerator utilizing single freewheel for random triggering motion energy harvesting and driver habits monitoring. *Nano Energy* 68: 104360.
82 Wu, Z., Ding, W., Dai, Y. et al. (2018). Self-powered multifunctional motion sensor enabled by magnetic-regulated triboelectric nanogenerator. *ACS Nano* 12: 5726–5733.

83 Lin, Z., Zhang, B., Guo, H. et al. (2019). Super-robust and frequency-multiplied triboelectric nanogenerator for efficient harvesting water and wind energy. *Nano Energy* 64: 103908.

84 Zhang, L., Meng, B., Xia, Y. et al. (2020). Galloping triboelectric nanogenerator for energy harvesting under low wind speed. *Nano Energy* 70: 104477.

85 Chen, B., Yang, Y., and Wang, Z.L. (2018). Scavenging wind energy by triboelectric nanogenerators. *Adv. Energy Mater.* 8: 1702649.

86 Liang, X., Jiang, T., Liu, G. et al. (2020). Spherical triboelectric nanogenerator integrated with power management module for harvesting multidirectional water wave energy. *Energy Environ. Sci.* 13 (1): 277–285.

87 Xu, M., Zhao, T., Wang, C. et al. (2019). High power density tower-like triboelectric nanogenerator for harvesting arbitrary directional water wave. *Energy* 13: 1932–1939.

88 Wang, Z.L., Chen, J., Lin, L., and Science, E. (2015). Progress in triboelectric nanogenerators as a new energy technology and self-powered sensors. *Energy Environ. Sci.* 8: 2250–2282.

89 Leng, Q., Chen, L., Guo, H. et al. (2014). Harvesting heat energy from hot/cold water with a pyroelectric generator. *J. Mater. Chem. A* 2: 11940–11947.

90 Ryu, H., Yoon, H.J., and Kim, S.W. (2019). Hybrid energy harvesters: toward sustainable energy harvesting. *Adv. Mater.* 31: 1802898.

91 Yan, C., Gao, Y., Zhao, S. et al. (2020). A linear-to-rotary hybrid nanogenerator for high-performance wearable biomechanical energy harvesting. *Nano Energy* 67: 104235.

92 Khan, A.A., Mahmud, A., and Ban, D. (2018). Evolution from single to hybrid nanogenerator: a contemporary review on multimode energy harvesting for self-powered electronics. *IEEE Trans. Nanotechnol.* 18: 21–36.

93 Lee, K.Y., Gupta, M.K., and Kim, S.-W. (2015). Transparent flexible stretchable piezoelectric and triboelectric nanogenerators for powering portable electronics. *Nano Energy* 14: 139–160.

94 Shi, B., Zheng, Q., Jiang, W. et al. (2016). A packaged self-powered system with universal connectors based on hybridized nanogenerators. *Adv. Mater.* 28: 846–852.

95 Huang, T., Wang, C., Yu, H. et al. (2015). Human walking-driven wearable all-fiber triboelectric nanogenerator containing electrospun polyvinylidene fluoride piezoelectric nanofibers. *Nano Energy* 14: 226–235.

96 Chowdhury, A.R., Abdullah, A.M., Hussain, I. et al. (2019). Lithium doped zinc oxide based flexible piezoelectric-triboelectric hybrid nanogenerator. *Nano Energy* 61: 327–336.

97 Shi, K., Huang, X., Sun, B. et al. (2019). Cellulose/batio3 aerogel paper based flexible piezoelectric nanogenerators and the electric coupling with triboelectricity. *Nano Energy* 57: 450–458.

98 Kim, D.H., Dudem, B., and Yu, J.S. (2018). High-performance flexible piezoelectric-assisted triboelectric hybrid nanogenerator via polydimethylsiloxane-encapsulated nanoflower-like ZnO composite films for scavenging energy from daily human activities. *ACS Sustain. Chem. Eng.* 6: 8525–8535.

99 Chen, X., Song, Y., Su, Z. et al. (2017). Flexible fiber-based hybrid nanogenerator for biomechanical energy harvesting and physiological monitoring. *Nano Energy* 38: 43–50.

100 Zhao, C., Zhang, Q., Zhang, W. et al. (2019). Hybrid piezo/triboelectric nanogenerator for highly efficient and stable rotation energy harvesting. *Nano Energy* 57: 440–449.

101 Wei, G., Bi, Y., Li, X. et al. (2018). Self-powered hybrid flexible nanogenerator and its application in bionic micro aerial vehicles. *Nano Energy* 54: 10–16.

102 Chen, C., Tsai, C., Xu, M. et al. (2019). A fully encapsulated piezoelectric–triboelectric hybrid nanogenerator for energy harvesting from biomechanical and environmental sources. *Express Polym. Lett.* 13: 533–542.

103 Singh, H.H. and Khare, N.J. (2018). Flexible ZnO-PVDF/PTFE based piezo-tribo hybrid nanogenerator. *Nano Energy* 51: 216–222.

104 Moezzi, A., McDonagh, A.M., and Cortie, M.B. (2012). Zinc oxide particles: synthesis, properties and applications. *Chem. Eng. J.* 185: 1–22.

105 Choi, M.Y., Choi, D., Jin, M.J. et al. (2009). Mechanically powered transparent flexible charge-generating nanodevices with piezoelectric ZnO Nanorods. *Adv. Mater.* 21: 2185–2189.

106 Xi, Y., Song, J., Xu, S. et al. (2009). Growth of ZnO nanotube arrays and nanotube based piezoelectric nanogenerators. *J. Mater. Chem.* 19: 9260–9264.

107 Stassi, S., Cauda, V., Ottone, C. et al. (2015). Flexible piezoelectric energy nanogenerator based on ZnO nanotubes hosted in a polycarbonate membrane. *Nano Energy* 13: 474–481.

108 Zhang, G., Liao, Q., Ma, M. et al. (2018). Uniformly assembled vanadium doped ZnO microflowers/bacterial cellulose hybrid paper for flexible piezoelectric nanogenerators and self-powered sensors. *Nano Energy* 52: 501–509.

109 Lee, Y., Kim, S., Kim, D. et al. (2020). Direct-current flexible piezoelectric nanogenerators based on two-dimensional ZnO nanosheet. *Appl. Surf. Sci.* 509: 145328.

110 Hu, F., Cai, Q., Liao, F. et al. (2015). Recent advancements in nanogenerators for energy harvesting. *Small* 11: 5611–5628.

111 Wu, Y., Qiu, J., Zhou, S. et al. (2018). A piezoelectric spring pendulum oscillator used for multi-directional and ultra-low frequency vibration energy harvesting. *Appl. Energy* 231: 600–614.

112 Park, J.Y., Salauddin, M., and Rasel, M.S. (2019). Nanogenerator for scavenging low frequency vibrations. *J. Micromech. Microeng.* 29: 053001.

113 Xue, C., Li, J., Zhang, Q. et al. (2015). A novel arch-shape nanogenerator based on piezoelectric and triboelectric mechanism for mechanical energy harvesting. *Nanomaterials* 5: 36–46.

114 Zhu, J., Hou, X., Niu, X. et al. (2017). The D-arched piezoelectric-triboelectric hybrid nanogenerator as a self-powered vibration sensor. *Sens. Actuators A Phys.* 263: 317–325.

115 Kim, H.S., Kim, D.Y., Kim, J.E. et al. (2019). Ferroelectric-polymer-enabled contactless electric power generation in triboelectric nanogenerators. *Adv. Funct. Mater.* 29: 1905816.

116 Shaw, S.J.J. (1985). Forced vibrations of a beam with one-sided amplitude constraint: theory and experiment. *J. Sound Vib.* 99: 199–212.

117 Qi, S., Shuttleworth, R., Oyadiji, S.O., and Wright, J. (2010). Design of a multiresonant beam for broadband piezoelectric energy harvesting. *Smart Mater. Struct.* 19: 094009.

118 Zhou, W., Penamalli, G.R., and Zuo, L. (2011). An efficient vibration energy harvester with a multi-mode dynamic magnifier. *Smart Mater. Struct.* 21: 015014.

119 Wang, H., Li, B., Liu, Y. et al. (2019). Exploiting elastically supported masses in cantilever for resonance frequencies down-shifted vibration energy harvester. *Energies* 12: 2207.

120 Nour, E., Nur, O., and Willander, M. (2017). Zinc oxide piezoelectric nano-generators for low frequency applications. *Semicond. Sci. Technol.* 32: 064005.

121 Nour, E., Khan, A., Nur, O., and Willander, M. (2014). A flexible sandwich nanogenerator for harvesting piezoelectric potential from single crystalline zinc oxide nanowires. *Nanomater. Nanotechnol.* 4: 24.

122 Nour, E.S., Chey, C.O., Willander, M., and Nur, O. (2016). Low frequency accelerator sensor based on piezoelectric ZnO nanorods grown by low temperature scalable process. *Phys. Status Solidi A* 213: 2503–2508.

123 Koka, A. and Sodano, H.A. (2013). High-sensitivity accelerometer composed of ultra-long vertically aligned barium titanate nanowire arrays. *Nat. Commun.* 4: 1–10.

124 Hsu, C.-L. and Chen, K.-C. (2012). Improving piezoelectric nanogenerator comprises ZnO nanowires by bending the flexible pet substrate at low vibration frequency. *J. Phys. Chem. C* 116: 9351–9355.

125 Deng, W., Jin, L., Chen, Y. et al. (2018). An enhanced low-frequency vibration ZnO nanorod-based tuning fork piezoelectric nanogenerator. *Nanoscale* 10: 843–847.

126 Deng, W., Jin, L., Zhang, B. et al. (2016). A flexible field-limited ordered ZnO nanorod-based self-powered tactile sensor array for electronic skin. *Nanoscale* 8: 16302–16306.

127 Wei, C. and Jing, X. (2017). A comprehensive review on vibration energy harvesting: modelling and realization. *Renew. Sust. Energ. Rev.* 74: 1–18.

128 Balpande, S.S., Pande, R.S., and Patrikar, R.M. (2016). Design and low cost fabrication of green vibration energy harvester. *Sens. Actuators A Phys.* 251: 134–141.

129 Williams, C. and Yates, R.B. (1996). Analysis of a micro-electric generator for microsystems. *Sens. Actuators A Phys.* 52 (1–3): 8–11.

130 Agrawal, R. and Espinosa, H.D. (2011). Giant piezoelectric size effects in zinc oxide and gallium nitride nanowires. A first principles investigation. *Nano Lett.* 11: 786–790.
131 Empizo, M.J.F., Zosa, M.A.H., Solibet, E.J.C. et al. (2019). Grazing incidence x-ray diffraction measurements of a deuterium-ion plasma-irradiated bulk zinc oxide single crystal. *Proc. Samahang Pisika Pilipinas Phys. Congr.* .
132 Zheng, G., Chen, M., Yin, J. et al. (2019). Metal organic frameworks derived nano materials for energy storage application. *Int. J. Electrochem. Sci.* 14: 2345–2362.
133 Yin, B., Qiu, Y., Zhang, H. et al. (2014). Low-frequency flexible piezoelectric nanogenerators based on ZnO nanorods grown on Cu wires. *CrystEngComm* 16: 6831–6835.
134 Yu, J., Huang, B., Qin, X. et al. (2011). Hydrothermal synthesis and characterization of ZnO films with different nanostructures. *Appl. Surf. Sci.* 257: 5563–5565.
135 Wang, Z., Zhan, X., Wang, Y. et al. (2012). A flexible Uv nanosensor based on reduced graphene oxide decorated ZnO nanostructures. *Nanoscale* 4: 2678–2684.
136 Carcia, P., McLean, R., Reilly, M., and Nunes, G. Jr., (2003). Transparent ZnO thin-film transistor fabricated by Rf magnetron sputtering. *Appl. Phys. Lett.* 82: 1117–1119.
137 Liu, N., Fang, G., Zeng, W. et al. (2011). Novel ZnO nanorod flexible strain sensor and strain driving transistor with an ultrahigh 107 scale "on"–"off" ratio fabricated by a single-step hydrothermal reaction. *J. Phys. Chem. C* 115: 570–575.
138 Zhu, G., Yang, R., Wang, S., and Wang, Z.L. (2010). Flexible high-output nanogenerator based on lateral ZnO nanowire array. *Nano Lett.* 10: 3151–3155.
139 Savarimuthu, K., Rajamanickam, G., Shankararajan, R. et al. (2017). Experimental study on flexible ZnO based nanogenerator using Schottky contact for energy harvesting applications. *IEEE Trans. Nanotechnol.* 16: 469–476.
140 Song, M., Zhang, Y., Peng, M., and Zhai, J. (2014). Low frequency wideband nano generators for energy harvesting from natural environment. *Nano Energy* 6: 66–72.
141 Wang, Z.L. (2011). *Nanogenerators for Self-Powered Devices and Systems.* Georgia Institute of Technology.
142 Han, M., Zhang, X., Liu, W. et al. (2013). Low-frequency wide-band hybrid energy harvester based on piezoelectric and triboelectric mechanism. *Sci. China Technol. Sci.* 56: 1835–1841.
143 Ferrari, M., Alghisi, D., Baù, M., and Ferrari, V. (2012). Nonlinear multi-frequency converter array for vibration energy harvesting in autonomous sensors. *Proc. Eng.* 47: 410–413.
144 Poulin-Vittrant, G., Oshman, C., Opoku, C. et al. (2015). Fabrication and characterization of ZnO nanowire-based piezoelectric nanogenerators for low frequency mechanical energy harvesting. *Phys. Procedia* 70: 909–913.

145 Lee, S., Hinchet, R., Lee, Y. et al. (2014). Ultrathin nanogenerators as self-powered/active skin sensors for tracking eye ball motion. *Adv. Funct. Mater.* 24: 1163–1168.

146 Zheng, Z., Lim, Z.S., Peng, Y. et al. (2013). General route to ZnO nanorod arrays on conducting substrates via galvanic-cell-based approach. *Sci. Rep.* 3: 1–5.

147 Lin, H.-I., Wuu, D.-S., Shen, K.-C., and Horng, R.-H. (2013). Fabrication of an ultra-flexible ZnO nanogenerator for harvesting energy from respiration. *ECS J. Solid State Sci. Technol.* 2: P400.

4

Investigation of Hot Carrier–induced Degradation in Nanoscale Junctionless MOSFETs: A Reliability-based Analysis

Nidhal Abdelmalek[1], Fayçal Djeffal[1,2], and Toufik Bentrcia[2]

[1] *University Mostefa Benboulaid – Batna 2, Faculty of Technology, LEA, Department of Electronics, Chahid Boukhlouf M. El Hadi Road, Batna 05000, Algeria*
[2] *University Hadj Lakhdar – Batna 1, Faculty of Matter Sciences, LEPCM, Department of Physics, 05 Avenue Chahid Boukhlouf, Batna 05000, Algeria*

4.1 Introduction

Reliability of MOSFET devices was a major concern in the electronics industry in the past decades, where the aging phenomenon severely degraded the device operation and lifetime. Extensive research has been done to identify the causes of aging and to seek viable solutions. Channel hot carrier injection (HCI)/substrate hot carrier injection (SCI), channel hot electron/hole (CHE/H), drain avalanche injection, and radiation are found to be the basic source mechanisms of degradation [1–7].

Hot carrier–induced degradation is of prime interest for designers as it aggravates with downscaling. The degradation is essentially due to the enhancement in an electric field with dimension reduction, in the lucky electron model representation, providing the carriers with enough energy to be injected through the oxide layer. These carriers are most likely to be captured by existing oxide traps or break silicon bounds and thus giving rise to localized and nonuniform pile-up of the interface states near the drain–channel junction [8]. An energy-driven paradigm proposed by Rauch et al. suggested that the electron–electron scattering mechanism, governed by energy rather than the electric field, is responsible for hot carrier degradation (HCD) aggravation in downscaled devices [9]. The HCD in the device properties is obvious, and degradation metrics such as drain current or transconductance determine the device lifetime [10]. However, assessing the aging process is more complex at the circuit level, and compact models or predefined power-law degradation models should be used [11]. These models do not consider the profile over the gate length of the interface states or their energy distribution. So far, capacitance of the interface traps is included in advanced compact models [12]. Therefore, new models are required to gain further insights into the associated physics and allow reasonable estimation of degradation in circuits. In this field, some works have proposed the introduction of interface trap charges in the surface potential expression [13, 14].

Nanotechnology in Electronics: Materials, Properties, Devices, First Edition.
Edited by Visakh P.M., Artem Semkin, Raneesh Balakrishnan, and Saša Lazović.

Nevertheless, to the best of our knowledge, the time dependence of the trap distribution is not included.

In this topic, it would be interesting to explore the possibilities of developing new models that include interface trap distributions in junctionless field-effect transistors (JLFETs). Especially, as experimental results of stress-induced degradation have demonstrated, the device exhibits high immunity against HCD and has a longer lifetime than its inversion mode (IM) counterpart [15–17]. It is essential to first elucidate the creation of interface states and proven reliability of the JLFET, particularly as high-impact ionization (II) rate is reached in this device, which would normally increase degradation. Thus, we propose an in-depth analysis of HCD in both junctionless (JL) and IM VSG-MOSFETs in the following sections.

The chapter is organized as follows. In Section 4.2, the principal aspects of the JL structure are described. In Section 4.3, the simulation framework adopted to model the HCD is discussed. Section 4.4 is devoted to interface trap creation. To highlight the impact of HCD, a comparative study is conducted in Sections 4.5 and 4.6 by assessing the main analog/radio frequency (RF) and digital performance criteria. Finally, in light of this study, some concluding remarks are given in Section 4.7.

4.2 Overview of the Junctionless Paradigm

The JL transistor concept was proposed by Lee et al. [18] as a simple way to overcome the process limitations. Indeed, as downscaling trends toward channel lengths of less than 10 nm, controlling process-induced variabilities becomes imperative to conserve transistor properties. At these scales, inherent scattering and diffusion of shallow-extension impurities into ultrashort channels poses a serious problem. Given that high doping concentration gradient is inherent, the formation of extensions with the actual process techniques implies more complex and expensive operations (additional millisecond annealing techniques) [19]. The first study suggested that the suppression of the junctions by setting a uniform one-type doping over the whole device yields better resilience to short channel effects (SCEs) compared to conventional MOSFETs and an excellent turnoff. Soon after, Colinge and coworkers successfully fabricated the first JL transistor with full complementary metal oxide semiconductor (CMOS) functionality, which was named gated resistor. The transistor exhibited a subthreshold slope close to 60 mV dec^{-1} and a very low leakage current [19].

Because of the absence of junctions to prevent the carrier flow, the JL is generally considered a closed device (on). However, the work function difference between the gate and the heavily doped channel forces depletion of carriers and thus prevents the conduction path formation, ensuring an efficient off-state. In the subthreshold regime, the JLFET operates as an accumulation-mode device in depletion and little current flows by diffusion. As the gate bias is increased, a conduction channel appears in the center of the body and keeps enlarging until it occupies the entire channel volume. At this point, the flat-band condition is reached and the carriers are slightly uniformly distributed in the depth direction. The channel is then said to be electrically neutral, and the transverse electric field tends toward zero. A higher gate voltage leads to the formation of an accumulation layer beneath the gate oxide. The saturation mechanism, on the other hand, is very similar to that of an IM device. As the drain bias is increased, partial depletion near the drain increases until a pinch-off of the conduction channel is formed. The pinch-off is located at the body center of the JL and is caused by the electrostatic control, while in the IM device, the pinch-off is situated beneath the gate and results from the extension of the p–n$^+$ junction's space charge region.

Overall, with a well-sized JLFET, MOSFET-like transfer and output characteristics can be obtained. The trigate resistor fabricated by Colinge and coworkers achieved an I_{on}/I_{off} ratio of greater than 10^6 and an subthreshold slope (*SS*) value of 64 mV dec^{-1} [19]. In comparison with a 20 nm trigate nMOSFET that exhibited an *SS* value of 92 mV dec^{-1} and a DIBL of 78 mV V^{-1}, equivalent JLFET performed an *SS* value of 75 mV dec^{-1} and a DIBL of 10 mV V^{-1} [20]. Otherwise, the same I_{on} and I_{on}/I_{off} ratio were reported for both devices. Furthermore, an enhanced saturation current of the JLFET than its IM counterpart was measured for the same "$V_{gs} - V_{th}$" voltage [20]. Conversely, Rios et al. obtained different results from their comparative study based on the matched off-state current [21]. In fact, major degradation in all JLFET metrics was reported. This degradation worsens with the increasing doping concentration. Moreover, the JLFET achieves relatively lower transconductance and cutoff frequency compared to conventional MOSFETs. The cutoff frequency degrades further above the threshold voltage with an increase in the total gate capacitance.

Additional transport mechanisms have been reported in JLFETs. The band-to-band tunneling from the channel to drain is found to be the source of leakage current enhancement with consequent deterioration in the subthreshold swing and DIBL, especially in narrow bandgap materials [22–24]. II was also identified in JL devices [25, 26]. Experimental results have demonstrated a lower II threshold energy (less than the conventional 1.5 times E_g [27]) and a higher multiplication factor compared to IM devices. This fact is explained by the improved impact generation rate and the electron temperature distributed over a larger volume as a result of a particular electrostatic profile. The II enhancement in JL devices can also be attributed to the bandgap narrowing induced by high body doping concentration [25, 26].

Furthermore, two mechanisms lead to creation of minority carriers. The accumulation of the generated holes in the channel results in a floating body effect and a parasitic bipolar junction transistor (BJT). The combination of the avalanche effect due to the high II multiplication factor and the enhancement of the BJT base formed by the excess generated holes creates a positive feedback loop mechanism. The triggering of this feedback loop is reflected by an instant current rise depicted by the kink effect in the output characteristics and a sharp subthreshold slope in the transfer characteristics [26]. Experimental results have shown that both IM and JL multi-gate FETs can reach a minimum *SS* value of 5 mV dec^{-1}. However, only 2.5 V of the drain supply is needed for the JLFET, while 5 V for the IM device to achieve the same subthreshold slope [25]. Moreover, the generation rate enhancement in narrow bandgap materials while maintaining a roughly similar recombination rate yields an impressive sharp subthreshold slope <1 mV dec^{-1} and a larger magnitude of current change [28]. Additionally, a memory effect that holds the BJT feedback loop mechanism is observed as the gate voltage is swept backward. This effect is depicted by a hysteresis in the transfer characteristics, which yields two threshold voltages corresponding to the forward and backward sweep processes. Therefore, the dynamic operation of logic circuits (gates or memories) is altered by changing the drive current, noise margin (NM), and propagation delay. It can also lead to undesirable and instable effects such as racing or bit reversal [29]. Nevertheless, the idea of exploiting the hysteresis characteristics as a single transistor dynamic memory is interesting, all the more so a large hysteresis window can be obtained at lower drain biases and even a larger one with narrow bandgap materials [30]. It is also important to notice that in the presence of the majority carriers of the channel, the minority carriers' recombination occurs continuously, which attenuates the positive feedback loop. Consequently, a lower amount of current change, delayed kink effect, and higher breakdown voltage are achieved [25, 31].

As mentioned earlier, JLFETs have better immunity against SCEs. Overall, experiments and simulations have demonstrated the scaling capability of JLFETs with a much lower subthreshold slope (often approaching 60 mV dec^{-1}) degradation and threshold voltage roll-off than equivalent IM transistors. JLFETs also exhibit lower DIBL, which is important for low-voltage CMOS applications [20, 31, 32]. This feature is explained by the fact that in the absence of junctions and dopant scattering, the channel effective length is sensibly equal to the gate physical length. Furthermore, the depletion at the periphery of the body is larger than at its center, yielding an effective length larger than the physical one, which can be advantageous to reduce SCEs [33]. Downscaling in JLFETs is not just a technological requirement but also a functionality condition. Indeed, JLFETs must have an enough thin body and oxide layer to permit a deep and efficient electrostatic control over the entire volume, ensuring the total carrier depletion of the channel in the off-state [19]. Multi-gate structures are also suitable for JLFETs due to their enhanced electrostatic controllability; gate all around (GAA) and nanowires are found to be the most efficient structures that provide full carrier depletion for thin bodies [20, 31, 33, 35, 36]. Moreover, studies and experiments have demonstrated that even at atomic scales, JLFETs still have a total functionality with good characteristics [31, 34, 36]. Furthermore, the reduction in the body thickness yields a significant reduction in the threshold voltage sensitivity toward variability in device dimensions. As reported by Colinge et al., JLFETs offer large flexibility in the threshold voltage by varying the body thickness and width while keeping the same doping and equivalent oxide thickness (EOT) [37]. If this feature represents a design advantage, it also reflects high sensitivity to process parameter variability as the line edge roughness (LER). For instance, a device with the p^+ polysilicon gate, 2×10^{19} cm^{-3} channel doping, 1 nm EOT, and 15 nm body thickness exhibit a high threshold voltage variability of 100 mV nm^{-1}. Likewise, variations in the metrics induced by random dopant fluctuation (RDF) are comparable to the LER and are roughly similar to that of IM devices [37, 38]. It was found that reduction in the channel thickness attenuates the RDF-induced variability. Even if the volume is shrunk (which normally should worsen the variability, the impact of length reduction is more relevant [39]), it is argued that the resulting enhanced controllability is responsible for this variability attenuation [38]. Moreover, the enhancement in screening of ionized impurities might reduce the variability further.

As explained, one of the most interesting characteristics of JL transistors is its very low transverse electric field, which is negligible when the transistor is in flat-band conditions [40]. Consequently, the channel carrier's mobility, especially at the surface, is supposed to be totally free from transverse electric field effects. Even before reaching the flat-band condition, the conduction path is essentially located in the center of the body where the transverse electric field is much lower compared to the peripheral, yielding lesser sensitivity to the bulk carrier's mobility against the gate voltage. It is worth mentioning that the impact of the electric field on mobility degradation is equally important to that induced by the increase in the doping concentration. Nevertheless, it was reported that degradation attains stationary levels beyond $10^{19}\ cm^{-3}$ of the doping concentration either for p- or n-type [41].

Conversely to IM MOSFETs, mobility measurements highlighted an enhancement in the effective mobility with the increase in the gate bias [21]. The enhancement is so significant that the effective mobility exceeds the corresponding bulk mobility especially in the accumulation regime [42]. A substantial improvement in the surface mobility was also observed in the accumulation layer of heavily doped devices [43], which indicate large current in the accumulation regime of JLFETs. This effect is attributed to the impact of the accumulation layer on the screening of ionized impurities, reducing Coulomb scattering [33, 43]. In heavily doped devices, measurements highlight the correlation between the carrier density and the Coulomb scattering limitation where a significant improvement in the effective mobility with respect to carrier density is observed (largely exceeding the bulk mobility) [43, 44]. Also, it was demonstrated that better electrostatic controllability provided by multi-gates and thinner EOT yields higher carrier density and improved mobility [44]. The improvement is more significant in thin channels as a consequence of better screening of the reduced dopant atoms number in shallow volumes [44]. Similarly, the additional subband filling due to quantum confinement effects in ultra-thin structures results in an increase in the carrier density [31]. Persson et al. proposed a more insightful explanation for the mobility enhancement [45]. They suggested that the majority carriers interact with the impurities as quantum wells while the transmission occurs above these wells. Therefore, increasing the screening will have no further impact. In contrast, the minority carriers in IM devices perceive charged impurities as tunnel barriers that can strongly impede their flow. In this case, the screening effect of the gate, the gate oxide, and the inversion layer effectively reduces the potential impurity barriers. Persson et al. concluded that the enhancement in the majority carrier mobility in thin bodies and nanowires is not due to the screening effect but rather to the complex interplay between the lateral confinements and impurity wells [45].

Mobility limitation in JLFETs, as in heavily doped accumulation devices, is largely governed by dopant Coulomb scattering, while phonon and surface roughness scattering have minor impacts [33, 43]. Indeed, the extracted mobility as a function of temperature variations clearly shows a weak impact of phonon scattering. In contrast to IM devices where the effective mobility drops by 36% for a temperature increase of about 200 °C, less than 7% decrease is observed for JLFETs [33]. Furthermore, as the transport below the accumulation regime occurs mainly in the body center, surface roughness scattering has no tangible effect on mobility. Also, the bulk carriers are not subject to Coulomb scattering induced by the potential of the oxide and interface state charges.

4.3 Simulation Framework of Hot Carrier Degradation

To investigate HCD in both JL and IM VSG-MOSFETs, two devices of the same dimensions under the same conditions were compared. In Table 4.1, different parameters used for the simulation are summarized [46].

Physical models are chosen to reflect the important mobility behaviors in both devices. Therefore, the Klaassen mobility model, which accounts for different scattering mechanisms, was incorporated. This model also includes an electric field and the doping-dependent mobility. Furthermore, the Shirahata mobility model, which considers the inversion layer screening effect, was combined with the aforementioned model. As the JL body was heavily doped, bandgap narrowing was considered. Besides, Fermi–Dirac statistics was used for both devices. More effects were activated in the simulation of HCD to consider additional phenomena such as hot electron injection, hot electron transport, II.

Figure 4.1 illustrates the simulated transfer characteristics of both devices. The JL gate work function was chosen to match the off-current of the IMFET. The JLFET configuration yielded a good switching property with an acceptable on-state current. In comparison to the IM device, where the I_{on}/I_{off} ratio (@$V_{gs} = V_{ds} + V_{off}$, $V_{off} = V_{gs}$ @ $I_{off} = 10$ pA) and the SS were, respectively, 5.3×10^5 and 60.05 mV dec^{-1}, the JLFET achieved 1.7×10^5 and 60.25 mV dec^{-1}.

The surface and center channel mobility variations with respect to the gate supply are depicted in Figure 4.2. In the IM device, both surface and body center mobilities remain relatively constant in the subthreshold regime. As the gate voltage

Table 4.1 Important simulation parameters and their values used in the simulation.

Parameter	Junctionless design	Inversion mode design
L (nm)	30–50	30–50
T_{si} (nm)	10	10
t_{ox} (nm)	2	2
$N_{D/S}$ (cm^{-3})	1×10^{19}	10^{20}
N_{Ch} (cm^{-3})	1×10^{19}	10^{17}
Φ_m (eV)	4.88	4.63

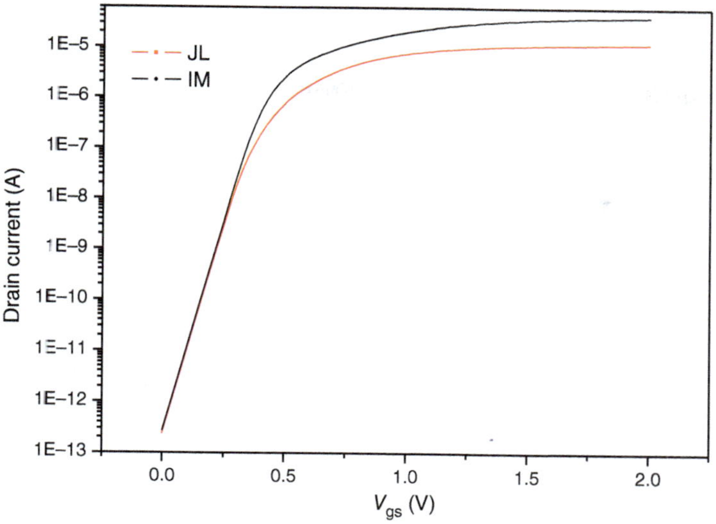

Figure 4.1 Transfer characteristics of a JL and an IM VSG-MOSFET at $V_{ds} = 0.5$ V.

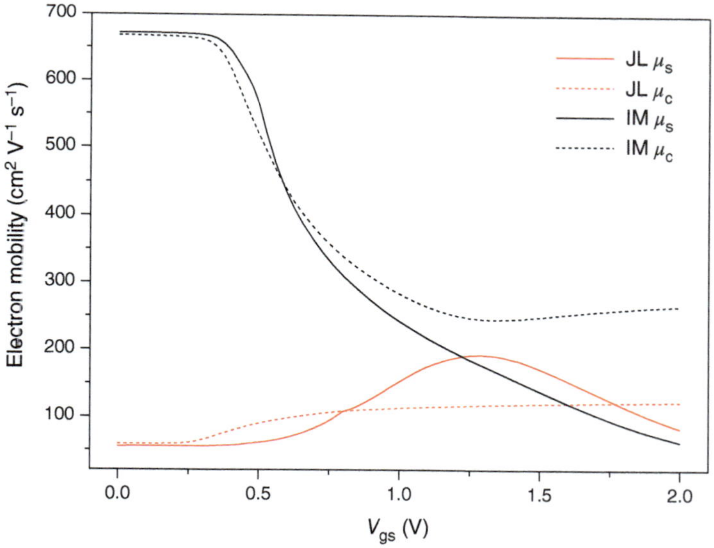

Figure 4.2 Electron mobility in the mid-channel at the surface (line) and body center (dashed) as a function of the gate bias for JL VSG-FET and IM VSG-FET.

is increased, the mobility degrades, the transverse electrical field is enhanced, and the inversion layer is formed. The screening effect is perceptible only at low gate voltages. As the inversion layer is completely formed above the threshold voltage, the combination of increasing scattering and the transverse electric field further degrades the surface mobility, which falls below that in the JLFET. In the JL device and as explained earlier, the reduction of the transverse electric field related to the screening of Coulomb scattering significantly improves the mobility. At the channel center, mobility enhancement is principally due to the low electric field rather than the screening effect. The mobility μ_c increases with respect to the gate bias and exceeds the bulk mobility, which is about 10^8 cm^2 Vs^{-1} for the aforementioned doping concentration (10^{19} cm^{-3}). Furthermore, the surface mobility follows the same increasing trend due to the enhancement in the accumulation layer screening effect. The highest value is reached above the flat-band condition. In these simulations, the surface mobility attains a maximum value of 192 cm^2 V^{-1} s^{-1} exceeding the IM surface mobility, which is 180 cm^2 V^{-1} s^{-1} at the same voltage. Nevertheless, this surface mobility improvement will have a tangible effect only if the conduction path is extended to the entire channel volume. Above this peak, the enhancement of surface scattering with the increase in the accumulation layer density degrades the surface mobility. The resulting current limitation and low transconductance suggest that the JLFET operation in the accumulation regime may not be recommended [19].

4.4 Creation of Interface Traps

Hot carrier effect is believed to be one of the most important causes of performance degradation in MOSFETs. In n-type silicon-on-insulator devices, there are two mechanisms: CHE injection and hot electron injection through the oxide layer. In the first mechanism, the accelerated carriers acquire adequate kinetic energy to break silicon bonds. If the interface is already H-passivated, hot electrons will depassivate the Si—H bond by collision in a single or multi-particle process [47]. The resulting dangling bonds act as interface states that can capture the channel electrons or reemit them back into the channel. Also, the predominance of the CHE on interface trap generation was demonstrated. This conclusion was made by the analysis of the deuterium generated by CHE and HCI. A large amount of deuterium is generated in the presence of CHE, resulting in higher interface trap density [6].

However, in this work, we consider only HCI as the main process responsible for interface trap creation. Although it has been proved that this widely accepted concept is physically incomplete, especially at the submicron scale, it provides an acceptable description of HCD [3–5]. The common model supposes that hot electrons can be injected into the oxide layer if they gain sufficient energy to surmount the Si—SiO_2 interface barrier evaluated at 3.2 eV [3]. Once the electrons cross the interface barrier, they can result in either interface state creation or can be captured by the oxide traps.

Moreover, the influence of the interface states on the oxide trap generation depends on the stress bias conditions. The interface state generation is predominant when $V_g \approx V_d/2$. The reverse situation is observed for $V_g \approx V_d$ [5, 7]. Furthermore, the injection of II and avalanche multiplication of excess carriers enhance the interface and oxide trap creation [1].

In the simulation framework, the Hänsch reliability model is used to generate HCD [4, 46]. The interface trap creation is calculated using Eq. (4.1)

$$\frac{dN(x,t)}{dt} = \frac{\sigma_e}{q} J_{inj,n}(x,t)\left(N_{acc}^0(x) - N(x,t)\right) \tag{4.1}$$

where N_{acc}^0 represents the initial acceptor-like trap density, σ_e is the electron capture cross section, and $N(x, t)$ is the time- and position-dependent traps.

Only the acceptor-like traps are considered with the initial density of $5 \times 10^{12}\,cm^{-2}$, uniformly distributed over the silicon–oxide interface, which is coherent with the experimental data reported for both device types [48, 49]. $J_{inj,n}$ represents the HEI current density based on the Tam lucky electron model [3, 46]. It is assumed that only HEI occurs under the present DC stress conditions, namely, $V_{ds} = 3\,V$ and $V_{gs} = 1.5\,V$. Also, these bias conditions are exaggerated to amplify the degradation and reduce the simulation time. It has been reported in literature that the power supply for 10 years lifetime of JL and IM nanowire structures varies around 1.5 V [17]. Furthermore, recovery of the interface traps is neglected, as well as the oxide trapping/detrapping process and the fixed oxide charge.

The probability that an electron gains sufficient energy to surmount the local barrier and create an interface trap depends essentially on the electric field and the surface carrier density. As these two factors are more important in the IM than in the JL device, the injected current follows the same trend as depicted in Figure 4.3. Consequently, the interface trap generation reaches higher densities in the IM than in the JL structure. Approximately, the total interface trap charge is 1.5 times greater in the IM device (Figure 4.4).

The effect of the II on interface trap creation is also depicted in Figure 4.4. In the first period, the excess carriers generated by II at the drain junction of the IM device

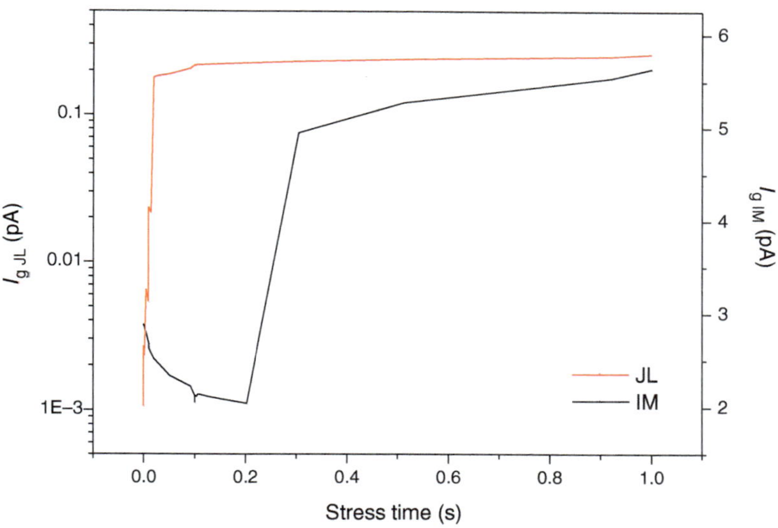

Figure 4.3 Gate leakage current as a function of stress time.

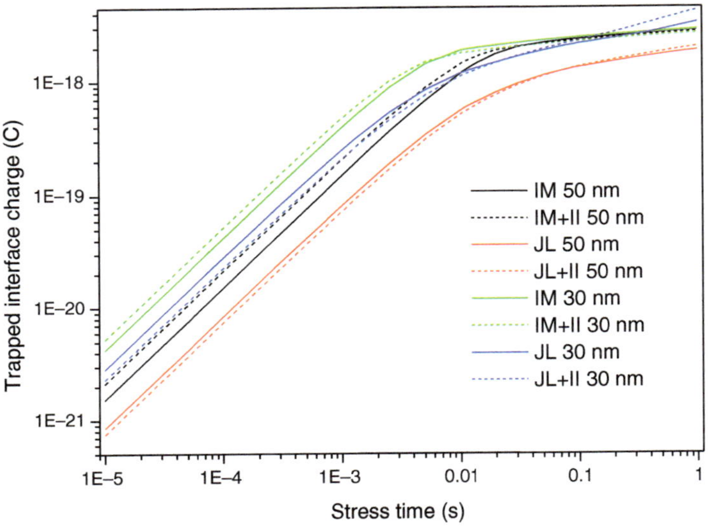

Figure 4.4 Trapped interface electron charge as a function of stress time.

enhance interface state creation. With extended stress time, the enhanced trapped electron charges lead to a reduction in the electric field and carrier density at the interface. Consequently, the second period of the HCD exhibits a decrease in the II process as well as trap creation that starts to saturate. The JLFET, on the other hand, exhibits a reverse behavior. As mentioned, the II process occurs mainly in the drain extension. Therefore, the generated electrons are evacuated toward the drain contact and have small impacts on the gated interface. These electrons would be implicated in the creation of interface traps if oxide spacers are considered, which is not the case in this study. Furthermore, the generated holes are injected in the channel where they recombine. Hence, the channel carrier density decreases, resulting in less interface trap creation in the first period. In the second period, however, the enlargement of the interface trap distribution attenuates II generation, resulting in less carrier recombination in the channel and high interface carrier densities. An enhancement in the electric field is also observed, particularly the perpendicular component beneath the damaged interface. Hence, the interface state creation and the trapped electron charge keep increasing as depicted in Figure 4.4.

Considering the HCD in shorter channels, interface trap creation is significantly enhanced in both device types due to the increase in the electric field. Such relevant trap creation results in larger degradation and reduction of the device lifetime as experimentally reported in [17]. A more pronounced enhancement of the II effect is observed in the JLFET. This fact can be explained by the reduction of carrier recombination in shorter channels, where a large amount of the generated holes reach the source before recombining and maintain the interface carrier density at high levels. The trapped interface charge of the IM device reaches approximately the same levels of saturation, while for the JLFET, the same ratio is observed between the linear and saturation region. This fact can be justified by exploring the temporal evolution

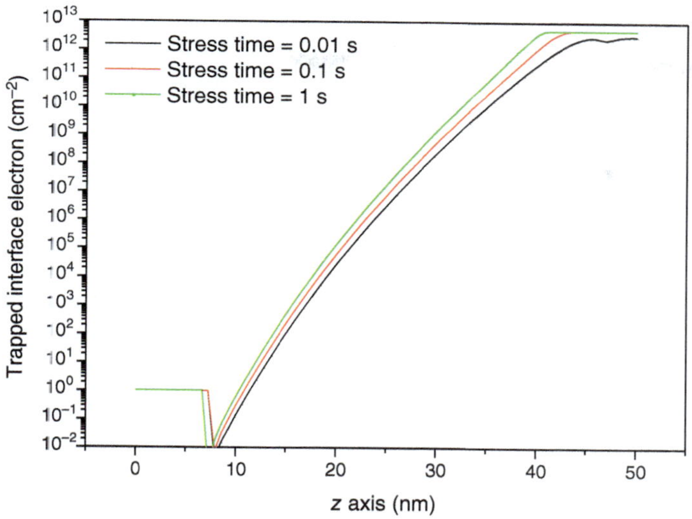

Figure 4.5 Trapped interface electron distribution for 50 nm IM device.

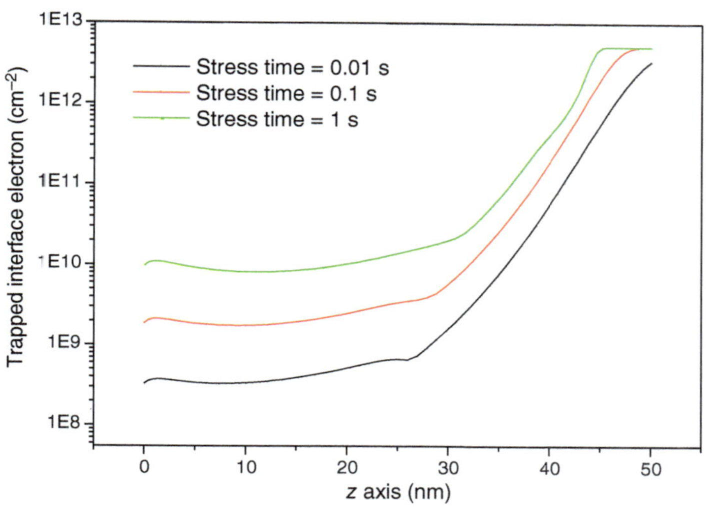

Figure 4.6 Trapped interface electron distribution for 50 nm JL device.

of the interface trap distribution over the gate length in both devices as illustrated in Figures 4.5 and 4.6. Indeed, the simulations performed on these two device types with a gate length of 50 nm elucidate a particular profile for each type. This specific interface trap distribution is elevated with extended stress time. When reaching the maximum allowable interface state density, the profile saturates and exhibits an extension toward the source [47]. Furthermore, the enhanced HCD in the IMFET is clearly depicted in Figure 4.5 with a larger saturated damaged region. The enhanced trapped interface charge of the JL device can also be explained by the increase in the interface trap densities in the remaining channel interface as shown in Figure 4.6.

Usually, the interface traps are limited to the drain side over a fixed length, neglecting the remaining interface. This approach results in misleading evaluation of the device properties degradation. Even if the interface state density decreases while receding from the drain junction, its impact on the surface potential and the electric field must be considered, especially for densities above 10^{10} cm^{-2}.

4.5 Performance Degradation Due to Hot Carrier Effect

In this section, the hot carrier effect on the device properties is assessed. For this purpose, the previous extracted interface trap distribution from the HCD simulation on IM VSG-MOSFET is used for both devices, rather to vary the trap density over a fixed length as generally used. Furthermore, the amphoteric property and density of state of interface traps is considered. In this manner, a more realistic behavior is obtained. Only acceptor-like interface traps are considered as the donor-like traps yield small variations in the IM device properties and negligible in the JLFET. The amphoteric property of acceptor-like interface traps means that the latter act as a negatively charged state when filled and neutral when empty. The level of trap filling depends on Fermi level position in the channel with respect to the energy distribution of traps. Thus, a dynamic response of the interface states is obtained for a different operating regime conversely to the static and fixed oxide charges.

The interface state energy distribution depends on many factors. Indeed, the energy distribution of interface states in ultrathin oxide layer was found to be dependent on the nature of the dangling bond (it interacts with Si or O atoms) and the temperature of the oxide layer formation. For instance, only one energy peak in the mid-gap was observed at 350 °C with ~0.1 eV width, while two peaks were detected above 550 °C [50]. A correlation between the paramagnetic centers P_b (which are dangling bonds) and the interface state density was demonstrated [48]. This type of interface trap depends on the Si/SiO_2 interface orientation. Indeed, only one P_b center appears in the (111) interface, yielding a U-shaped density profile [51]. In the (100) interface orientation, two centers are observed, P_{b0} and P_{b1}. The P_{b0} centers represent the dominant part of the density of states (DOS), they are well centered in each half of the bandgap. The P_{b1} yields lower trap density and are located near the mid-gap [52]. Also, higher interface trap density was measured for n-type doping [48]. Moreover, the oxide material, thickness, body doping concentration, and characterization methods result in different interface state energy distributions [49, 53].

Without a real consensus, an arbitrarily large Gaussian distribution of interface trap DOS is incorporated into a peak energy/peak distribution of 0.49 eV from the silicon conduction band and a characteristic decay energy of 0.2 eV. The peak distribution increases from 10^{12} to a maximum of 10^{13}, which corresponds to the first phase of interface trap creation. The saturation phase is described by an enlargement of the peak distribution length toward the source.

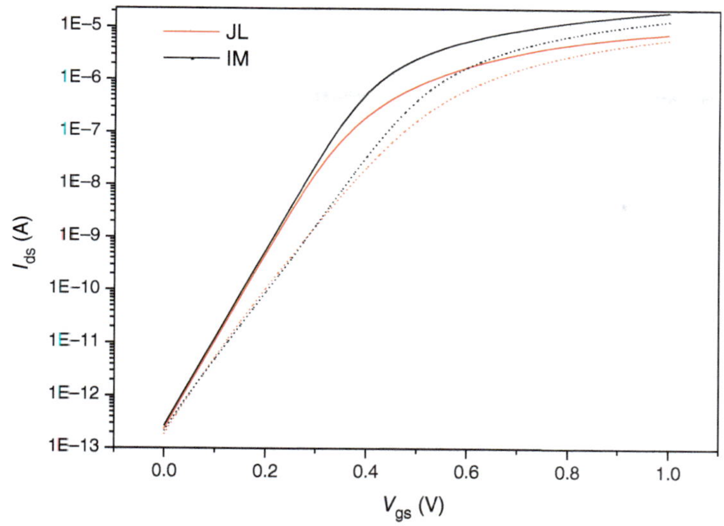

Figure 4.7 Drain current for fresh (line) and damaged (dot) JLFET and IM VSG-FET with $L = 50\,\text{nm}$, $V_{ds} = 0.5\,\text{V}$, $N_{it} = 7.9\text{E}^{11}\,\text{cm}^{-2}$.

Figure 4.7 illustrates the drain current degradation for both JL and IM devices. The interface trap density (N_t) in the rest of this study corresponds to the distribution mean value computed by $1/L \int_0^L N_{it}(z)\text{d}z$. The obtained results corroborate the reported HCD on IM devices [5, 54]. It is shown that degradation of transfer characteristics occurs mainly in the depletion and weak inversion regimes. In fact, the negatively charged traps lead to an elevation of the channel barrier, which further reduces the subthreshold current and delays the device onset. As the gate voltage increases, the formation of the inversion layer causes a bending of the channel conduction band at the interface. Consequently, the Fermi level with respect to the trap energy level is modified, thereby increasing the interface charge. The impact of the interface states on the JL is very similar, the degradation being relevant during full and partial depletion. The negative charge behavior is expected to be at its maximum during full depletion as the band profile at the interface yields filling of maximum traps. The effect decreases progressively as the device switches to partial depletion and almost vanishes above the flat-band condition.

Figure 4.8 illustrates the impact of interface state density enhancement on the subthreshold slope degradation of both devices. The IM device exhibits relative stability in the presence of low trap densities. As the density increases and gets deeper in the channel, the impact on the barrier elevation is enlarged, yielding major subthreshold degradation. On the other hand, a moderate impact of the interface states on the JL device, compared to its IM counterpart, is observed. This relatively low degradation is mainly due to the location of the conduction path. It is clear that the bulk conduction mode is less affected by the interface charge than the surface conduction mode. As the interface state distribution increases, a deeper and larger depletion area is formed, which delays the creation of the conduction path creation. The JL

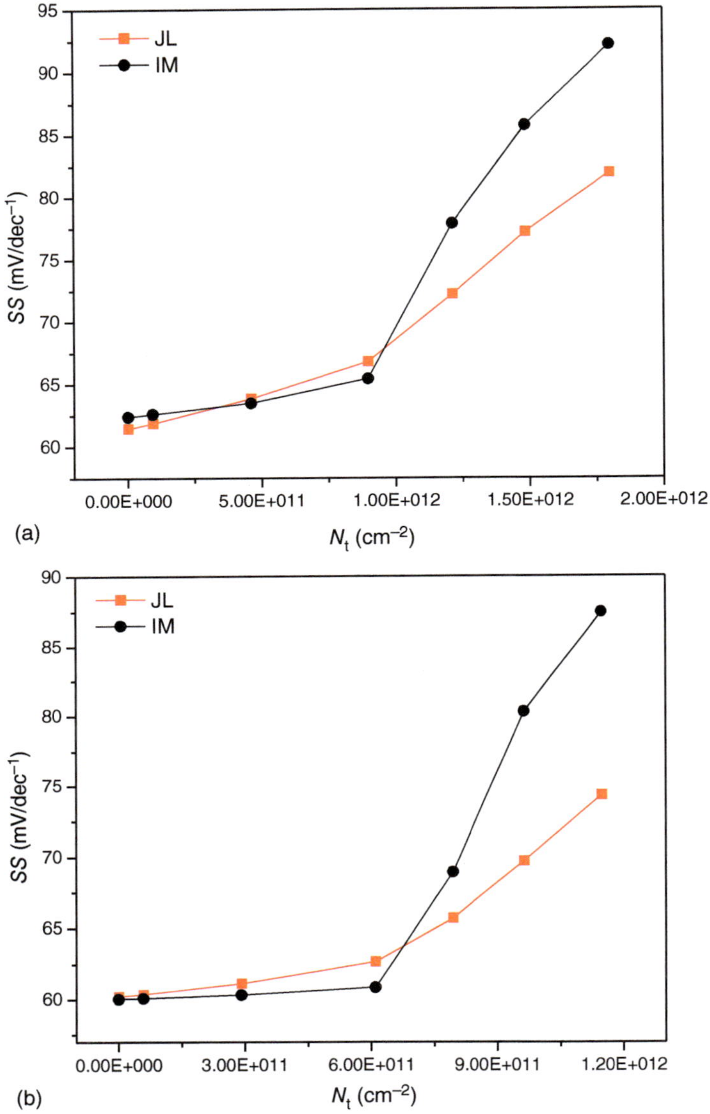

Figure 4.8 Subthreshold slope degradation for JLFET and IM VSG-FET for (a) $L = 30$ nm and (b) $L = 50$ nm ($V_{ds} = 0.5$ V).

device immunity against SCEs is also observed in Figure 4.8a, yielding lower SS for fresh devices. The degradation follows the same evolution as that of longer channel lengths. However, relatively larger degradation is observed.

The channel potential shift induced by the charged interface states yields a sensible variation in the threshold voltage as depicted in Figure 4.9. The extraction was made using the maximum transconductance derivative method. It appears that

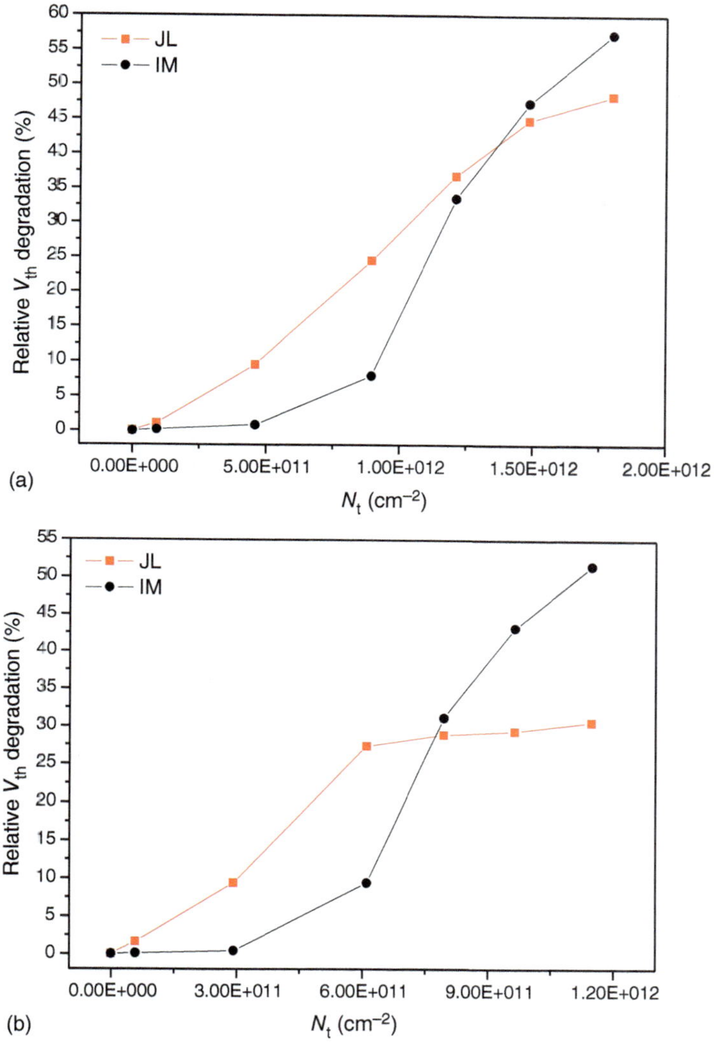

Figure 4.9 Relative threshold voltage degradation for JLFET and IM VSG-FET for (a) $L = 30$ nm and (b) $L = 50$ nm ($V_{ds} = 0.5$ V).

V_{th} of the IM device shifts gradually for low interface trap densities and increases exponentially for higher values. For the JL device, relative degradation increases quasi-linearly with the increasing interface trap density. However, an attenuation is observed for high densities, suggesting that degradation might reach a saturation level with the longer aging phenomenon. Furthermore, larger V_{th} variation is observed for shorter channel lengths. For 30 nm JLFET, degradation reaches 50%, while for 50 nm, it does not exceed 30%. Conversely, high degradation is observed in the IM device for both channel lengths. Overall, for low and moderate interface state densities, the IMFET exhibits much lower V_{th} degradation compared to its JL counterpart. Nevertheless, this tendency is reversed for high trap densities.

The on-current variation and its relative degradation are shown in Figure 4.10. A quasi-linear degradation is observed for the JL device, whereas for the IM type, the degradation evolves linearly for low trap densities and increases with the enlargement of the interface state distribution. The I_{on} is measured for 1 V of gate bias. At this bias condition, the IM device is in strong inversion and the JL device is

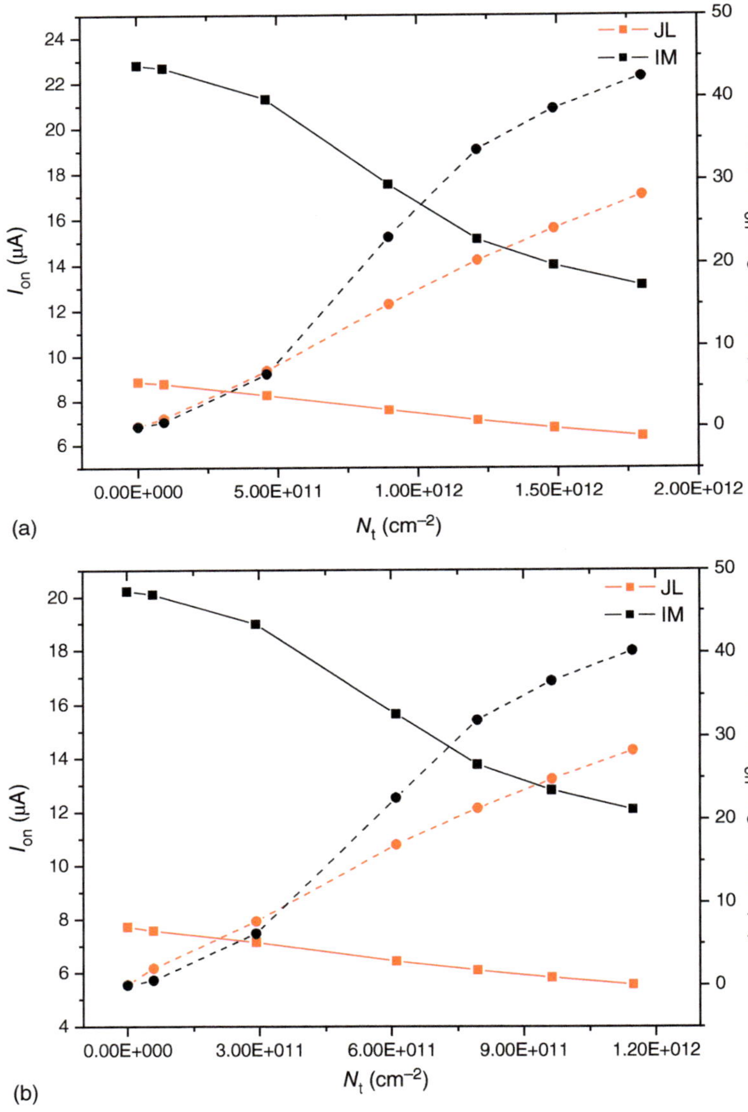

Figure 4.10 I_{on} (line) and relative I_{on} degradation (dashed) for JLFET and IM VSG-FET for (a) $L = 30$ nm and (b) $L = 50$ nm ($V_{ds} = 0.5$ V and $V_{gs} = 1$ V).

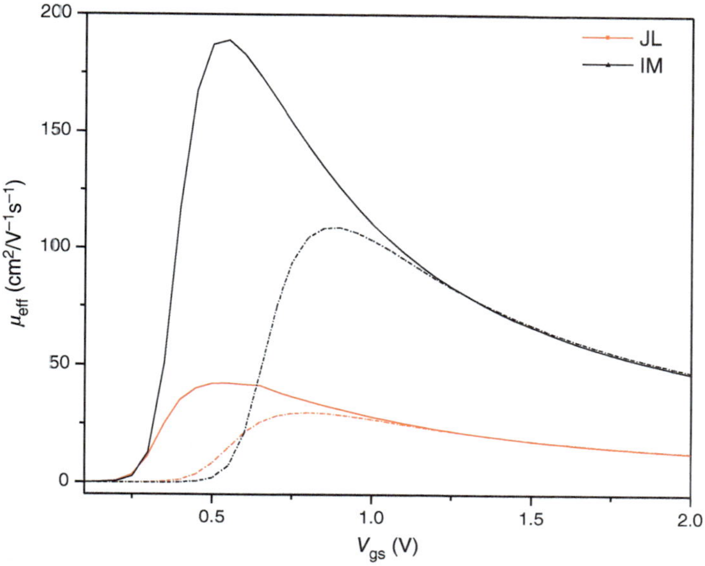

Figure 4.11 Effective mobility as a function of the gate bias for fresh (line) and damaged (dashed) JLFET and IM VSG-FET with $L = 50$ nm and $V_{ds} = 0.01$ V.

in the partial depletion regime. The amount of trapped electrons in the acceptor-like interface states is substantial in the JL device, yielding larger interface charges. Nevertheless, the difference in the conduction path location leads to a different impact of interface charges on the carrier transport. It is clear that the inversion layer, by its position, is more affected by the interface states, especially at high densities and thus despite the low level of trapped charge. Although there is a small improvement in the current for shorter lengths, the relative degradation reaches approximately the same level for both devices.

The large current degradation in the IMFET can also be attributed to the enhancement of Coulomb scattering due to the local potential of the charged states [2]. Obviously, this additional scattering causes a degradation in the surface mobility and has no direct impact on the carrier transport in the channel center. Simulations show that the presence of interface traps results in a shift and lowering of the effective mobility peak as depicted in Figure 4.11. With Coulomb scattering being dominant for low carrier density, i.e. low gate bias, the impact of interface traps on the effective mobility is observable only for weak inversion (for the IM device) and partial depletion (for the JL device). The effective mobility was extracted using the following formula [42]:

$$\mu_{\text{eff}} = \frac{LI_{\text{d}}(V_{\text{g}})}{qn_{\text{s}}(V_{\text{g}})V_{\text{d}}W} = \frac{L^2 I_{\text{d}}(V_{\text{g}})}{V_{\text{d}} \int C_{\text{g}}(V_{\text{g}})\text{d}V_{\text{g}}} \tag{4.2}$$

where n_s is sheet density of mobile carriers in the channel obtained by integration of the total gate capacitance C_g extracted from C–V curves. V_{ds} was set at 0.01 V to eliminate the effect of the lateral high electric field.

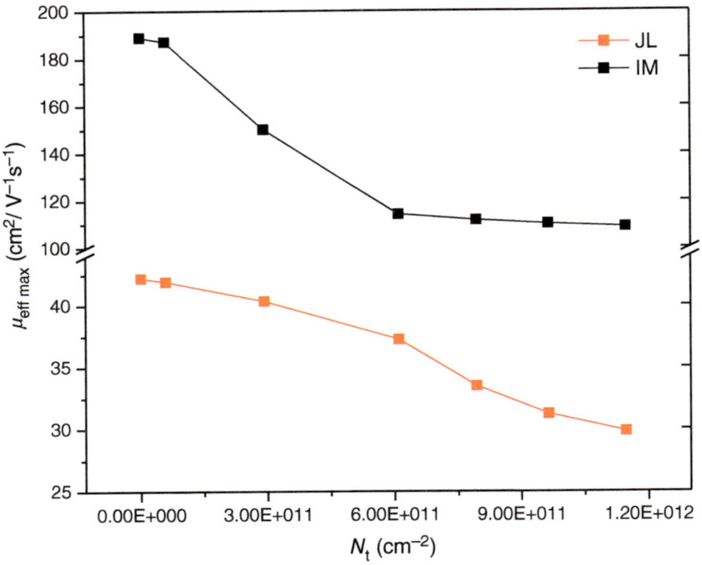

Figure 4.12 Maximum effective mobility degradation with respect to N_{it} for JLFET and IM VSG-FET with $L = 50$ nm and $V_{ds} = 0.01$ V.

Figure 4.12 shows the maximum effective mobility degradation as a function of the interface state density. As expected, major degradation of the IM effective mobility exceeding 40% is observed. The JL device, on the other hand, exhibits much lower degradation that reaches 20%. It is believed that this degradation is not due to Coulomb scattering of the interface states but rather the enhancement in the perpendicular electric field.

The shift in the transfer characteristics also yields a relative decrease in transconductance. Similar to the drain current, a linear degradation in the transconductance is observed in the JL device (Figure 4.13). The IM type, on the other hand, exhibits small variations in low trap densities and much less relative degradation. As the interface state density extends toward the source, major degradation in transconductance is observed, attaining the JLFET values. Relatively, the degradation approaches 100% and largely exceeds its JL counterpart. This major degradation in the IM device can be attributed to the same reasons of drain current degradation, i.e. the enhanced effect of the charged interface states on the conduction path and surface mobility. Although the sensible improvement of transconductance results from the channel length reduction for both devices (Figure 4.13a), the same degradation evolution is observed for both lengths.

Figure 4.14 illustrates the intrinsic gain degradation for JL and IM devices. The gain of the fresh IM device reaches the value twice that of its JL counterpart. Furthermore, the length reduction reduces the gain by half, which shows a large improvement in output conductance. Consequent to the drop in transconductance in the presence of interface states, the intrinsic gain degrades as well. However, an improvement in the gain is observed above a certain amount of the interface

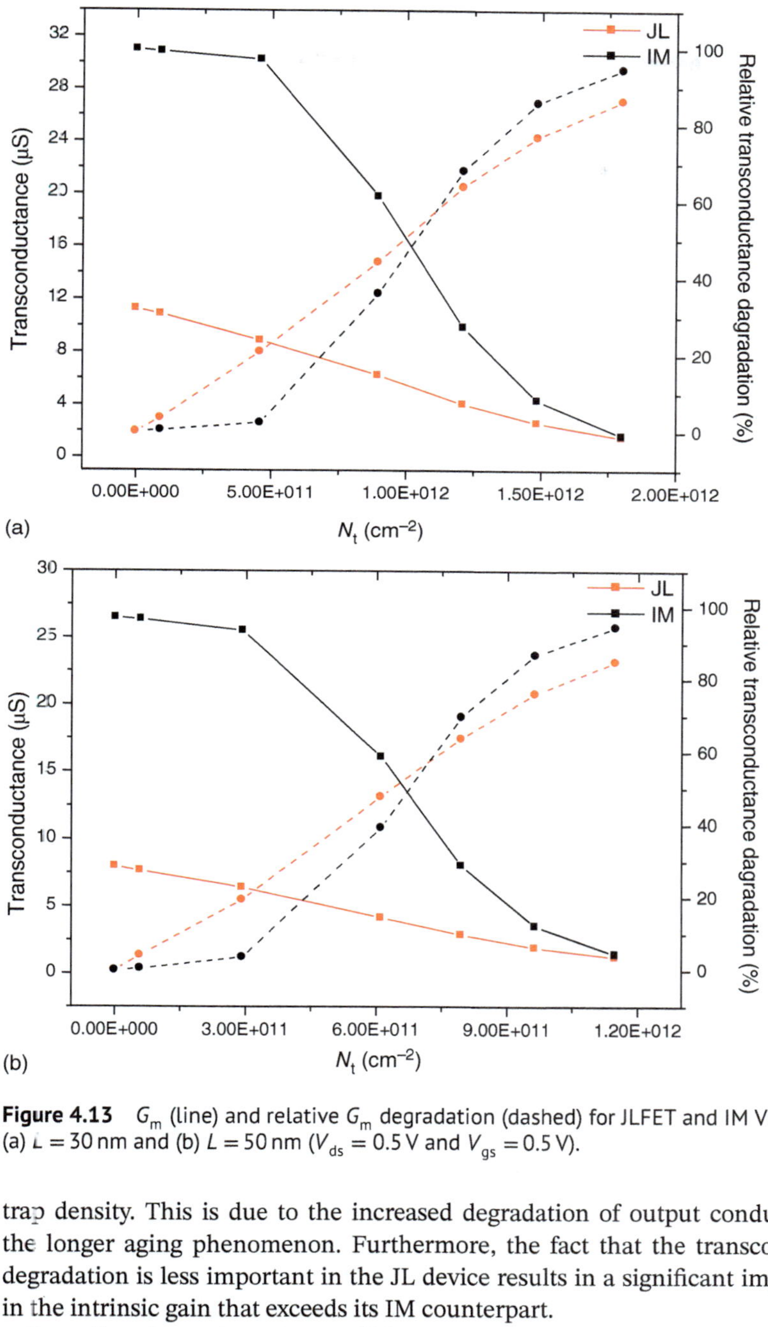

Figure 4.13 G_m (line) and relative G_m degradation (dashed) for JLFET and IM VSG-FET (a) $L = 30$ nm and (b) $L = 50$ nm ($V_{ds} = 0.5$ V and $V_{gs} = 0.5$ V).

trap density. This is due to the increased degradation of output conductance for the longer aging phenomenon. Furthermore, the fact that the transconductance degradation is less important in the JL device results in a significant improvement in the intrinsic gain that exceeds its IM counterpart.

Regarding the unity gain cutoff frequency (Figure 4.15), the simulations have provided acceptable values for the JL device, exceeding the half of the fresh IM values. Moreover, the channel length reduction by 20 nm yields a significant improvement in the JLFET cutoff frequency. As for the previous parameters, the

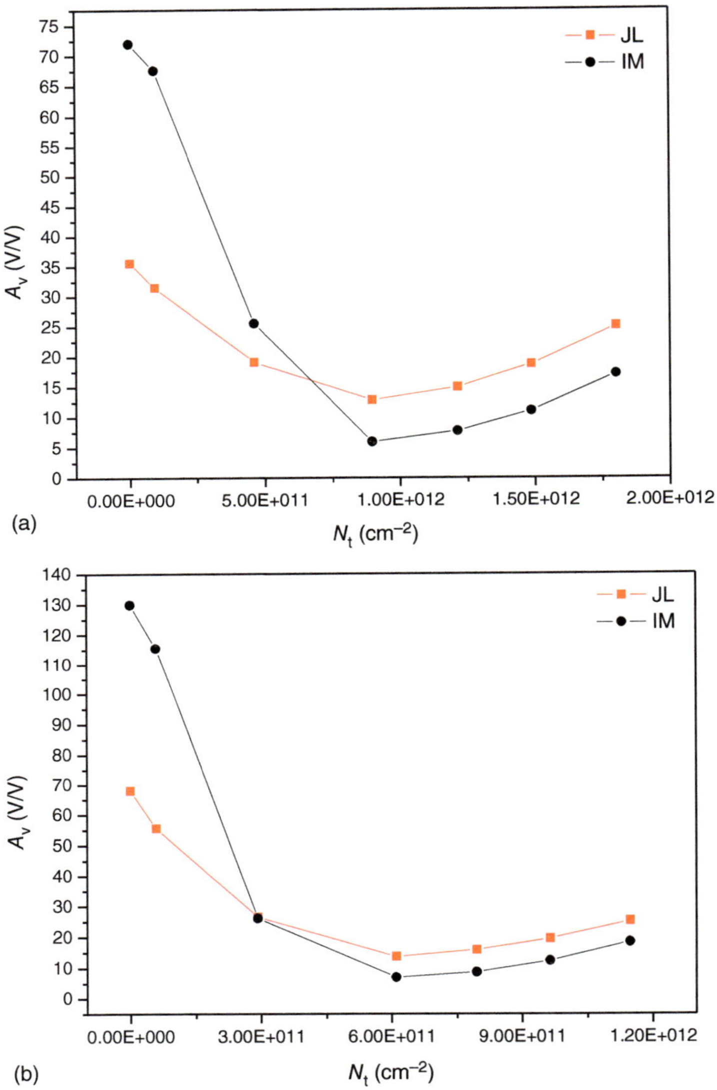

Figure 4.14 Intrinsic gain degradation for JLFET and IM VSG-FET (a) $L = 30$ nm and (b) $L = 50$ nm ($V_{ds} = 0.5$ V and $V_{gs} = 0.5$ V).

IM exhibits moderate cutoff frequency degradation for low interface trap densities. However, the extended HCD yields major degradation that approaches 100% and exceeds the JLFET value for 30 nm gate length. In the JL device, a quasi-linear degradation is observed. A sensible attenuation is observed for higher interface state densities, which can be attributed to the degradation of the total gate capacitance accentuated by the increasing interface charges.

Summarizing the previous results, it appears that HCD is very harmful to the operation of both devices either in analog/RF or digital applications. Considering

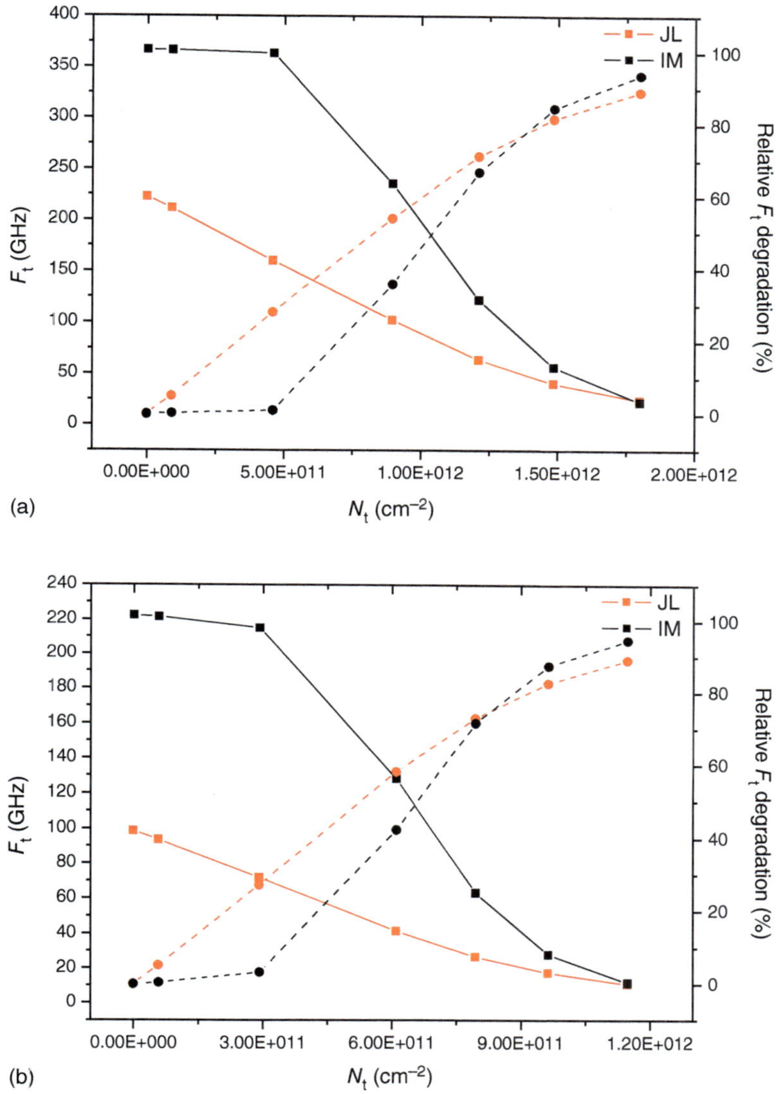

Figure 4.15 Cutoff frequency (line) and its relative degradation (dashed) for JLFET and IM VSG-FET (a) $L = 30$ nm and (b) $L = 50$ nm ($V_{ds} = 0.5$ V and $V_{gs} = 0.5$ V).

analog/RF applications, drain current and transconductance degradation will reduce the upper limit of the operating range. Signal distortion, compression/extension, and corruption might also occur. In amplifiers, gain degradation will result in a loss in the original circuit performance. The cutoff frequency drop will reduce the bandwidth, which can be disadvantageous for RF transceivers. If the inversion mode device exhibits relative stability for low interface trap densities, the major degradation revealed for higher densities suggests that the JL transistor will have better resilience to the longer aging phenomenon.

4.6 Hot Carrier Degradation in Digital Applications

In this section, the impact of HCD on digital applications is discussed, where both static and transient analyses are performed.

4.6.1 Static Analysis

In this subsection, the impact of HCD on CMOS inverter operation is considered. The degradation of the conception keys, threshold voltage and transconductance, implies deterioration in the inverter properties. Figure 4.16 illustrates the voltage transfer characteristics (VTCs) of a fully functional complementary JLFET-based inverter. The p-type JL is identical to the n-type with a channel doping of 2.5×10^{19} cm^{-3} and a gate work function of 4.26 eV; the latter values are chosen to obtain a switching threshold of $V_{dd}/2$ ($V_{dd} = 1$ V).

The NM is a key feature of logic application design. Largest NM values ensure correct circuit operation and stability against signal noise that may lead to a bit reversal in cascade configurations [55]. The NM for the two inverter states is defined as $NM_H = V_{OH} - V_{IH}$ and $NM_L = V_{IL} - V_{OL}$. $V_{I,O/L,H}$ are the operational points and correspond to a voltage gain of −1. For fresh devices, the JL inverter achieved large

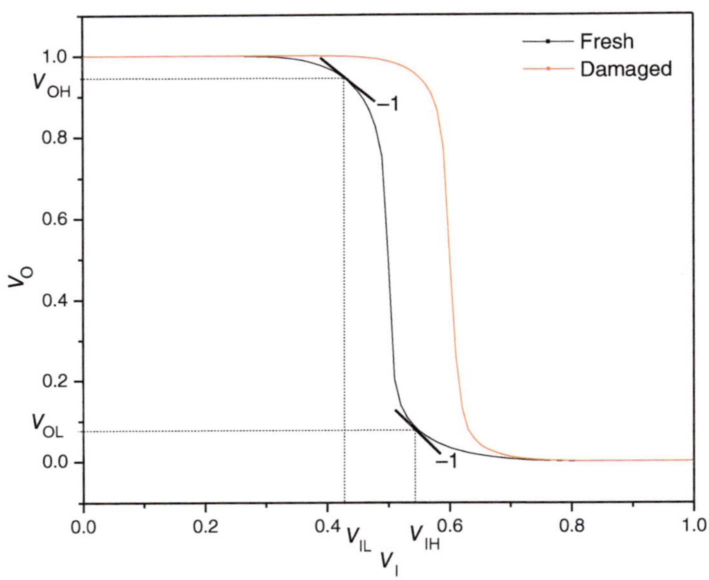

Figure 4.16 VTCs of 50-nm VSG-JL-based inverter for fresh and damaged n-type devices ($N_{it} = 7.9 \times 10^{11}$ cm^{-2}).

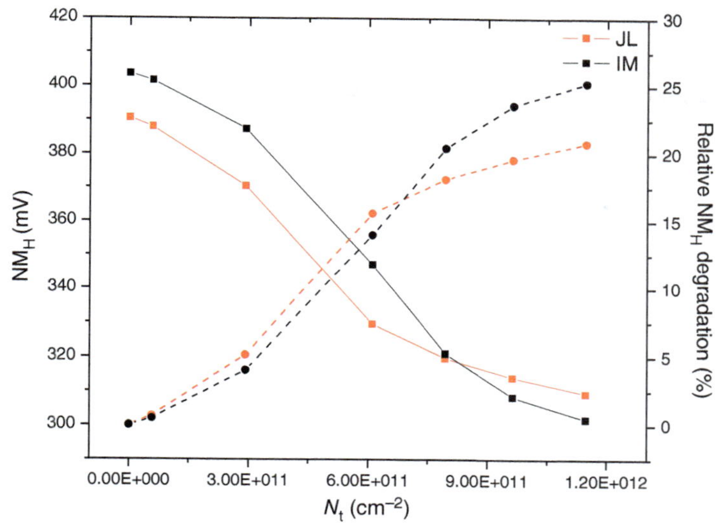

Figure 4.17 NM_H (line) and relative NM_H degradation (dashed) for VSG-JL- and IM-based inverters.

NMs with $NM_H = 390$ mV and $NM_L = 359$ mV. On the other hand, the IM inverter achieved $NM_H = 403$ mV and $NM_L = 340$ mV.

As shown in Figure 4.16, the presence of interface states in the n-type device yields a shift toward the right of the output voltage and the switching threshold. As a result, the high NM decreases with a relative increase in the low NM. Considering the worst-case scenario (i.e. the lowest NM), the NM_H degradation is depicted in Figure 4.17. For low interface trap densities, the degradation follows the same profile for both devices with relatively larger values for the JL-based inverter. As the trap density further increases, the relative NM degradation of the IM-based inverter exceeds that of the JL type, yielding lower NM values for the former circuit than for the latter one.

4.6.2 Transient Analysis

Figure 4.18 presents the transient response of the VSG-JL- and IM-based inverters. The output is lumped with an external capacitance of 2 fF for the IM-based inverter. Because of the large equivalent resistance (lower saturation current) and C_{gd} of the JLFET, the external capacitance is set to 0.2 fF to obtain rise and fall times close to those offered by the IMFET. The oxide capacitance is estimated at 34 aF for both devices. However, the total internal capacitance is lower in the JLFET due to the absence of junction capacitances. Simulation outcomes have demonstrated an increase in the fall time of the damaged device–based inverter. This increase results from the degradation of many parameters such as transconductance, drain current, capacitance. Despite the fact that the trapped electrons yield a negative charge that decreases C_{gd} during saturation, the impact of the drain current degradation on the

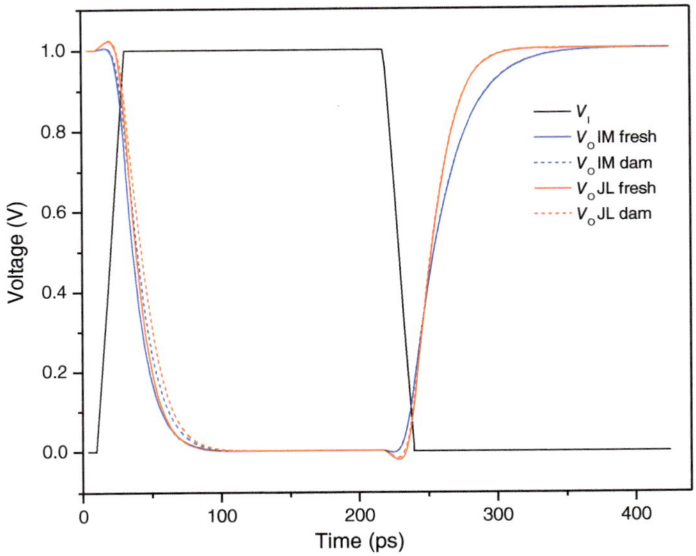

Figure 4.18 Transient response of fresh and damaged VSG-JL and IM-based inverters.

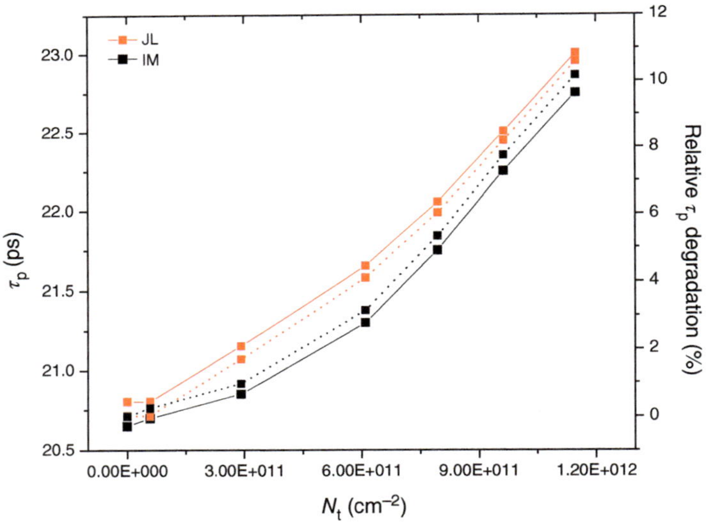

Figure 4.19 τ_p (line) and relative τ_p degradation (dashed) for VSG-JL- and IM-based inverters.

discharging time is significant. On the other hand, the rise time remains unaffected as it depends essentially on the p-type transistor characteristics.

Degradation of propagation delay is illustrated in Figure 4.19. It is defined as $\tau_p = (\tau_{pH} + \tau_{pL})/2$, where τ_{pH} and τ_{pL} are, respectively, the high-to-low and low-to-high transition propagation delays that reflect the elapsed time for the

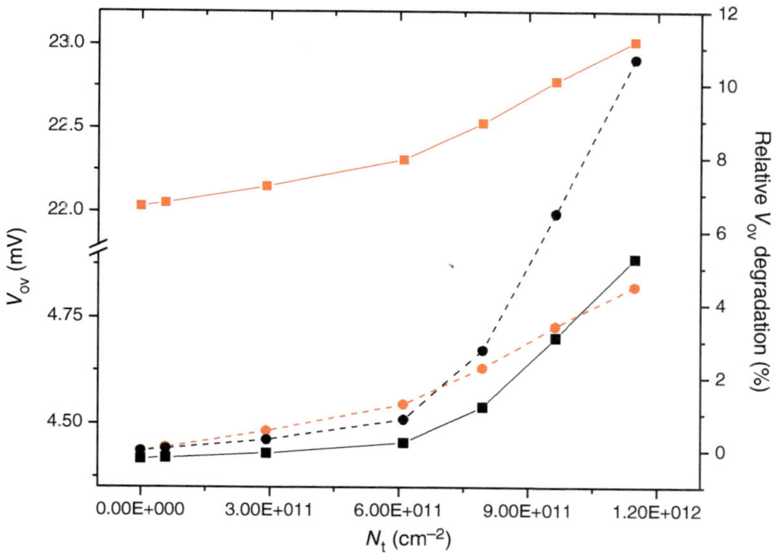

Figure 4.20 V_{ov} (line) and relative V_{ov} degradation (dashed) for VSG-JL and IM-based inverter.

output to reach 50% of V_{dd} with respect to the input voltage at the same level. As expected from the large equivalent resistance of the JLFET, relatively longer propagation delay and higher relative degradation are obtained. Degradation for both devices follows the same increasing profile. Nevertheless, the discrepancy tends to decrease with extended stress time.

Moreover, the transient analysis reveals an amplification of the over-/undershoots voltage peaks. These peaks are inherent in the inverter operation and result from the coupling of the input ramp to the output while the change in the transistor states has not started yet. The coupling occurs via the C_{gd} (Miller capacitance) of the two transistors. Also, the over-/undershooting time adds up to the transition one, resulting in an enhancement in the propagation delay and power dissipation. In Figure 4.20, the overshoot voltage V_{ov} is extracted for various interface state densities of both JL- and IM-based inverters. For the fresh JLFET, V_{ov} is almost five times higher than its IM counterpart. This large value results from the lower C_L as the peak voltage can be computed by $V_{ov} = V_{dd} \times C_M/(C_M + C_L)$ [56]. Considering the relative degradation of V_{ov}, better values are observed for the IM-based inverter for low interface trap densities. Such a trend reverses with a longer aging phenomenon, attaining more than two times the relative degradation of the JLFET-based circuit.

Overall, the simulations performed show that the inverters based on the two device types still have their total functionality after a long aging process even if it loses its symmetry. In a single stage, degradation in parameters will essentially increase the total power dissipation. For cascade topologies such as ring oscillators, propagation delay degradation will result in a decrease in oscillation frequency. The results also highlight better resilience of the JLFET-based inverter for the longer aging process.

4.7 Concluding Remarks

In this chapter, a comparative study of HCD in the JL and IM VSG-MOSFET devices was proposed. The analysis focused on interface state creation. Using the appropriate stress voltage conditions, the simulations showed that the created interface charge is 1.5 times higher than in the IM device. Despite of the enhanced II in the JLFET, the continuous recombination in the channel attenuates interface trap creation. Also, a deeper penetration into the channel of the interface state profile with respect to stress time is observed for the IM device.

Based on the interface trap profile evolution with stress time and their amphoteric properties, a more realistic response of the interface states in different operating regimes is obtained. HCD in many electrical characteristics of these two devices was evaluated. Major degradation in all the metrics in the subthreshold and superthreshold regimes were observed. Enhanced degradation was noticed in the IMFET for high interface state densities, leading, in most cases, to the underperformance of the device compared to its JL counterpart. The enhanced immunity of the JLFET against SCEs was demonstrated. For equivalent interface state densities, improved immunity of shorter channel devices was observed. Accordingly, the combination of lower interface trap creation and low degradation exhibited by the JLFET demonstrated enhanced immunity against HCD and confirms the longer lifetime of this device.

Besides, HCD on both JL- and IM VSG-based CMOS inverters was assessed. It can be deduced that the presence of interface states yields degradation in the majority of the circuit metrics. Nevertheless, the JLFET-based inverter exhibits better immunity against HCD. In addition to elucidating the superior reliability of the JLFET at the device and circuit levels, this study emphasized the importance of interface state creation and distribution for HCD evaluation. Therefore, it is necessary to develop new compact models that support a wide range of parasitic phenomena in order to provide a more realistic view of the aging process.

References

1 Takeda, E. (1993). Hot-carrier effects in scaled MOS devices. *Microelectron. Reliab.* 33: 1687–1711.

2 Roblin, P., Samman, A., and Bibyk, S. (1988). Simulation of hot-electron trapping and aging of nMOSFETs. *IEEE Trans. Electron Devices* 35: 2229–2237.

3 Tam, S., Ko, P.K., and Hu, C. (1984). Lucky-electron model of channel hot-electron injection in MOSFET's. *IEEE Trans. Electron Devices* 31: 1116–1125.

4 Hänsch, W., Schwerin, A.V., and Hofmann, F. (1990). A new self-consistent modeling approach to investigating MOSFET degradation. *IEEE Electron Device Lett.* 11: 362–364.

5 Doyle, B., Bourcerie, M., Marchetaux, J.C., and Boudou, A. (1990). Interface state creation and charge trapping in the medium-to-high gate voltage range

($V_d/2 \geq V_g \geq V_d$) during hot-carrier stressing of n-MOS transistors. *IEEE Trans. Electron Devices* 37: 744–754.

6 Chen, Z., Hess, K., Lee, J. et al. (2000). On the mechanism for interface trap generation in MOS transistors due to channel hot carrier stressing. *IEEE Electron Device Lett.* 21: 24–26.

7 Bravaix, A., Guerin, C., Huard, V. et al. (2009). Hot-carrier acceleration factors for low power management in DC-AC stressed 40 nm NMOS node at high temperature. Reliability Physics Symposium, 2009 IEEE International, Montreal, Quebec, Canada, 26–30 April 2009.

8 Djeffal, F., Ghoggali, Z., Dibi, Z., and Lakhdar, N. (2009). Analytical analysis of nanoscale multiple gate MOSFETs including effects of hot-carrier induced interface charges. *Microelectron. Reliab.* 49: 377–381.

9 Rauch, S., La Rosa, G., and Guarin, F. (2001). Role of E–E scattering in the enhancement of channel hot carrier degradation of deep-submicron nMOSFETs at high V_{gs} conditions. *IEEE Trans. Device Mater. Reliab.* 1: 113–119.

10 Djeffal, F., Guessasma, S., Benhaya, A., and Chahdi, M. (2005). An analytical approach based on neural computation to estimate the lifetime of deep submicron MOSFETs. *Semicond. Sci. Technol.* 20: 158–164.

11 Quader, K.N., Minami, E.R., Ko, W.J. et al. (1994). Hot-carrier-reliability design guidelines for CMOS logic circuits. *IEEE J. Solid State Circuits* 29: 253–262.

12 http://bsim.berkeley.edu/.

13 Sanchez Esqueda, I., Barnaby, H.J., and King, M.P. (2015). Compact modeling of total ionizing dose and aging effects in MOS technologies. *IEEE Trans. Nucl. Sci.* 62: 1501–1515.

14 Ghoggali, Z. and Djeffal, F. (2010). Analytical analysis of nanoscale fully depleted Double-Gate MOSFETs including the hot-carrier degradation effects. *Int. J. Electron.* 97 (2): 119–127.

15 Kim, S.Y., Lee, B.H., Hur, J. et al. (2018). A comparative study on hot-carrier injection in 5-story vertically integrated inversion-mode and junctionless-mode gate-all-around MOSFETs. *IEEE Electron Device Letters* 39: 4–7.

16 Lee, S.M., Kim, J.Y., Yu, C.G., and Park, J.T. (2013). A comparative study on hot carrier effects in inversion-mode and junctionless MuGFETs. *Solid State Electron.* 79: 253–257.

17 Park, C.H., Ko, M.D., Kim, K.H.et al. (2012). Investigation on hot carrier effects in n-type short-channel junctionless nanowire transistors. 12th IEEE Conference on Nanotechnology (IEEE-NANO), Birmingham, UK, 20–23 August 2012.

18 Lee, C.W., Afzalian, A., Dehdashti Akhavan, N. et al. (2009). Junctionless multigate field-effect transistor. *Appl. Phys. Lett.* 94 (1-2): 053511.

19 Colinge, J.P., Lee, C.W., Afzalian, A. et al. (2010). Nanowire transistors without junctions. *Nat. Nanotechnol.* 5: 225–229.

20 Park, C.H., Ko, M.D., Kim, K.H. et al. (2012). Electrical characteristics of 20 nm junctionless Si nanowire transistors. *Solid State Electron.* 73: 7–10.

21 Rios, R., Cappellani, A., Armstrong, M. et al. (2011). Comparison of junctionless and conventional trigate transistors with L_g down to 26 nm. *IEEE Electron Device Letters* 32: 1170–1172.

22 Yu, R., Georgiev, Y.M., Das, S. et al. (2013). Junctionless nanowire transistor fabricated with high mobility Ge channel. *Phys. Status Solidi RRL* 8: 65–68.

23 Kamata, Y., Kamata, Y., Kamimuta, Y. et al. Superior cut-off characteristics of $L_g = 40$ nm Wfi $n = 7$ nm poly Ge junctionless tri-gate FET for stacked 3D circuits integration. Symposium on VLSI Technology Digest of Technical Papers, T94–T95, 2013, Kyoto, Japan.

24 Sahay, S. and Kumar, M.J. (2017). Diameter dependence of leakage current in nanowire junctionless field effect transistors. *IEEE Trans. Electron Devices* 64: 1330–1335.

25 Yu, R., Nazarov, A.N., Lysenko, V.S. et al. (2013). Impact ionization induced dynamic floating body effect in junctionless transistors. *Solid State Electron.* 90: 28–33.

26 Lee, C.W., Nazarov, A.N., Ferain, I. et al. (2010). Low subthreshold slope in junctionless multigate transistors. *Appl. Phys. Lett.* 96: 102106.1–102106.3.

27 Sze, S.M. (1981). *Physics of Semiconductor Devices*, 2e.e. New York: Wiley.

28 Gupta, M. and Kranti, A. (2015). Germanium junctionless MOSFET with steep subthreshold swing. *Electrochem. Soc. Trans.* 66: 79–86.

29 Krishnan, S. and Fossum, J.G. (1998). Grasping SOI floating-body effects. *IEEE Circuits Devices Mag.* 14: 32–37.

30 Gupta, M. and Kranti, A. (2017) "Suppressing single transistor latch effect in energy efficient steep switching junctionless MOSFETs. 2017 30th International Conference on VLSI Design and 2017 16th International Conference on Embedded Systems (VLSID), Hyderabad, India, 7–11 January 2017.

31 Yeh, M.S., Yeh, M.S., Wu, Y.C.et al. 2014 "High performance ultra-thin body (2.4 nm) poly-Si junctionless thin film transistors with a Trench structure. IEEE International Electron Devices Meeting (IEDM), San Francisco, CA, USA, 15–17 December 2014.

32 Colinge, J.P. (2014). Chapter 6: Silicon-on-insulator (SOI) junctionless transistors. In: *Silicon-On-Insulator (SOI) Technology* (ed. O. Kononchuk and B.-Y. Nguyen), 167–194. UK: Woodhead Publishing.

33 Colinge, J.P., Lee, C.W., Dehdashti Akhavan, N. et al. (2012). Junctionless transistors: physics and properties. In: *Semiconductor-On-Insulator Materials for Nanoelectronics Applications* (ed. A. Nazarov, J.-P. Colinge, F. Balestra, et al.), 187–200. Berlin Heidelberg: Springer.

34 Thirunavukkarasu, V., Jhan, Y.R., Liu, Y.B. et al. (2017). Gate-all-around junctionless silicon transistors with atomically thin nanosheet channel (0.65 nm) and record sub-threshold slope (43 mV/dec). *Appl. Phys. Lett.* 110: 032101.1–032101.5.

35 Chiang, T.K. (2012). A quasi-two-dimensional threshold voltage model for short-channel junctionless double-gate MOSFETs. *IEEE Trans. Electron Devices* 59: 2284–2289.

36 Ansari, L., Feldman, B., Fagas, G. et al. (2010). Simulation of junctionless Si nanowire transistors with 3 nm gate length. *Appl. Phys. Lett.* 97: 062105.

37 Colinge, J.P., Kranti, A., Yan, R. et al. (2011). Junctionless nanowire transistor (JNT): properties and design guidelines. *Solid State Electron.* 65-66: 33–37.

38 Leung, G. (2015). Variability and heterogeneous integration of emerging device technologies. PhD dissertation, University of California, Los Angeles, CA.

39 Asenov, A., Brown, A.R., Roy, G. et al. (2009). Simulation of statistical variability in nano-CMOS transistors using drift-diffusion, Monte Carlo and non-equilibrium Green's function techniques. *J. Comput. Electron.* 8: 349–373.

40 Colinge, J.P., Lee, C.W., Ferain, I. et al. (2010). Reduced electric field in junctionless transistors. *Appl. Phys. Lett.* 96 (1–3): 073510.

41 Jacoboni, C., Canali, C., Ottaviani, G., and Alberigi Quaranta, A. (1977). A review of some charge transport properties of silicon. *Solid State Electron.* 20: 77–89.

42 Jeon, D.Y. (2013). Revisited parameter extraction methodology for electrical characterization of junctionless transistors. *Solid State Electron.* 90: 86–93.

43 Goto, K.I., Yu, T.H., Wu, J. et al. (2012). Mobility and screening effect in heavily doped accumulation-mode metal-oxide-semiconductor field-effect transistors. *Appl. Phys. Lett.* 101: 073503.

44 Rudenko, T., Nazarov, A., Yu, R. et al. (2013). Electron mobility in heavily doped junctionless nanowire SOI MOSFETs. *Microelectron. Eng.* 109: 326–329.

45 Persson, M.P., Mera, H., Niquet, Y.M. et al. (2010). Charged impurity scattering and mobility in gated silicon nanowires. *Phys. Rev. B* 82: 115318.

46 ATLAS User Manual: Device Simulation Software (2012).

47 Tyaginov, S. (2015). Physics-based modeling of hot-carrier degradation. chapter. In: *Hot Carrier Degradation in Semiconductor Devices* (ed. T. Grasser). Cham, Switzerland: Springer International Publishing.

48 Poindexter, E.H., Caplan, P.J., Deal, B.E., and Razouk, R.R. (1981). Interface states and electron spin resonance centers in thermally oxidized (111) and (100) silicon wafers. *J. Appl. Phys.* 52: 879–884.

49 Trevisoli, R., Trevisoli Doria, R., Michelly de Souza, D., and Pavanello, M.A. (2015). Extraction of the interface trap density energetic distribution in SOI junctionless nanowire transistors. *Microelectron. Eng.* 147: 23–26.

50 Kubota, T. and Asano, A. (1999). Theoretical and spectroscopic studies of gap-states at ultrathin silicon oxide/silicon interfaces. *J. Chem. Phys.* 111: 8136–8143.

51 Lenahan, P.M. and Dressendorfer, P.V. (1984). Hole traps and trivalent silicon centers in metal/oxide/silicon devices. *J. Appl. Phys.* 55: 3495–3499.

52 Campbell, J.P. and Lenahan, P.M. (2002). Density of states of Pb_1 Si/SiO_2 interface trap centers. *Appl. Phys. Lett.* 80: 1945–1947.

53 Kim, S., Choi, S.J., Moon, D.I., and Choi, Y.K. (2011). An extraction method of the energy distribution of interface traps by an optically assisted charge pumping technique. *IEEE Trans. Electron Devices* 58: 3667–3673.

54 Hu, C., Tam, S.C., Hsu, F.C. et al. (1985). Hot-electron-induced MOSFET degradation-model, monitor, and improvement. *IEEE Trans. Electron Devices* 32: 375–385.

55 Rabaey, J.M., Chandrakasan, A.P., and Nikolić, B. (2003). *Digital Integrated Circuits: A Design Perspective*. New Jersey, USA: Pearson Education.

56 Mookerjea, S., Krishnan, R., Datta, S., and Narayanan, V. (2009). On enhanced miller capacitance effect in interband tunnel transistors. *IEEE Electron Device Letters* 30: 1102–1104.

5

Study of Electrostatic and Dispersion Forces in Nanoelectromechanical Systems (NEMS)

Masoud SoltanRezaee and Amin Farrokhabadi

Tarbiat Modares University, Department of Mechanical Engineering, Jalal AleAhmad, Nasr, Tehran 14115-111, Iran

5.1 Introduction

In recent years, due to the outstanding electrical and mechanical properties of nanostructures, scientists are interested in novel generations of ultrasmall devices. As a result, micro- and nanoelectromechanical systems (MEMS and NEMS, respectively) have been developed, which are used in several applications, such as switches, actuators, sensors, atomic force microscopes as well as mass and force detectors.

Several important building blocks in NEMS consist of at least two conductive electrodes, one of which is movable (Figure 5.1). The deformable electrode deflects toward the fixed substrate due to the electrostatic Coulomb and/or dispersion forces. When this attraction becomes larger than the corresponding restoring force, it results in pull-in instability. In other words, when the applied voltage exceeds a critical value, the electrostatic and/or molecular forces become larger than the corresponding restoring forces, it leads to unstable collapsing (Figure 5.2). Investigation of the pull-in instability phenomenon has attracted industrial and research attention. While pull-in instability confines electromechanical systems, determining the instability characteristics is essential in the design procedure of various micro- and nano-electronic devices.

The appearance of new effects through the scale alteration in geometry can include several additional terms into the theory of nanoscale structures, while their influence on other scales is negligible. One such effect is dispersion attractions. Dispersion forces or quantum interactions, including the Casimir and van der Waals (vdW) regimes, play a significant role in the behavior of nanodevices [1–3], for example, structure dimensions/configurations and the gap between electrically actuated electrodes [4].

Neglecting these attractions in ultrasmall scales can cause significantly incorrect results. For instance, Figure 5.3 shows that neglecting the effect of the Casimir regime in the simulation of nanostructures can cause inaccurate prediction of the system behavior, where the predicted pull-in voltage will be overestimated [5–7].

Nanotechnology in Electronics: Materials, Properties, Devices, First Edition.
Edited by Visakh P.M., Artem Semkin, Raneesh Balakrishnan, and Saša Lazović.

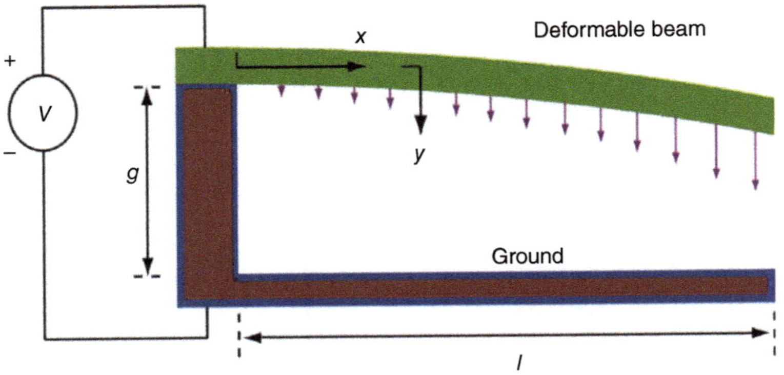

Figure 5.1 Schematic of an actuated cantilever beam where the deformable electrode deflects toward the substrate.

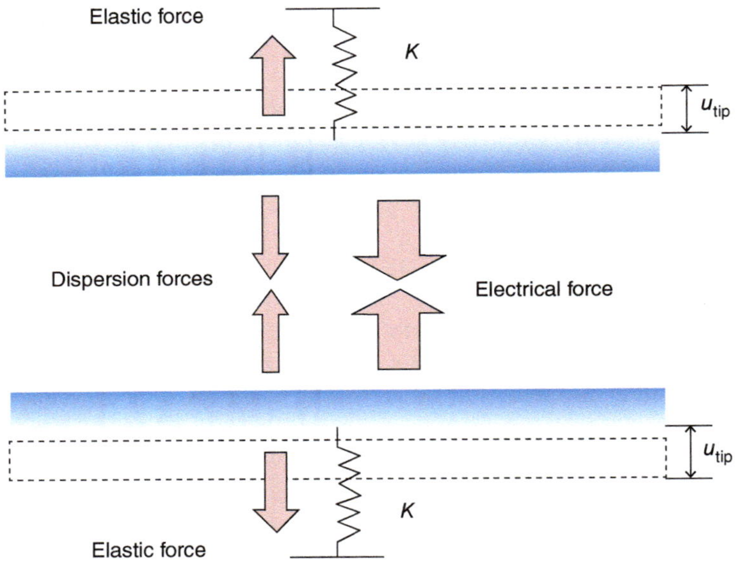

Figure 5.2 Schematic configuration of a nanosystem model by considering different forces. Source: Farrokhabadi et al. [13] / with permission of Elsevier.

Furthermore, in some circumstances, the instability conditions cannot be predicted without considering the Casimir force [8]. According to the reliable physical concepts, miniature devices can usually be under either Casimir or vdW regime. Therefore, it is imperative to mention that influence of the Casimir or vdW regime depends on the conditions considered and cannot be generalized for any arbitrary system, where other geometries can cause different results. Figure 5.4 illustrates the Casimir force is dominant due to the system circumstances considered [9, 10]. It is important to mention that there is a transition region where both regimes can be applied to the system [1, 11].

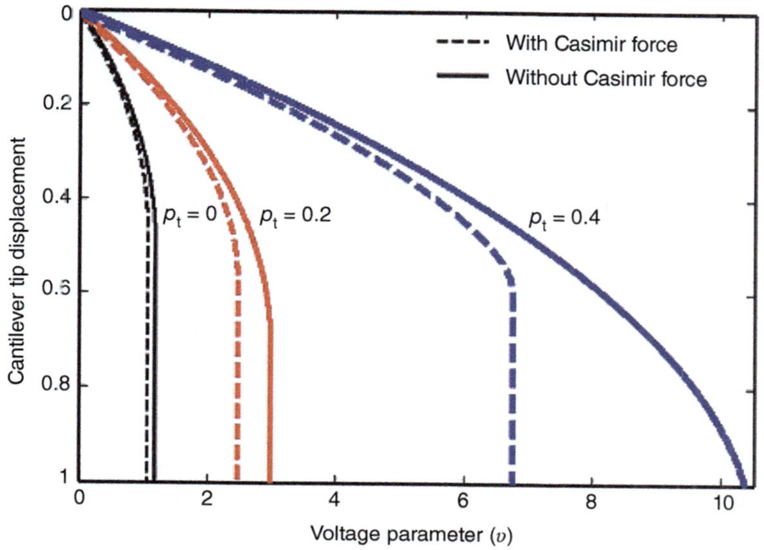

Figure 5.3 Effect of neglecting the dispersion forces in the simulation of nanostructures. Source: SoltanRezaee et al. [9]/with permission of Elsevier.

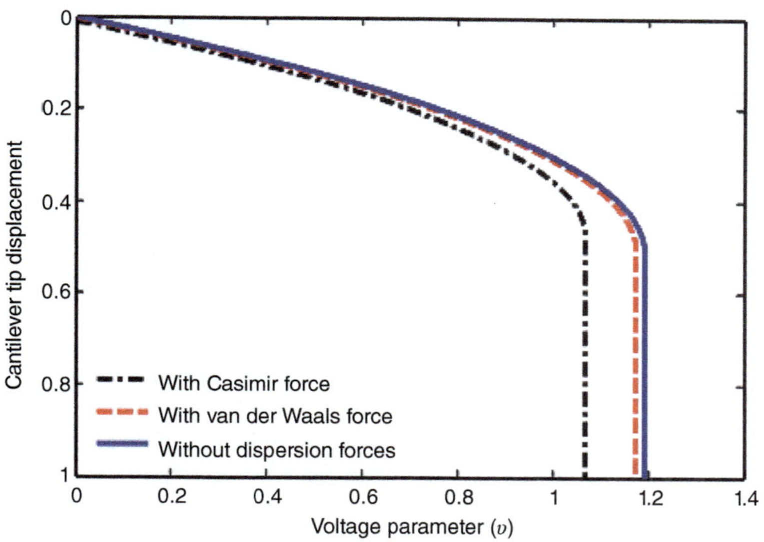

Figure 5.4 Effect of the dispersion forces on the nanosystem behavior. Source: SoltanRezee et al. [9] / with permission of Elsevier.

For more accurate evaluation, one force is effective at small distance separation, while the other at larger separation. In general, efficient electric actuation at reasonable actuation voltages is obtained by considering the gap to beam length, which is typically on the order of 10^{-2}–10^{-3} [12].

Indeed, both Casimir and vdW attractions are connected to the existence of zero-point vacuum oscillations of an electromagnetic field. They can be expressed in terms of quantum confinement effects when a pair of interacting bodies is brought close together [14]. For close conductive bodies, a virtual photon emitted by an atom of one surface reaches the atom of the other surface during its lifetime. The vdW force arises from the correlated oscillations of instantaneously induced dipole moments of these atoms. However, when the gap between two surfaces is sufficiently large, the virtual photon emitted by the atom of the first surface cannot reach the second one during its lifetime [4, 14]. In this case, the dispersion energy starts to decay faster, and the interaction between the two surfaces is described by the Casimir force relation. This retardation effect is the result of finite propagation speed of the electromagnetic field. If the separation is larger than the plasma (for metals) or absorption (for dielectrics) wavelength of the surface material, then the retardation becomes significant [4, 15]. Furthermore, emphasizing the gap limitation is crucial to the validity of the dominant regime assumption. It should be noted that other geometrical parameters can also affect the dominant force.

Beneficial discussions on the comparison between the vdW and Casimir attractions and the dominancy range of each regime can be found in the literature for planar geometries [16]. However, a few researchers have studied the instability of nanodevices with other configurations, for example, cylindrical elements. Farrokhabadi et al. [17] have illustrated the difference between these regimes by comparing the Casimir and vdW forces for typical nano-tweezers. It has been demonstrated that without considering the valid range of operation for dissipation forces, using the vdW or Casimir formula leads to inaccurate results [17].

Another interesting point is that under some considerations, the provided models for circular (shaft) tweezers can be used for approximating the system behavior in nanostructures made of conductive nanotubes. It should be noted that the local buckling of nanotubes is not usually considered in such models. Moreover, when the nanotube wall is sufficiently thin, using shaft tweezer models requires more corrections [18].

In this study, we intend to categorize the well-known relations of electrostatic, Casimir, and vdW forces for different configurations. Here, the most frequently used configurations for beam-based structures are considered: rectangular beam–plate, wire–plate or cylinder–plate, carbon nanotube (CNT) sheets, rectangular tweezers, circular tweezers, and CNT tweezers. The presented relations are valid for cantilever and doubly clamped boundary conditions (BCs). Moreover, the effects of the fringing field, other famous theories related to the Casimir force, and the freestanding phenomenon are considered separately. Owing to the lack of similar categorized relations in the specialized literature, they are expected to contribute to a better understanding about the actuation of nanodevices and would be beneficial to simulation and development of such structures.

5.2 Electrostatic Forces

Various structural configurations can be considered in the NEMS. The device responses completely depend on the system configuration, which should be considered to accurately derive the governing equations. In the following sections, classic expressions of the electrostatic forces for different geometrical configurations of actuated electrodes are presented, neglecting the electric fringing field and dispersion forces, hence valid in macro systems. It should be noted that all the geometrical parameters have been shown in the related figures in all cases.

5.2.1 Rectangular Beam–Plate

The parallel-plate model for the electric force is the most common configuration in numerous applications (Figure 5.5). Considering two parallel plates (a moveable beam-based electrode and a substrate), the electrostatic force per unit length of the beam can be approximated as [19–21]

$$F_{\text{els}} = \frac{\varepsilon_0 b V^2}{2(g - w)^2} \tag{5.1}$$

where V is the external applied voltage (the electrical potential difference within the substrate and electrode) and $\varepsilon_0 = 8.854 \times 10^{-12}\ \text{F m}^{-1}$ is the permittivity constant of free space.

5.2.2 Wire–Plate

It is essential to point out that in this relation, the wire and plate are considered perfect conductors (Figure 5.6). For a conductive cylinder of infinite length, the electrostatic energy per unit length of the wire, E_{els}, is given by [22]

$$E_{\text{els}} = \frac{CV^2}{2} = \frac{\pi V^2 \varepsilon_0}{\text{arc}\ \cosh(1 + D/R)} \tag{5.2}$$

where C is the capacitance per unit length of the rod.

Therefore, the electrostatic force per unit length F_{els} will be obtained by differentiating the electrostatic energy as follows [23]:

$$F_{\text{els}} = \frac{\mathrm{d}E_{\text{els}}}{\mathrm{d}D} = \frac{\pi V^2 \varepsilon_0}{\sqrt{D(D + 2R)}\text{arc}\ \cosh^2(1 + D/R)} \tag{5.3}$$

Note that by applying the voltage, the wire deflects and the distance between the wire and the substrate reduces from D to $(D - w)$. Hence, with the assumption that

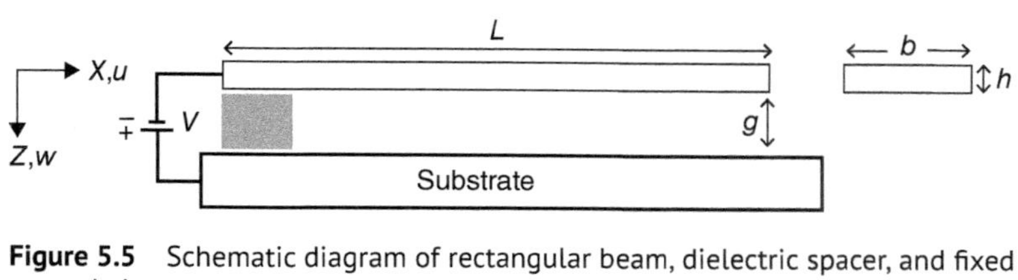

Figure 5.5 Schematic diagram of rectangular beam, dielectric spacer, and fixed ground plane.

R is much less than the distance between wires, which leads to $(D \pm R) \approx D$, the electrostatic force per unit length of the deflected wire can be further simplified as

$$F_{els} \approx \frac{\pi V^2 \varepsilon_0}{(D - w)\text{arc}\cosh^2\left(\frac{D-w}{R}\right)} \approx \frac{\pi V^2 \varepsilon_0}{(D - w)\ln^2\left(2\frac{D-w}{R}\right)} \tag{5.4}$$

5.2.3 Carbon Nanotube (CNT) Sheets

When a conductive nanotube is placed over a conductive substrate (in the presence of an applied potential difference between the tube and electrode), the electrostatic charge is induced on both the tube and substrate [25, 26]. To calculate the electrical forces acting on the tube, a capacitance model is used (Figure 5.6). In this case, the relation of the electrostatic force is similar to that for a typical wire–plate system [25–27].

5.2.4 Rectangular Tweezers

The electrical force in rectangular tweezers (Figure 5.7) can be computed by differentiating energy per unit length stored in the capacitor [20]:

$$F_{els} = -\frac{V^2}{2}\frac{dC_g}{dg} \tag{5.5}$$

where C_g is the capacitance per unit length of the system.

Batra et al. [20] showed that for two parallel narrow beams the capacitance can be estimated as follows:

$$C_g = 2\varepsilon_0\varepsilon_r\left(0.77 + \frac{b}{g} + 1.06\left(\left(\frac{b}{g}\right)^{0.25} + \left(\frac{h}{g}\right)^{0.5}\right)\right) \tag{5.6}$$

where $_r$ is the relative permittivity of the dielectric.

It should be noted that by applying an external voltage to the tweezers, the arms will deflect toward each other to reduce their gap from g to $(g - w_1 - w_2)$. If both

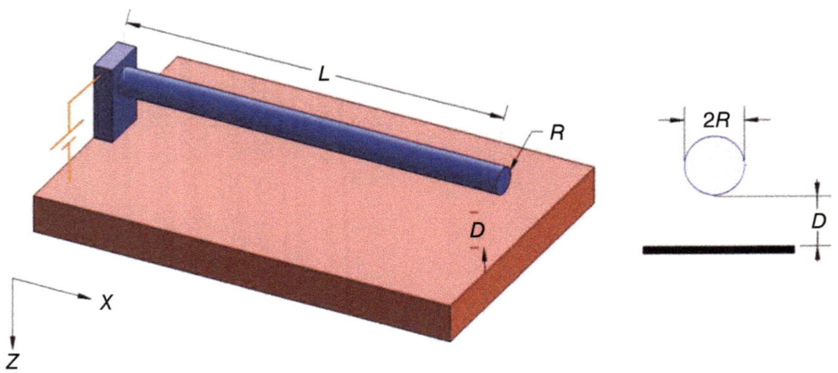

Figure 5.6 Schematic representation of typical cantilever wire-based nanostructures. Source: Mohebshahedin et al. [24] / with permission of Elsevier.

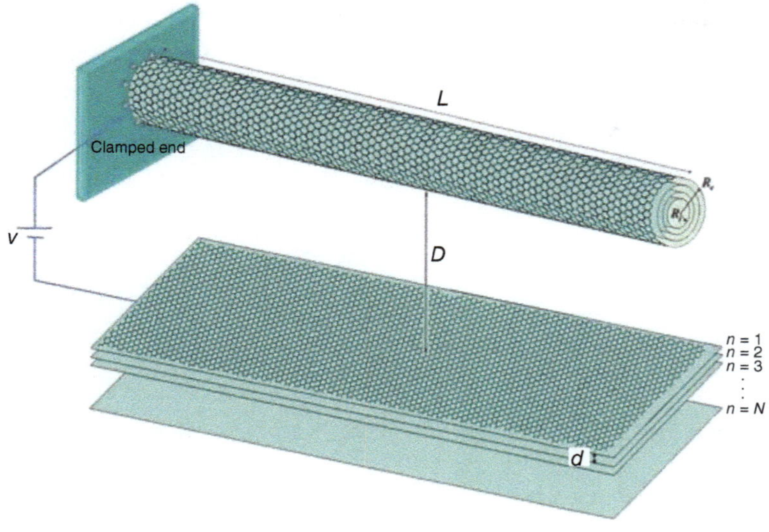

Figure 5.7 Schematic representation of cantilever CNT. Source: Koochi et al. [26] / SciELO / CC BY 4.0.

Figure 5.8 Schematic representation of rectangular tweezers.

arms have the same geometry and material properties, their deflections will be equal ($w = w_1 = w_2$). Thus, by replacing g with $(g - 2w)$, one can write the electrostatic force per unit length of the arm as follows [28]:

$$F_{\text{els}} = -\frac{\mathrm{d}C_g}{\mathrm{d}g} = \varepsilon_0 \varepsilon_r \frac{bV^2}{(g-2w)^2}\left(1 + 0.265\left(\frac{g-2w}{b}\right)^{0.75} + 0.53\frac{h}{b}\left(\frac{g-2w}{h}\right)^{0.5}\right) \tag{5.7}$$

For this type of structure, nanowires are considered perfect cylindrical conductors (Figure 5.9). Therefore, the electrostatic energy per unit length is [13, 23]

$$E_{\text{els}} = \frac{CV^2}{2} = \frac{\pi V^2 \varepsilon_0}{\ln\left(1 + \frac{D}{R} + \sqrt{\left(1 + \frac{D}{R}\right)^2 - 1}\right)} \tag{5.8}$$

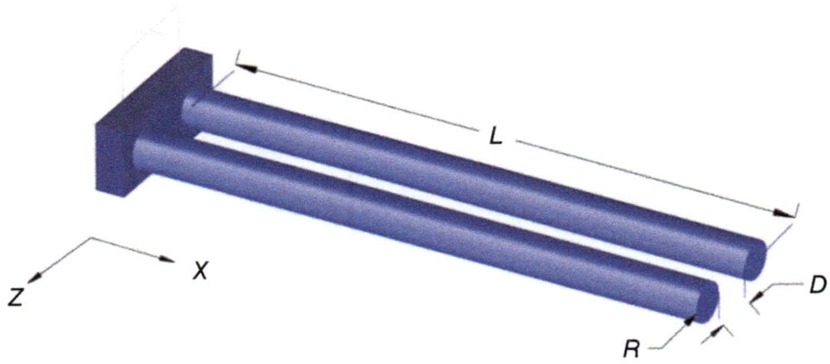

Figure 5.9 Schematic representation of typical wire nano-tweezers. Source: Mohebshahedin [24] / with permission of Elsevier.

where C is the capacitance per unit length of two parallel cylindrical beams that are oppositely charged.

Hence, the electrostatic force per unit length can be defined by

$$F_{\text{els}} = \frac{dE_{\text{els}}}{dD} = \frac{\pi V^2 \varepsilon_0}{\sqrt{D(D+2R)}\left(\ln\left(1+\frac{D}{R}+\sqrt{\left(1+\frac{D}{R}\right)^2-1}\right)\right)^2} \tag{5.9}$$

In a similar deduction to M/NEMS, by applying an external voltage to the structure, the moveable arms will deflect toward each other to reduce their gap from D to $(D - w_1 = -w_2)$. If both wires have the same geometry and material properties, their deflections will be equal $(w_1 = w_2 = w)$. Thus, by replacing D with $(D - 2w)$ in the above relation, one can obtain

$$F_{\text{els}} = \frac{\pi V^2 \varepsilon_0}{\sqrt{(D-2w)(D-2w+2R)}\left(\ln\left(1+\frac{D-2w}{R}+\sqrt{\left(1+\frac{D-2w}{R}\right)^2-1}\right)\right)^2} \tag{5.10}$$

Using the Taylor-series expansion, the above relation can be written as [13]

$$F_{\text{els}} = \frac{\pi V^2 \varepsilon_0}{(D-2w+R)(\ln((R+D-2w)2/R))^2} \tag{5.11}$$

Considering that R is much less than the distance between the wires,

$$F_{\text{els}} = \frac{\pi V^2 \varepsilon_0}{(D-2w+R)(\ln((R+D-2w)2/R))^2} \tag{5.12}$$

5.2.5 Carbon Nanotube (CNT) Tweezers

A typical CNT-based tweezer is composed of two CNT arms with length L, outer radius r, and effective thickness t as presented in Figure 5.10. The initial distance between two CNT arms is g. The arms deflect toward each other due to the presence of the electrostatic Coulomb force. The electrostatic force is computed by using

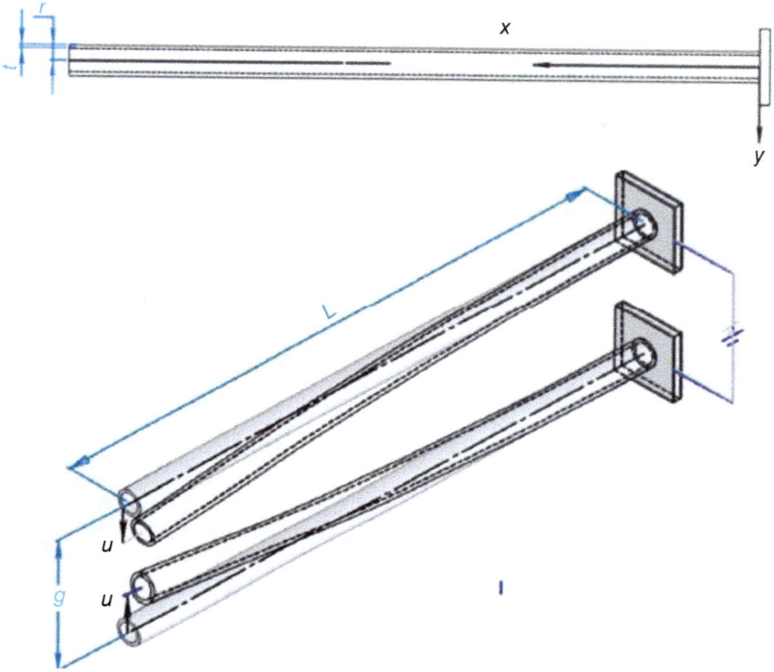

Figure 5.10 Schematic configuration of CNT tweezers. Source: Farrokhabadi et al. [13] / with permission of Elsevier.

a standard capacitance model. The nanotubes are considered perfect cylindrical conductors. In this case, the relations are similar to that of typical circular tweezers [13, 29].

5.3 Fringing Field Effects

Researchers have demonstrated that electric fringing field is an important and effective nonlinearity in investigating electrostatically actuated M/NEMS devices. This particular electrostatic force is more significantly acting on a suspended rectangular beam over a fixed plate (Figure 5.11). In this situation, it is a function of the actuator geometrical properties (width, thickness, and gap) and air permittivity. Due to the nonlinearity of the fringing field relations, governing equations of a system by considering this correction cannot generally be solved in a closed-form way. Hence, numerical analyses are usually involved to obtain responses.

For a narrow-width charged beam, various higher-order corrections such as Batra correction [21] have been proposed for the fringing field effect. Ouakad et al. [30] have studied this effect as a driving actuation in their research studies in detail. Furthermore, SoltanRezaee et al. [9, 31] have illustrated significant effects of fringing field correction in the system responses.

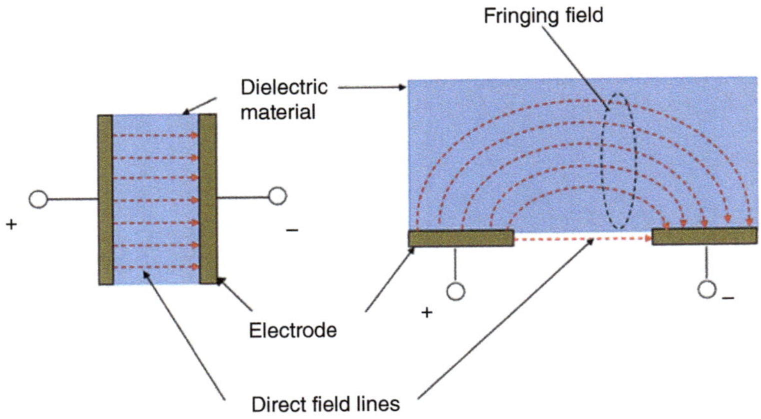

Figure 5.11 Schematic of fringing field effect (top and side views).

Two well-known models for the electric fringing field effect, namely, Palmer's and the Mejis–Fokkema models have been suggested and investigated in the literature, which are presented next.

5.3.1 Palmer's Model

Considering the most frequently used fringing field relation, which takes just the first-order correction into account, the electric force is given by [19]

$$F_{\mathrm{els,P}} = \frac{\varepsilon_0 b V^2}{2(G - w(x,t))^2} + \frac{0.65\varepsilon_0 V^2}{2(G - w(x,t))} \tag{5.13}$$

5.3.2 Mejis–Fokkema Model

The Mejis–Fokkema model is another well-known correction that adjusts the electric force [20, 21]:

$$F_{\mathrm{els,MF}} = \frac{\varepsilon_b b V^2}{2(G - w(x,t))^2}\left(1 + 0.265\left(\frac{G - w(x,t)}{b}\right)^{0.75} + 0.53\frac{h}{b}\left(\frac{G - w(x,t)}{b}\right)^{0.5}\right) \tag{5.14}$$

In this model, not only the impact of the beam width but also its thickness are considered.

The effects of electric fringing field corrections on pull-in instability and threshold voltage of a beam-based nanodevice are displayed in Figure 5.12 It can be found that the Mejis–Fokkema model largely affects the behavior of the actuated ultrasmall system.

5.3.3 Other Models

Other expressions are available for fringing field corrections, which are not widely used. For example, Kacem et al. [33] used a coefficient considering the electrode

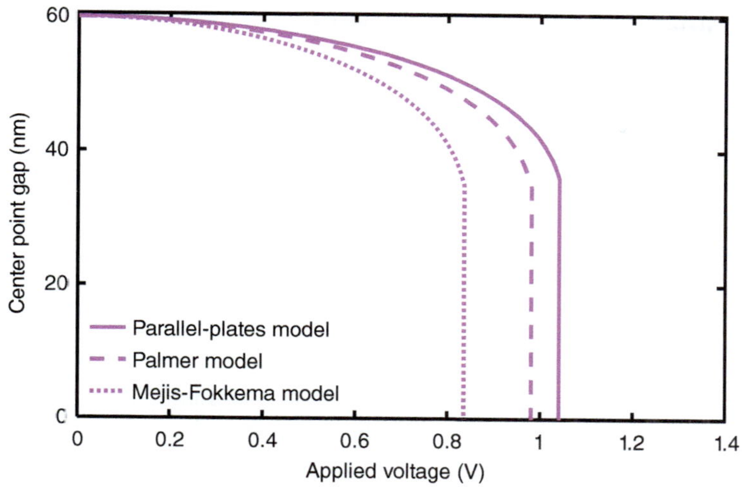

Figure 5.12 Effect of different fringing field corrections on pull-in instability of a nanostructure. Source: SoltanRezaee et al. [32] / Springer Nature / CC BY 4.0.

length. As they have stated [33], for weak g/b ratios, the fringing field effect is not considered. However, in typical cases of g/b ratios higher than 0.05, this effect can significantly increase the capacitance value compared to that of the parallel-plate model. Thus, the corrected electric force can be written as

$$F_{\mathrm{els,MF}} = \frac{\varepsilon_b b V^2}{2(G - w(x,t))^2}\left(1 + 0.265\left(\frac{G - w(x,t)}{b}\right)^{0.75} + 0.53\frac{h}{b}\left(\frac{G - w(x,t)}{b}\right)^{0.5}\right) \tag{5.15}$$

In this correction, which can be employed in special conditions, the effect of electrode thickness is not reflected. Moreover, a few other expressions have been developed in the open literature that have specific applications in electrostatically actuated nanodevices [34, 35].

5.4 Van der Waals Force

The vdW effect stands for the electrostatic interaction among a pair of magnetic poles in the atomic dimension. As mentioned earlier, when the separation is well below the plasma (for metals) or the absorption (for dielectrics) wavelength of the surface material (typically less than several tens of nanometers), the Casimir force should be replaced by the vdW force [4, 15]. In this case, retardation is not significant [36].

To obtain a more accurate expression for the vdW force between finite interacting bodies, one should integrate appropriate molecular potentials over the volume of the interacting bodies.

5.5 Rectangular Beam–Sheets

The intermolecular vdW force (per unit length) of a typical rectangular nano-beam over a fixed plate is written as [37] (which is generally related to the third-order inverse of the spacing of conductors)

$$F_{\text{vdW}} = \frac{\underline{A}b}{6\pi(D-w)^3} \tag{5.16}$$

where $\underline{A}$ is the Hamaker constant in the range of $0.4–4 \times 10^{19}$ J [38].

5.5.1 Wire–Plate

For a suspended wire, by integrating over two volumes and assuming ($R < D$), the Lennard–Jones potential per unit length [39] can be calculated as [38]

$$E_{\text{vdW}} = -\frac{\pi^2 C_{1,2}\rho_1\rho_2 R}{3D^3} = -\frac{\underline{A}R^2}{3D^3} \tag{5.17}$$

Now the molecular force per unit length F_{vdW} can be obtained:

$$F_{\text{vdW}} = \frac{\text{d}E_{\text{vdW}}}{\text{d}D} = \frac{\underline{A}R^2}{D^4} \tag{5.18}$$

By deflecting the wire, the initial gap between the moveable electrode and the fixed ground reduces from D to $(D-w)$. Therefore, the vdW force due to deflection of the wire can be written as

$$F_{\text{vdW}} = \frac{\underline{A}R^2}{(D-w)^4} \tag{5.19}$$

5.5.2 Carbon Nanotube (CNT) Sheets

It should be noted that the relations for CNTs are generally similar to those of cylindrical beams, for example, for the electrostatic and Casimir forces. However, the expression of the wire–plate configuration cannot be generalized for the vdW force.

For a CNT over a graphene surface and for distances larger than 5 Å, the difference between two values of the vdW energy, as specified by the continuum model and molecular dynamics, is less than 1% [40]. The Lennard–Jones potential is a suitable model to describe the vdW interaction between bodies. It defines the potential between atoms i and j by

$$\varphi_{ij} = \frac{C_{12}}{r_{ij}^{12}} - \frac{C_6}{r_{ij}^6} \tag{5.20}$$

where r_{ij} is the distance between atoms, while C_6 and C_{12} are the attractive and repulsive constants, respectively. For distances greater than 3.4 Å, the repulsive term decays extremely fast and can be neglected [41]. For the carbon–carbon interaction, $C_6 = 15.2\,\text{eV}\,\text{Å}^6$ [42]. A reliable continuum model has been established to compute the vdW energy by a double-volume integral of the Lennard–Jones potential [37, 39, 40]. This method provides acceptable results for explaining the CNT–graphene substrate attraction compared to that of direct pairwise summation through the molecular dynamics [25], that is

$$\begin{aligned} E_{\mathrm{vdW}} &= \int_{v_1}\int_{v_2} -\frac{C_6\rho_1\rho_2}{r^6(v_1,v_2)}\mathrm{d}v_2\mathrm{d}v_1 \\ &= -C_6\sigma^2\pi^2\sum_{R=R_\mathrm{i}}^{R_\mathrm{o}}\sum_{r=D}^{D+(N-1)d}\frac{R(R+r)(3R^2+2(R+r)^2)}{2\sqrt{(R+r)^2-R^2}^{7}} \end{aligned} \tag{5.21}$$

where v_1 and v_2 represent two domains of integration (CNTs) and the distance between any two points on v_1 and v_2 is $r(v_1,v_2)$. In this relation, R_i and R_o are the internal and external radii of the CNT, respectively. N is the number of graphene sheets and $\sigma = 38\,\mathrm{nm}^{-2}$ is the graphene surface density [25].

Once the vdW energy is computed, the corresponding energy terms are used to derive the component of the intermolecular force per unit length along the r-direction. In most applications, it is practically assumed that the diameter of the tube is much smaller than the distance between the nanotube and the graphene surface (i.e., $2R \ll D$). According to this assumption, the (5.21) is simplified as

$$\begin{aligned} F_{\mathrm{vdW}} &= \frac{\mathrm{d}E_{\mathrm{vdW}}}{\mathrm{d}r} \\ &= C_6\sigma^2\pi^2\sum_{R=R_\mathrm{i}}^{R_\mathrm{o}}\sum_{r=D}^{D+(N-1)d}\frac{8r^4R+32r^3R^2+72r^2R^3+80rR^4+35R^5}{2\sqrt{r(2R+r)}^{9}} \\ &\approx 4C_6\sigma^2\pi^2N_\mathrm{W}R_\mathrm{W}\sum_{r=D}^{D+(N-1)d}\frac{1}{r^5} \end{aligned} \tag{5.22}$$

where N_W is the number of walls of the nanotube and R_W is the mean value of their radii. For a large number of layers (i.e., $D+(N-1)d \gg D$), replacing the summation with an integral yields

$$\sum_{r=D}^{D+(N-1)d}\frac{1}{r^5} \approx \frac{1}{d}\int_D^{D+(N-1)d}\frac{1}{r^5}\mathrm{d}r = \frac{1}{4d}\left(\frac{1}{D^4}-\frac{1}{(D+(N-1)d)^4}\right) \approx \frac{1}{4dD^4} \tag{5.23}$$

Finally,

$$F_{\mathrm{vdW}} \approx C_6\sigma^2\pi^2N_\mathrm{W}R_\mathrm{W}\frac{1}{dD^4} \tag{5.24}$$

5.5.3 Rectangular Tweezers

Again, the reliable continuum model has been established to compute the vdW energy [39], which can be obtained as

$$E_{\mathrm{vdW}} = \int_{v_1}\int_{v_2} -\frac{C_{1,2}\rho_1\rho_2}{r^6(v_1,v_2)}\mathrm{d}v_2\mathrm{d}v_1 \tag{5.25}$$

where $C_{1,2}$ is the London constant for the material of types 1, 2. The terms v_1 and v_2 represent two domains of integration; ρ_1 and ρ_2 are the number of atoms per unit volume (densities of atoms) in two bodies, respectively. Thus, the vdW energy of two parallel beams with the same geometry and material properties can be defined as

$$E_{\mathrm{vdW}} = -\frac{\pi C_{1,2}\rho^2}{12}\left(\frac{1}{g^2} + \frac{1}{(g+2h)^2} - \frac{2}{(g+h)^2}\right) \tag{5.26}$$

By differentiating with respect to the gap and replacing g with (g − 2w), one can obtain the vdW force per unit length as follows:

$$F_{\mathrm{vdW}} = -\frac{\mathrm{d}E_{\mathrm{vdW}}}{\mathrm{d}g} = -\frac{Ab}{6\pi}\left(\frac{1}{(g-2w)^3} + \frac{1}{(g+2h-2w)^3} - \frac{2}{(g+h-2w)^3}\right) \tag{5.27}$$

where $A = \pi^2 C_{1,2}\rho_2$ is the Hamaker constant.

5.5.4 Circular Tweezers

For two suspended wires with distance D, where ($R < D$), the Lennard–Jones potential [39] can be calculated as [38]

$$E_{\mathrm{vdW}} = -\frac{A\sqrt{R}}{24\sqrt{D^3}} \tag{5.28}$$

Now the intermolecular force per unit length is given by

$$F_{\mathrm{vdW}} = \frac{\mathrm{d}E_{\mathrm{vdW}}}{\mathrm{d}D} = \frac{A\sqrt{R}}{16\sqrt{D^5}} \tag{5.29}$$

As mentioned, by deflecting the arms, the gap reduces from D to $(D - w_1 - w_2)$. By assuming the identical geometry of the arms, the displacements will be the same. Therefore, similarly to the electrical force, the vdW force is derived by replacing D with $(D - 2w)$:

$$F_{\mathrm{vdW}} = \frac{A\sqrt{R}}{16\sqrt{(D-2w)^5}} \tag{5.30}$$

5.5.5 Carbon Nanotube (CNT) Tweezers

Considering the explanations of the CNT–plate, the Lennard–Jones potential [39] can be written as [13, 29]

$$\begin{aligned} E_{\mathrm{vdW}} &= \int_{v_1}\int_{v_2} -\frac{C_6\rho_1\rho_2}{r^6(v_1,v_2)}\mathrm{d}v_2\mathrm{d}v_1 \\ &= \frac{AL}{12\sqrt{2}}\left(\frac{1}{\sqrt{g+2t}^3}\sqrt{\frac{r_1}{2}} + \frac{2}{\sqrt{g+t}^3}\sqrt{\frac{r_1 r_2}{r_1+r_2}} - \frac{1}{\sqrt{g}^3}\sqrt{\frac{r_2}{2}}\right) \end{aligned} \tag{5.31}$$

where the distance between any two points on v_1 and v_2 is $r(v_1, v_2)$ and r is the distance between atoms, while C_6 is the attractive constant. r_1 and r_2 are the internal and external radii of the CNT arms, respectively. Now, the molecular force per unit length can be obtained as

$$F_{\text{vdW}} = \frac{\mathrm{d}E_{\text{vdW}}}{\mathrm{d}g} = -\frac{\mathcal{A}}{8\sqrt{2}}\left(\frac{1}{\sqrt{g+2t}^{5}}\sqrt{\frac{r_1}{2}} + \frac{2}{\sqrt{g+t}^{5}}\sqrt{\frac{r_1 r_2}{r_1+r_2}} - \frac{1}{\sqrt{g}^{5}}\sqrt{\frac{r_2}{2}}\right) \tag{5.32}$$

By assuming the similarity of the deformable arms and equal geometry for two nanotubes, their displacements will be the same. Therefore, the vdW force can be derived by replacing g with $(g - 2w)$:

$$F_{\text{vdW}} = -\frac{\mathcal{A}}{8\sqrt{2}}\left(\frac{1}{\sqrt{g+2t-2w}^{5}}\sqrt{\frac{r_1}{2}} + \frac{2}{\sqrt{g+t-2w}^{5}}\sqrt{\frac{r_1 r_2}{r_1+r_2}} - \frac{1}{\sqrt{g-2w}^{5}}\sqrt{\frac{r_2}{2}}\right) \tag{5.33}$$

It is possible to further simplify the formula of the vdW force to obtain a more practical relation in the case of $(r_2 = r_1 = r)$ as [13]

$$F_{\text{vdW}} = -\frac{\mathcal{A}}{8\sqrt{2}\sqrt{g-2w}^{5}}\sqrt{\frac{r}{2}}\left(1 - \frac{2\sqrt{\dfrac{1+t/r}{1+0.5t/r}}}{\sqrt{1+\dfrac{t}{g-2w}}^{5}} + \frac{\sqrt{1+\frac{t}{r}}}{\sqrt{1+\dfrac{2t}{g-2w}}^{5}}\right) \tag{5.34}$$

Using the appropriate Taylor-series expansion and considering $g \gg r$, the obtained relation can be rewritten as [13]

$$\begin{aligned} F_{\text{vdW}} &= -\frac{\mathcal{A}t^2}{8\sqrt{2}r^2\sqrt{g-2w}^{5}}\sqrt{\frac{r}{2}}\left(\frac{3}{16} - \frac{5}{4}\frac{r}{g-2w} + \frac{35}{4}\left(\frac{r}{g-2w}\right)^2\right) \\ &\approx -\frac{3\mathcal{A}t^2}{128\sqrt{2}\sqrt{r}^{3}\sqrt{g-2w}^{5}} \end{aligned} \tag{5.35}$$

5.6 Casimir Force

The Casimir effect arises from quantum fluctuations in the ground state of an electromagnetic field. At micro-separations (typically less than several tens of micrometers), the attraction between two surfaces can be described by the Casimir interaction [43].

The starting point for precisely calculating the Casimir energy in large separations is a path integral representation of the effective action, which yields a trace formula for the density of states. The Casimir energy of two disconnected conductive surfaces is precisely determined by the change in the photon density of states caused by moving the surfaces from infinity to a finite distance in vacuum.

The total electromagnetic Casimir energy is the sum of energies of Dirichlet and Neumann modes. However, in large separations, the Dirichlet mode is dominant and thus the Neumann mode can be neglected [44]. It should be noted that the large separation assumption is acceptable for most real nano-tweezers since the gap between the arms is extremely larger than the radius of the arms ($D \gg R$). Typically, for ($D < 100R$), the contribution of the Neumann mode is negligible compared to the Dirichlet mode, and the asymptotic approximation provides an excellent estimation of the Casimir attraction [45]. In this section, practical mathematical models by assuming the large separation approximation (LSA) (Dirichlet mode) are developed. Other theories related to the Casimir regime based on the small separation approximation (SSA) and the Neumann mode are mentioned in the next section.

In general, by considering the first-order approximation for a system with a relatively large initial gap, the Casimir energy based on the Dirichlet mode is evaluated [31]. Then, by differentiating the energy and considering the deflection of the nanoelectrodes, the Casimir force per unit length for the LSA can be obtained.

5.6.1 Rectangular Beam–Plate

The Casimir regime is important in micro- and nanodevices, which can be disregarded beyond these scales. The Casimir force for a rectangular beam over a fixed substrate can be given by [4] (which is generally related to the fourth-order inverse of spacing of the conductors)

$$F_{\text{cas}} = \frac{\pi^2 bcH}{240(G-w)^4} \tag{5.36}$$

where H is the reduced Planck's constant (Planck's constant divided by 2π) equal to 1.05457×10^{-34} J s and $c = 2.998 \times 10^8$ m s^{-1} is the speed of light.

5.6.2 Wire–Plate

In nanowires, when $D \gg R$, the Dirichlet condition is quite significant compared to the Neumann condition. Therefore, the asymptotic expression of the attractive interaction energy for structures with a large separation gap should be considered [45, 46]:

$$E_{\text{cas}} = -\frac{HcL}{D^2}\frac{1}{16\pi\ln(D/R)} \tag{5.37}$$

By differentiating the energy, the Casimir force can be obtained:

$$F_{\text{cas}} = \frac{\mathrm{d}E_{\text{cas}}}{\mathrm{d}D} = \frac{Hc}{8\pi\ln(D/R)D^3}\left(1 + \frac{1}{2\ln(D/R)}\right) \tag{5.38}$$

It should be remarked that by applying an external voltage to the wire, it will deflect toward the ground to reduce the gap between them from D to $(D - w)$.

$$F_{\text{cas}} = \frac{Hc}{16\pi\ln^2\left(\frac{D-w}{R}\right)(D-w)^3}\left(1 + 2\ln\left(\frac{D-w}{R}\right)\right) \tag{5.39}$$

5.6.3 Carbon Nanotube (CNT) Sheets

The Casimir force per unit length for the CNT–plate configuration with a large separation gap, i.e. $D \gg R$, is generally similar to that of a typical wire–plate configuration [27]. Moreover, the Casimir force in small separation gaps are derived at the end of this chapter based on the proximity force approximation (PFA) and the Lifshitz theory. Furthermore, there are other relations, for example, for single-walled CNTs (SWCNTs) suspended over graphene sheets [47], which are not widely used.

5.6.4 Rectangular Tweezers

In this section, the modified expression of the Casimir force for rectangular nano-tweezers will be derived. By considering the finite dimension and edge effects, the Casimir force per unit length between two finite parallel plates can be explained as [48]

$$F_{\text{cas}} = F_{\text{cas,inf}} + \gamma_{\text{mod}} Hc/g^3 \tag{5.40}$$

where $\gamma_{\text{mod}} = 0.00231$ is the universal coefficient [48] and the Casimir force between two infinite plates can be expressed as

$$F_{\text{cas,inf}} = \frac{Hcb\pi^2}{240g^4} \tag{5.41}$$

Now by replacing g with $(g - 2w)$, which is practical in most nano-tweezers, the modified Casimir force per unit length of the beam is obtained as

$$F_{\text{cas}} = \frac{Hcb\pi^2}{240(g - 2w)^4} + 0.00231\frac{Hc}{(g - 2w)^3} \tag{5.42}$$

5.6.5 Circular Tweezers

Considering the first-term approximation for two wires with a large distance, the asymptotic expression of the attractive interaction energy at a large distance can be obtained as [45]

$$E_{\text{cas}} = -\frac{HcL}{D^2}\frac{1}{8\pi\ln^2(D/R)} \tag{5.43}$$

By differentiating the energy and considering the deflection of wires, w_1 and w_2, we can obtain the Casimir force per unit length for the LSA as follows:

$$F_{\text{cas}} = \frac{dE_{\text{cas}}}{dD} = \frac{HcL}{4\pi\ln^3(D/R)D^3}\left(1 + \ln\left(\frac{D}{R}\right)\right) \tag{5.44}$$

It should be noted that by applying an external voltage to the wires, the arms deflect toward each other to reduce the gap in between them from D to $(D - w_1 - w_2)$. If both arms of the tweezers have the same geometry and material properties, then their deflections will be equal ($w_1 = w_2 = w$). Therefore, the Casimir force can be derived as follows:

$$F_{\text{cas}} = \frac{HcL}{4\pi\ln^3\left(\frac{D}{R}\left(1 - \frac{2w}{D}\right)\right)(D - 2w)^3}\left(1 + \ln\left(\frac{D}{R}\left(1 - \frac{2w}{D}\right)\right)\right) \tag{5.45}$$

5.6.6 Carbon Nanotube (CNT) Tweezers

Based on the attractive interaction energy [46], the Casimir force per unit length for the CNT tweezers with a large separation gap, i.e. $D \gg R$, is generally similar to that of typical circular tweezers [49].

5.7 Other Theories Related to the Casimir Force

5.7.1 Proximity Force Approximation (PFA)

It should be noted that relations of dispersion forces can be derived based on different theories for some cases. Therefore, several corrections, for instance, for circular or curved electrodes, have been presented to improve the prediction of the system response. It should be noted that considering various theories may yield dissimilar results. For example, PFA is a simplistic method for approximating the Casimir interaction between nonplanar geometries [50], which will be described.

Based on the Lifshitz theory, the Casimir energy per unit area for conducting parallel infinite flat plates (E_{pp}) separated by a distance D is [50]

$$E_{pp} = -\frac{\pi^2 Hc}{720D^3} \tag{5.46}$$

This equation is the fundamental starting point for the PFA. According to the PFA, any complex configuration of interacting surfaces can be treated as a sum of infinitesimal parallel plates. For very short separations, the PFA provides the zeroth-order approximation for the Casimir energy as

$$E^{PFA} = \int_S E_{pp}(D)\mathrm{d}S = -\frac{\pi^2 Hc}{720}\iint \frac{\mathrm{d}S}{D^3} \tag{5.47}$$

where S is the surface defining the gap.

Generally, the PFA might be suggested for modeling quantum vacuum fluctuations between interacting bodies due to its simplicity and convenience. Unfortunately, the PFA provides acceptable results only for interacting flat bodies in proximity; hence, this approximation is not valid for modeling the Casimir attraction for real objects with considerable separation distances. A powerful theory for approximating the vacuum fluctuations between real nonplanar structures is the multi-scattering method [46, 50], which has also been presented in this chapter. This method was successfully applied to approximate the Casimir force between different structures, e.g. spheres, sphere–plane, tweezers, cylinder–plane geometries [46, 50].

Farrokhabadi et al. [17] analyzed the effect of considering the scattering theory versus the PFA for circular tweezers. The variation of the ratio of F_{cas} obtained from the scattering theory and from the PFA [45] is plotted in Figure 5.13 as a function of the parameter $D/2R$. Similarly, Farrokhabadi et al. [51] plotted the variation for a cylinder–plate configuration (Figure 5.14). Their results demonstrated that by increasing the initial gap, the difference between the Casimir force values obtained from the PFA and the scattering theory becomes more significant [17].

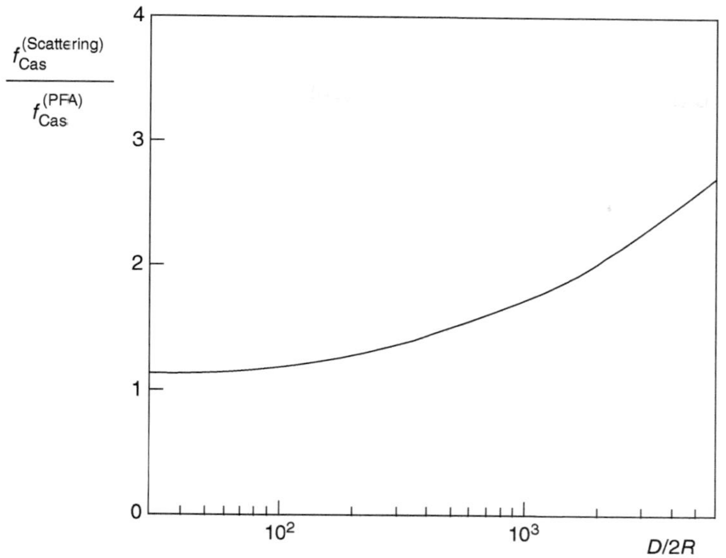

Figure 5.13 Variation of the ratio of the Casimir force obtained from the scattering theory and the PFA for cantilever circular tweezers. Source: Farrakhabadi et al. [51] / with permission of Elsevier.

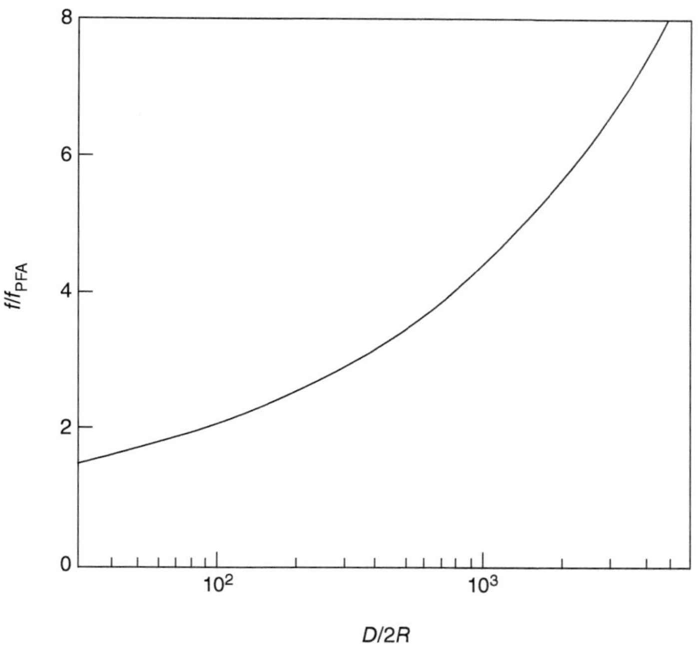

Figure 5.14 Variation of the ratio of the Casimir force obtained from the scattering theory and the PFA for a wire–plate configuration. Source: Farrakhabadi et al. [51] / with permission of Elsevier.

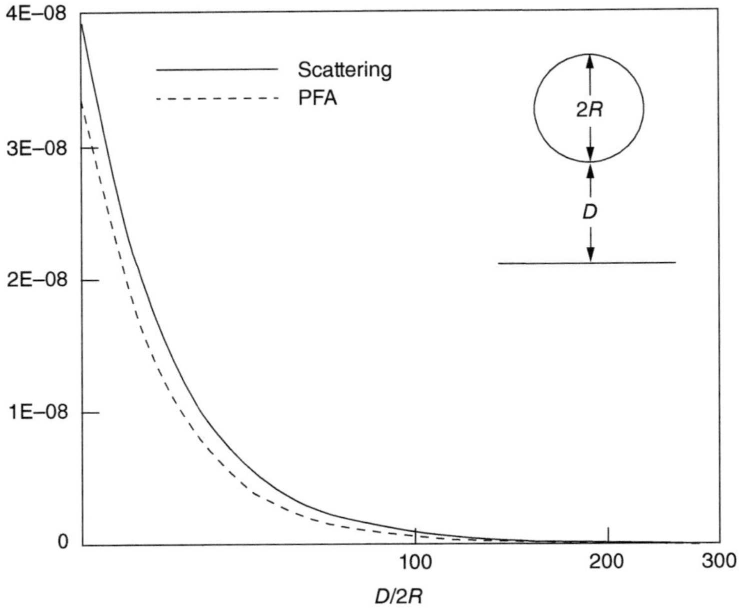

Figure 5.15 Variation of the dimensionless Casimir force. (PFA and scattering theory) for the cylinder–plate geometry. Source: Farrokhabadi et al. [51] / with permission of Elsevier.

For the cylinder–plate configuration, the variation of the Casimir force obtained from different theories is illustrated in Figure 5.15 [51]. In this case, for a small distance gap, the PFA can be used; hence, the Casimir energy can be evaluated as follows [50]:

$$E^{\text{PFA}} = -\frac{\pi^3 HcL}{960}\sqrt{\frac{R}{2D^5}} \tag{5.48}$$

By differentiating the energy, the Casimir force for the SSA can be obtained as follows:

$$F_{\text{cas}} = \frac{\pi^3 HcL}{768}\sqrt{\frac{2R}{D^7}} \tag{5.49}$$

The relation of the Casimir force based on the scattering theory was derived earlier (Casimir force in the wire–plate case). The results reveal the difference between the Casimir force values obtained from the scattering theory and PFA for the curved configuration.

Furthermore, for the doubly clamped cylinder–cylinder configuration, the variations of the Casimir force obtained from different theories are illustrated in Figure 5.16 [51]. By means of the PFA for this case, the Casimir energy can be evaluated as [45]

$$E^{\text{PFA}} = -\frac{\pi^3 HcL}{1920}\sqrt{\frac{R}{D^5}} \tag{5.50}$$

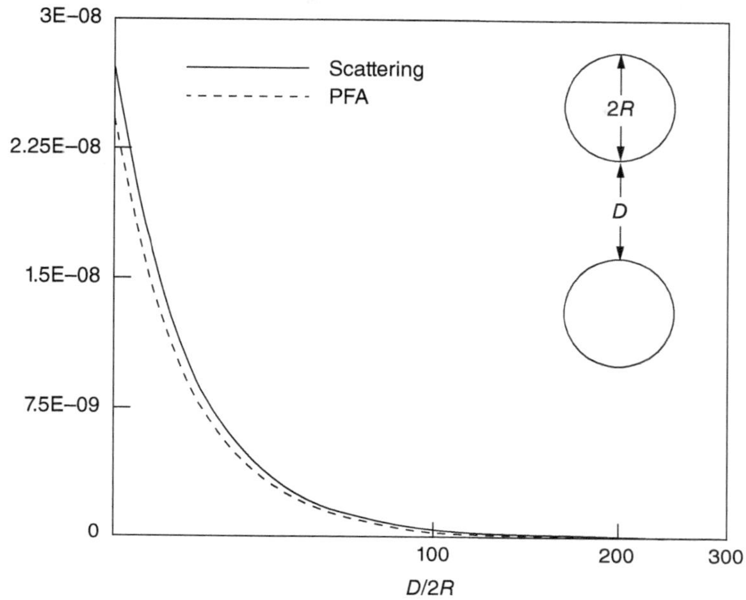

Figure 5.16 Variation of the dimensionless Casimir force (PFA and scattering theory) for the double clamped cylinder–cylinder geometry. Source: Farrokhabadi et al. [51] / with permission of Elsevier.

Again, by differentiating the energy, the Casimir force for the SSA can be obtained as

$$F_{cas} = \frac{\pi^3 HcL}{768}\sqrt{\frac{R}{D^7}} \tag{5.51}$$

The relation of Casimir force based on the scattering theory was derived earlier (Casimir force in the circular tweezers case). Moreover, the abovementioned relations are also valid for cantilever circular tweezers [18]. The obtained results illustrate the difference between the Casimir forces calculated from scattering theory and the PFA, which completely depends on the initial gap of actuated electrodes.

Figure 5.17 depicts the variation of the system stable length as a function of the wire diameter and the initial gap [51]. The curves are numerically obtained for a nano-resonator by scattering theories for LSA and SSA, respectively. As Farrokhabadi et al. [51] reported, the difference between values obtained via the theories increases with the increasing initial gap and the decreasing nanowire diameter. As shown, there is a substantial difference between the predicted values for these two cases. This figure shows that scattering theory is more conservative, i.e. it predicts lower L_{max} and higher D_{min} values for the structure. This is because the LSA leads to lower critical values of the Casimir force and deflection compared to the SSA, i.e. the PFA overestimates the critical values. This should be considered in the theoretical modeling of nanostructures to achieve precise and reliable simulation results. It demonstrates that the effect of the Casimir attraction on the instability of NEMS is very important in separations below 1 μm. Finally, it

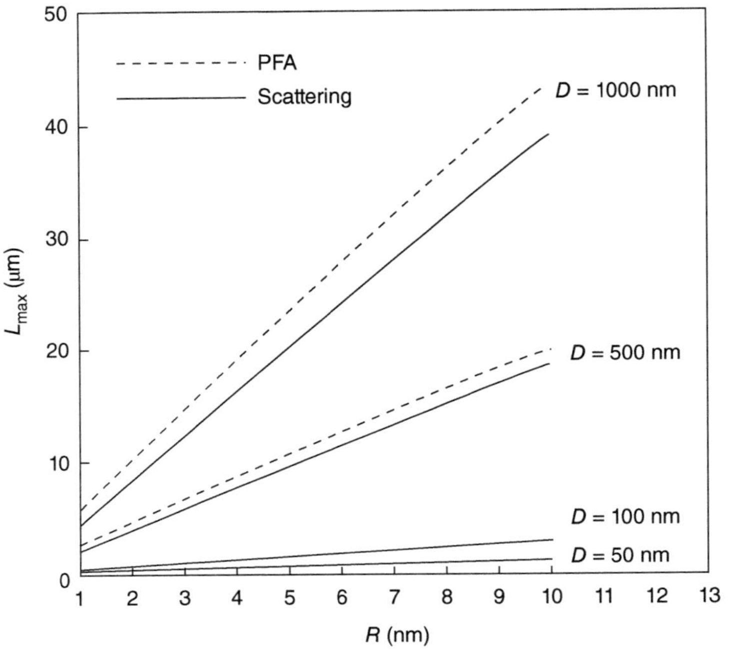

Figure 5.17 Variation of the system stable length as a function of the wire diameter and the initial gap using PFA and scattering theories for the cylinder–plate configuration. Source: Farrokhabadi et al. [51] / with permission of Elsevier.

should be noted that deformation of nanowires is a nonlinear phenomenon, and the solution is only valid for small deformation instability behavior.

5.7.2 Dirichlet and Neumann Modes

The electromagnetic energy consists of both transverse magnetic (TM) and electric (TE) modes described by a scalar field that satisfies both Dirichlet and Neumann modes of BCs. For large distances $g \gg R$ (R is the radius of wires or CNT), the Casimir energy between surfaces including Dirichlet and Neumann energies of the electromagnetic field is [45, 46, 49]

$$E^{\mathrm{Dir}} = -\frac{HcL}{8\pi g^2 \ln^2(g/R)}\left(1 - \frac{2}{\ln(g/R)}\right)$$
$$E^{\mathrm{Neu}} = -HcL\frac{7R^4}{5\pi g^6} \tag{5.52}$$

The electromagnetic Casimir energy comes from both electric and magnetic energies. It is interesting to note that the assumption of large separations for tweezers is usually applicable because the distance of the electrodes is considerably larger than the diameter. It should be remarked that by applying an external voltage to the system, the electrodes will deflect toward each other to reduce the gap from g to $(g - u_1 - u_2)$. To derive the Casimir force, the energy is differentiated with respect to the initial gap. Therefore, it can be calculated as [49]

$$F_{cas} = \frac{dE_{cas}}{dg} = \frac{Hc}{4\pi \ln^4\left(\frac{g-u_1-u_2}{R}\right)(g-u_1-u_2)^3}$$
$$\times \left(3 + \ln\left(\frac{g-u_1-u_2}{R}\right) - \ln^2\left(\frac{g-u_1-u_2}{R}\right)\right) - \frac{42HcR^4}{5\pi(g-u_1-u_2)^7} \quad (5.53)$$

By assuming the same material properties and geometry for both electrodes, due to the considered symmetry in the configuration of CNT tweezers, the deflections of the electrodes are assumed equal ($u_1 = u_2 = u$). Therefore, the Casimir force by considering both Dirichlet and Neumann modes on the conducting surfaces will be derived as

$$F_{cas} = \frac{Hc}{4\pi \ln^4\left(\frac{g-2u}{R}\right)(g-2u)^3}\left(3 + \ln\left(\frac{g-2u}{R}\right) - \ln^2\left(\frac{g-2u}{R}\right)\right) - \frac{42HcR^4}{5\pi(g-2u)^7} \quad (5.54)$$

5.8 Freestanding Phenomenon

In micro- and nanoscales, dispersion forces can strongly affect the operation of miniature devices. When the gap between electrodes is sufficiently small, then even without an external applied voltage, the deformable arms can adhere to each other due to the intermolecular attractions. Computing threshold values, which are the electrode(s) maximum length and minimum initial gap, is essential. The critical values of the dispersion force and the corresponding arm critical deflection can be obtained by ignoring the electrostatic attraction and then solving the governing equation of the system.

5.8.1 Detachment Length

The maximum length L_{max} of arms to prevent stiction, where electrodes do not stick together without the application of a voltage differential, is called the detachment length [52]. When the electrode length exceeds this amount, the electrode sticks to the substrate or the other electrode [51]. The detachment length is the maximum permissible length of the freestanding system. On the other hand, if the length of the movable arms is known, there is a minimum gap g_{min}, which prevents stiction between the arms due to dispersion forces. The maximum length and minimum gap are very important in the design of miniature systems.

A comprehensive study about the detachment length and minimum gap of the freestanding for wire–plate and cylindrical tweezers with cantilever and doubly clamped BCs was presented by Farrokhabadi et al. [51]. They derived the governing equations and calculated the critical values as some numerical case studies. They considered the Casimir regime and obtained the system responses using numerical finite difference method (FDM) and semianalytical modified adomian decomposition (MAD) methods for both SSA and LSA cases by the PFA and scattering theories, respectively. However, they did not consider the stretching

Table 5.1 General formulas for computing the detachment length (L_{max}) and minimum gap (g_{min}) of freestanding due to the Casimir force (parameters a_i generally depend on the configuration, BCs, electrodes distance, and solution methods).

Detachment length	Minimum gap
$a_1 g \sqrt[4]{E_{eff} I}$	$a_2 L \sqrt[4]{E_{eff} I}$

Source: Based on Farrokhabadi et al. [51].

Table 5.2 Formulas for computing the detachment length (L_{max}) and minimum gap (g_{min}) of freestanding cantilever CNT tweezers due to the vdW force using several methods.

Method	Detachment length	Minimum gap
MAD	$2.56\sqrt[4]{\dfrac{E_{eff} I \sqrt{g^7}\sqrt{r^3}}{At^2}}$	$0.340\sqrt[7]{\dfrac{A^2 t^4 L^8}{r^3 (E_{eff} I)^2}}$
FDM	$2.55\sqrt[4]{\dfrac{E_{eff} I \sqrt{g^7}\sqrt{r^3}}{At^2}}$	$0.343\sqrt[7]{\dfrac{A^2 t^4 L^8}{r^3 (E_{eff} I)^2}}$
HPM	$2.48\sqrt[4]{\dfrac{E_{eff} I \sqrt{g^7}\sqrt{r^3}}{At^2}}$	$0.355\sqrt[7]{\dfrac{A^2 t^4 L^8}{r^3 (E_{eff} I)^2}}$

Source: Based on Farrokhabadi et al. [13, 29].

effect of clamped–clamped structures in double clamped structures. A brief of their results has been reported in Table 5.1, where the parameters a_i of different systems are not the same. They depend on the wire–plate or circular tweezer structures, cantilever or doubly clamped BCs, considering SSA or LSA, and numerical (FDM) or analytical (MAD) solutions.

It should be noted that the freestanding phenomenon may happen due to both dispersion forces. Therefore, both Casimir and vdW regimes should be examined. Keivani et al. [53] have investigated the freestanding phenomenon for both cantilever cylinder–plate and cylinder–cylinder nanosystems under the Casimir and vdW effects separately. They have reported their results graphically using several solution methods. In addition, Table 5.2 shows the comparison between the L_{max} and g_{min} values for a cantilever CNT tweezers subjected to the vdW force obtained by various methods [13, 29].

5.8.2 Surface Layer and Size-dependent Effects

Other issues that can affect threshold parameters of the freestanding phenomenon are the surface layer energy and nonclassical continuum mechanics theories. Investigating different nonclassical size theories and the surface layer in addition to its aspects, theories, and effects is beyond the scope of this study because the surface layer and size-dependent parameters do not affect the Coulomb and dispersion forces. However, their impact on threshold parameters in the freestanding phenomenon is considerable. As a result, to predict the detachment length and the minimum gap accurately, the surface layer energy and nonclassical continuum theories should be essentially considered. Mohebshahedin and Farrokhabadi [24] derived the detachment length for cantilever cylinder–plate and cylinder–cylinder nanosystems. They considered surface elasticity and residual surface stress and presented the relations based on both Casimir and vdW attractions. In another study, Farrokhabadi et al. [18] investigated the freestanding phenomenon due to the Casimir energy for the cantilever wire–plate and circular tweezers considering the size-dependent parameters based on the modified couple stress theory.

5.9 Summary

This chapter derived in brief the well-known relations of electrostatic and dispersion forces for beam-based nanostructures. The cases were classified for different structural configurations, i.e. rectangular beam–plate, wire–plate, cylinder–plate, CNT sheets, rectangular tweezers, circular tweezers, and CNT tweezers. In addition, the effects of different electric fringing field corrections, the freestanding phenomenon owing to the Casimir or vdW attractions, and other famous theories related to the Casimir force were studied separately. Due to the lack of similar classic relations in the literature, they have been expected to contribute to the development of the nanotechnology and would be beneficial to the modeling of actuated nanodevices.

References

1 Tavakolian, F., Farrokhabadi, A., SoltanRezaee, M. et al. (2019). Dynamic pull-in of thermal cantilever nanoswitches subjected to dispersion and axial forces using nonlocal elasticity theory. *Microsyst. Technol.* 25 (1): 19–30.

2 Esfahani, S., Khadem, S.E., and Mamaghani, A.E. (2019). Size-dependent nonlinear vibration of an electrostatic nanobeam actuator considering surface effects and inter-molecular interactions. *Int. J. Mech. Mater. Des.* 15 (3): 489–505.

3 SoltanRezaee, M., Bodaghi, M., Farrokhabadi, A. et al. (2019). Nonlinear stability analysis of piecewise actuated piezoelectric microstructures. *Int. J. Mech. Sci.* 160: 200–208.

4 Lamoreaux, S.K. (2005). The Casimir force: background, experiments, and applications. *Rep. Prog. Phys.* 68 (1): 201.

5 SoltanRezaee, M. and Ghazavi, M.R., Thermal, size and surface effects on the nonlinear pull-in of small-scale piezoelectric actuators. *Smart Mater. Struct.*, 2017. 26(9): p. 095023.

6 SoltanRezaee, M., Ghazavi, M.R., and Najafi, A. Mathematical modelling for vibration evaluation of powertrain systems. in Zhangxing John Chen, M.H. Hamza *853: Modelling, Simulation and Identification/854: Intelligent Systems and Control*. 2017. Canada: Acta Press.

7 SoltanRezaee, M. and Bodaghi, M. (2020). Simulation of an electrically actuated cantilever as a novel biosensor. *Sci. Rep.* 10 (1): 1–14.

8 SoltanRezaee, M. and Afrashi, M. (2016). Modeling the nonlinear pull-in behavior of tunable nano-switches. *Int. J. Eng. Sci.* 109: 73–87.

9 SoltanRezaee, M., Farrokhabadi, A., and Ghazavi, M.R. (2016). The influence of dispersion forces on the size-dependent pull-in instability of general cantilever nano-beams containing geometrical non-linearity. *Int. J. Mech. Sci.* 119: 114–124.

10 SoltanRezaee, M., Afrashi, M., and Rahmanian, S. (2018). Vibration analysis of thermoelastic nano-wires under coulomb and dispersion forces. *Int. J. Mech. Sci.* 142–143: 33–43.

11 Dowlati, S. and Rezazadeh, G. (2018). A new approach to the evaluation of Casimir and van der Waals forces in the transition region. *Chin. J. Phys.* 56 (3): 1133–1146.

12 Chaterjee, S. and Pohit, G. (2009). A large deflection model for the pull-in analysis of electrostatically actuated microcantilever beams. *J. Sound Vib.* 322 (4–5): 969–986.

13 Farrokhabadi, A., Rach, R., and Abadyan, M. (2013). Modeling the static response and pull-in instability of CNT nanotweezers under the Coulomb and van der Waals attractions. *Phys. E: Low-Dimens. Syst. Nanostruct.* 53: 137–145.

14 Klimchitskaya, G., Mohideen, U., and Mostepanenko, V. (2000). Casimir and van der Waals forces between two plates or a sphere (lens) above a plate made of real metals. *Phys. Rev. A* 61 (6): 062107.

15 Ramezani, A., Alasty, A., and Akbari, J. (2007). Closed-form solutions of the pull-in instability in nano-cantilevers under electrostatic and intermolecular surface forces. *Int. J. Solids Struct.* 44 (14): 4925–4941.

16 Zhao, Y.P., Wang, L., and Yu, T. (2003). Mechanics of adhesion in MEMS – a review. *J. Adhes. Sci. Technol.* 17 (4): 519–546.

17 Farrokhabadi, A., Mokhtari, J., Rach, R. et al. (2015). Modeling the influence of the Casimir force on the pull-in instability of nanowire-fabricated nanotweezers. *Int. J. Mod. Phys. B* 29 (02): 1450245.

18 Farrokhabadi, A., Abadian, N., Kanjouri, F. et al. (2014). Casimir force-induced instability in freestanding nanotweezers and nanoactuators made of cylindrical nanowires. *Int. J. Mod Phys. B* 28 (19): 1450129.

19 Gupta, R.K. (1997). Electrostatic pull-in test structure design for mechanical property characterization of microelectromechanical systems (MEMS). Massachusetts Institute of Technology (MIT), Cambridge.

20 Batra, R.C., Porfiri, M., and Spinello, D. (2006). Electromechanical model of electrically actuated narrow microbeams. *J. Microelectromech. Syst.* 15 (5): 1175–1189.

21 Batra, R., Porfiri, M., and Spinello, D. (2006). Capacitance estimate for electrostatically actuated narrow microbeams. *Micro Nano Lett.* 1 (2): 71–73.

22 Li, M., Bhiladvala, R.B., Morrow, T.J. et al. (2008). Bottom-up assembly of large-area nanowire resonator arrays. *Nat. Nanotechnol.* 3 (2): 88.

23 Hayt, W.H. and Buck, J.A. (1981). *Engineering Electromagnetics*, vol. 6. New York: McGraw-Hill.

24 Mohebshahedin, A. and Farrokhabadi, A. (2015). The influence of the surface energy on the instability behavior of NEMS structures in presence of intermolecular attractions. *Int. J. Mech. Sci.* 101: 437–448.

25 Fazli, N., Koochi, A., Kazemi, A.S. et al. (2013). Influence of electrostatic force and the van der Waals attraction on the pull-in instability of the CNT-based probe–actuator. *Can. J. Phys.* 92 (9): 1047–1057.

26 Koochi, A., Fazli, N., and Rach, R. (2014). Modeling the pull-in instability of the CNT-based probe/actuator under the Coulomb force and the van der Waals attraction. *Lat. Am. J. Solids Struct.* 11 (8): 1315–1328.

27 Sedighi, H.M. and Farjam, N. (2017). A modified model for dynamic instability of CNT based actuators by considering rippling deformation, tip-charge concentration and Casimir attraction. *Microsyst. Technol.* 23 (6): 2175–2191.

28 Koochi, A., Hosseini-Toudeshky, H., and Abadyan, M. (2018). A corrected model for static and dynamic electromechanical instability of narrow nanotweezers: incorporation of size effect, surface layer and finite dimensions. *Int. J. Mod. Phys. B* 32 (08): 1850089.

29 Farrokhabadi, A., Koochi, A., and Abadyan, M. (2014). Modeling the instability of CNT tweezers using a continuum model. *Microsyst. Technol.* 20 (2): 291–302.

30 Ouakad, H.M. (2014). Static response and natural frequencies of microbeams actuated by out-of-plane electrostatic fringing-fields. *Int. J. Non-Linear Mec.* 63: 39–48.

31 SoltanRezaee, M., Bodaghi, M., and Farrokhabadi, A. (2019). A thermosensitive electromechanical model for detecting biological particles. *Sci. Rep.* 9 (1): 11706.

32 SoltanRezaee, M. and Bodaghi, M. (2020). Nonlinear dynamic stability of piezoelectric thermoelastic electromechanical resonators. *Sci. Rep.* 10 (1): 1–14.

33 Kacem, N., Hentz, S., Pinto, D. et al. (2009). Nonlinear dynamics of nanomechanical beam resonators: improving the performance of NEMS-based sensors. *Nanotechnology* 20 (27): 275501.

34 Krylov, S., Ilic, B.R., and Lulinsky, S. (2011). Bistability of curved microbeams actuated by fringing electrostatic fields. *Nonlinear Dynamics* 66 (3): 403.

35 Linzon, Y., Ilic, B., Lulinsky, S. et al. (2013). Efficient parametric excitation of silicon-on-insulator microcantilever beams by fringing electrostatic fields. *J. Appl. Phys.* 113 (16): 163508.

36 Sedighi, H.M., Daneshmand, F., and Zare, J. (2014). The influence of dispersion forces on the dynamic pull-in behavior of vibrating nano-cantilever based NEMS including fringing field effect. *Arch. Civ. Mech. Eng.* 14 (4): 766–775.

37 Israelachvili, J. (1992). *Intermolecular and Surface Forces*. London: Academic Press.

38 Guo, J.G. and Zhao, Y.P. (2006). Dynamic stability of electrostatic torsional actuators with van der Waals effect. *Int. J. Solids Struct.* 43 (3): 675–685.

39 Lennard Jones, J.E. (1930). Perturbation problems in quantum mechanics. *Proc. R. Soc. London* 129 (811): 598–615.

40 Tserpes, K. (2007). Role of intertube spacing in the pullout forces of double-walled carbon nanotubes. *Mater. Des.* 28 (7): 2197–2201.

41 Dequesnes, M., Rotkin, S., and Aluru, N.R. (2002). Calculation of pull-in voltages for carbon-nanotube-based nanoelectromechanical switches. *Nanotechnology* 13 (1): 120.

42 Girifalco, L., Hodak, M., and Lee, R.S. (2000). Carbon nanotubes, buckyballs, ropes, and a universal graphitic potential. *Phys. Rev. B* 62 (19): 13104.

43 Gusso, A. and Delben, G.J. (2008). Dispersion force for materials relevant for micro-and nanodevices fabrication. *J. Phys. D. Appl. Phys.* 41 (17): 175405.

44 Chan, H., Bao, Y., Zou, J. et al. (2008). Measurement of the Casimir force between a gold sphere and a silicon surface with nanoscale trench arrays. *Phys. Rev. Lett.* 101 (3): 030401.

45 Rahi, S.J., Emig, T., Jaffe, R.L. et al. (2008). Casimir forces between cylinders and plates. *Phys. Rev. A* 78 (1): 012104.

46 Emig, T., Jaffe, R.L., Kardar, M. et al. (2006). Casimir interaction between a plate and a cylinder. *Phys. Rev. Lett.* 96 (8): 080403.

47 Rasekh, M., Khadem, S., and Tatari, M. (2010). Nonlinear behaviour of electrostatically actuated carbon nanotube-based devices. *J. Phys. D. Appl. Phys.* 43 (31): 315301.

48 Gies, H. and Klingmüller, K. (2006). Casimir edge effects. *Phys. Rev. Lett.* 97 (22): 220405.

49 Zare, J. and Shateri, A. (2017). Instability threshold of rippled carbon nanotube nanotweezers in the low vacuum gas flow incorporating Dirichlet and Neumann modes of Casimir energy. *Phys. E: Low-Dimens. Syst. Nanostruct.* 90: 67–75.

50 Bordag, M., Mohideen, U., and Mostepanenko, V.M. (2001). New developments in the Casimir effect. *Phys. Rep.* 353 (1–3): 1–205.

51 Farrokhabadi, A., Abadian, N., Rach, R. et al. (2014). Theoretical modeling of the Casimir force-induced instability in freestanding nanowires with circular cross-section. *Phys. E: Low-Dimens. Syst. Nanostruct.* 63: 67–80.

52 Lambrecht, A. and Reynaud, S. (2000). Casimir force between metallic mirrors. *Eur. Phys. J. D* 8 (3): 309.

53 Keivani, M., Koochi, A., Sedighi, H.M. et al. (2016). Effect of surface layer on electromechanical stability of tweezers and cantilevers fabricated from conductive cylindrical nanowires. *Surf. Rev. Lett.* 23 (02): 1550101.

6

Nanomaterials for Wearable, Flexible, and Stretchable Strain/Pressure Sensors

Ali Sedighi[1] *and Majid Montazer*[2]

[1] *Graphene and Advanced Materials Laboratory (GAM lab), Textile Department, Amirkabir University of Technology, Tehran, Iran*
[2] *Textile Department, Functional Fibrous Structures & Environmental Enhancement (FFSEE), Amirkabir University of Technology, Tehran, Iran*

6.1 Introduction

Hardness and rigidity of microelectronics and smart electronics cause mechanical deformation because of the brittleness and inflexibility of their components. This restricted the application of non-flexible electronics, so they were replaced with electronic devices containing polymer and plastic substrates as flexible electronics in the last decades. In electronic devices, flexibility implies lightweight, elasticity, non-brittle, bending ability, and easy fabrication [1]. Flexible electronics has aroused great interest among researchers, industries, organizations, and military sectors to design and manufacture hybrid composite structures with lightweight flexible electronic devices [2]. Furthermore, the rapid growth of technology, especially the Internet, in everyday life has resulted in the advent of wearable technology in recent years [3].

Interesting possibilities have become practical by integrating processors, sensors, and information analyzers into the human body [4, 5]. Wearable electronics are designed to be worn comfortably and flexibly on the user's body. These devices have recently attracted remarkable attention in industry and research areas because of their wide applications in health monitoring, navigation, wireless communications, and entertainment [2, 6–14].

A smartwatch on the wrist, a head-mounted gaming display, a smart garment that tracks the physiological functions, and a glucose sensor tattoo accomplish at least one of the required target users [15]. Our lifestyle has changed in recent years in such a way that it is impossible to imagine the world without wearable/portable electronics [16]. Different companies worldwide are interested in wearable patient monitoring, fitness trackers, and entertainment devices with the market requirement growth and research development [17].

Wearable and flexible sensors and biosensors are widely explored in research and industry [18–20]. Sensors convert one form of energy into another form for

Nanotechnology in Electronics: Materials, Properties, Devices, First Edition.
Edited by Visakh P.M., Artem Semkin, Raneesh Balakrishnan, and Saša Lazović.

processing [17]. A sensor detects the environmental changes through conductive materials and sends the data to the computer for processing [21]. Conventional detecting methods have been changed by the advent of wearable and stretchable sensing electronics accompanied by nanomaterials [22]. Structures and designs of wearable sensors and nanomaterials determine the applications and performance of sensors. Depending on their properties, shape, size, structure, porosity, and fabrication method, nanomaterials and nanocomposites may impart different sensing properties to wearable substrates [17, 19].

Smartwatches, smartphones, and smart electronics as wearable sensors are widely used for monitoring health and environmental changes with wireless communications [17, 23]. Many challenging features, including multifunctional integration, stability and durability, adverse environmental impacts, mechanical flexibility, bending ability, stretchability, userfriendliness, communication, real-time monitoring, sustainability, and power source supply, are still in development [2]. Various flexible wearable sensors have been developed using nanomaterials and nanocomposites with different sensing properties, including strain, pressure, chemical, optical, humidity, temperature, and biochemical. This chapter summarizes nanomaterials and structure developments for various types of wearable stretchable strain/pressure sensors. Furthermore, approaches to improve the sensitivity and performance of wearable strain/pressure sensors and their applications in wearable electronics are discussed.

6.2 Wearable Strain/Pressure Sensors

Strain/pressure sensors measure deformation and pressure stimuli and collect data and mechanical properties [24, 25]. Metallic and semiconducting films were conventionally utilized as strain sensors and exhibited brittle and rigid characterizations with low stretchability [26]. These bulk materials could not attach to wearable substrates. Nanomaterials, including metal and metal-oxide nanoparticles (NPs) and nanowires (NWs), conducting polymers (CPs), graphene, and carbon nanotubes (CNTs), were explored due to their larger surface area compared to bulk materials [17]. Higher mechanical and electrical properties accompanied with usages of lower active materials are advantages of integrating nanostructures and hybrid nanomaterials with flexible supports. The supports can also maintain their stretchability, wearability, and flexibility [27–32]. Wearable sensors have many applications in health-care monitoring, physical activity continuous capturing, soft robotics, and artificial intelligence systems [17, 24, 33–35]. Monitoring body motions, including detection of large-scale motions such as bending of hands and legs as well small-scale motions such as subtle movements of the face and chest during breathing and speaking, requires strain sensors to diagnose the damage to vocal cord, respiratory disorders, determine spinal posture degree of deformation, and monitor Parkinson's disease [34, 36].

A wearable stretchable sensor generally consists of active sensing nanomaterials embedded on or integrated into a stretchable substrate [17, 37]. Stretchable intrinsic materials and stretchable structural designs as wearable substrates or additives are requisite for mechanical sensors [38, 39]. Various sensing nanomaterials were selected to fabricate high-performance sensors with tunable conductivity, high sensitivity, low power consumption, linearity, and a wide range of applied force [40–44]. Different types of flexible and wearable strain/pressure sensors with various sensing mechanisms can be categorized into piezoresistive, piezoelectric, capacitive, and triboelectric sensors [24, 25, 45].

6.2.1 Piezoresistive Sensors

Piezoresistive strain/pressure sensors rely on an electrical resistance change of a conductive material subjected to mechanical deformation in response to externally applied pressure or strain [17, 24, 36]. Resistance-type sensors with simple systems for signal read-out, simplicity in fabrication, and a broad test range are desirable for the realization of electronic skin (E-skin), health monitoring, and human–machine interfaces [24, 25, 46–50]. Traditional piezoresistive sensors are fixed directional sensors with the possibility of one specific direction measurement that cannot be embedded in structural materials [51]. New types of piezoresistive tactile sensors with high flexibility, sensitivity, and stretchability are developed that can be easily attached to the human body for monitoring health and physical motions [36].

Flexible wearable piezoresistive strain/pressure sensors are composed of conductive sensing elements in addition to elastic polymers or other flexible/stretchable substrates. The contact resistance of the conductive part is changed with strain/pressure stimuli, which is detected by electrical measuring systems [25]. Due to their desirable electrical and mechanical properties, different conductive nanomaterials were explored as conductive fillers of elastomer composites, including metal NPs, NWs, and thin films, CPs, and carbon nanomaterials [36, 46, 50, 52–58]. Moreover, PDMS, polyurethane (PU), silicon rubber, Ecoflex, and Lycra were explored as elastic substrates or components for conductive nanomaterials [45, 59, 60]. On the other hand, incorporating metallic nanomaterials, CNTs, and graphene with flexible wearable substrates proposes desirable gauge factor (GF) ranges, introduces different structures, and results in complex strains inside the matrix during stretching [61, 62]. Disconnection between nanomaterials decreases their contact area and leads to higher electrical resistance. Furthermore, the cracked structures of nanocomposites cause a different density distribution and can improve the sensitivity of wearable strain sensors [17].

Conductive nanomaterials also enable the matrix to deform under pressure and cause alteration in electrical conductivity [22]. Conductive carbon-based nanomaterials have received attention because of their unique chemical, physical, and electrical properties [63–66]. CNTs with different shapes, including aligned arrays, films, networks, yarns, fibers, fabrics, and aerogels, have been explored for wearable flexible strain/pressure sensors [66–72]. Due to desirable electrical conductivity, promising flexibility, and good mechanical properties, single-walled carbon nanotubes (SWCNTs) and multi-walled carbon nanotubes (MWCNTs) can be embedded into flexible and stretchable polymer materials to develop piezoresistive strain sensors [36, 45].

CNTs were exploited alone or in combination with elastic polymers as conductive fillers or a hybrid system to detect tactile and pressure [47, 50, 56, 70, 72–74]. Li et al. demonstrated extensively twisted, thin, and flexible CNT yarns indicating potential applications for super-elastic strain sensors [73]. The yarn entanglements were formed based on the dry-spinning technique, and large-strain stretching and rotational actuation were achieved as a consequence of random entanglements. However, randomness caused the local fluctuations and instability of the flexible sensors. CNT fibers were also grown on the Ecoflex substrate to develop an ultrahigh stretchable and wearable strain sensor [72]. The sensor could be stretched over 900% and showed high sensitivity, responsiveness, and durability [72]. CNTs have intrinsic flexibility due to their structure. Thus, hierarchical CNT structures were combined with microstructure elastic substrates to enhance the performance of sensors [25].

CNTs were coated or covered onto elastomers to form a conductive layer through various methods such as dipping, chemical vapor deposition (CVD), and printing [75–77]. However, the stretching mismatch between the conductive layer and the polymer substrate is usually possible [45]. A wearable and stretchable CNT strain sensor was fabricated with high durability, fast response, and low creep using the CVD method [68]. The obtained piezoresistive strain sensor could withstand strains up to 280% with a detection limit of 14 ms. Tadakaluru et al. developed a strain sensor by sandwiching conducting films between natural rubber layers [75]. The randomly distributed, entangled CNTs were used to devise a high sensitivity and durable multifunctional sensor to measure strain and pressure. Moreover, a scalable approach was explored to generate wearable strain sensors using nano- to microscale wrinkled CNT thin films [78]. The high-sensitive wearable sensor showed up to 750% strain with the capability to detect human motions by mounting onto the joint areas of fingers, elbows, and knees.

Some researchers explored porous and microstructures besides hierarchical and hollow architectures to improve tunneling piezoresistance for enhancing sensitivity [47, 79, 80]. Han et al. presented a flexible, compressible CNT/PDMS sponge with piezoresistive strain sensitivity [79]. The porosity of the sponge ranged from 200 to 800 μm. Further, Park et al. designed a flexible E-skin based on composite elastomer films containing interlocked microdome arrays [47]. They fabricated a CNT/PDMS liquid mixture by micromolding with the array position and enabled a maximum variation in the contact area for applied pressure. The working principle of the devised E-skin is illustrated in Figure 6.1a. Similarly, interlocked microdome arrays were explored for piezoresistive stretchable sensitive E-skins by Park et al. [56]. The CNT elastomer films can detect normal, shear, stretch, bending, and twisting forces. Another E-skin sensor consisting of sunflower pollen–based microcapsules with biomimetic architecture functionalized with CNT was designed for wearable strain sensor applications [74]. The functionalized microcapsules were embedded within an interlocked hierarchy into a rubber matrix composite with a low detection limit of 1.6 Pa for applied pressure.

Although CNTs offer high sensitivity, their interface between the elastic polymer substrate is not strong enough. Combining 2D materials such as graphene may

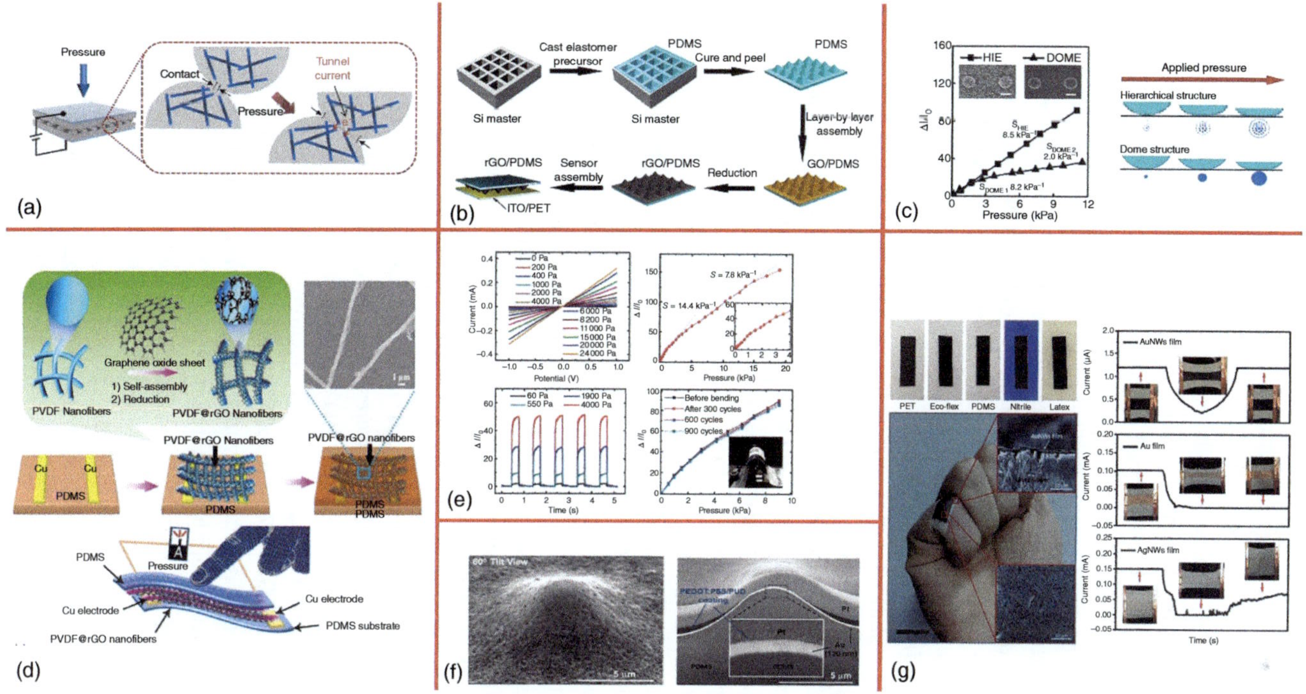

Figure 6.1 (a) Schematic of the working mechanism of the E-skin. The external pressure concentrates stress, deforms the microdomes, and causes an increase in the contact area and the tunneling currents, respectively. Source: Park et al. [56]/with permission of American Chemical Society. (b) The fabrication schematic of a tactile sensor device. A sensor was devised through a sandwich structure composed of the microstructured rGO/PDMS film and ITO-coated PET film. Source: Zhu et al. [54]/with permission of John Wiley & Sons. (c) Pressure–response curves exhibit the geometrical structure effects. Insets are SEM images of the hierarchical and dome-structured samples (scale bar: 50 μm). Source: [82] / with permission of John Wiley & Sons. (d) Schematic for the formation mechanism of PVDF fibers coated by rGO nanosheets (top-left) and FESEM image of coated fibers (top-right). The fabrication schematic of the flexible pressure sensor (bottom). Source: [112] / with permission of Elsevier. (e) I-V curves of the sensor for various applied pressures (top-left), recorded current variation–pressure relationship of three pressure sensors (top-right), plot of current profiles as a function of time at different pressures (bottom-left), and the performance before and after 900 bending cycles Source: [71] / with permission of John Wiley & Sons. (f) Pyramids are coated by drop-casting that grafts the pyramid with the increasing coating thickness downward along the slanted wall. Source: [57] / with permission John Wiley & Sons. (g) Photographs of Au NW film on different soft substrates (top-left), photographs of bending of the strain sensor ring attached on the little finger (scale bar: 1 cm); the inset is the top-view SEM image of the Au NW film on a latex rubber substrate (bottom-left). (Right-top to down) Electrical resistance changes of three different films on latex rubber under strain changes of 0%–100%–0% with a stage moving speed of 3 mm s^{-1}. Source: [61] / with permission of John Wiley & Sons.

improve stability, mechanical strength, and performance due to their synergistic effect [67, 69]. A chemically hybridized film of CNTs and graphene was presented in a flexible strain sensor by Shi et al. [69]. The hybrid structures improved the strength and load transfer at the nanotube joints and maintained the high electrical conductivity of CNTs. The buckling deformation of CNTs under cycling strain was resisted, and the films showed a linear and reliable response to strain as a wearable strain sensor to monitor human motions. Similarly, a high-performance flexible pressure sensor was fabricated using aligned CNT/graphene hybrid films [67]. The microstructured PDMS was molded from *Epipremnum Aureum* leaves as the flexible substrate. Jian et al. reported good performance of the sensor, including the sensitivity of 19.8 kPa^{-1} with a low detection limit of 0.6 Pa and 16.7 ms response time that could be attributed to the hierarchical structures of both CNT/graphene and PDMS accompanied by the excellent conformation of the active materials on the substrate. Moreover, Tsui et al. developed graphene-coated CNT aerogel for piezoresistive strain sensor [81]. The sensor exhibited stable, ultrafast, fatigue- and creep-resistant responses for static and dynamic applications [81].

Graphene is another impressive carbon-based nanomaterial for resistance-based strain sensors due to its high electrical conductivity and mechanical properties [82–84]. Applied strain can change and partially destroy the hexagonal structure of the edges of graphene films [85]. Graphene can be combined with elastic polymers, generally grown through CVD or derived from exfoliated graphite [33, 86–93]. The shape and structure of graphene determine the flexibility and stretchability of a sensor [25]. To increase the stretchability and sensitivity of graphene in strain/pressure sensors, particular structures and engineering strategies were designed [33, 36, 52, 94, 95]. For instance, Wang et al. reported graphene on PDMS to increase the tensile strain of graphene (>30%) for piezoresistive strain sensors [86]. Pre-strains of the elastic substrates and the width of graphene thickness determined the geometrical configurations of graphene ribbons. In another work, macroscopic flat, uniform graphene ribbons were fabricated through a modified wet spinning method [96]. The flexible ribbons exhibited good mechanical properties for potential application in sensors and actuators. Conversely, 3D macroporous nanopapers composed of crumpled graphene and nanocellulose were embedded into a PDMS matrix to increase the sensitivity of strain sensors to 100% compared to graphene [97]. The graphene nanopaper sensor fabricated through vacuum filtration improved the stretchability to 100% with a GF of 7.1 (at 100% strain) [97]. A paper-like rGO foam pressure sensor with open porous networks was presented by Hou et al. with excellent mechanical, electrical, and flexible properties as well as the ability to locate and measure finger pressure points and the pressure levels without any power supply [52]. Samad et al. developed a two-step method to create graphene foams with tunable densities for strain/pressure sensors [89]. The graphene foam/PDMS composite demonstrated high sensitivity (90% resistance change with 10% compressive strain). Furthermore, a high-performance strain sensor with a fish-skin-like graphene layer was reported with a large stretchability and a sensing range up to 82% strain, high sensitivity, and low detection limit (<0.1% strain) [33].

A graphene-woven fabric can also be another ideal candidate for piezoresistive sensors for strain/pressure sensing due to the special framework structure of microribbon components [98]. Li et al. embedded graphene woven fabric into PDMS [51]. High sensitivity and significant resistance changes of 10 times at 2% and 10 000 times at 8% enabled the composite to be an ideal structure for sensing tensile deformation by strain changing. Another graphene-woven structure was reported as a wearable sensor for sound signal acquisition and recognition [99]. For this purpose, graphene was coated on Cu meshes through CVD and then transferred to PDMS and a double-side tape. Every tone or phrase can be distinguished based on the resistance, changes, and characteristic peaks of each vocalization due to high electrical performance [36, 99].

Planar elastomer composites usually suffer from poor sensing performance [24]. 3D networks, porous structures, and microstructure patterns have been explored to overcome this issue [53, 54, 100]. Microstructured sensors demonstrate high performance due to their anisotropic configuration. A resistive tactile sensor based on graphene arrays on pyramid microstructured PDMS was fabricated with high sensitivity at a low-pressure range and a low detection limit of 1.5 Pa [54]. Microstructured graphene film was coated on PDMS by layer-by-layer (LBL) assembly to produce great stability and an ultrafast response of 0.2 ms (Figure 6.1b). Also, a high-performance piezoresistive pressure sensor with a high sensitivity of 8.5 kPa^{-1} for a wide pressure range was presented by Bae et al. [82]. A bioinspired hierarchical structure composed of PDMS coated with monolayer graphene resulted in a transparent sensor with a durability of 10 000 cycles and a detection limit of 1 Pa. The effects of the morphology of microstructures on sensitivity and linearity were studied as a function of applied pressure (Figure 6.1c). The contact phenomena of the hierarchical and dome structures under pressure are also illustrated in Figure 6.1c.

Conductive porous sponges or foams are ideal candidates for fabricating piezoresistive strain/pressure sensors due to their mechanical flexibility and porosity. A 3D random fiber network was devised with a microstructured design for wearable sensors [53]. PU sponge was dip-coated in GO solution and then reduced to generate a highly sensitive and durable sensor [53]. Modulating the graphene bandgap by tweaking the width and specific edge of the strained graphene, breaking the symmetry of the graphene lattice, and doping nitrogen or boron can open the graphene bandgap [101–105]. The effect of NH_3/Ar plasma treatment on the graphene piezoresistive behavior was studied to investigate the graphene sheet changes [106]. The plasma doping increased the piezoresistivity of the graphene sensor. Pyridinic-*N*, pyrrolic-*N*, and graphitic-*N* bonds formed inside the graphene structure exhibited high sensitivity in the lowest range of human perception (0–10 kPa^{-1}).

In addition to CNTs and graphene, other conductive carbon-based materials, including graphite and carbon black (CB), were exploited for flexible strain/pressure sensors [106, 107]. A strain gauge sensor using CB-doped PDMS was designed by Lu et al. [108]. The flexible sensor was composed of molded, straight resistors of CB-PDMS joined by CNT-PDMS interconnects, both in a matrix of PDMS substrate. The strain was measured between 11.2 and 22.6% on the wrist. In another demonstration, a sensitive graphite-based strain sensor was devised by pencil traces on flexible printing paper [109]. The prototype can be fabricated in less than 5 minutes with minimal equipment and exhibit high GFs of 150.5, 60.7, and 536.6 in 3 strain ranges from −0.62 to +0.62%. Different GFs are possible because of the sliding of graphene slices and the separation and rearranging between the graphite slices on the paper. Amjadi et al. also used graphite thin films on a soft elastomer film to fabricate a wearable strain sensor with high sensitivity and stretchability based on the formation of reversible micro-cracks [110]. The strain sensor showed good stability, reliability, and fast response to both tensile and compressive strain with low hysteresis.

Textiles, including fabrics, yarns, fibers, and nanofibers, have been explored as wearable components to incorporate graphene and CNTs for strain/pressure sensors [50, 77, 111, 112]. To develop e-textiles to sense strain and pressure, fiber sensors are woven into electronic fabrics [45, 113, 114]. They can also be integrated into conventional textiles or clothes [115–117]. Furthermore, conventional textiles can be used as substrates and/or functionalized with active nanomaterials [118–121]. On the other hand, nanofibers can be electrospun with conductive materials as electronic textile sensors [122–124]. Silk fabrics with microstructured surfaces were used as a mold to construct patterned flexible PDMS films integrated with freestanding ultrathin CNTs using the LBL exfoliation method [50]. The obtained E-skin sensor presented high sensitivity (1.8 kPa^{-1}), low detection limit (0.6 Pa), a fast response time (<10 ms), and high stability in detecting pressure. Similarly, Park et al. prepared graphene-coated stretchable yarns incorporated with PDMS using the LBL assembly technique [34]. The highly sensitive piezoresistive sensor with high stretchability (up to 150%) could detect even small motions in the throat and chest, and there was a linear relationship between strain and the resistance change. On the other hand, Lou et al. presented a wearable pressure sensor with high sensitivity (15.6 kPa^{-1}), a fast response time (5 ms), and long-term stability (100 000 cycles) utilizing rGO-encapsulated P(VDF-TrFE) nanofibers sandwiched between two PDMS films (Figure 6.1d) [112]. A textile-based pressure sensor composed of bottom integrated textile electrodes and a top bridge of cotton fabric coated with CNTs was fabricated by Liu et al. [71]. Laser-scribing masking and electroless nickel deposition on PET and/or nylon fabric were exploited for the interdigitated conductive electrode. Response to static pressure, sensitivity, current profiles ($\Delta I/I_0$) at different pressures, and mechanical stability of the textile sensor are shown in Figure 6.1e. This wearable piezoresistive sensor with high sensitivity, stability (1000 cycles), a fast response time (≈24 ms), and a low detection limit of 2 Pa can be attached to human skin to detect various vibrations and forces and monitor real-time pulse waves with low power consumption [71].

CB and graphite have also been integrated with textiles to show piezoresistivity [117, 125–127]. Natural biomaterials, including natural textiles such as silk and cotton, can be converted into conductive carbon materials via hydrothermal and pyrolysis processes [128, 129]. Wang et al. generated a highly sensitive strain sensor composed of a carbonized silk woven fabric to detect human motions [129]. The wearable and stretchable sensor exhibited a tolerable strain of greater than 500% and responded in less than 70 ms [129]. Carbonized silk nanofiber membranes integrated with PDMS film were also reported as skin-like pressure sensors [130]. The sensor was fabricated using electrospinning and carbonization and exhibited high sensitivity, good transparency, and excellent flexibility with a detection limit of 0.8 Pa.

CPs are conjugated organic polymers with unpaired π-electrons per carbon atom [131]. The electrons are mobile along the backbone of the polymer chain and cause the conduction of electricity. This leads to semiconductor or metallic behavior in CPs depending on the difference between their valence and conduction bands [132]. CPs, especially polypyrrole (PPy) and PEDOT:PSS, have also been investigated as active nanomaterials for piezoresistive tactile sensors [118]. One of the common approaches to achieving wearable sensors with elastic and conductive behavior is coating textiles with CPs. For example, Yamashita et al. coated polyethylene terephthalate (PET) ribbons with poly(3,4-ethylenedioxythiophene) polystyrene sulfonate (PEDOT:PSS) using the die-coating method [115]. Spandex fabric was also deposited on PEDOT:PSS as active nanomaterials [133]. Similarly, PPy was coated on Lycra, Lycra/cotton, and nylon/Lycra fabrics to detect strain and pressure [119, 120, 134]. However, the CPs often have poor adhesion to elastomer substrates and fabric polymers. Therefore, chemical functional groups and/or structures and defects are produced on the substrates [135–137]. This Problem can be addressed by combining conductive nanomaterials with elastomers in composite formation, this problem can be addressed [138]. Pan et al. designed an elastic conductive microstructure polymer consisting of hollow-sphere structures of PPy through a multiphase synthesis technique [139]. A spherical shell geometry enables inherently brittle nanostructures to show tunable effective elastic modulus withstanding large strains and stresses. Further, a flexible temperature–pressure sensor based on microstructured organic thermoelectric materials was developed using PEDOT:PSS-deposited PU with a sandwich structure [60]. The combination of piezoresistive and thermoelectric mechanisms enabled simultaneous detection of pressure stimuli and temperature for health-monitoring wearable devices. Choong et al. presented a resistive pressure sensor using micro-pyramid PDMS arrays incorporated with PEDOT:PSS to detect compression and tension (Figure 6.1f) [57]. The wearable sensor showed a high sensitivity of 10.3 kPa^{-1} when stretched by 40%. CPs can also be employed as conductive components of fibers or nanofibers. Seyedin et al. demonstrated a method to devise homogeneous PEDOT:PSS/PU fibers for wearable strain sensors through wet spinning [138]. The synthesized fibers responded to a wide range of strains, showing the potential of motion sensing and biomedical monitoring. Also, a stretchable strain sensor was reported based on electrospun PEDOT:PSS/PVA nanofibers [140]. The strain sensor exhibited high durability, fast response, and high sensitivity for monitoring even small human motions.

Metallic NWs and NPs have been extensively used for imparting electrical conductance and fabricating flexible wearable electronics [58, 141–143]. Metals with excellent electrical conductivity have been introduced into polymer matrices and stretchable textile substrates to exhibit high sensitivity for the detection of strain and pressure stimuli [144–148]. An Au NW-impregnated tissue paper sandwiched between blank, and patterned PDMS sheets was fabricated as a wearable pressure sensor with important properties including low energy consumption (<30 μW), high sensitivity (>1.14 kPa^{-1}), a fast response time (<17 ms), and good detection limit (13 Pa) [58]. Gong et al. used black silver NW films in designing flexible strain sensors through drop-casting on some elastomeric sheets including latex, nitrile rubber, Ecoflex, PDMS, and PET reported a GF of 6.9–9.9 and a fast response time of less than 22 ms (Figure 6.1g) [61]. On the other hand, a highly stretchable strain sensor based on silver NWs was devised by Amjadi et al. [141]. Sandwich structured nanocomposites consisting of Ag NW thin films embedded between two PDMS layers were fabricated to produce an excellent performance for both static and dynamic loads with high linearity and negligible hysteresis. Moreover, Ag or Cu NWs were studied as conductive components on PI substrate matrices in skin-mountable electronic sensors [144]. The network nanocomposite elastomers showed remarkable mechanical properties against fracture and delamination, crack resistance, contact adhesion, and normal/shear strength due to the fabrication strategy. Although high-performance tactile sensors can mimic and surpass the human skin sensation, they cannot retain the sensation information after stimuli removal. In this regard, Zhu et al. prepared haptic memory devices through the integration of resistive switching memory with resistive pressure sensors [145]. Microstructured PDMS films embedded on Ag NWs provided the sensor with high sensitivity in low-pressure regimes (<1 kPa) for the advancement of next-generation high-performance sensing systems in E-skins and humanoid robots (Figure 6.2a).

The introduction of metallic nanoscale cracks into elastic microporous layers disconnected/reconnected between adjacent crack junctions causes small changes in resistivity due to external stimuli [149]. A stretchable strain sensor based on microcracks of metal NPs was designed through the direct transfer patterning of Ag NP inks on microstructured PDMS and exhibited high sensitivity and fast response/recovery under tensile and compressive strain [150]. Another piezoresistive sensor based on channel crack–based gold/PU sponges was created through ion sputtering [151]. The channel cracks perpendicular to 3D microporous Au/PU sponge skeletons considerably improved sensitivity at low pressure (<10 kPa) as well as endowed the capability to detect large human motions for the sensor. Similarly, Park et al. evaluated the effect of geometric factors on nanoscale crack-based sensors by modulation crack geometry and reported high sensitivity [151]. The sensor containing a cracked 20-nm-thick Pt top layer and a 10-μm-thick PU layer showed reversibility, reliability, and no hysteresis as a mechanosensor.

Metal NWs and NPs are also integrated with or coated on fibers and nanofibers to detect pressure/strain [146, 147, 152]. A stretchable pressure sensor was proposed using Au-deposited PDMS micropillars on the top surface, and PANI nanofibers are grown on a PET substrate at the bottom [146]. A decrease in contact resistance between the top and bottom surfaces under applied pressure and a large contact area of the nanofibers resulted in high sensitivity. On the other hand, Ge et al. reported fabric sensors able to map and quantify mechanical stress [147]. An elastic thread consisting of PU fibers as core and nylon fiber as sheath was modified with (3-aminopropyl)triethoxysilane (APTES) and coated with Ag NWs. It was then coated with piezoresistive rubber composed of PDMS and CB weaved as a

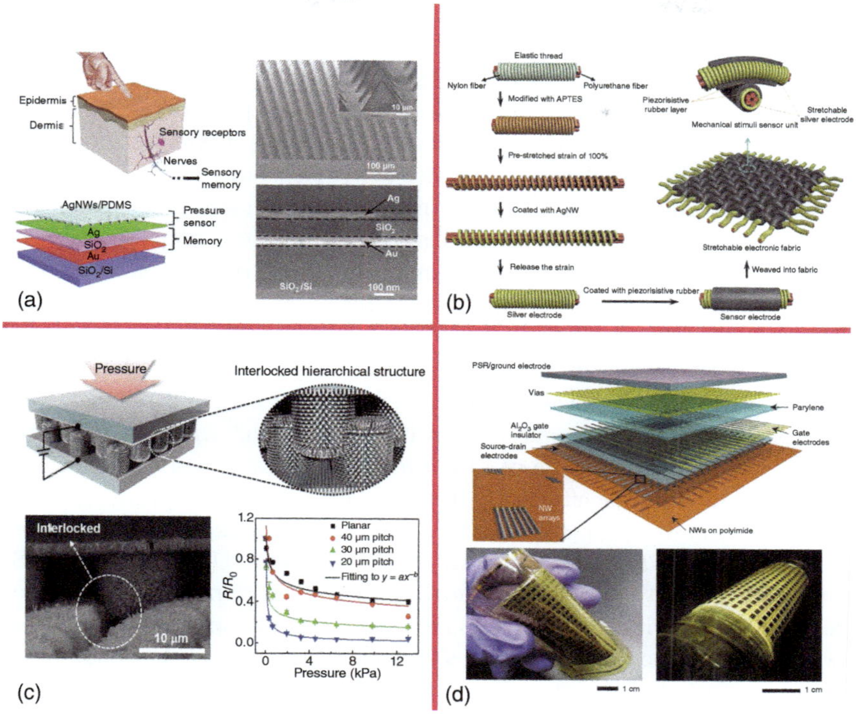

Figure 6.2 (a) Haptic memory devices consist of resistive switching memory devices and resistive pressure sensors inspired by the human sensory memory. (a) Schematic illustration of human skin responding to applied pressure. When a finger touches the skin, the sensory receptors perceive the force and send the sensation information to the brain through nerves (top-left). Schematic of the haptic memory device comprising pressure sensor and resistive switching memory devices (bottom-left). Tilted SEM image of the pyramid microstructures of Ag NWs/PDMS film (top-right). The inset shows an enlarged image. Cross-sectional SEM image with the LBL configuration on a silicon wafer (Au: bottom electrode; SiO_2: resistive switching layer; Ag: top electrode). Source: [145] / with permission of John Wiley & Sons. (b) Fabrication schematic of the electronic fabric artificial skin with fibrous sensor units. Source: Ge et al. [147] / with permission of John Wiley & Sons. (c) Static-pressure-sensing capability of e-skins based on the interlocking geometry of hierarchical ZnO NW arrays. (a) Schematic of the piezoresistive E-skin device utilizing the interlocking geometry of hierarchical ZnO NW arrays (top). A cross-sectional SEM image of interlocked geometry on the PDMS micropillar arrays (bottom-left). Relative resistance changes as a function of normal pressure for different pitch sizes of E-skins (bottom-right). Source: [161] / with permission of John Wiley & Sons. (d) Schematic of the passive and active layers of nanowire-based macroscale flexible E-skin (top). Optical photographs of a fabricated E-skin device under bending (bottom-left) and rolling (bottom-right) conditions. Source: [13] / with permission of Springer Nature.

stretchable fabric in which mechanical sensor units formed at each cross contact (Figure 6.2b). The sensor was repeatable and reproducible with excellent stability that could be used for biomedical prostheses and physiological analysis devices. A stretchable fiber strain sensor based on P(VDF-TrFE) polymer nanofiber mats and silver NW layer was designed by Chen et al. [153]. An elastic rubber latex thread core and PU fiber yarn winding around the shell acted as the elastic scaffold. High GF of 5.326, a fast response time of 20 ms, and considerable durability (10 000 strain cycles) were reported for the resistive strain sensor.

Poor stretchability and heterogeneous electromechanical properties prevent the widespread use of single nanotubes or NWs as strain sensors. Therefore, some new approaches have been developed to enhance the stretchability and sensitivity of strain sensors by integrating functional nanomaterials with stretchable polymers [154]. Several mechanisms, including disconnection, crack propagation, and tunneling effects, have been proposed to improve the performance. However, the piezoresistivity contribution of nanomaterials is still low because a large elastic mismatch accompanied by weak interfacial adhesion strength between nanomaterials and polymers results in insignificant deformations of nanomaterials after stretching [141, 154, 155]. Flexible and wearable strain sensors using carbon-based materials exhibit a large sensing range and low sensitivity or a small sensing range with high sensitivity. The introduction of CPs or metal/metal-oxide nanomaterials is a practical approach to enhancing the performance of carbon-based pressure/strain sensors [25, 40, 76, 156–159]. A transparent, stretchable, and tunable strain sensor based on CNTs and PEDOT:PSS sandwiched between the PU layers and PDMS substrate exhibited 63% GF of 62.3 and high sensing performance due to the interaction of CNTs with PU/PEDOT:PSS in a sandwich-like structure [40]. Silver NPs were also combined with graphene sheets integrated with PU to exhibit high sensitivity besides a low detection limit as a wearable strain sensor [160]. Metal oxides can also be integrated with conductive nanomaterials and elastic polymers to prevent stacking and aggregation. For instance, a piezoresistive nanocomposite sensor was developed on a PU matrix using magnetite-decorated graphene to detect tactile [156]. Similarly, micro- and nanostructured Pt coated-ZnO NWs were grown on a PDMS micropillar substrate to detect static and dynamic tactile stimuli through piezoresistive and piezoelectric transduction modes [161]. An interlocked geometry between the hierarchical ZnO NW arrays allows the contact area between the NWs to change in response to stimuli, leading to a highly sensitive piezoresistive static sensor (Figure 6.2c).

Field-effect transistor (FET)-type strain/pressure sensors with high sensing performance and high resolution have been developed recently, requiring some steps to be fabricated [162–164]. These steps include fabrication of transistor, encapsulation of device via hole opening, and integration with the sensor [164]. Some researchers have reported flexible strain/pressure sensors based on FET active matrices [13, 142, 165, 166]. Accordingly, Ge/Si core–shell NW arrays on a PI substrate were followed by device fabrication for low-voltage macroscale artificial skins (Figure 6.2d) [13]. Optical images comprising a pixel matrix connected to an NW-array FET maintained the pixel state while other pixels were addressed. The schematic and SEM images of the E-skin device with an integrated NW active-matrix backplane are shown in Figure 6.2d. Similarly, the FET was integrated with a graphite-based pressure sensor layer to form a very wide structure with a low-cost processing technology [167]. Sun et al. described a transparent graphene-based FET pressure sensor for E-skin applications [164]. The applied pressure to the substrate induced the contact between the square-type graphene on the top cover and interdigitated-type graphene on the FET backplane, leading to lower resistance between the sources and drain electrodes and higher transconductance. A flexible MoS_2 FET was developed in another work, and the properties were reported under applied strain [166]. The high sensitivity of the sensor was attributed to the strain-induced bandgap change. Also, a bioinspired strategy was reported for a flexible pressure sensor inspired by mimosa, a plant that closes its leaves under external stimuli [142]. A mimosa leaflet was used as the template for the fabrication of microstructured PDMS thin films coated with gold/titanium NPs. Conductive wires were anchored on the edge of both films as the source–drain electrodes, leading to a bioinspired flexible pressure sensor. Yeom et al. demonstrated a CNT-based pressure sensor array capable of mapping out the applied pressure utilizing a printed active-matrix thin film transistor backplane [165].

Piezoresistive strain sensors are highly stretchable and sensitive and, however, generally exhibit nonlinear response, large hysteresis, and irreversibility. After releasing the pressure, it is hard for the nanomaterials to slide back to the initial position or entirely close the cracks, leading to the time delay between the electrical output and the mechanical input [27, 154].

6.2.2 Capacitive Sensor

A piezocapacitive strain, pressure, or tactile sensor comprises a dielectric layer sandwiched between two electrically conductive layers [27, 45, 168]. The mechanical, physical, or chemical stimuli cause measurable changes in physical size or the dielectric constant of a capacitor [24, 27, 45, 169, 170]. A capacitor is a passive electronic that stores energy electrostatically [171]. The capacitance of a parallel-plate capacitor is defined according to Eq. (6.1) as follows:

$$C = \varepsilon_0 \varepsilon_r A/d \tag{6.1}$$

where ε_0 and ε_r are relative permittivity of the dielectric material and vacuum permittivity, respectively, and A and d are the capacitor surface area and the distance between two layers, respectively [172].

The elastic polymer is displaced upon external pressure, and A, d, and ε_r are changed consequently due to normal/shear force and strain. The capacitor change is linearly related to tensile strain [24]. Small modulus stretchable elastomers such as PU, Ecoflex, and PDMS are appropriate choices for dielectric materials [11, 173–175]. Piezocapacitive sensors exhibit high sensitivity and low energy consumption [169]. They show low hysteresis, fast response, and appropriate availability despite the viscoelasticity of the polymer matrix and friction under deformation between nanomaterials and the polymer [171, 176, 177].

Carbon-based capacitive strain/pressure sensors have widely been studied in wearable devices. Graphene and CNTs, as flexible/stretchable conductive electrodes, exhibit high sensitivity [173, 176–179]. A stretchable strain gauge with high sensitivity, optical transparency, durability, and stability was fabricated using CNTs to measure strains as large as 300% [176]. The capacitive sensor used CNT films as electrodes and PDMS as the dielectric layer and exhibited a fast response (<100 ms) without hysteresis or relaxation [179]. A graphene-based capacitive pressure sensor fabricated using PMMA elastic polymer exhibited a high sensitivity of 15.15 aF/Pa [173]. Ag NW composites with different structures have been intensively explored in stretchable capacitance-based sensors due to their good stretchability, electrical conductivity, and transparency [180–184]. A multifunctional wearable sensor based on Ag NWs was presented to detect strain, pressure, and finger touch [184]. In another study, Ag NW percolation networks were used to design a real-time skin-like multidimensional strain sensor with a large GF (20) and high stretchability (35%) [180]. The sensor exhibited highly polarized and decoupled electrical resistance change in the axial and perpendicular directions while measuring principal strain. Ho and coworkers demonstrated a transparent and stretchable capacitive sensor employing a microfluidic system improved by microcracks into Ag NW–embedded rubber channels for E-skin applications [183]. While measuring principal strain, the sensor with two laminated PDMS films bearing Ag NW–embedded microfluidic channels formed a cross-bar-type capacitive sensor with an air gap sandwiched between two electrodes.

The capacitance change highly depends on the geometrical structure of the electrodes and dielectric layers [162]. In a typical parallel-plate-type capacitive strain sensor, strain and pressure can induce changes in capacitance, while the minimum thickness of the sensor cannot be achieved due to the stacked layer structure for conformal contact in wearable electronics [162, 185, 186]. The crosstalk problem (effect of applied pressure on the adjacent pixels) between strain and pressure sensing may also occur in parallel-plate types [171, 186]. Therefore, interdigitated capacitive sensors have been widely explored due to better linearity, high sensitivity, and lower hysteresis [17, 186–189]. Hu et al. presented a molding process to realize a flexible capacitive sensor utilizing 3D CNT interdigitated finger electrodes to detect strain, bending curvature, tactile force, and proximity distance [187]. Similarly, a flexible paper-based interdigitated capacitive touch sensor was fabricated using a graphite pencil [189]. Moreover, interdigitated capacitive sensors were fabricated based on patterned Ag NW electrodes with micrometer dimensions and spacing embedded onto the surface [186]. The strain sensor showed a GF of −2, which is larger than the GF limit of 1 for parallel-plate capacitive sensors.

High stretchable dielectric elastomers with smaller mechanical modulus, such as PDMS, PU, and Ecoflex, are typically employed for higher sensitivity in capacitive sensors [17]. Moreover, researchers have attempted to improve the sensitivity of the capacitive sensor by changing the dielectric layer with different materials and structures using compressible elastomeric materials and microstructured dielectric layers [49, 190, 191]. Highly flexible wearable bandage-like pressure sensors with signal amplification were fabricated using PDMS microhair-structured interfacial dielectric layers [49]. Figure 6.3a, b illustrates a schematic view of detecting pulse on a human's neck with a microhair sensor and a cross-sectional diagram of the fabricated sensor. These portable sensors with wireless transmitters are advantageous for diagnosing cardiovascular and cardiac disease. Luo et al. presented a highly

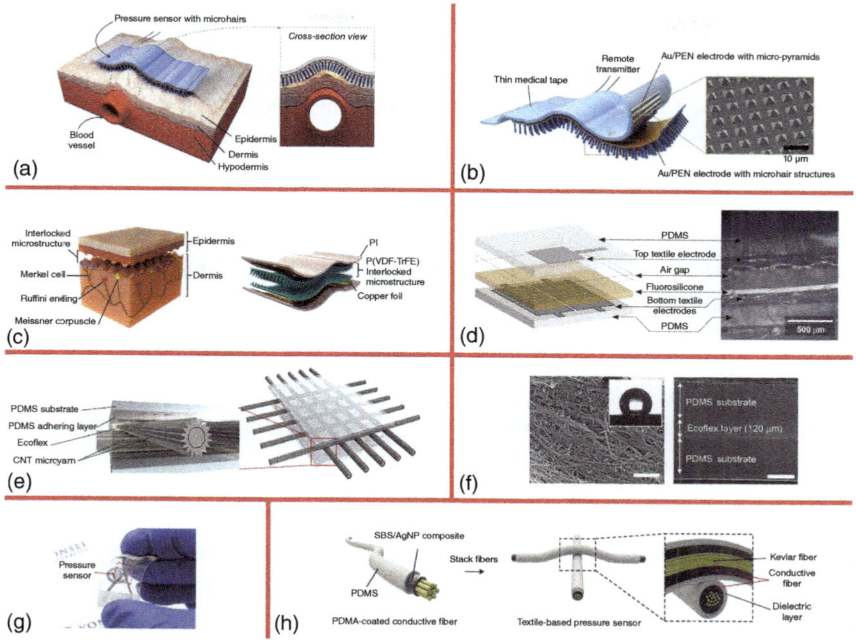

Figure 6.3 (a) Schematic illustration of pulse-detectable pressure sensor to detect pulse on a human's neck with microhair sensor. (b) Cross-sectional diagram of the microhair-structured sensor. Source: [49] / with permission of John Wiley & Sons. (c) Schematic of human skin structure with interlocked epidermal–dermal layers and sensory neurons and schematic of the capacitive tactile sensor based on P(VDF-TrFE) films with the interlocked microstructure. Source: Niu et al. [199] / with permission of John Wiley & Sons. (d) Schematic illustration of the sensor architecture with the corresponding cross-sectional view. Source: [172] / with permission of John Wiley & Sons. (e) Hierarchically engineered CNT microyarns and piezocapacitive stretchable E-skin devices. (f) Schematic of the CNT microyarn-based sensor skin. (f) FESEM of CNT microyarn surface (right side). Inset shows the water contact angle of CNT microyarn surface and cross-sectional electron micrograph of layered structure without CNT. Source: [57] / with permission John Wiley & Sons. (g) Photograph of the fabricated pressure sensor on a PET substrate using conductive fibers. Source: [61] / with permission of John Wiley & Sons. (h) Schematic view of the fabrication of the pressure sensor. Source: Takamatsu et al. [206] / with permission of John Wiley & Sons.

sensitive, flexible capacitive pressure sensor based on a PDMS dielectric layer structure with tilted micropillar arrays, strongly bonded to Au-coated electrodes [24]. The bending deformation of the tilted micropillar arrays replaced compression deformation, facilitating the distance changes between the electrodes even after small pressure and resulting in high sensing performance (0.42 kPa^{-1}). On the other hand, utilizing the air gap can improve the sensitivity [24, 192]. A microstructure-like randomly rough surface graphite-coated paper was employed to design a capacitive pressure sensor with 3D touch capability [179]. The air gap was formed when two graphite-coated papers faced each other to develop a high-sensitivity capacitive sensor.

To address the unstable capacitance under large strains and the complicated fabrication process of stretchable capacitive touch sensors, Choi et al. reported a selective-patterning process for direct coating of stretchable electrodes on the elastomeric polymer by choosing materials considering Poisson's ratio to reduce the electric deformation and, thereby, minimal changes in capacitance [193]. The other promising strategy to enhance the sensitivity and sensing range of the dielectric layer is microporous structures [25, 194–196]. A wearable piezocapacitive pressure sensor based on 3D microporous dielectric elastic polymers with high sensitivity and stability was generated using CNT-Ecoflex [194]. The micropores were opened and closed without viscoelastic behavior, resulting in lower hysteresis. Solid elastomeric PDMS foams filled with conductive CB were fabricated as piezocapacitive strain sensors [196]. CB particles increased the dielectric constant by improving the interfacial polarization by generating local microcapacitors. Compression of microcapacitors led to large deformation and a significant change in permittivity; therefore, adding conductive fillers into the dielectric below the percolation threshold can enhance the electric constant under pressure [27, 197]. Shi et al. proposed a capacitive pressure sensor with a diaphragm structure that can easily deform under both positive and negative pressure [198]. The sensor composed of PEDOT:PSS electrodes, an Ecoflex dielectric layer, and PDMS encapsulation layers exhibited good sensitivity and great repeatability under compressive loads, vacuum suction, bending, and stretching [198]. Interlocked microstructures located at the interface between the dermis and epidermis of human skin play an essential role in recovering the afferent stimuli in pressure perception (Figure 6.3c) [199, 200]. Inspired by this phenomenon, Niu et al. developed a capacitive tactile sensor with interlocked asymmetric nanocones. P(VDF-TrFE) showed low viscoelasticity, medium modulus, and high dielectric constant suitable for fabricating miniaturized and highly integrated tactile sensors (Figure 6.3c) [199].

One of the fundamental functions of touch sensors is detecting normal and tangential forces raising the level of encoded information by an artificial touch system close to natural touch [172]. Some capacitive skin-like sensors were reported to have three-dimensional sensing for wearable applications [162, 172, 201]. Viry et al. reported a flexible capacitive three-axial force sensor made of conductive-coated fabrics and fluorosilicone film as the electrode and dielectric layer (Figure 6.3d) [172]. Similarly, a wearable mutual capacitance touch sensor based on graphene electrodes and an acrylic dielectric layer on PET was devised for multi-touch sensing as well as 3D sensing in a highly deformed state [201]. Organic field-effect transistors (OFETs) are also explored to amplify to improve sensitivity. They are combined with capacitive wearable tactile sensors to improve the performance [11, 196, 202]. Mannsfeld et al. and Schwartz reported active OFET pressure sensor devices employing microstructured elastomeric rubber and PDMS dielectric films, respectively [11, 203].

Similar to other types of strain and pressure sensors, capacitive textile-based sensors were explored, consisting of two compliant conductive fabrics as electrodes and a dielectric layer such as foam, spacer, and soft polymer [172, 204, 205]. A highly sensitive, wearable, and multimodal all-carbon skin sensor was designed using hierarchical CNT fabrics and the Ecoflex dielectric layer to sense tactile, touch, temperature, humidity, and biological variables in a single pixel (Figure 6.3e–h) [206]. A textile-based pressure sensor with high sensitivity, durability, and the fast response was reported based on conductive Ag-coated Kevlar fibers and dielectric PDMS films coated on the fibers [48].

In conclusion, piezocapacitive strain/pressure sensors have smaller GFs than resistive sensors. However, their GFs remains constant in the entire range, making them suitable for large strain/pressure monitoring [17, 27].

6.2.3 Piezoelectric Sensors

Piezoelectricity is the ability of non-centrosymmetric materials to generate a voltage in response to external pressure/strain due to reorientation under stimuli to form an internal polarization [36, 169, 207]. In other words, a particular type of solid exhibits ultrafast response time to transfer mechanical energy into output electricity under applied force [17, 22]. Due to high sensitivity to dynamic pressures, piezoelectric sensors are widely used to measure vibrations of sound and slip [169, 208, 209]. They are not well suited for static pressures because of their transient nature, similar to the RA receptors in human skin [169, 208, 210]. Piezoelectric materials are also pyroelectric, meaning that a temperature change can induce a voltage. This property may affect the reliability of pressure detection without temperature sensor compensation [141]. Moreover, the stretchability and flexibility limitations of piezoelectric sensors make them unstable in the human body for health-monitoring applications [36].

ZnO is one of the important piezoelectric inorganic materials for pressure/strain sensors [211–213]. Textured ZnO NW film was grown on the surface of CF for flexible piezotronic strain sensors and showed a fast response and current real-time response upon both static and dynamic mechanical loads [212]. Lead titanate and lead zirconate titanate (PZT) are ferroelectric/piezoelectric materials explored as pressure and strain sensors [214–217]. A large-area PZT thin film on a flexible PET substrate was designed with high output piezoelectric performance as a sensor and nanogenerator [216]. On the other hand, a pressure sensor for static measurements based on $PbTiO_3$ NWs/graphene heterostructures was presented by Chen et al. [217]. The flexible piezoelectric sensor showed a sensitivity of up to 9.4×10^{-3} kPa^{-1} due to high carrier mobility of graphene.

PVDF as a piezoelectric polymer has been developed so far due to its flexibility in pressure/strain sensor applications than inorganic materials. PVDF and its co-polymers can exploit deformations induced by small forces through pressure, compression, bending, and twisting [218]. Lower coefficients and ease of processing of PVDF than conventional ceramics cause widespread usage in wearable piezoelectronics [219–223]. For instance, Garain reported a flexible pressure sensor using electrospun Ce^{3+}-doped PVDF/graphene composite nanofibers [220]. The combination of high sensing performance with mechanical flexibility and robustness of nanofibers enabled real-time monitoring of sound waves as well as detection of different types of musical vibration. Wang et al. presented a sensor based on single-electrode piezoelectric nanogenerators utilizing PVDF nanofibers to detect static pressure by integrating cold/heat sensing [222]. The P(VDF-TrFE) is one of the most promising piezoelectric materials proposed for wearable piezoelectric strain/pressure sensors [218, 224–227]. A flexible piezoelectric material consisting of P(VDF-TrFE) fiber was produced to enable ultrahigh sensitivity for pressure measurement even at small values (0.1 Pa) [218]. Similarly, a pulse monitoring piezoelectric sensor was reported by employing P(VDF-TrFE) in the form of the electrospun nanofiber mat sandwiched between two PDMS sheets coated with Au [225].

In addition to the mentioned piezoelectric materials, other materials have been explored with appropriate piezoelectric properties. For instance, $BaTiO_3$, MoS_2, GaAs, $ZnSnO_3$, and poly-L-lactic acid (PLLA) were reported as piezoelectric materials in strain/pressure sensors [228–232]. A strain sensor based on $ZnSnO_3$ NWs/microwires exhibited a high piezoelectric effect to detect tension and compression [231]. Lim et al. reported a transparent interactive human–machine interface using PLLA as a piezoelectric polymer [232]. The graphene-based wearable system had the capability to monitor human motions, convert recorded data into signals, deliver triggering signals to machines, and transfer feedback to operators through electrotactile devices.

To improve the performance of piezoelectric sensors, inorganic or polymer piezoelectric materials are combined with amplifier parts such as transistors to read out signals from sensing elements [36, 215, 227, 233]. Degdeviren et al. formulated a conformable metal-oxide semiconductor field-effect transistor (MOSFET)-integrated piezoelectric pressure sensor based on PZT with high sensitivity and fast response (0.1 ms) [215]. On the other hand, a piezopotential-powered active-matrix strain sensor consisting of a piezoelectric nanogenerator and a coplanar-gate graphene transistor was fabricated using a photolithography process [227]. Direct integration of piezoelectric polymer as gate dielectric into the transistor is employed to address complicated fabrication and high power consumption of FET-based piezoelectric sensors and nanogenerators [226, 234]. For instance, Tien et al. reported direct integration of the piezoelectric and pyroelectric gate dielectric and the piezo-thermoresistive organic semiconductor channel into FET platforms to detect pressure and temperature simultaneously [226].

Most sensors need external power supplies and additional signal processing amplifications. Integrating a battery or other energy supplies to the sensor is required for applicable wearable electronics such as wearable body monitoring, smart tags, and implants [235, 236]. However, conventional batteries and other integrated energy supplies greatly restricted wearable sensors, especially E-skins. Hence, self-powered wearable flexible electronics have been recently considered due to their small nano-, microscale devices, and extremely low power consumption. Many researchers focused on integrated nanogenerators for wearable piezoelectric sensors [216, 220, 222–224, 227, 229, 233, 235, 236]. A flexible triboelectric–piezoelectric nanogenerator was devised based on P(VDF-TrFE) nanofibers and PDMS/MWCNT membrane [222]. In another work, a multifunctional self-powered E-skin was developed utilizing rGO-on-PVDF-nanofibers [223]. Three types of sensors, including a pressure sensor, a photodetector, and a gas sensor, besides on-chip microcapacitors in parallel, were integrated into a single pixel to detect environmental conditions and physiological health monitoring [235]. On the other hand, the ferroelectric

$BaTiO_3$ with excellent pyroelectric and piezoelectric properties was integrated with Ag and PDMS to obtain high sensitivity to both strain and temperature within a self-powered system [229]. The organic/inorganic hybrid composed of piezoelectric polymers such as PVDF with inorganic piezoelectric nanostructures demonstrated the capability to convert the mechanical energy of motion into electrical energy and was found to be highly suited for self-powered E-skins [226, 236, 237]. A flexible self-powered ZnO/PVDF/fabric E-skin was designed by He et al. with multifunctional properties, including tactile perception, atmosphere detection, and self-cleaning behavior [236]. Another flexible piezoelectric nanogenerator composed of P(VDF-TrFE) nanofibers and $BaTiO_3$ NPs on PDMS was presented with high stability and durability along with stable operation during harsh working conditions [237].

Although piezoelectric sensors possess advantages such as wide frequency response range, high sensitivity to dynamic pressure, fast response time, and low power consumption, the voltage decay after the load's removal makes them unreliable for static sensing [169, 208, 238, 239].

6.2.4 Triboelectric Sensors

Triboelectric nanogenerators (TENGs) are developed based on the triboelectric effect, and they harvest mechanical energy from the environment and convert it into electricity [240]. Due to the requirement of self-powered sensor network systems in next-generation wearable and implantable devices, especially health-care applications, self-powered wearable tactile sensors based on triboelectric effect have been generated [241–244]. When two contacted materials are electrically charged upon separation through friction, called the triboelectric effect [245]. The coupling effect of contact electrification and electrostatic induction is the main mechanism of TENG-based devices [242]. A tactile sensor is a transducer that transforms physical stimuli such as touch or pressure into measurable electronic signals [246, 247]. TENG-based sensors act as energy collectors, harvest all kinds of energy, convert it into electricity as well as supply the energy for sensing [240].

In recent years, much focus has been on wearable self-powered triboelectric sensors for health care, motion detection, and human–machine interface applications [224, 244, 248–250]. A self-powered triboelectric sensor based on flexible thin films composed of ITO and fluorinated ethylene propylene (FEP) was reported by Zhu et al. [241]. Similarly, a membrane-based self-powered triboelectric sensor was fabricated using latex/FEP for surveillance and health monitoring [248]. The output voltage signals can be employed for air pressure sensing and frequency monitoring due to their good linearity, high sensitivity, and stability. Also, elastomeric materials and an inner helix electrode sticking on a tube with the Ecoflex dielectric layer and an outer electrode for TENG-based sensors are appropriate for wearable electronics such as smartwatches and fitness trackers [224]. On the other hand, a graphene tribotronic touch sensor based on the coplanar coupling of a single-electrode-mode TENG and a graphene FET was designed by Khan et al. [244]. The touch TENG-based sensor showed excellent sensing performance with a low detection limit (<1 kPa), touch sensitivity of 2% kPa^{-1}, and response time of 30 ms. Another self-powered triboelectric tactile sensor was fabricated, consisting of Ag nanofiber electrodes on the elastomeric substrate to detect and map trajectory profiles [243]. Utilizing the cross-location technology, the device can execute more rapid tactile mapping with a response time of 70 ms.

Although self-powered tactile sensors based on the triboelectric effect demonstrate many significant advantages over other tactile sensors, further improvement in pressure mapping resolution is still needed by the integration of other mechanisms [240].

6.3 Applications

6.3.1 Movement Monitoring and Daily Performance Tracking

Wearable strain pressure sensors have extensively been used to monitor human movements and body motions in daily life [68, 251–258]. Body motions, including fingers, elbow, wrist, knee, arm, feet, and hand movements, can be detected by wearable sensors [107, 259–266]. For instance, the APTES bonding layer and MWCNT/graphene on the PDMS substrate sandwich structure decorated with Ag NPs and heptadecafluoro-1, 1, 2, 2-tetradecyl trimethoxysilane strain sensor was reported [267]. A wearable sensor coated with the bacteria *S. aureus* was used to detect routine human motions, including wrist, finger, and knee bending (Figure 6.4a). The sensor exhibited ultra-sensitivity with a maximum GF of 1989, broad strain ranges of 0.1–170% as well as a fast response time of 150 ms. Moreover, Su et al. presented a selective wearable sensor with bending/stretching force differentiation

Figure 6.4 (a) Relative resistance responses of the (heptadecafluoro-1,1,2,2-tetradecyl) trimethoxysilane/Ag/MWCNG/ graphene–PDMS strain sensor during the pulse, wrist, finger, and knee bending detection with a droplet of bacteria on the sensor surface. Source: [267] / with permission of John Wiley & Sons. (b) The wearable device based on real-time monitoring of the arterial pulse (top-left). Schematic illustration of the monitoring device system (top-right). The signal of real-time artery pulse and the plotted interpreted waveforms measured during cycling by the sensor mounted onto the radial artery of the wrist. Source: [283] / with permission of Elsevier. (c) Schematic illustration of human joint motions (spine, wrist, finger, elbow, and knee). (d) Relative resistance changes detected by textile strain sensors attached to the finger (top-left), the elbow (middle-left), and the wrist (bottom-left). The Bluetooth messages that output "Get the phone" and "Hang up" resulting from the folding motions of the finger joint (top-right), Bluetooth message output of the number of different folding motions of the elbow joint in the angular range 45° and 100° (middle-right), and Bluetooth messages "Channel up" and "Channel down" output from the different rotational motions of the elbow joint in the angular range (bottom-right). Source: [284] / with permission of John Wiley & Sons. (e) Changes in resistance monitored by the textile strain sensor attached to the knee joint (top-left), changes from the different folding and rotational motions of the knee joint in the following angular ranges of motion of −38° and 30° and 45° and 90°, and the messages of "Walking" and "Running" from different folding motions in the angular range of 30° and 45° (top-right). Changes in resistance from toe-out gait and normal gait motions (bottom). Source: [284] / with permission of John Wiley & Sons.

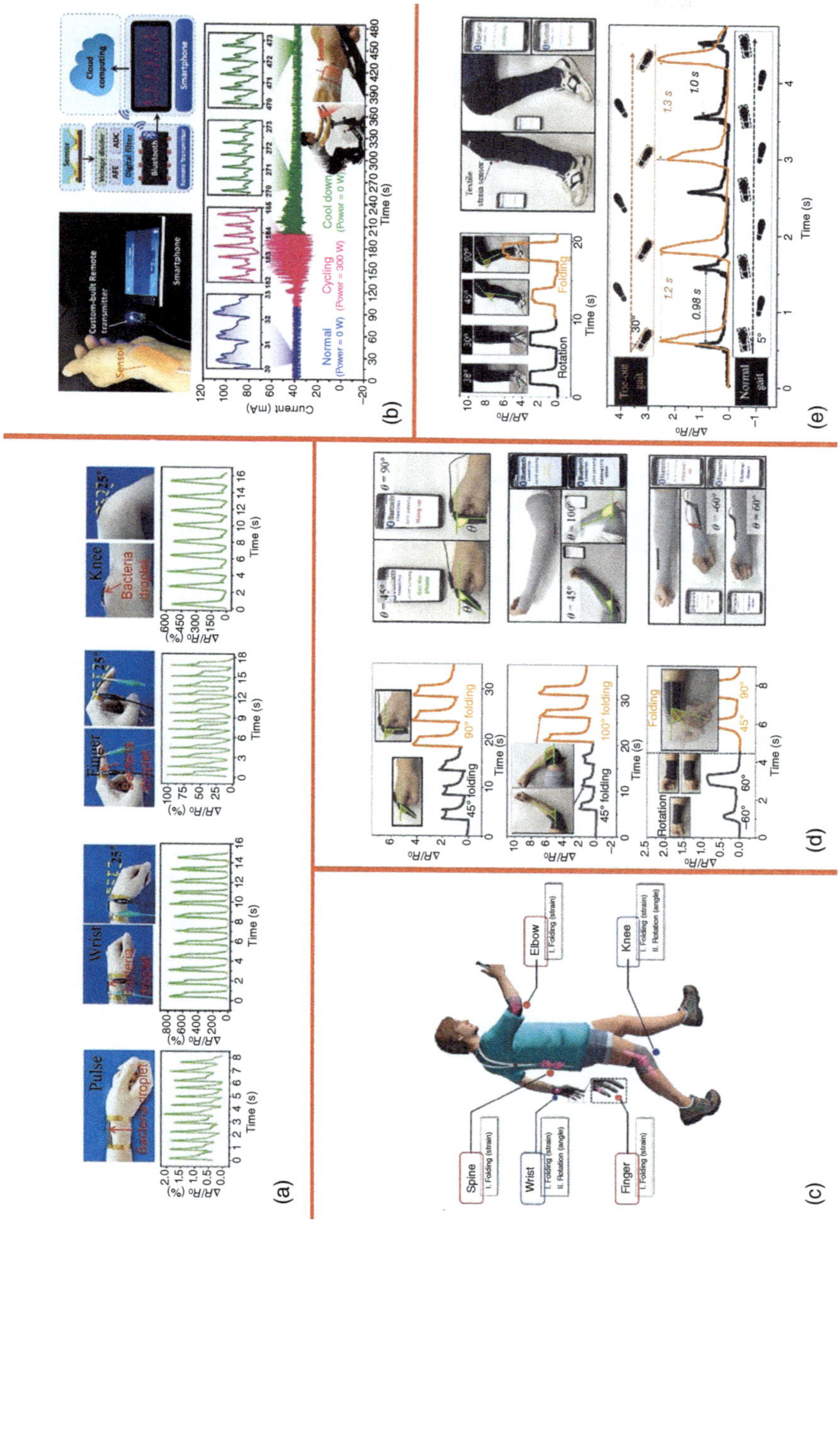
(a)
Pulse
Wrist
Finger
Knee
Bacteria droplet
ΔR/R₀ (%)
Time (s)
(b)
Custom-built Remote transmitter
Sensor
Smartphone
Cloud computing
Bluetooth
Current (mA)
Normal (Power = 0 W)
Cycling (Power = 300 W)
Cool down (Power = 0 W)
Time (s)
(c)
Spine
I. Folding (strain)
Wrist
I. Folding (strain)
II. Rotation (angle)
Finger
I. Folding (strain)
Elbow
I. Folding (strain)
Knee
I. Folding (strain)
II. Rotation (angle)
(d)
45° folding
90° folding
100° folding
Rotation
Folding
ΔR/R₀
Time (s)
(e)
Textile strain sensor
Rotation
Folding
Toe-off gait
Normal gait
1.2 s
1.3 s
0.98 s
1.0 s
ΔR/R₀
Time (s)

using NaCl-doped agarose gel as a biocompatible filler and a 3D holding elastomer matrix [268]. Different arm bending states, curl motions, and arm exercises were detected to investigate real-life applications of the sensor.

In addition to large-scale movements, subtle motions such as breathing, voice vibrations and phonation, respiration, and pulse detections have widely been explored for monitoring human motions and daily activities [269–274]. On this basis, a patterned graphene strain sensor was fabricated by Lee et al. to monitor subtle body motions [275]. A wearable sensor was attached to the neck to monitor throat motions during speech. Furthermore, strain sensors were utilized to take the pulse from the wrist, confirming high sensitivity in detecting subtle motions. Facial expressions, including frowning, laughing, crying, surprising, fearing, blinking, and eye movements, can be detected by wearable strain/pressure sensors [40, 271, 273, 276, 277]. As such, Ag NP–based curve arrays on a PDMS substrate were used for multi-recognition facial expressions, including eyeball movements at four orientations, relaxation, anger, disgust, fear, laughter, sadness, smile, and surprise [278]. Similarly, braided composites of yarn@polydopamine@PPy were employed as wearable strain sensors to monitor smiling, surprising, and frowning expressions in addition to large-scale and subtle motions [279]. A large workable strain range (up to 105%), high sensitivity with a GF of 51.2 in 0–40% strain, and long-term stability were achieved for the wearable sensor.

A wearable sensor can also be mounted on body joints such as the knee to recognize walking, running, jumping, and squatting to check the well-being and fitness sports performance of athletes or others during sports activities [265, 271, 280–282]. Using a graphene pressure sensor, He et al. reported a wearable pulse monitoring system with high sensitivity of 1875.53 kPa^{-1} and large linearity of 0–40 kPa to detect human pulse signals during cycling, walking, and running, according to Figure 6.4b [283]. Articular types of motions of body joints, including the spine, wrist, finger, elbow, and knee, are also illustrated in Figure 6.4c [284].

Attaching strain/pressure sensors to textiles has been explored due to the motion limitation of other skin-mountable sensors and the repeatability and shareability of clothing-integrated sensors [17]. Flexible strain/pressure sensors were assimilated with textiles and shoes to record real-time sport or daily activity data [34, 273, 279]. The textile substrate selection is vital because of the forces applied to textile-based strain sensors by human body motions. Park et al. devised an all-textile-based, highly stretchable structure capable of sensing, folding, and rotation of human finger, wrist, elbow, spine, and knee on textile substrates, including braided elastic fabric, knitted and woven fabrics (Figure 6.4d, e) [284]. Similarly, Jeerapan et al. fabricated a self-powered sensor on commercial soft substrates such as socks, underwear, and textile straps using screen printing [285]. The wearable textile-based pressure sensor can monitor the foot pressure of athletes or diabetic patients. It can be integrated with mobility sport tracking to collect data on vital signs via wireless communications.

6.3.2 Health Monitoring

Strain/pressure sensors are mounted on the skin or embedded/integrated with textiles for real-time health monitoring of body motions, heart rate, respiration rate, blood pressure, and pulse oxygenation [27, 44, 286]. Because of the increasing population, difficulties/complexities, and high cost of hospital-centered healthcare and medical systems, personalized long-distance health monitoring is desirable to improve early diagnosis and treatments and long-term monitoring of patients [27, 262]. Parkinson's disease, epilepsy, stroke, and arthritis can cause abnormal body motions and affect activities of daily living and working styles [262]. A wearable pressure sensor has been demonstrated to monitor the cervical region to identify cervical postures to tackle the issue [287]. The high-stretchable capacitive sensor based on GaIn liquid metal electrodes indicated high sensitivity of 39% kPa^{-1}. Body movement monitoring, including fingers, wrist and knee joint bending, voice vibrations, tissue swelling, and fitness, were explored to evaluate body functions [61, 184, 288–291].

Heart rate or pulse is the frequency of cardiac cycles, which changes with a person's physical and mental state [291, 292]. Heart rate is usually detected through radial, carotid, and tibial artery pulses to monitor rates at the wrist, neck, and ankle-brachial [58, 61, 148, 225, 293–295]. Gao et al. fabricated a high-sensitive tactile diaphragm pressure sensor based on microchannels capable of real-time pulse monitoring [296]. A PDMS wrist band with an embedded pressure sensor was located at the position of the radial artery during exercise to test the sensor performance. Similarly, a flexible and wearable wrist band based on MEMS elements array was developed to address the need for a low-cost and easy-to-use wearable cardiovascular health-monitoring device [297]. The capacitive wrist band device produces a high-quality signal with healthy subjects and a good ability to detect heartbeats. Furthermore, a clustering analysis indicated an ideal classification accuracy of sinus rhythm and arterial fibrillation patients. On the other hand, a piezoresistive pressure sensor based on rGO was designed to monitor the heart rate through radial artery pulse and carotid artery pulse detection attached to a bandage (Figure 6.5a) [298]. The E-skin showed a sensitivity of 0.82 kPa^{-1} in the low-pressure region (0–0.6 kPa), a short response time of 24 ms, and a detection limit of 7 Pa.

Respiration rate is another critical vital sign, and its abnormality may lead to sleep apnea, asthma, and anemia [299–301]. Wearable strain/pressure sensors were placed on the abdomen or ribcage, or response the airflow and breath were investigated to monitor blood flow pulses [262, 291, 302]. The respiratory activities were monitored by attaching a flexible self-powered piezoelectric sensor based on PZT thin films onto a conventional mask, as illustrated in Figure 6.5b, c [303]. The sensor distinguished deep (red) and normal (blue) respiration modes with consistent output voltage. Additionally, radial/carotid artery pulse and trachea movements were detected by the pressure sensor to detect the early onset of cardiovascular disease. Similarly, both respiration rate and volume were measured by Chu et al. utilizing a wearable piezoresistive sensor [304]. The respiration sensor could measure the local strain of the ribcage and abdomen during breathing; it can be integrated with wireless Bluetooth units to monitor patients with chronic diseases in everyday settings.

Blood pressure is another important human physiological parameter widely monitored with wearable strain sensors [305–309]. For instance, a wearable piezoresistive pressure sensor fabricated by Fan et al. showed the capability for continuous tracking

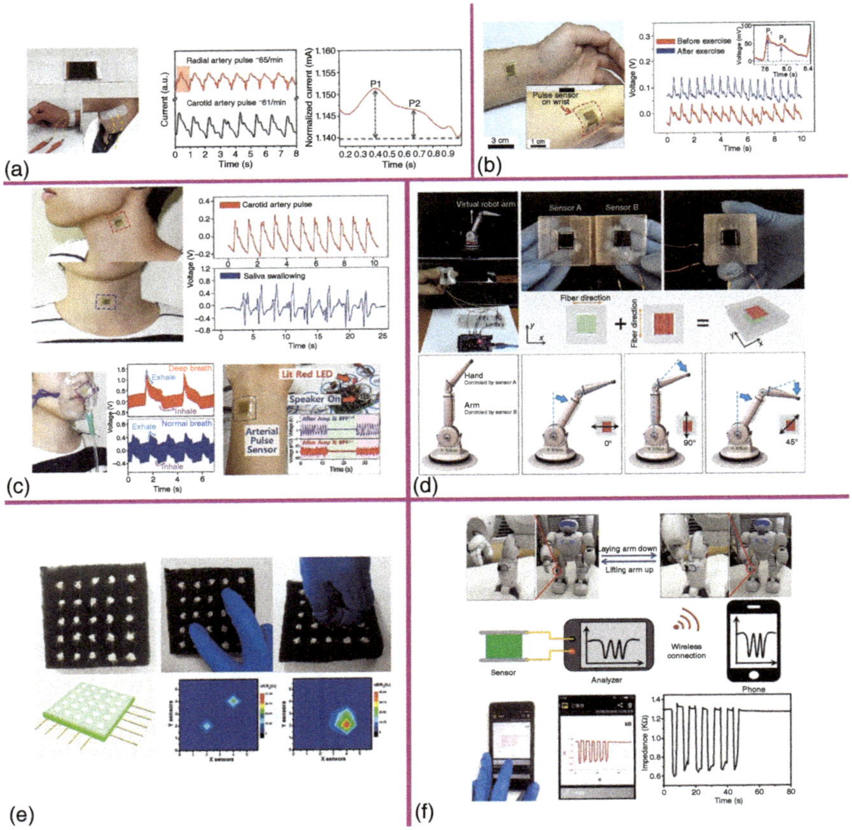

of both diastolic and systolic blood pressure with a mean absolute difference of less than 3 mmHg [305]. The liquid-capsule-based sensor can be used for physiological monitoring and the diagnosis and prognosis of cardiovascular diseases as wearable health-monitoring devices. In another work, a self-powered pressure sensor was introduced with the capability to monitor blood pressure in a wide pressure range in addition to radial or carotid artery pulses [306].

6.3.3 Human–machine Interface, Soft Robotics, and Artificial Skin

Wearable strain/pressure sensors are utilized in soft robotics and human–machine interfaces to sense the environment and surrounding objects. Human body movements, especially finger, hand, and arm motions, can actuate and control smart robots to perform different tasks. They cause changes in electrical signals by movement, bending, or touching [62, 293, 310–314]. Chen et al. fabricated a carbonized crepe paper strain sensor with high flexibility and fast response time for human–machine interfacing, controlling multiaxial robot manipulators as an input device [315]. Figure 6.5d shows the integrated sensor connection to a virtual simulator through a microcontroller. The integrated sensor on the hand or

Figure 6.5 (a) Detection of radial and carotid artery pulse using the assembled adhesive bandage (left). The radial artery pulse and carotid artery pulse signals (middle) and a single peak magnification of radial artery pulse. Source: [298] / with permission of John Wiley & Sons. (b) The photograph of a piezoelectric pulse sensor attached to the wrist by the biocompatible liquid bandage. The inset shows the deformed sensor consistent with the movement of the blood vessel (left). The signals of radial artery pulse monitored using the self-powered pulse sensor, exhibiting different heart rates and output voltages before and after exercise. The inset shows the magnified voltage signal of the radial artery pulse before physical exercise (right). Source: [303] / with permission of John Wiley & Sons. (c) Photograph of the pulse sensor attached to a carotid artery position (top-left) and the middle of the throat (middle-left) with the output voltage in response to carotid arterial pressure (top-right) and saliva swallowing actions (middle-left). Medical mask photographs integrated with the flexible pressure sensor for monitoring human respirations (bottom-left) and generated output voltage response of the sensor to deep and normal breathing (bottom-middle). Photograph of the LED and speaker unit operating synchronously, corresponding to the radial artery pulse. The inset shows the generated output voltage from the first (red) and second (blue) amplifier stages (bottom-right). Source: [303] / with permission of John Wiley & Sons. (d) The photograph of the integrated carbonized crepe paper strain sensor, controlling virtual robot arm by microcontroller (top), schematic illustration of the integration of two carbonized crepe paper strain sensors (middle-right), and the virtual robot images including hand and arm controlled by the integrated strain sensor by different angles of tensile bending (bottom). Source: [315] / with permission of John Wiley & Sons. (e) Photograph of the sponge-based E-skin assembled (top-left) and the scheme of the E-skin for touch sensing (bottom-left), the fingers touching photographs (top-middle), and the resistance change of the corresponding sensing map (bottom-middle) of the E-skin. The joint of the index finger pressing onto the E-skin (top-right) and the resistance change of the corresponding sensing map (bottom-right). Source: [321] / with permission of John Wiley & Sons. (f) The photograph of the sensor attached to the femorotibial joint of a robot with the repeated arm lifting along with the hand touching the wearable sensor (top) for remote monitoring by wireless communications (middle). The wirelessly received sensing performance for monitoring the reversible lifting of the arm (bottom). Source: [321] / with permission of John Wiley & Sons.

arm controls the movement of the robot manipulator by bending or movement. Also, a wearable strain sensor detects to interpret the kinematics of deformation in human fingernails [316]. Simple motions, gestures, finger writing, and grip strength are identified or quantified as a human–computer interface. Another wearable strain sensor was applied as the human–machine interface controller for a robot to convert resistance changes into signals for robot activation [263]. Bending the wrist also activates the robot movements as other experiments of strain sensors, including graphene platelets and silicone rubber. E-skins and prostheses have mimicked the skin's ability to accommodate body movements during bending and stretching [317]. A number of researchers have recently developed artificial E-skins with wearable sensor systems by mimicking the function of mechanoreceptors in the skin [56, 87, 180, 227, 318–320]. For instance, a wearable sensor was assembled as a smart artificial E-skin to simultaneously map the pressure distribution, shape, and location of touching sensing (Figure 6.5e) [321]. A hierarchically microstructured PANI nanohair/rGO@PU-based sensor bioinspired by the crack-shaped mechanosensor of spiders and the wing-locking sensing systems of beetles exhibited high sensitivity and excellent reproducibility. Moreover, the resistance variations for each pixel under the corresponding strain were collected for plotting the color contrast map, and the specific locations plotted, according to Figure 6.5e, were in good arrangement with the locations of the fingers. The local pressure distribution, shape, and location were recognized when the joint of the index finger pressed the E-skin (Figure 6.5f). Similarly, a flexible and degradable pressure sensor based on MXene, tissue paper, and PLA acted as an E-skin for mapping tactile stimuli [322]. The smart fabricated E-skin can recognize the contact points touched by fingers by collecting the sensing response of each pixel and plotting the pressure distribution and location in a response map.

Smart glove systems, an example of skin-mountable and wearable sensors, have numerous applications to control robots, performing surgical procedures and delicate works that are not accessible to humans [262, 323–325]. A flexible pneumatic actuator was integrated with a stretchable strain sensor embedded in a hand glove to analyze finger kinematics, stiffness, or dexterity [326]. Furthermore, a $CaCu_3Ti_4O_{12}$@PU hybrid sponge-based sensor for robotic tactile sensing was embedded into a glove to create an intelligent tactile glove [327]. The capacitive sensor was attached to the thumb of the glove, showing the capability to distinguish between the grasping of different weights.

Multimodal sensing capability is essential to mimic the human skin or construct a prosthetic and artificial skin. Bimodal and multimodal wearable strain/pressure sensors have been reported for human–machine interactions and soft robotic applications [314, 328]. Large sensing area, complicated device design, fabrication process, and developed interconnections are requisites for multimodal interactivity [329, 330]. For instance, a transparent, stretchable touch-pressure sensor capable of detecting and distinguishing touch and pressure input was presented by Huang et al. as a multimodal input device for future wearable electronics [328]. The capacitive-type bimodal device consisting of an upper piezoresistive PEDOT:PSS electrode was mounted on a forearm for human–machine interfacing by wireless controlling of movements of a small toy car. Similarly, Baumgartner et al. devised a resilient gelatin-based biogel derived entirely from natural and food constituents for sustainable soft robot units of multimodal sensors (Figure 6.6a) [331]. A set of durable, biodegradable soft actuators and autonomous electronic platforms with multimodal temperature, humidity, and strain sensing capabilities were created (Figure 6.6b). The E-skin monitored the deformation of the skin (Figure 6.6c, d), a load on a specific sensor, and the objects with complex shapes (Figure 6.6e, f).

TENGs have been recently developed to solve size and/or power problems in flexible motion transducers for wearable human–machine interfaces [332, 333]. A self-powered pressure sensor with high sensitivity of $0.101\,kPa^{-1}$ and a fast response was designed to detect flexion/extension of human fingers for the manipulation of a prosthetic robot finger and monitoring of wind speed and direction [333]. On the other hand, a new field of triboelectric sensors is introduced in which FETs are coupled for robotics and human–machine interactive systems [244, 334]. The reported performances of wearable strain/pressure sensors are summarized in Table 6.1 [32, 278, 335–352].

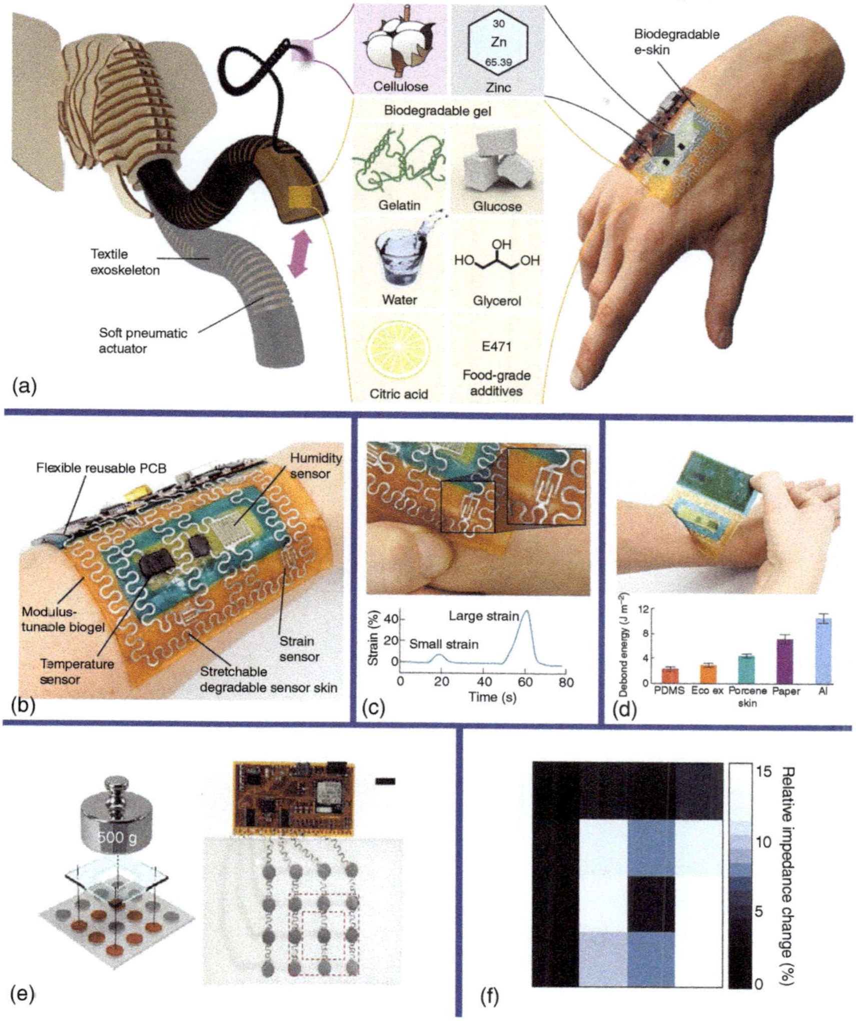

Figure 6.6 (a) Naturally derived ingredients, such as gelatin, citric acid, cellulose fibers, and zinc, together with soft and durable pneumatic actuators for the elastic and degradable biogel as multifunctional E-skins. (b) The soft E-skin on a human arm. (c) Mechanical deformation of the strain sensors (top) and output recordings for small and large strain levels (bottom). (d) The electronic patch peeling-off (top). The debond energy was measured for the biogel on different substrates (bottom). Scale bar, 3 cm. Error bars, s.d. for $n > 3$ measurements. (e) Photograph of a 500 g weight distributed to a part of a pressure sensor array. Scale bar, 1 cm. (f) Signals obtained by the pressure sensor patch using PDMS. Source: [331] / with permission of Springer Nature.

Table 6.1 Summary of the reported performances of representative wearable strain/pressure sensors.

Sensing mechanism	Materials	Sensitivity/ GF (kPa^{-1})	Sensing range (kPa)	Stretch-ability	Application	Ref.
Piezoresistive	MXene/ P(VDF-TrFE)	817.4 kPa^{-1} 2213.68 kPa^{-1}	(0.072–0.74) (0.74–3.083)	–	Monitoring speech recognition, muscle movements, and real-time pulses	[335]
Piezoelectric	Ag-3D textile-PVDF	0.006 $VkPa^{-1}$	0.05–100	–	Monitoring bio-signals from voice printing, respiration and blood pulse and body movement based on pressure of fingertip and gait, movement of neck, and joint	[336]
Piezoresistive	PEDOT : PSS/Au-PDMS@PET	35.8	0.42–1.08%	–	Monitoring subtle movements of fingers, wrist, and throat swallowing	[337]
Piezoresistive	PVA/phytic acid/amino-polyhedral oligomeric silsesquioxane	3.44	0–125%	363%	Monitoring bending of finger and wrist, clenching	[338]
Piezoresistive	PU@polymethacrylate@GaIn composite fiber	1.5	0–500%	500%	Monitoring bending of elbow, finger, wrist, knee	[339]
Piezoresistive	PVA/glycerol/NaCl ionic hydrogel	4.01	100–300%	570.7%	Finger, wrist, elbow, and knee joints motions monitoring	[340]
Piezoresistive	Au NP densely packed micro-/nanowires	0.048 kPa^{-1}	1.5 kPa	–	Monitoring artery pulse, movements of hand, finger, wrist, throat and voice vibrations	[341]
Piezoresistive	CB-PDMS/ collagen fiber/leather	21.49	0–2.22%	37.5%	Finger and wrist motions monitoring	[342]
Piezoresistive	rGO@PET/PDMS	−1.7 in x-direction −26 in y-direction	15% 8%	16% 65%	Monitoring both subtle and large human motions	[343]
Capacitive	parylene/Au/PDMS	3.05	0–140%	140%	Monitoring finger bending	[344]

Piezoresistive	Ag/CNT-P3HT/PU/PDMS	6.29 3.27	135% 150%	160%	Monitoring human motions (finger bending)	[278]
Piezoresistive	PU/PET fabric	4.13	0–30%	60%	Monitoring human physiological signals (wrist pulse, ECG) and body motions and speech vibrations	[345]
Piezoresistive	Ag/PU fibers	9.3×10^5 659	450% 200%	3000% and 900 for 1 and 3 reduction	Monitoring bladder volume and finger bending	[32]
Piezoresistive	CNT/Ecoflex	42 300	125–145%	≥145%	Human motion detection including finger and arm bending, swallowing and blood pulses monitoring	[346]
Piezoresistive	Annealed PAN nanofiber/graphene	$44.5\,kPa^{-1}$	1.2 kPa	—	Acoustic vibrations and pulse wave detection	[347]
Piezoresistive	Au/PDMS	$23\,kPa^{-1}$ $0.7\,kPa^{-1}$	(<600 Pa) (>600 Pa)	—	Artery pulse detection	[348]
Capacitive	PEDOT : PSS/PDMS (electrode) MXene/PVDF-TrFE (dielectric)	$0.51\,kPa^{-1}$	0–400 kPa	≈60%	Monitoring physiological signals (pulse rate, respiration, muscle movements, and eye twitching) for human–machine interface applications	[349]
Piezoresistive	Carbon ink/Ecoflex	3.8 ± 0.6	100%	700–800%	Human–machine interface, soft robotics	[350]
Triboelectric	Silk/Ag NW/PDMS	30	5%	15%	human–machine interface	[351]
Capacitive	Ag NW/PDMS (electrode) micro cilia array (dielectric)	$0.28\,kPa^{-1}$ $0.02\,kPa^{-1}$	0–10 kPa 50–200 kPa	–	Pulse sensing, voice recognition, gas-flow monitoring, and loading mapping, and large motions for health control, body motion feedback, and human–machine interfaces	[352]

6.4 Conclusion and Outlook

Recent investigation and advances in pressure/strain sensors were summarized based on nanomaterials and nanostructures were summarized, and their applications were elaborated in wearable devices and flexible electronics. Various nanomaterials, including carbon-based and metal-based nanostructures, and CPs were used to fabricate wearable and stretchable sensors. High sensitivity and conductivity, good linearity, wide sensing range, long-term stability and durability, and remarkable stretchability are the most repeated and important parameters for wearable strain/pressure sensors. Different types of strain/pressure sensors based on their sensing mechanisms were also studied and discussed. Different conductive nanomaterials and stretchable components with different architectures, microstructured patterns, and porosity were employed so far to develop wearable piezoresistive, capacitive, and piezoelectric sensors. A number of strategies used by researchers were deeply investigated to improve stretchability and sensitivity. Although resistive sensors demonstrate nonlinear response and large hysteresis, capacitive sensors employ small modulus stretchable elastomers exhibiting lower hysteresis and sensitivity. Piezoelectric and triboelectric sensors were also discussed in detail, along with their advantages and disadvantages. Recent developments in wearable strain/pressure sensor applications, including movement monitoring and daily performance, health monitoring, human–machine interface, and soft robotics, were also considered.

Despite the advances in the field of sensors, some issues should be addressed. Solutions and suggestions need to be considered for enhancing the performance and commercialization of wearable strain/pressure sensors. Simplicity and cost-effectiveness of manufacturing methods and materials, as well as packaging can increase wearable device availability and mass production. Biocompatible, environmentally friendly, and skin-friendly materials, especially for health-care and skin-mountable devices, are necessary for next-generation wearable sensors. Roll-to-roll processing and 3D printing methods should be explored, and more attention to manufacturing approaches. New materials and composites (nanomaterials, elastic polymers, and substrates) accompanied by new structures/architectures need to be created for development and replacement to improve the performance, comfort, and safety of wearable sensors. Moreover, multimodal and multifunctional sensors with more complicated designs are required to mimic human skin functions for advanced robotic and prosthetic applications. E-skins should have the ability to be easily and conformally attached to the human body to activate/control soft robotics or monitor human physiological health parameters or electrophysiological activities.

On the other hand, integrating wearable functional components into wearable strain/pressure sensors can improve the functional area and scope. Storage, transmission, and analysis systems with long-term sensing can enhance healthcare quality and treatments to a higher level. Furthermore, self-powered sensing devices have been developed to provide long-term continuous outputs for health monitoring and E-skins. Self-healable and self-cleaning protection coatings can also improve durability and fulfil the long-term monitoring demands. Last but not least, the side effects of electromagnetic waves from wireless systems should be considered in skin-mountable and prosthetic devices, which are supposed to be in constant contact with the body.

References

1 Wong, W.S., and Salleo, A. (Eds.). (2009). *Flexible electronics: materials and applications* (Vol. 11). Springer Science & Business Media.

2 Zuber, K., Fabretto, M., Hall, C., and Murphy, P. (2008). Improved PEDOT conductivity via suppression of crystallite formation in Fe (III) tosylate during vapor phase polymerization. *Macromol. Rapid Commun.* 29 (18): 1503–1508.

3 McCann, J. (2009). The garment design process for smart clothing: from fibre selection through to product launch. In Smart clothes and wearable technology (pp. 70–94). Woodhead Publishing.

4 Seymour, S. (2008). *Fashionable Technology: The Intersection of Design, Fashion, Science, and Technology*. Springer.

5 Seymour, S. (2010). *Functional Aesthetics: Visions in Fashionable Technology*. Springer.

6 Zeng, W., Shu, L., Li, Q. et al. (2014). Fiber-based wearable electronics: a review of materials, fabrication, devices, and applications. *Adv. Mater.* 26 (31): 5310–5336.

7 Wang, X., Gu, Y., Xiong, Z. et al. (2014). Electronic skin: silk-molded flexible, ultrasensitive, and highly stable electronic skin for monitoring human physiological signals. *Adv. Mater.* 26 (9): 1309–1309.

8 Wilson, J., Walker, B.N., Lindsay, J. et al. (2007). *Swan: System for Wearable Audio Navigation*, 91–98. 2007 11th IEEE international symposium on wearable computers, 2007; IEEE.

9 Honda, W., Harada, S., Arie, T. et al. (2014). Wearable, human-interactive, health-monitoring, wireless devices fabricated by macroscale printing techniques. *Adv. Funct. Mater.* 24 (22): 3299–3304.

10 Briseno, A.L., Tseng, R.J., Ling, M.M. et al. (2006). High-performance organic single-crystal transistors on flexible substrates. *Adv. Mater.* 18 (17): 2320–2324.

11 Mannsfeld, S.C., Tee, B.C., Stoltenberg, R.M. et al. (2010). Highly sensitive flexible pressure sensors with microstructured rubber dielectric layers. *Nat. Mater.* 9 (10): 859–864.

12 Kim, S., Kwon, H.J., Lee, S. et al. (2011). Low-power flexible organic light-emitting diode display device. *Adv. Mater.* 23 (31): 3511–3516.

13 Takei, K., Takahashi, T., Ho, J.C. et al. (2010). Nanowire active-matrix circuitry for low-voltage macroscale artificial skin. *Nat. Mater.* 9 (10): 821–826.

14 Wu, H., Huang, Y., Xu, F. et al. (2016). Energy harvesters for wearable and stretchable electronics: from flexibility to stretchability. *Adv. Mater.* 28 (45): 9881–9919.

15 Sazonov, E. (2014). *Wearable Sensors: Fundamentals, Implementation and Applications*. Elsevier.

16 Dubal, D.P., Chodankar, N.R., Kim, D.-H., and Gomez-Romero, P. (2018). Towards flexible solid-state supercapacitors for smart and wearable electronics. *Chem. Soc. Rev.* 47 (6): 2065–2129.

17 Jayathilaka, W.A.D.M., Qi, K., Qin, Y. et al. (2019). Significance of nanomaterials in wearables: a review on wearable actuators and sensors. *Adv. Mater.* 31 (7): 1805921.

18 Patel, S., Park, H., Bonato, P. et al. (2012). A review of wearable sensors and systems with application in rehabilitation. *J. Neuroeng. Rehabil.* 9 (1): 21.

19 Sedighi, A., Montazer, M., and Mazinani, S. (2019). Synthesis of wearable and flexible NiP0. 1-SnOx/PANI/CuO/cotton towards a non-enzymatic glucose sensor. *Biosens. Bioelectron.* 135: 192–199.

20 Sedighi, A., Montazer, M., and Samadi, N. (2014). Synthesis of nano Cu_2O on cotton: morphological, physical, biological and optical sensing characterizations. *Carbohydr. Polym.* 110: 489–498.

21 Wu, W. (2017). Inorganic nanomaterials for printed electronics: a review. *Nanoscale* 9 (22): 7342–7372.

22 Wang, X., Liu, Z., and Zhang, T. (2017). Flexible sensing electronics for wearable/attachable health monitoring. *Small* 13 (25): 1602790.

23 Gao, W., Emaminejad, S., Nyein, H.Y.Y. et al. (2016). Fully integrated wearable sensor arrays for multiplexed in situ perspiration analysis. *Nature* 529 (7587): 509–514.

24 Lou, Z., Li, L., Wang, L., and Shen, G. (2017). Recent progress of self-powered sensing systems for wearable electronics. *Small* 13 (45): 1701791.

25 Wang, C., Xia, K., Wang, H. et al. (2019). Advanced carbon for flexible and wearable electronics. *Adv. Mater.* 31 (9): 1801072.

26 Wagner, S. and Bauer, S. (2012). Materials for stretchable electronics. *MRS Bull.* 37 (3): 207–213.

27 Yao, S., Swetha, P., and Zhu, Y. (2018). Nanomaterial-enabled wearable sensors for healthcare. *Adv. healthc. mater.* 7 (1): 1700889.

28 Wang, Z., Zhou, H., Chen, W. et al. (2018). Dually synergetic network hydrogels with integrated mechanical stretchability, thermal responsiveness, and electrical conductivity for strain sensors and temperature alertors. *ACS Appl. Mater. Interfaces* 10 (16): 14045–14054.

29 Wang, Y., Hao, J., Huang, Z. et al. (2018). Flexible electrically resistive-type strain sensors based on reduced graphene oxide-decorated electrospun polymer fibrous mats for human motion monitoring. *Carbon* 126: 360–371.

30 Heo, Y., Hwang, Y., Jung, H.S. et al. (2017). Secondary sensitivity control of silver-nanowire-based resistive-type strain sensors by geometric modulation of the elastomer substrate. *Small* 13 (23): 1700070.

31 Zhong, J., Zhong, Q., Hu, Q. et al. (2015). Stretchable self-powered fiber-based strain sensor. *Adv. Funct. Mater.* 25 (12): 1798–1803.

32 Lee, J., Shin, S., Lee, S. et al. (2018). Highly sensitive multifilament fiber strain sensors with ultrabroad sensing range for textile electronics. *ACS Nano* 12 (5): 4259–4268.

33 Liu, Q., Chen, J., Li, Y., and Shi, G. (2016). High-performance strain sensors with fish-scale-like graphene-sensing layers for full-range detection of human motions. *ACS Nano* 10 (8): 7901–7906.

34 Park, J.J., Hyun, W.J., Mun, S.C. et al. (2015). Highly stretchable and wearable graphene strain sensors with controllable sensitivity for human motion monitoring. *ACS Appl. Mater. Interfaces* 7 (11): 6317–6324.

35 Ding, Y., Yang, J., Tolle, C.R., and Zhu, Z. (2018). Flexible and compressible PEDOT: PSS@ Melamine conductive sponge prepared via one-step dip coating as piezoresistive pressure sensor for human motion detection. *ACS Appl. Mater. Interfaces* 10 (18): 16077–16086.

36 Trung, T.Q. and Lee, N.E. (2016). Flexible and stretchable physical sensor integrated platforms for wearable human-activity monitoring and personal healthcare. *Adv. Mater.* 28 (22): 4338–4372.

37 Matsuhisa, N., Chen, X., Bao, Z., and Someya, T. (2019). Materials and structural designs of stretchable conductors. *Chem. Soc. Rev.* 48 (11): 2946–2966.

38 Liu, Z., Qi, D., Guo, P. et al. (2015). Thickness-gradient films for high gauge factor stretchable strain sensors. *Adv. Mater.* 27 (40): 6230–6237.

39 Trung, T.Q. and Lee, N.E. (2017). Recent progress on stretchable electronic devices with intrinsically stretchable components. *Adv. Mater.* 29 (3): 1603167.

40 Roh, E., Hwang, B.-U., Kim, D. et al. (2015). Stretchable, transparent, ultrasensitive, and patchable strain sensor for human–machine interfaces comprising a nanohybrid of carbon nanotubes and conductive elastomers. *ACS Nano* 9 (6): 6252–6261.

41 Tang, Y., Zhao, Z., Hu, H. et al. (2015). Highly stretchable and ultrasensitive strain sensor based on reduced graphene oxide microtubes–elastomer composite. *ACS Appl. Mater. Interfaces* 7 (49): 27432–27439.

42 Yin, G., Hu, N., Karube, Y. et al. (2011). A carbon nanotube/polymer strain sensor with linear and anti-symmetric piezoresistivity. *J. Compos. Mater.* 45 (12): 1315–1323.

43 Zhu, J., Fox, J.J., Yi, N., and Cheng, H. (2019). Structural design for stretchable microstrip antennas. *ACS Appl. Mater. Interfaces* 11 (9): 8867–8877.

44 Dinh, T., Nguyen, T., Phan, H.P. et al. (2020). Advances in rational design and materials of high-performance stretchable electromechanical sensors. *Small* 16 (14): 1905707.

45 Weng, W., Chen, P., He, S. et al. (2016). Smart electronic textiles. *Angew. Chem. Int. Ed.* 55 (21): 6140–6169.

46 Tee, B.C., Wang, C., Allen, R., and Bao, Z. (2012). An electrically and mechanically self-healing composite with pressure-and flexion-sensitive properties for electronic skin applications. *Nat. Nanotechnol.* 7 (12): 825.

47 Park, J., Lee, Y., Hong, J. et al. (2014). Giant tunneling piezoresistance of composite elastomers with interlocked microdome arrays for ultrasensitive and multimodal electronic skins. *ACS Nano* 8 (5): 4689–4697.

48 Lee, J., Kwon, H., Seo, J. et al. (2015). Conductive fiber-based ultrasensitive textile pressure sensor for wearable electronics. *Adv. Mater.* 27 (15): 2433–2439.

49 Pang, C., Koo, J.H., Nguyen, A. et al. (2015). Highly skin-conformal microhairy sensor for pulse signal amplification. *Adv. Mater.* 27 (4): 634–640.

50 Wang, X., Gu, Y., Xiong, Z. et al. (2014). Silk-molded flexible, ultrasensitive, and highly stable electronic skin for monitoring human physiological signals. *Adv. Mater.* 26 (9): 1336–1342.

51 Li, X., Zhang, R., Yu, W. et al. (2012). Stretchable and highly sensitive graphene-on-polymer strain sensors. *Sci. Rep.* 2: 870.

52 Hou, C., Wang, H., Zhang, Q. et al. (2014). Highly conductive, flexible, and compressible all-graphene passive electronic skin for sensing human touch. *Adv. Mater.* 26 (29): 5018–5024.

53 Yao, H.B., Ge, J., Wang, C.F. et al. (2013). A flexible and highly pressure-sensitive graphene–polyurethane sponge based on fractured microstructure design. *Adv. Mater.* 25 (46): 6692–6698.

54 Zhu, B., Niu, Z., Wang, H. et al. (2014). Microstructured graphene arrays for highly sensitive flexible tactile sensors. *Small* 10 (18): 3625–3631.

55 Jung, S., Kim, J.H., Kim, J. et al. (2014). Reverse-micelle-induced porous pressure-sensitive rubber for wearable human–machine interfaces. *Adv. Mater.* 26 (28): 4825–4830.

56 Park, J., Lee, Y., Hong, J. et al. (2014). Tactile-direction-sensitive and stretchable electronic skins based on human-skin-inspired interlocked microstructures. *ACS Nano* 8 (12): 12020–12029.

57 Choong, C.L., Shim, M.B., Lee, B.S. et al. (2014). Highly stretchable resistive pressure sensors using a conductive elastomeric composite on a micropyramid array. *Adv. Mater.* 26 (21): 3451–3458.

58 Gong, S., Schwalb, W., Wang, Y. et al. (2014). A wearable and highly sensitive pressure sensor with ultrathin gold nanowires. *Nat. Commun.* 5 (1): 1–8.

59 Bae, S.-H., Lee, Y., Sharma, B.K. et al. (2013). Graphene-based transparent strain sensor. *Carbon* 51: 236–242.

60 Zhang, F., Zang, Y., Huang, D. et al. (2015). Flexible and self-powered temperature–pressure dual-parameter sensors using microstructure-frame-supported organic thermoelectric materials. *Nat. Commun.* 6 (1): 1–10.

61 Gong, S., Lai, D.T., Su, B. et al. (2015). Highly stretchy black gold E-skin nanopatches as highly sensitive wearable biomedical sensors. *Adv. Electron. Mater.* 1 (4): 1400063.

62 Gong, S., Lai, D.T., Wang, Y. et al. (2015). Tattoolike polyaniline microparticle-doped gold nanowire patches as highly durable wearable sensors. *ACS Appl. Mater. Interfaces* 7 (35): 19700–19708.

63 Geim, A.K. and Novoselov, K.S. (2010). The rise of graphene. In: *Nanoscience and Technology: A Collection of Reviews from Nature Journals* (ed. P. Rodgers), 11–19. World Scientific.

64 Wei, W. and Qu, X. (2012). Extraordinary physical properties of functionalized graphene. *Small* 8 (14): 2138–2151.

65 Pang, S., Hernandez, Y., Feng, X., and Müllen, K. (2011). Graphene as transparent electrode material for organic electronics. *Adv. Mater.* 23 (25): 2779–2795.

66 Bae, S., Kim, H., Lee, Y. et al. (2010). Roll-to-roll production of 30-inch graphene films for transparent electrodes. *Nat. Nanotechnol.* 5 (8): 574.

67 Jian, M., Xia, K., Wang, Q. et al. (2017). Flexible and highly sensitive pressure sensors based on bionic hierarchical structures. *Adv. Funct. Mater.* 27 (9): 1606066.

68 Yamada, T., Hayamizu, Y., Yamamoto, Y. et al. (2011). A stretchable carbon nanotube strain sensor for human-motion detection. *Nat. Nanotechnol.* 6 (5): 296.

69 Shi, J., Li, X., Cheng, H. et al. (2016). Graphene reinforced carbon nanotube networks for wearable strain sensors. *Adv. Funct. Mater.* 26 (13): 2078–2084.

70 Zhao, H., Zhang, Y., Bradford, P.D. et al. (2010). Carbon nanotube yarn strain sensors. *Nanotechnology* 21 (30): 305502.

71 Liu, M., Pu, X., Jiang, C. et al. (2017). Large-area all-textile pressure sensors for monitoring human motion and physiological signals. *Adv. Mater.* 29 (41): 1703700.

72 Ryu, S.; Lee, P.; Chou, J. B.; Xu, R.; Zhao, R.; Hart, A. J.; Kim, S.-G., Fabrication of Extremely Elastic Wearable Strain Sensor Using Aligned Carbon Nanotube Fibers for Monitoring Human Motion.

73 Li, Y., Shang, Y., He, X. et al. (2013). Overtwisted, resolvable carbon nanotube yarn entanglement as strain sensors and rotational actuators. *ACS Nano* 7 (9): 8128–8135.

74 Wang, L., Jackman, J.A., Tan, E.-L. et al. (2017). High-performance, flexible electronic skin sensor incorporating natural microcapsule actuators. *Nano Energy* 36: 38–45.

75 Tadakaluru, S., Thongsuwan, W., and Singjai, P. (2014). Stretchable and flexible high-strain sensors made using carbon nanotubes and graphite films on natural rubber. *Sensors* 14 (1): 868–876.

76 Lee, K., Lee, S.S., Lee, J.A. et al. (2010). Carbon nanotube film piezoresistors embedded in polymer membranes. *Appl. Phys. Lett.* 96 (1): 013511.

77 Kang, T.J., Choi, A., Kim, D.-H. et al. (2010). Electromechanical properties of CNT-coated cotton yarn for electronic textile applications. *Smart Mater. Struct.* 20 (1): 015004.

78 Park, S.J., Kim, J., Chu, M., and Khine, M. (2016). Highly flexible wrinkled carbon nanotube thin film strain sensor to monitor human movement. *Adv. Mater. Technol.* 1 (5): 1600053.

79 Han, J.-W., Kim, B., Li, J., and Meyyappan, M. (2013). Flexible, compressible, hydrophobic, floatable, and conductive carbon nanotube-polymer sponge. *Appl. Phys. Lett.* 102 (5): 051903.

80 Slobodian, P., Riha, P., Benlikaya, R. et al. (2013). A flexible multifunctional sensor based on carbon nanotube/polyurethane composite. *IEEE Sensors J.* 13 (10): 4045–4048.

81 Tsui, M.N. and Islam, M.F. (2017). Creep-and fatigue-resistant, rapid piezoresistive responses of elastomeric graphene-coated carbon nanotube aerogels over a wide pressure range. *Nanoscale* 9 (3): 1128–1135.

82 Bae, G.Y., Pak, S.W., Kim, D. et al. (2016). Linearly and highly pressure-sensitive electronic skin based on a bioinspired hierarchical structural array. *Adv. Mater.* 28 (26): 5300–5306.

83 Boland, C.S., Khan, U., Backes, C. et al. (2014). Sensitive, high-strain, high-rate bodily motion sensors based on graphene–rubber composites. *ACS Nano* 8 (9): 8819–8830.

84 Yang, T., Jiang, X., Zhong, Y. et al. (2017). A wearable and highly sensitive graphene strain sensor for precise home-based pulse wave monitoring. *ACS sensors* 2 (7): 967–974.

85 Li, Y., Jiang, X., Liu, Z., and Liu, Z. (2010). Strain effects in graphene and graphene nanoribbons: the underlying mechanism. *Nano Res.* 3 (8): 545–556.

86 Wang, Y., Yang, R., Shi, Z. et al. (2011). Super-elastic graphene ripples for flexible strain sensors. *ACS Nano* 5 (5): 3645–3650.

87 Jeong, Y.R., Park, H., Jin, S.W. et al. (2015). Highly stretchable and sensitive strain sensors using fragmentized graphene foam. *Adv. Funct. Mater.* 25 (27): 4228–4236.

88 Zhao, J., Wang, G., Yang, R. et al. (2015). Tunable piezoresistivity of nanographene films for strain sensing. *ACS Nano* 9 (2): 1622–1629.

89 Samad, Y.A., Li, Y., Schiffer, A. et al. (2015). Graphene foam developed with a novel two-step technique for low and high strains and pressure-sensing applications. *Small* 11 (20): 2380–2385.

90 Li, J., Zhao, S., Zeng, X. et al. (2016). Highly stretchable and sensitive strain sensor based on facilely prepared three-dimensional graphene foam composite. *ACS Appl. Mater. Interfaces* 8 (29): 18954–18961.

91 Xia, K., Wang, C., Jian, M. et al. (2018). CVD growth of fingerprint-like patterned 3D graphene film for an ultrasensitive pressure sensor. *Nano Res.* 11 (2): 1124–1134.

92 Fu, X.-W., Liao, Z.-M., Zhou, J.-X. et al. (2011). Strain dependent resistance in chemical vapor deposition grown graphene. *Appl. Phys. Lett.* 99 (21): 213107.

93 Sedighi, A., Taheri, R.A., and Montazer, M. (2020). High-performance electromagnetic interference shielding electrodes/substrates for wearable electronics. *Ind. Eng. Chem. Res.*

94 Tian, H., Shu, Y., Cui, Y.-L. et al. (2014). Scalable fabrication of high-performance and flexible graphene strain sensors. *Nanoscale* 6 (2): 699–705.

95 Hempel, M., Nezich, D., Kong, J., and Hofmann, M. (2012). A novel class of strain gauges based on layered percolative films of 2D materials. *Nano Lett.* 12 (11): 5714–5718.

96 Sun, J., Li, Y., Peng, Q. et al. (2013). Macroscopic, flexible, high-performance graphene ribbons. *ACS Nano* 7 (11): 10225–10232.

97 Yan, C., Wang, J., Kang, W. et al. (2014). Highly stretchable piezoresistive graphene–nanocellulose nanopaper for strain sensors. *Adv. Mater.* 26 (13): 2022–2027.

98 Li, X., Sun, P., Fan, L. et al. (2012). *Multifunctional graphene woven fabrics. Sci. Rep.* 2: 395.

99 Wang, Y., Yang, T., Lao, J. et al. (2015). Ultra-sensitive graphene strain sensor for sound signal acquisition and recognition. *Nano Res.* 8 (5): 1627–1636.

100 Liu, H., Dong, M., Huang, W. et al. (2017). Lightweight conductive graphene/thermoplastic polyurethane foams with ultrahigh compressibility for piezoresistive sensing. *J. Mater. Chem. C* 5 (1): 73–83.

101 Peng, X. and Velasquez, S. (2011). Strain modulated band gap of edge passivated armchair graphene nanoribbons. *Appl. Phys. Lett.* 98 (2): 023112.

102 Zhou, S.Y., Gweon, G.-H., Fedorov, A. et al. (2007). Substrate-induced bandgap opening in epitaxial graphene. *Nat. Mater.* 6 (10): 770–775.

103 Zhang, Y., Tang, T.-T., Girit, C. et al. (2009). Direct observation of a widely tunable bandgap in bilayer graphene. *Nature* 459 (7248): 820–823.

104 Usachov, D., Vilkov, O., Gruneis, A. et al. (2011). Nitrogen-doped graphene: efficient growth, structure, and electronic properties. *Nano Lett.* 11 (12): 5401–5407.

105 Chang, C.-K., Kataria, S., Kuo, C.-C. et al. (2013). Band gap engineering of chemical vapor deposited graphene by in situ BN doping. *ACS Nano* 7 (2): 1333–1341.

106 Haniff, M., Hafiz, S., Huang, N. et al. (2017). Piezoresistive effect in plasma-doping of graphene sheet for high-performance flexible pressure sensing application. *ACS Appl. Mater. Interfaces* 9 (17): 15192–15201.

107 Kong, J.-H., Jang, N.-S., Kim, S.-H., and Kim, J.-M. (2014). Simple and rapid micropatterning of conductive carbon composites and its application to elastic strain sensors. *Carbon* 77: 199–207.

108 Lu, N., Lu, C., Yang, S., and Rogers, J. (2012). Highly sensitive skin-mountable strain gauges based entirely on elastomers. *Adv. Funct. Mater.* 22 (19): 4044–4050.

109 Liao, X., Liao, Q., Yan, X. et al. (2015). Flexible and highly sensitive strain sensors fabricated by pencil drawn for wearable monitor. *Adv. Funct. Mater.* 25 (16): 2395–2401.

110 Amjadi, M., Turan, M., Clementson, C.P., and Sitti, M. (2016). Parallel microcracks-based ultrasensitive and highly stretchable strain sensors. *ACS Appl. Mater. Interfaces* 8 (8): 5618–5626.

111 Liu, Y., Tao, L.-Q., Wang, D.-Y. et al. (2017). Flexible, highly sensitive pressure sensor with a wide range based on graphene-silk network structure. *Appl. Phys. Lett.* 110 (12): 123508.

112 Lou, Z., Chen, S., Wang, L. et al. (2016). An ultra-sensitive and rapid response speed graphene pressure sensors for electronic skin and health monitoring. *Nano Energy* 23: 7–14.

113 Ehrmann, A., Heimlich, F., Brücken, A. et al. (2014). Suitability of knitted fabrics as elongation sensors subject to structure, stitch dimension and elongation direction. *Text. Res. J.* 84 (18): 2006–2012.

114 Zysset, C., Kinkeldei, T., Münzenrieder, N. et al. (2013). Combining electronics on flexible plastic strips with textiles. *Text. Res. J.* 83 (11): 1130–1142.

115 Yamashita, T., Takamatsu, S., Miyake, K., and Itoh, T. (2013). Fabrication and evaluation of a conductive polymer coated elastomer contact structure for woven electronic textile. *Sensors Actuators A Phys.* 195: 213–218.

116 Atalay, O. and Kennon, W.R. (2014). Knitted strain sensors: impact of design parameters on sensing properties. *Sensors* 14 (3): 4712–4730.

117 Shyr, T.-W., Shie, J.-W., Jiang, C.-H., and Li, J.-J. (2014). A textile-based wearable sensing device designed for monitoring the flexion angle of elbow and knee movements. *Sensors* 14 (3): 4050–4059.

118 Lorussi, F., Rocchia, W., Scilingo, E.P. et al. (2004). Wearable, redundant fabric-based sensor arrays for reconstruction of body segment posture. *IEEE Sensors J.* 4 (6): 807–818.

119 Campbell, T.E., Munro, B.J., Wallace, G.G., and Steele, J.R. (2007). Can fabric sensors monitor breast motion? *J. Biomech.* 40 (13): 3056–3059.

120 Li, Y., Cheng, X., Leung, M. et al. (2005). A flexible strain sensor from polypyrrole-coated fabrics. *Synth. Met.* 155 (1): 89–94.

121 Wei, Y., Chen, S., Lin, Y. et al. (2016). Silver nanowires coated on cotton for flexible pressure sensors. *J. Mater. Chem. C* 4 (5): 935–943.

122 Greiner, A. and Wendorff, J.H. (2007). Electrospinning: a fascinating method for the preparation of ultrathin fibers. *Angew. Chem. Int. Ed.* 46 (30): 5670–5703.

123 Agarwal, S., Greiner, A., and Wendorff, J.H. (2013). Functional materials by electrospinning of polymers. *Prog. Polym. Sci.* 38 (6): 963–991.

124 Li, D. and Xia, Y. (2004). Electrospinning of nanofibers: reinventing the wheel? *Adv. Mater.* 16 (14): 1151–1170.

125 Wang, F., Zhu, B., Shu, L., and Tao, X. (2013). Flexible pressure sensors for smart protective clothing against impact loading. *Smart Mater. Struct.* 23 (1): 015001.

126 Shyr, T.-W., Shie, J.-W., and Jhuang, Y.-E. (2011). The effect of tensile hysteresis and contact resistance on the performance of strain-resistant elastic-conductive webbing. *Sensors* 11 (2): 1693–1705.

127 Cochrane, C., Koncar, V., Lewandowski, M., and Dufour, C. (2007). Design and development of a flexible strain sensor for textile structures based on a conductive polymer composite. *Sensors* 7 (4): 473–492.

128 Li, Y., Samad, Y.A., and Liao, K. (2015). From cotton to wearable pressure sensor. *J. Mater. Chem. A* 3 (5): 2181–2187.

129 Wang, C., Li, X., Gao, E. et al. (2016). Carbonized silk fabric for ultrastretchable, highly sensitive, and wearable strain sensors. *Adv. Mater.* 28 (31): 6640–6648.

130 Wang, Q., Jian, M., Wang, C., and Zhang, Y. (2017). Carbonized silk nanofiber membrane for transparent and sensitive electronic skin. *Adv. Funct. Mater.* 27 (9): 1605657.

131 Heeger, A.J. (2001). Semiconducting and metallic polymers: the fourth generation of polymeric materials. *ACS Publications* 40 (14): 2591–2611.

132 Sedighi, A., Montazer, M., and Mazinani, S. (2018). Fabrication of electrically conductive superparamagnetic fabric with microwave attenuation, antibacterial properties and UV protection using PEDOT/magnetite nanoparticles. *Mater. Des.* 160: 34–47.

133 Ding, Y., Invernale, M.A., and Sotzing, G.A. (2010). Conductivity trends of PEDOT-PSS impregnated fabric and the effect of conductivity on electrochromic textile. *ACS Appl. Mater. Interfaces* 2 (6): 1588–1593.

134 Scilingo, E.P., Lorussi, F., Mazzoldi, A., and De Rossi, D. (2003). Strain-sensing fabrics for wearable kinaesthetic-like systems. *IEEE Sensors J.* 3 (4): 460–467.

135 Wu, J., Zhou, D., Too, C.O., and Wallace, G.G. (2005). Conducting polymer coated lycra. *Synth. Met.* 155 (3): 698–701.

136 Xue, P., Tao, X., and Tsang, H. (2007). In situ SEM studies on strain sensing mechanisms of PPy-coated electrically conducting fabrics. *Appl. Surf. Sci.* 253 (7): 3387–3392.

137 Xue, P., Wang, J., and Tao, X. (2014). Flexible textile strain sensors from polypyrrole-coated XLA™ elastic fibers. *High Perform. Polym.* 26 (3): 364–370.

138 Seyedin, M.Z., Razal, J.M., Innis, P.C., and Wallace, G.G. (2014). Strain-responsive polyurethane/PEDOT: PSS elastomeric composite fibers with high electrical conductivity. *Adv. Funct. Mater.* 24 (20): 2957–2966.

139 Pan, L., Chortos, A., Yu, G. et al. (2014). An ultra-sensitive resistive pressure sensor based on hollow-sphere microstructure induced elasticity in conducting polymer film. *Nat. Commun.* 5 (1): 1–8.

140 Liu, N., Fang, G., Wan, J. et al. (2011). Electrospun PEDOT: PSS–PVA nanofiber based ultrahigh-strain sensors with controllable electrical conductivity. *J. Mater. Chem.* 21 (47): 18962–18966.

141 Amjadi, M., Pichitpajongkit, A., Lee, S. et al. (2014). Highly stretchable and sensitive strain sensor based on silver nanowire–elastomer nanocomposite. *ACS Nano* 8 (5): 5154–5163.

142 Su, B., Gong, S., Ma, Z. et al. (2015). Mimosa-inspired design of a flexible pressure sensor with touch sensitivity. *Small* 11 (16): 1886–1891.

143 Sedighi, A., Montazer, M., and Hemmatinejad, N. (2014). Copper nanoparticles on bleached cotton fabric: in situ synthesis and characterization. *Cellulose* 21 (3): 2119–2132.

144 Han, S., Kim, M.K., Wang, B. et al. (2016). Mechanically reinforced skin-electronics with networked nanocomposite elastomer. *Adv. Mater.* 28 (46): 10257–10265.

145 Zhu, B., Wang, H., Liu, Y. et al. (2016). Skin-inspired haptic memory arrays with an electrically reconfigurable architecture. *Adv. Mater.* 28 (8): 1559–1566.

146 Park, H., Jeong, Y.R., Yun, J. et al. (2015). Stretchable array of highly sensitive pressure sensors consisting of polyaniline nanofibers and Au-coated polydimethylsiloxane micropillars. *ACS Nano* 9 (10): 9974–9985.

147 Ge, J., Sun, L., Zhang, F.R. et al. (2016). A stretchable electronic fabric artificial skin with pressure-, lateral strain-, and flexion-sensitive properties. *Adv. Mater.* 28 (4): 722–728.

148 Pang, C., Lee, G.-Y., Kim, T.-i. et al. (2012). A flexible and highly sensitive strain-gauge sensor using reversible interlocking of nanofibres. *Nat. Mater.* 11 (9): 795–801.

149 Park, B., Kim, J., Kang, D. et al. (2016). Dramatically enhanced mechanosensitivity and signal-to-noise ratio of nanoscale crack-based sensors: effect of crack depth. *Adv. Mater.* 28 (37): 8130–8137.

150 Lee, J., Yang, D., Kim, S., and Park, I. (2013). *Stretchable Strain Sensor Based on Metal Nanoparticle Thin Film for Human Motion Detection & Flexible Pressure Sensing Devices*, 2624–2627. 2013 Transducers & Eurosensors XXVII: The 17th International Conference on Solid-State Sensors, Actuators and Microsystems (TRANSDUCERS & EUROSENSORS XXVII), 2013; IEEE.

151 Wu, Y.-H., Liu, H.-Z., Chen, S. et al. (2017). Channel crack-designed gold@ PU sponge for highly elastic piezoresistive sensor with excellent detectability. *ACS Appl. Mater. Interfaces* 9 (23): 20098–20105.

152 Ghosh, S.K., Adhikary, P., Jana, S. et al. (2017). Electrospun gelatin nanofiber based self-powered bio-e-skin for health care monitoring. *Nano Energy* 36: 166–175.

153 Chen, S., Lou, Z., Chen, D. et al. (2016). Polymer-enhanced highly stretchable conductive fiber strain sensor used for electronic data gloves. *Adv. Mater. Technol.* 1 (7): 1600136.

154 Amjadi, M., Kyung, K.U., Park, I., and Sitti, M. (2016). Stretchable, skin-mountable, and wearable strain sensors and their potential applications: a review. *Adv. Funct. Mater.* 26 (11): 1678–1698.

155 Hu, N., Itoi, T., Akagi, T. et al. (2013). Ultrasensitive strain sensors made from metal-coated carbon nanofiller/epoxy composites. *Carbon* 51: 202–212.

156 Tung, T., Robert, C., Castro, M. et al. (2016). Enhancing the sensitivity of graphene/polyurethane nanocomposite flexible piezo-resistive pressure sensors with magnetite nano-spacers. *Carbon* 108: 450–460.

157 Ma, Y., Liu, N., Li, L. et al. (2017). A highly flexible and sensitive piezoresistive sensor based on MXene with greatly changed interlayer distances. *Nat. Commun.* 8 (1): 1207.

158 Han, S., Liu, C., Xu, H. et al. (2018). Multiscale nanowire-microfluidic hybrid strain sensors with high sensitivity and stretchability. *npj Flex. Electron.* 2 (1): 1–10.

159 Zhang, Y., Anderson, N., Bland, S. et al. (2017). All-printed strain sensors: building blocks of the aircraft structural health monitoring system. *Sensors Actuators A Phys.* 253: 165–172.

160 Chen, S., Wei, Y., Yuan, X. et al. (2016). A highly stretchable strain sensor based on a graphene/silver nanoparticle synergic conductive network and a sandwich structure. *J. Mater. Chem. C* 4 (19): 4304–4311.

161 Ha, M., Lim, S., Park, J. et al. (2015). Bioinspired interlocked and hierarchical design of ZnO nanowire arrays for static and dynamic pressure-sensitive electronic skins. *Adv. Funct. Mater.* 25 (19): 2841–2849.

162 Lipomi, D.J., Vosgueritchian, M., Tee, B.C. et al. (2011). Skin-like pressure and strain sensors based on transparent elastic films of carbon nanotubes. *Nat. Nanotechnol.* 6 (12): 788.

163 Lin, P. and Yan, F. (2012). Organic thin-film transistors for chemical and biological sensing. *Adv. Mater.* 24 (1): 34–51.

164 Sun, Q., Kim, D.H., Park, S.S. et al. (2014). Transparent, low-power pressure sensor matrix based on coplanar-gate graphene transistors. *Adv. Mater.* 26 (27): 4735–4740.

165 Yeom, C., Chen, K., Kiriya, D. et al. (2015). Large-area compliant tactile sensors using printed carbon nanotube active-matrix backplanes. *Adv. Mater.* 27 (9): 1561–1566.

166 Tsai, M.-Y., Tarasov, A., Hesabi, Z.R. et al. (2015). Flexible MoS2 field-effect transistors for gate-tunable piezoresistive strain sensors. *ACS Appl. Mater. Interfaces* 7 (23): 12850–12855.

167 Someya, T., Sekitani, T., Iba, S. et al. (2004). A large-area, flexible pressure sensor matrix with organic field-effect transistors for artificial skin applications. *Proc. Natl. Acad. Sci.* 101 (27): 9966–9970.

168 Mitrakos, V., Macintyre, L., Denison, F.C. et al. (2017). Design, manufacture and testing of capacitive pressure sensors for low-pressure measurement ranges. *Micromachines* 8 (2): 41.

169 Hammock, M.L., Chortos, A., Tee, B.C.K. et al. (2013). 25th anniversary article: the evolution of electronic skin (e-skin): a brief history, design considerations, and recent progress. *Adv. Mater.* 25 (42): 5997–6038.

170 Eaton, W.P. and Smith, J.H. (1997). Micromachined pressure sensors: review and recent developments. *Smart Mater. Struct.* 6 (5): 530.

171 Yao, S. and Zhu, Y. (2015). Nanomaterial-enabled stretchable conductors: strategies, materials and devices. *Adv. Mater.* 27 (9): 1480–1511.

172 Viry, L., Levi, A., Totaro, M. et al. (2014). Flexible three-axial force sensor for soft and highly sensitive artificial touch. *Adv. Mater.* 26 (17): 2659–2664.

173 Chen, Y.-M., He, S.-M., Huang, C.-H. et al. (2016). Ultra-large suspended graphene as a highly elastic membrane for capacitive pressure sensors. *Nanoscale* 8 (6): 3555–3564.

174 Chun, S., Hong, A., Choi, Y. et al. (2016). A tactile sensor using a conductive graphene-sponge composite. *Nanoscale* 8 (17): 9185–9192.

175 Nie, B., Li, R., Cao, J. et al. (2015). Flexible transparent iontronic film for interfacial capacitive pressure sensing. *Adv. Mater.* 27 (39): 6055–6062.

176 Cai, L., Song, L., Luan, P. et al. (2013). Super-stretchable, transparent carbon nanotube-based capacitive strain sensors for human motion detection. *Sci. Rep.* 3 (1): 1–9.

177 Cohen, D.J., Mitra, D., Peterson, K., and Maharbiz, M.M. (2012). A highly elastic, capacitive strain gauge based on percolating nanotube networks. *Nano Lett.* 12 (4): 1821–1825.

178 Habibi, M., Darbari, S., Rajabali, S., and Ahmadi, V. (2016). Fabrication of a graphene-based pressure sensor by utilising field emission behavior of carbon nanotubes. *Carbon* 96: 259–267.

179 Wang, D., Li, H., Li, M. et al. (2013). Stretchable conductive polyurethane elastomer in situ polymerized with multi-walled carbon nanotubes. *J. Mater. Chem. C* 1 (15): 2744–2749.

180 Kim, K.K., Hong, S., Cho, H.M. et al. (2015). Highly sensitive and stretchable multidimensional strain sensor with prestrained anisotropic metal nanowire percolation networks. *Nano Lett.* 15 (8): 5240–5247.

181 Xu, F. and Zhu, Y. (2012). Highly conductive and stretchable silver nanowire conductors. *Adv. Mater.* 24 (37): 5117–5122.

182 Hu, W., Niu, X., Zhao, R., and Pei, Q. (2013). Elastomeric transparent capacitive sensors based on an interpenetrating composite of silver nanowires and polyurethane. *Appl. Phys. Lett.* 102 (8): 38.

183 Ho, D.H., Song, R., Sun, Q. et al. (2017). Crack-enhanced microfluidic stretchable E-skin sensor. *ACS Appl. Mater. Interfaces* 9 (51): 44678–44686.

184 Yao, S. and Zhu, Y. (2014). Wearable multifunctional sensors using printed stretchable conductors made of silver nanowires. *Nanoscale* 6 (4): 2345–2352.

185 Hu, W., Chen, S., Zhuo, B. et al. (2016). Highly sensitive and transparent strain sensor based on thin elastomer film. *IEEE Electron Device Lett.* 37 (5): 667–670.

186 Kim, S.-R., Kim, J.-H., and Park, J.-W. (2017). Wearable and transparent capacitive strain sensor with high sensitivity based on patterned Ag nanowire networks. *ACS Appl. Mater. Interfaces* 9 (31): 26407–26416.

187 Hu, C.-F., Wang, J.-Y., Liu, Y.-C. et al. (2013). Development of 3D carbon nanotube interdigitated finger electrodes on polymer substrate for flexible capacitive sensor application. *Nanotechnology* 24 (44): 444006.

188 Cui, Z., Poblete, F.R., Cheng, G. et al. (2015). Design and operation of silver nanowire based flexible and stretchable touch sensors. *J. Mater. Res.* 30 (1): 79–85.

189 Kanaparthi, S. and Badhulika, S. (2017). Low cost, flexible and biodegradable touch sensor fabricated by solvent-free processing of graphite on cellulose paper. *Sensors Actuators B Chem.* 242: 857–864.

190 Luo, Y., Shao, J., Chen, S. et al. (2019). Flexible capacitive pressure sensor enhanced by tilted micropillar arrays. *ACS Appl. Mater. Interfaces* 11 (19): 17796–17803.

191 Woo, S.-J., Kong, J.-H., Kim, D.-G., and Kim, J.-M. (2014). A thin all-elastomeric capacitive pressure sensor array based on micro-contact printed elastic conductors. *J. Mater. Chem. C* 2 (22): 4415–4422.

192 Lee, K., Lee, J., Kim, G. et al. (2017). Rough-surface-enabled capacitive pressure sensors with 3D touch capability. *Small* 13x (44): 1700368.

193 Choi, T.Y., Hwang, B.-U., Kim, B.-Y. et al. (2017). Stretchable, transparent, and stretch-unresponsive capacitive touch sensor array with selectively patterned silver nanowires/reduced graphene oxide electrodes. *ACS Appl. Mater. Interfaces* 9 (21): 18022–18030.

194 Kwon, D., Lee, T.-I., Shim, J. et al. (2016). Highly sensitive, flexible, and wearable pressure sensor based on a giant piezocapacitive effect of three-dimensional microporous elastomeric dielectric layer. *ACS Appl. Mater. Interfaces* 8 (26): 16922–16931.

195 Wan, S., Bi, H., Zhou, Y. et al. (2017). Graphene oxide as high-performance dielectric materials for capacitive pressure sensors. *Carbon* 114: 209–216.

196 Pruvost, M., Smit, W.J., Monteux, C. et al. (2019). Polymeric foams for flexible and highly sensitive low-pressure capacitive sensors. *npj Flex. Electron.* 3 (1): 1–6.

197 Wang, J., Jiu, J., Nogi, M. et al. (2015). A highly sensitive and flexible pressure sensor with electrodes and elastomeric interlayer containing silver nanowires. *Nanoscale* 7 (7): 2926–2932.

198 Shi, H., Al-Rubaiai, M., Holbrook, C.M. et al. (2019). Screen-printed soft capacitive sensors for spatial mapping of both positive and negative pressures. *Adv. Funct. Mater.* 29 (23): 1809116.

199 Niu, H., Gao, S., Yue, W. et al. (2019). Highly morphology-controllable and highly sensitive capacitive tactile sensor based on epidermis-dermis-inspired interlocked asymmetric-nanocone arrays for detection of tiny pressure. *Small* 16 (4): 1904774.

200 Cao, Y., Li, T., Gu, Y. et al. (2018). Fingerprint-inspired flexible tactile sensor for accurately discerning surface texture. *Small* 14 (16): 1703902.

201 Kang, M., Kim, J., Jang, B. et al. (2017). Graphene-based three-dimensional capacitive touch sensor for wearable electronics. *ACS Nano* 11 (8): 7950–7957.

202 Zang, Y., Zhang, F., Huang, D. et al. (2015). Flexible suspended gate organic thin-film transistors for ultra-sensitive pressure detection. *Nat. Commun.* 6 (1): 1–9.

203 Schwartz, G., Tee, B.C.-K., Mei, J. et al. (2013). Flexible polymer transistors with high pressure sensitivity for application in electronic skin and health monitoring. *Nat. Commun.* 4 (1): 1–8.

204 Takamatsu, S., Kobayashi, T., Shibayama, N. et al. (2012). Fabric pressure sensor array fabricated with die-coating and weaving techniques. *Sensors Actuators A Phys.* 184: 57–63.

205 Takamatsu, S., Yamashita, T., Imai, T., and Itoh, T. (2014). Lightweight flexible keyboard with a conductive polymer-based touch sensor fabric. *Sensors Actuators A Phys.* 220: 153–158.

206 Kim, S.Y., Park, S., Park, H.W. et al. (2015). Highly sensitive and multimodal all-carbon skin sensors capable of simultaneously detecting tactile and biological stimuli. *Adv. Mater.* 27 (28): 4178–4185.

207 Kim, H., Yoon, J., Lee, G. et al. (2016). Encapsulated, high-performance, stretchable array of stacked planar micro-supercapacitors as waterproof wearable energy storage devices. *ACS Appl. Mater. Interfaces* 8 (25): 16016–16025.

208 Gong, S. and Cheng, W. (2017). One-dimensional nanomaterials for soft electronics. *Adv. Electron. Mater.* 3 (3): 1600314.

209 Cotton, D.P., Chappell, P.H., Cranny, A. et al. (2007). A novel thick-film piezoelectric slip sensor for a prosthetic hand. *IEEE Sensors J.* 7 (5): 752–761.

210 Dargahi, J. and Najarian, S. (2004). Human tactile perception as a standard for artificial tactile sensing–a review. *Int. J. Med. Robot.* 1 (1): 23–35.

211 Zhou, J., Gu, Y., Fei, P. et al. (2008). Flexible piezotronic strain sensor. *Nano Lett.* 8 (9): 3035–3040.

212 Liao, Q., Mohr, M., Zhang, X. et al. (2013). Carbon fiber–ZnO nanowire hybrid structures for flexible and adaptable strain sensors. *Nanoscale* 5 (24): 12350–12355.

213 Zhang, W., Zhu, R., Nguyen, V., and Yang, R. (2014). Highly sensitive and flexible strain sensors based on vertical zinc oxide nanowire arrays. *Sensors Actuators A Phys.* 205: 164–169.

214 Graz, I., Krause, M., Bauer-Gogonea, S. et al. (2009). Flexible active-matrix cells with selectively poled bifunctional polymer-ceramic nanocomposite for pressure and temperature sensing skin. *J. Appl. Phys.* 106 (3): 034503.

215 Dagdeviren, C., Su, Y., Joe, P. et al. (2014). Conformable amplified lead zirconate titanate sensors with enhanced piezoelectric response for cutaneous pressure monitoring. *Nat. Commun.* 5 (1): 1–10.

216 Park, K.I., Son, J.H., Hwang, G.T. et al. (2014). Highly-efficient, flexible piezoelectric PZT thin film nanogenerator on plastic substrates. *Adv. Mater.* 26 (16): 2514–2520.

217 Chen, Z., Wang, Z., Li, X. et al. (2017). Flexible piezoelectric-induced pressure sensors for static measurements based on nanowires/graphene heterostructures. *ACS Nano* 11 (5): 4507–4513.

218 Persano, L., Dagdeviren, C., Su, Y. et al. (2013). High performance piezoelectric devices based on aligned arrays of nanofibers of poly (vinylidenefluoride-co-trifluoroethylene). *Nat. Commun.* 4 (1): 1–10.

219 Joshi, A.B., Kalange, A.E., Bodas, D., and Gangal, S. (2010). Simulations of piezoelectric pressure sensor for radial artery pulse measurement. *Mater. Sci. Eng. B* 168 (1-3): 250–253.

220 Garain, S., Jana, S., Sinha, T.K., and Mandal, D. (2016). Design of in situ poled Ce^{3+}-doped electrospun PVDF/graphene composite nanofibers for fabrication of nanopressure sensor and ultrasensitive acoustic nanogenerator. *ACS Appl. Mater. Interfaces* 8 (7): 4532–4540.

221 Li, B., Zhang, F., Guan, S. et al. (2016). Wearable piezoelectric device assembled by one-step continuous electrospinning. *J. Mater. Chem. C* 4 (29): 6988–6995.

222 Wang, X., Song, W.-Z., You, M.-H. et al. (2018). Bionic single-electrode electronic skin unit based on piezoelectric nanogenerator. *ACS Nano* 12 (8): 8588–8596.

223 Ai, Y., Lou, Z., Chen, S. et al. (2017). All rGO-on-PVDF-nanofibers based self-powered electronic skins. *Nano Energy* 35: 121–127.

224 Wang, X., Yang, B., Liu, J. et al. (2016). A flexible triboelectric-piezoelectric hybrid nanogenerator based on P (VDF-TrFE) nanofibers and PDMS/MWCNT for wearable devices. *Sci. Rep.* 6: 36409.

225 Park, S.-H., Lee, H.B., Yeon, S.M. et al. (2016). Flexible and stretchable piezoelectric sensor with thickness-tunable configuration of electrospun nanofiber mat and elastomeric substrates. *ACS Appl. Mater. Interfaces* 8 (37): 24773–24781.

226 Tien, N.T., Jeon, S., Kim, D.I. et al. (2014). A flexible bimodal sensor array for simultaneous sensing of pressure and temperature. *Adv. Mater.* 26 (5): 796–804.

227 Sun, Q., Seung, W., Kim, B.J. et al. (2015). Active matrix electronic skin strain sensor based on piezopotential-powered graphene transistors. *Adv. Mater.* 27 (22): 3411–3417.

228 Tao, S., Chen, D., Wang, J. et al. (2016). A high sensitivity pressure sensor based on two-dimensional photonic crystal. *Photonic Sens.* 6 (2): 137–142.

229 Song, K., Zhao, R., Wang, Z.L., and Yang, Y. (2019). Conjuncted pyro-piezoelectric effect for self-powered simultaneous temperature and pressure sensing. *Adv. Mater.* 31 (36): 1902831.

230 Wu, W., Wang, L., Li, Y. et al. (2014). Piezoelectricity of single-atomic-layer MoS 2 for energy conversion and piezotronics. *Nature* 514 (7523): 470–474.

231 Wu, J.M., Chen, C.-Y., Zhang, Y. et al. (2012). Ultrahigh sensitive piezotronic strain sensors based on a $ZnSnO_3$ nanowire/microwire. *ACS Nano* 6 (5): 4369–4374.

232 Lim, S., Son, D., Kim, J. et al. (2015). Transparent and stretchable interactive human machine interface based on patterned graphene heterostructures. *Adv. Funct. Mater.* 25 (3): 375–383.

233 Wu, W., Wen, X., and Wang, Z.L. (2013). Taxel-addressable matrix of vertical-nanowire piezotronic transistors for active and adaptive tactile imaging. *Science* 340 (6135): 952–957.

234 Trung, T.Q., Tien, N.T., Seol, Y.G., and Lee, N.-E. (2012). Transparent and flexible organic field-effect transistor for multi-modal sensing. *Org. Electron.* 13 (4): 533–540.

235 Hu, K., Xiong, R., Guo, H. et al. (2016). Self-powered electronic skin with biotactile selectivity. *Adv. Mater.* 28 (18): 3549–3556.

236 He, H., Fu, Y., Zang, W. et al. (2017). A flexible self-powered T-ZnO/PVDF/fabric electronic-skin with multi-functions of tactile-perception, atmosphere-detection and self-clean. *Nano Energy* 31: 37–48.

237 Siddiqui, S., Kim, D.-I., Roh, E. et al. (2016). A durable and stable piezoelectric nanogenerator with nanocomposite nanofibers embedded in an elastomer under high loading for a self-powered sensor system. *Nano Energy* 30: 434–442.

238 Xu, S., Qin, Y., Xu, C. et al. (2010). Self-powered nanowire devices. *Nat. Nanotechnol.* 5 (5): 366.

239 Forman, C. (1996). *High-Performance Films*. CRC Press.

240 Tao, J., Bao, R., Wang, X. et al. (2019). Self-powered tactile sensor array systems based on the triboelectric effect. *Adv. Funct. Mater.* 29 (41): 1806379.

241 Zhu, G., Yang, W.Q., Zhang, T. et al. (2014). Self-powered, ultrasensitive, flexible tactile sensors based on contact electrification. *Nano Lett.* 14 (6): 3208–3213.

242 Lin, L., Xie, Y., Wang, S. et al. (2013). Triboelectric active sensor array for self-powered static and dynamic pressure detection and tactile imaging. *ACS Nano* 7 (9): 8266–8274.

243 Wang, X., Zhang, Y., Zhang, X. et al. (2018). A highly stretchable transparent self-powered triboelectric tactile sensor with metallized nanofibers for wearable electronics. *Adv. Mater.* 30 (12): 1706738.

244 Khan, U., Kim, T.H., Ryu, H. et al. (2017). Graphene tribotronics for electronic skin and touch screen applications. *Adv. Mater.* 29 (1): 1603544.

245 Wang, Z.L. (2013). Triboelectric nanogenerators as new energy technology for self-powered systems and as active mechanical and chemical sensors. *ACS Nano* 7 (11): 9533–9557.

246 Yang, T., Xie, D., Li, Z., and Zhu, H. (2017). Recent advances in wearable tactile sensors: materials, sensing mechanisms, and device performance. *Mater. Sci. Eng. R. Rep.* 115: 1–37.

247 Wan, Y., Wang, Y., and Guo, C.F. (2017). Recent progresses on flexible tactile sensors. *Mater. Today Phys.* 1: 61–73.

248 Bai, P., Zhu, G., Jing, Q. et al. (2014). Membrane-based self-powered triboelectric sensors for pressure change detection and its uses in security surveillance and healthcare monitoring. *Adv. Funct. Mater.* 24 (37): 5807–5813.

249 Wang, J., Li, S., Yi, F. et al. (2016). Sustainably powering wearable electronics solely by biomechanical energy. *Nat. Commun.* 7 (1): 1–8.

250 Xia, X., Chen, J., Liu, G. et al. (2017). Aligning graphene sheets in PDMS for improving output performance of triboelectric nanogenerator. *Carbon* 111: 569–576.

251 Liao, M., Wan, P., Wen, J. et al. (2017). Wearable, healable, and adhesive epidermal sensors assembled from mussel-inspired conductive hybrid hydrogel framework. *Adv. Funct. Mater.* 27 (48): 1703852.

252 Joo, Y., Byun, J., Seong, N. et al. (2015). Silver nanowire-embedded PDMS with a multiscale structure for a highly sensitive and robust flexible pressure sensor. *Nanoscale* 7 (14): 6208–6215.

253 Pan, J., Hao, B., Song, W. et al. (2020). Highly sensitive and durable wearable strain sensors from a core-sheath nanocomposite yarn. *Compos. Part B* 183: 107683.

254 Wu, J., Wu, Z., Lu, X. et al. (2019). Ultrastretchable and stable strain sensors based on antifreezing and self-healing ionic organohydrogels for human motion monitoring. *ACS Appl. Mater. Interfaces* 11 (9): 9405–9414.

255 Xu, H., Zhang, M.K., Lu, Y.F. et al. (2020). Dual-mode wearable strain sensor based on graphene/colloidal crystal films for simultaneously detection of subtle and large human motions. *Adv. Mater. Technol.* 5 (2): 1901056.

256 Xia, S., Song, S., Jia, F., and Gao, G. (2019). A flexible, adhesive and self-healable hydrogel-based wearable strain sensor for human motion and physiological signal monitoring. *J. Mater. Chem. B* 7 (30): 4638–4648.

257 Li, J., Wang, L., Wang, X. et al. (2019). Highly conductive PVA/AG coating by aqueous in situ reduction and its stretchable structure for strain sensor. *ACS Appl. Mater. Interfaces* 12 (1): 1427–1435.

258 Yang, Y., Shi, L., Cao, Z. et al. (2019). Strain sensors with a high sensitivity and a wide sensing range based on a $Ti_3C_2T_x$ (MXene) nanoparticle–nanosheet hybrid network. *Adv. Funct. Mater.* 29 (14): 1807882.

259 Shin, U.-H., Jeong, D.-W., Park, S.-M. et al. (2014). Highly stretchable conductors and piezocapacitive strain gauges based on simple contact-transfer patterning of carbon nanotube forests. *Carbon* 80: 396–404.

260 Li, D., Fan, Y., Han, G., and Guo, Z. (2020). Superomniphobic silk fibroin/AG nanowires membrane for flexible and transparent electronic sensor. *ACS Appl. Mater. Interfaces* 12 (8): 10039–10049.

261 Chen, Y., Wang, Z., Xu, R. et al. (2020). A highly sensitive and wearable pressure sensor based on conductive polyacrylonitrile nanofibrous membrane via electroless silver plating. *Chem. Eng. J.* 394: 124960.

262 Amjadi, M., Yoon, Y.J., and Park, I. (2015). Ultra-stretchable and skin-mountable strain sensors using carbon nanotubes–Ecoflex nanocomposites. *Nanotechnology* 26 (37): 375501.

263 Shi, G., Zhao, Z., Pai, J.H. et al. (2016). Highly sensitive, wearable, durable strain sensors and stretchable conductors using graphene/silicon rubber composites. *Adv. Funct. Mater.* 26 (42): 7614–7625.

264 Sun, S., Liu, Y., Chang, X. et al. (2020). A wearable, waterproof, and highly sensitive strain sensor based on three-dimensional graphene/carbon black/Ni sponge for wirelessly monitoring human motions. *J. Mater. Chem. C* 8 (6): 2074–2085.

265 Zheng, Y., Li, Y., Zhou, Y. et al. (2019). High-performance wearable strain sensor based on graphene/cotton fabric with high durability and low detection limit. *ACS Appl. Mater. Interfaces* 12 (1): 1474–1485.

266 Bai, M., Zhai, Y., Liu, F. et al. (2019). Stretchable graphene thin film enabled yarn sensors with tunable piezoresistivity for human motion monitoring. *Sci. Rep.* 9 (1): 1–10.

267 Lin, J., Cai, X., Liu, Z. et al. (2020). Anti-liquid-interfering and bacterially anti-adhesive strategy for highly stretchable and ultrasensitive strain sensors based on cassie-baxter wetting state. *Adv. Funct. Mater.* 30 (23): 2000398.

268 Su, X., Borayek, R., Li, X. et al. (2020). Integrated wearable sensors with bending/stretching selectivity and extremely enhanced sensitivity derived from agarose-based ionic conductor and its 3D-shaping. *Chem. Eng. J.* 389: 124503.

269 Huang, J., Zeng, J., and Yan, Z. (2020). *Multi-Arches Structured All-Carbon Aerogels with Super Elasticity and High Fatigue Resistance As Sensitive Wearable Sensors*. 237th ECS Meeting with the 18th International Meeting on Chemical Sensors (IMCS 2020)(May 10-14, 2020), 2020; ECS.

270 He, X., Liu, Q., Zhong, W. et al. (2019). Strategy of constructing light-weight and highly compressible graphene-based aerogels with an ordered unique configuration for wearable piezoresistive sensors. *ACS Appl. Mater. Interfaces* 11 (21): 19350–19362.

271 Sun, H., Dai, K., Zhai, W. et al. (2019). A highly sensitive and stretchable yarn strain sensor for human motion tracking utilizing a wrinkle-assisted crack structure. *ACS Appl. Mater. Interfaces* 11 (39): 36052–36062.

272 Wang, Z., Si, Y., Zhao, C. et al. (2019). Flexible and washable poly (Ionic Liquid) nanofibrous membrane with moisture proof pressure sensing for real-life wearable electronics. *ACS Appl. Mater. Interfaces* 11 (30): 27200–27209.

273 Xia, K., Chen, X., Shen, X. et al. (2019). Carbonized Chinese art paper-based high-performance wearable strain sensor for human activity monitoring. *ACS Appl. Electron. Mater.* 1 (11): 2415–2421.

274 Huang, T., He, P., Wang, R. et al. (2019). Porous fibers composed of polymer nanoball decorated graphene for wearable and highly sensitive strain sensors. *Adv. Funct. Mater.* 29 (45): 1903732.

275 Lee, S.W., Park, J.J., Park, B.H. et al. (2017). Enhanced sensitivity of patterned graphene strain sensors used for monitoring subtle human body motions. *ACS Appl. Mater. Interfaces* 9 (12): 11176–11183.

276 Wang, Y., Wang, L., Yang, T. et al. (2014). Wearable and highly sensitive graphene strain sensors for human motion monitoring. *Adv. Funct. Mater.* 24 (29): 4666–4670.

277 Lv, Z., Liu, J., Yang, X. et al. (2020). Naturally derived wearable strain sensors with enhanced mechanical properties and high sensitivity. *ACS Appl. Mater. Interfaces* 12 (19): 22163–22169.

278 Su, M., Li, F., Chen, S. et al. (2016). Nanoparticle based curve arrays for multi-recognition flexible electronics. *Adv. Mater.* 28 (7): 1369–1374.

279 Pan, J., Yang, M., Luo, L. et al. (2019). Stretchable and highly sensitive braided composite yarn@ polydopamine@ polypyrrole for wearable applications. *ACS Appl. Mater. Interfaces* 11 (7): 7338–7348.

280 Chen, S., Zhuo, B., and Guo, X. (2016). Large area one-step facile processing of microstructured elastomeric dielectric film for high sensitivity and durable sensing over wide pressure range. *ACS Appl. Mater. Interfaces* 8 (31): 20364–20370.

281 Nie, B., Huang, R., Yao, T. et al. (2019). Textile-based wireless pressure sensor array for human-interactive sensing. *Adv. Funct. Mater.* 29 (22): 1808786.

282 Xu, J., Wang, G., Wu, Y. et al. (2019). Ultrastretchable wearable strain and pressure sensors based on adhesive, tough, and self-healing hydrogels for human motion monitoring. *ACS Appl. Mater. Interfaces* 11 (28): 25613–25623.

283 He, J., Xiao, P., Lu, W. et al. (2019). A Universal high accuracy wearable pulse monitoring system via high sensitivity and large linearity graphene pressure sensor. *Nano Energy* 59: 422–433.

284 Park, S., Ahn, S., Sun, J. et al. (2019). Highly bendable and rotational textile structure with prestrained conductive sewing pattern for human joint monitoring. *Adv. Funct. Mater.* 29 (10): 1808369.

285 Jeerapan, I., Sempionatto, J.R., Pavinatto, A. et al. (2016). Stretchable biofuel cells as wearable textile-based self-powered sensors. *J. Mater. Chem. A* 4 (47): 18342–18353.

286 Hwang, B.-U., Lee, J.-H., Trung, T.Q. et al. (2015). Transparent stretchable self-powered patchable sensor platform with ultrasensitive recognition of human activities. *ACS Nano* 9 (9): 8801–8810.

287 Zhang, Y., Liu, S., Miao, Y. et al. (2020). A highly stretchable and sensitive pressure sensor array based on icicle-shaped liquid metal film electrodes. *ACS Appl. Mater. Interfaces* 12 (25): 27961–27970.

288 Choi, M.K., Park, I., Kim, D.C. et al. (2015). Thermally controlled, patterned graphene transfer printing for transparent and wearable electronic/optoelectronic system. *Adv. Funct. Mater.* 25 (46): 7109–7118.

289 Liu, X., Liu, D., Lee, J.-H. et al. (2018). Spider-web-inspired stretchable graphene woven fabric for highly sensitive, transparent, wearable strain sensors. *ACS Appl. Mater. Interfaces* 11 (2): 2282–2294.

290 Sun, R., Carreira, S.C., Chen, Y. et al. (2019). Stretchable piezoelectric sensing systems for self-powered and wireless health monitoring. *Adv. Mater. Technol.* 4 (5): 1900100.

291 Khan, Y., Ostfeld, A.E., Lochner, C.M. et al. (2016). Monitoring of vital signs with flexible and wearable medical devices. *Adv. Mater.* 28 (22): 4373–4395.

292 Evans, D., Hodgkinson, B., and Berry, J. (2001). Vital signs in hospital patients: a systematic review. *Int. J. Nurs. Stud.* 38 (6): 643–650.

293 Chen, M., Li, K., Cheng, G. et al. (2018). Touchpoint-tailored ultrasensitive piezoresistive pressure sensors with a broad dynamic response range and low detection limit. *ACS Appl. Mater. Interfaces* 11 (2): 2551–2558.

294 Jason, N.N., Wang, S.J., Bhanushali, S., and Cheng, W. (2016). Skin inspired fractal strain sensors using a copper nanowire and graphite microflake hybrid conductive network. *Nanoscale* 8 (37): 16596–16605.

295 Yan, C., Deng, W., Jin, L. et al. (2018). Epidermis-inspired ultrathin 3D cellular sensor array for self-powered biomedical monitoring. *ACS Appl. Mater. Interfaces* 10 (48): 41070–41075.

296 Gao, Y., Ota, H., Schaler, E.W. et al. (2017). Wearable microfluidic diaphragm pressure sensor for health and tactile touch monitoring. *Adv. Mater.* 29 (39): 1701985.

297 Kaisti, M., Panula, T., Leppänen, J. et al. (2019). Clinical assessment of a non-invasive wearable MEMS pressure sensor array for monitoring of arterial pulse waveform, heart rate and detection of atrial fibrillation. *NPJ digit. med.* 2 (1): 1–10.

298 Huang, C.B., Witomska, S., Aliprandi, A. et al. (2019). Molecule–graphene hybrid materials with tunable mechanoresponse: highly sensitive pressure sensors for health monitoring. *Adv. Mater.* 31 (1): 1804600.

299 Guo, L., Berglin, L., Wiklund, U., and Mattila, H. (2013). Design of a garment-based sensing system for breathing monitoring. *Text. Res. J.* 83 (5): 499–509.

300 Min, S.D., Yun, Y., and Shin, H. (2014). Simplified structural textile respiration sensor based on capacitive pressure sensing method. *IEEE Sensors J.* 14 (9): 3245–3251.

301 Atalay, O., Kennon, W.R., and Demirok, E. (2014). Weft-knitted strain sensor for monitoring respiratory rate and its electro-mechanical modeling. *IEEE Sensors J.* 15 (1): 110–122.

302 Shuai, X., Zhu, P., Zeng, W. et al. (2017). Highly sensitive flexible pressure sensor based on silver nanowires-embedded polydimethylsiloxane electrode with microarray structure. *ACS Appl. Mater. Interfaces* 9 (31): 26314–26324.

303 Park, D.Y., Joe, D.J., Kim, D.H. et al. (2017). Self-powered real-time arterial pulse monitoring using ultrathin epidermal piezoelectric sensors. *Adv. Mater.* 29 (37): 1702308.

304 Chu, M., Nguyen, T., Pandey, V. et al. (2019). Respiration rate and volume measurements using wearable strain sensors. *NPJ digit. med.* 2 (1): 1–9.

305 Fan, X., Huang, Y., Ding, X. et al. (2018). Alignment-free liquid-capsule pressure sensor for cardiovascular monitoring. *Adv. Funct. Mater.* 28 (44): 1805045.

306 Chen, S., Wu, N., Lin, S. et al. (2020). Hierarchical elastomer tuned self-powered pressure sensor for wearable multifunctional cardiovascular electronics. *Nano Energy* 70: 104460.

307 Kim, J., Chou, E.F., Le, J. et al. (2019). Soft wearable pressure sensors for beat-to-beat blood pressure monitoring. *Adv. healthc. mater.* 8 (13): 1900109.

308 Kim, K., Choi, J., Jeong, Y. et al. (2019). Highly sensitive and wearable liquid metal-based pressure sensor for health monitoring applications: integration of a 3D-printed microbump array with the microchannel. *Adv. Healthc. Mater.* 8 (22): 1900978.

309 Sekine, T., Sugano, R., Tashiro, T. et al. (2018). Fully printed wearable vital sensor for human pulse rate monitoring using ferroelectric polymer. *Sci. Rep.* 8 (1): 1–10.

310 Gao, L., Zhu, C., Li, L. et al. (2019). All paper-based flexible and wearable piezoresistive pressure sensor. *ACS Appl. Mater. Interfaces* 11 (28): 25034–25042.

311 Liu, P., Liu, J., Zhu, X. et al. (2019). A highly adhesive flexible strain sensor based on ultra-violet adhesive filled by graphene and carbon black for wearable monitoring. *Compos. Sci. Technol.* 182: 107771.

312 Wang, B., Thukral, A., Xie, Z. et al. (2020). Flexible and stretchable metal oxide nanofiber networks for multimodal and monolithically integrated wearable electronics. *Nat. Commun.* 11 (1): 1–11.

313 Zhang, F., Wu, S., Peng, S., and Wang, C.H. (2018). The effect of dual-scale carbon fibre network on sensitivity and stretchability of wearable sensors. *Compos. Sci. Technol.* 165: 131–139.

314 Zhou, Z., Li, Y., Cheng, J. et al. (2018). Supersensitive all-fabric pressure sensors using printed textile electrode arrays for human motion monitoring and human–machine interaction. *J. Mater. Chem. C* 6 (48): 13120–13127.

315 Chen, S., Song, Y., Ding, D. et al. (2018). Flexible and anisotropic strain sensor based on carbonized crepe paper with aligned cellulose fibers. *Adv. Funct. Mater.* 28 (42): 1802547.

316 Sakuma, K., Abrami, A., Blumrosen, G. et al. (2018). Wearable nail deformation sensing for behavioral and biomechanical monitoring and human-computer interaction. *Sci. Rep.* 8 (1): 1–11.

317 Chortos, A., Liu, J., and Bao, Z. (2016). Pursuing prosthetic electronic skin. *Nat. Mater.* 15 (9): 937–950.

318 Lee, Y. and Ahn, J.-H. (2020). Biomimetic tactile sensors based on nanomaterials. *ACS Nano* 14 (2): 1220–1226.

319 Zhang, Q., Liu, X., Duan, L., and Gao, G. (2020). Nucleotide-driven skin-attachable hydrogels toward visual human–machine interfaces. *J. Mater. Chem. A* 8 (8): 4515–4523.

320 Wang, C., Hwang, D., Yu, Z. et al. (2013). User-interactive electronic skin for instantaneous pressure visualization. *Nat. Mater.* 12 (10): 899–904.

321 Guo, Y., Guo, Z., Zhong, M. et al. (2018). A flexible wearable pressure sensor with bioinspired microcrack and interlocking for full-range human–machine interfacing. *Small* 14 (44): 1803018.

322 Guo, Y., Zhong, M., Fang, Z. et al. (2019). A wearable transient pressure sensor made with MXene nanosheets for sensitive broad-range human–machine interfacing. *Nano Lett.* 19 (2): 1143–1150.

323 Xu, J., Wang, H., Ma, T. et al. (2020). A graphite nanoplatelet-based highly sensitive flexible strain sensor. *Carbon* 166: 316–327.

324 Li, X., Koh, K.H., Farhan, M., and Lai, K.W.C. (2020). An ultraflexible polyurethane yarn-based wearable strain sensor with a polydimethylsiloxane infiltrated multilayer sheath for smart textiles. *Nanoscale* 12 (6): 4110–4118.

325 Choi, J., Kwon, D., Kim, K. et al. (2019). Synergetic effect of porous elastomer and percolation of carbon nanotube filler toward high performance capacitive pressure sensors. *ACS Appl. Mater. Interfaces* 12 (1): 1698–1706.

326 Yeo, J.C., Yap, H.K., Xi, W. et al. (2016). Flexible and stretchable strain sensing actuator for wearable soft robotic applications. *Adv. Mater. Technol.* 1 (3): 1600018.

327 Chhetry, A., Sharma, S., Yoon, H. et al. (2020). Enhanced sensitivity of capacitive pressure and strain sensor based on $CaCu_3Ti_4O_{12}$ wrapped hybrid sponge for wearable applications. *Adv. Funct. Mater.* 30 (31): 1910020.

328 Hwang, B.-U., Zabeeb, A., Trung, T.Q. et al. (2019). A transparent stretchable sensor for distinguishable detection of touch and pressure by capacitive and piezoresistive signal transduction. *NPG Asia Mater.* 11 (1): 1–12.

329 Xu, B., Akhtar, A., Liu, Y. et al. (2016). An epidermal stimulation and sensing platform for sensorimotor prosthetic control, management of lower back exertion, and electrical muscle activation. *Adv. Mater.* 28 (22): 4462–4471.

330 Zhao, S. and Zhu, R. (2017). Electronic skin with multifunction sensors based on thermosensation. *Adv. Mater.* 29 (15): 1606151.

331 Baumgartner, M., Hartmann, F., Drack, M. et al. (2020). Resilient yet entirely degradable gelatin-based biogels for soft robots and electronics. *Nat. Mater.* 19 (10): 1–8.

332 Anaya, D.V., He, T., Lee, C., and Yuce, M.R. (2020). Self-powered eye motion sensor based on triboelectric interaction and near-field electrostatic induction for wearable assistive technologies. *Nano Energy* 72: 104675.

333 Choi, J., Kwon, D., Kim, B. et al. (2020). Wearable self-powered pressure sensor by integration of piezo-transmittance microporous elastomer with organic solar cell. *Nano Energy* 74: 104749.

334 Li, J., Zhang, C., Duan, L. et al. (2016). Flexible organic tribotronic transistor memory for a visible and wearable touch monitoring system. *Adv. Mater.* 28 (1): 106–110.

335 Li, L., Fu, X., Chen, S. et al. (2020). Hydrophobic and stable mxene–polymer pressure sensors for wearable electronics. *ACS Appl. Mater. Interfaces* 12 (13): 15362–15369.

336 Ahn, S., Cho, Y., Park, S. et al. (2020). Wearable multimode sensors with amplified piezoelectricity due to the multi local strain using 3D textile structure for detecting human body signals. *Nano Energy* 74: 104932.

337 Tang, N., Zhou, C., Qu, D. et al. (2020). A highly aligned nanowire-based strain sensor for ultrasensitive monitoring of subtle human motion. *Small* 16 (24): 2001363.

338 Shao, L., Li, Y., Ma, Z. et al. (2020). A highly sensitive strain sensor based on stretchable and conductive poly(vinyl alcohol)/Phytic Acid/NH_2-POSS hydrogel with a 3D microporous structure. *ACS Appl. Mater. Interfaces* 12 (23): 26496–26508.

339 Chen, G., Wang, H., Guo, R. et al. (2020). Superelastic EGaIn composite fibers sustaining 500% tensile strain with superior electrical conductivity for wearable electronics. *ACS Appl. Mater. Interfaces* 12 (5): 6112–6118.

340 Pan, S., Xia, M., Li, H. et al. (2020). Transparent, high-strength, stretchable, sensitive and anti-freezing poly(vinyl alcohol) ionic hydrogel strain sensors for human motion monitoring. *J. Mater. Chem. C* 8 (8): 2827–2837.

341 Li, S.-X., Xia, H., Xu, Y.-S. et al. (2019). Gold nanoparticle densely packed micro/nanowire-based pressure sensors for human motion monitoring and physiological signal detection. *Nanoscale* 11 (11): 4925–4932.

342 Xie, R., Hou, S., Chen, Y. et al. (2019). Leather-based strain sensor with hierarchical structure for motion monitoring. *Adv. Mater. Technol.* 4 (10): 1900442.

343 Yang, Z., Pang, Y., Han, X.-L. et al. (2018). Graphene textile strain sensor with negative resistance variation for human motion detection. *ACS Nano* 12 (9): 9134–9141.

344 Nur, R., Matsuhisa, N., Jiang, Z. et al. (2018). A highly sensitive capacitive-type strain sensor using wrinkled ultrathin gold films. *Nano Lett.* 18 (9): 5610–5617.

345 Chun, S., Son, W., Kim, D.W. et al. (2019). Water-resistant and skin-adhesive wearable electronics using graphene fabric sensor with octopus-inspired microsuckers. *ACS Appl. Mater. Interfaces* 11 (18): 16951–16957.

346 Lee, J., Pyo, S., Kwon, D.S. et al. (2019). Ultrasensitive strain sensor based on separation of overlapped carbon nanotubes. *Small* 15 (12): 1805120.

347 Ren, H., Zheng, L., Wang, G. et al. (2019). Transfer-medium-free nanofiber-reinforced graphene film and applications in wearable transparent pressure sensors. *ACS Nano* 13 (5): 5541–5548.

348 Zhu, B., Ling, Y., Yap, L.W. et al. (2019). Hierarchically structured vertical gold nanowire array-based wearable pressure sensors for wireless health monitoring. *ACS Appl. Mater. Interfaces* 11 (32): 29014–29021.

349 Sharma, S., Chhetry, A., Sharifuzzaman, M. et al. (2020). Wearable capacitive pressure sensor based on MXene composite nanofibrous scaffolds for reliable human Physiological signal acquisition. *ACS Appl. Mater. Interfaces* 12 (19): 22212–22224.

350 Muth, J.T., Vogt, D.M., Truby, R.L. et al. (2014). Embedded 3D printing of strain sensors within highly stretchable elastomers. *Adv. Mater.* 26 (36): 6307–6312.

351 Gogurla, N., Roy, B., Park, J.-Y., and Kim, S. (2019). Skin-contact actuated single-electrode protein triboelectric nanogenerator and strain sensor for biomechanical energy harvesting and motion sensing. *Nano Energy* 62: 674–681.

352 Zhou, Q., Ji, B., Wei, Y. et al. (2019). A bio-inspired cilia array as the dielectric layer for flexible capacitive pressure sensors with high sensitivity and a broad detection range. *J. Mater. Chem. A* 7 (48): 27334–27346.

7

Conductive Nanomaterials for Printed and Flexible Electronics Application

K. Thiyagarajan[1], S.G. Rahul[2], G.K. Rajini[1], Debashis Maji[3], and Arunkumar Chandrasekhar[3]

[1] *Vellore Institute of Technology, School of Electrical Engineering, Vellore, 632014, India*
[2] *Vel Tech Rangarajan Dr Sagunthala R&D Institute of Science and Technology, Department of Electronics & Communication Engineering, Chennai, 600062, India*
[3] *Vellore Institute of Technology, School of Electronics Engineering, Nanosensors and Nanoenergy Lab, Department of Sensor and Biomedical Technology, Vellore, 632014, India*

7.1 Introduction

Printed electronics (PE) involves a collection of electronic devices produced through conventional printing technology. The significant features of printing technology are high-volume production and low cost. Compared with traditional electronic devices, PE technology easily offers the fabrication of microlevel and nanoscale electronic devices in a faster and simpler techno-economical approach. Considering an example of radio-frequency identification (RFID) antennas, they are manufactured by both conventional production and printing technology. Similarly, several devices are being manufactured conventionally and now by printing technology. In terms of performance, devices manufactured by printed technology possess more advantages. The conventional manufacturing procedure involves etching of metal film to create the pattern of the antenna, resulting in huge solvent wastage. Printing techniques save labor and energy. Recently, several studies are carried out and explored in large-scale fabrication of printed and flexible electronic devices using nanomaterial inks. The increasing technological impacts reveal that the field of wearable and flexible electronics and the Internet of Things (IoT) is gaining significant consideration, which is tremendously changing the lifestyle of the globe.

Therefore, the rising domain of PE tremendously offers a favorable technique to fabricate and produce novel and highly efficient electronics. This chapter mainly focuses on the advancement of nanomaterial inks in flexible and thin printed electronic devices. As shown in Figure 7.1, PE technology is deployed in wide-ranging fields such as energy conversion devices, storage devices, lighting devices, flexible display materials, wearable sensors, biomedical engineering. Revolution in printing technology has made it possible mass production of electronic devices at a cheaper rate on flexible substrates such as plastic, paper, or textile materials.

Nanotechnology in Electronics: Materials, Properties, Devices, First Edition.
Edited by Visakh P.M., Artem Semkin, Raneesh Balakrishnan, and Saša Lazović.

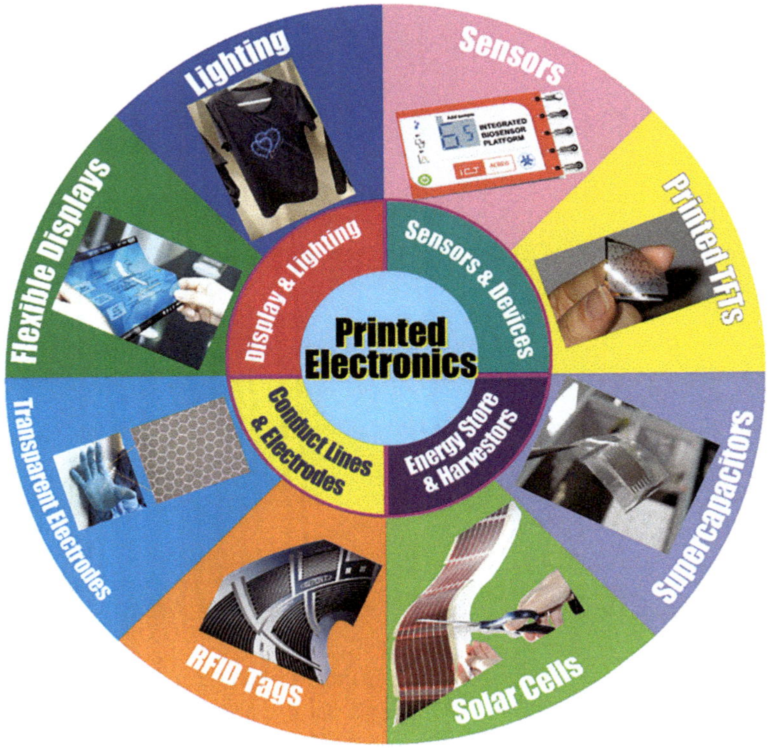

Figure 7.1 The different fields of implementation of the printed electronics technique. Source: Reproduced with permission [1], Royal Society of Chemistry.

This flexibility has also made the materials to be biodegradable, thereby reducing e-waste production. In this chapter, various advances and recent applications of nanomaterial inks and their synthesis in the field of PE are discussed. The initial discussions involve the fundamentals and features of different printing technologies and their requirements and features. Several kinds of literature are reported involving different types of nanomaterials such as silver-based metallic inks and 1D, 2D carbon-based inks along with their applications. Subsequently, several research activities such as identification of finding new printable nanomaterials, their functionality and practical applications are still being carried out with a hope of achieving much more efficient nanomaterials to serve the fled of PE.

7.2 Printing Technology and Challenges with Fabrication of Electronics

The different types and categories of printing technologies are presented in Figure 7.2. Printing is a process whereby formulated inks are deposited on top of the substrate material for data or information transfer just as graphics, images, text, and

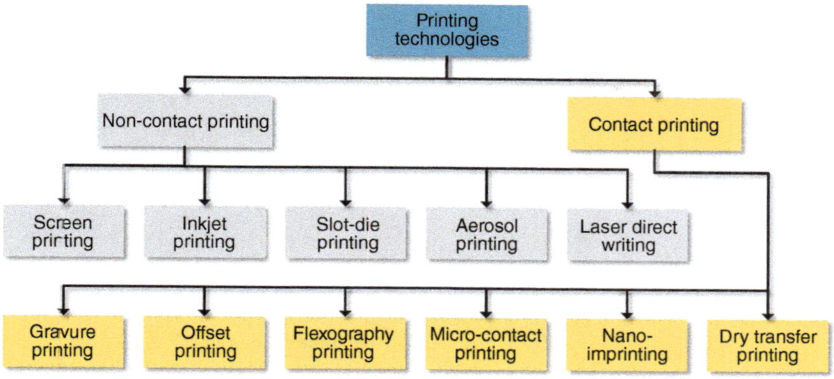

Figure 7.2 Classification of the common printing technologies.

the forms reappear as an image-carrying medium called a substrate. Accordingly, thus, PE and flexible electronics follow a set of printing methodologies that are used to build electronic devices on flexible substrates. The printing process generally uses a printing apparatus to define patterns of various materials over substrates using either contact or non-contact techniques as highlighted in Figure 7.2.

7.2.1 Inkjet Printing

In inkjet printing, upon receiving a digital signal, the drop-on-demand (DOD) printer discharges essential ink droplets as shown in the Figure 7.3b. Based on the nozzle, there are two methods of jet printing: (i) bubble-jet printing model and (ii) piezoelectric printing model. In the bubble-jet printing model, there is a print head with a heating unit. This heat unit forms ink bubbles and also forces the ink drops out of the nozzle. In the piezoelectric printing model, the nozzle has a ceramic piezoelectric tile that gets expanded into the ink reservoir inside the print head and forces the ink drops out of the nozzle. DOD inkjet printing requires good technical expertise in the field of PE. It is a non-contact printing methodology without mask utilization. Different thin layer patterns are deposited through inkjet printing on the top of the substrate and can be easily modified with a computer. Reduced material wastage and cost-effectiveness are some of its promising features.

The complementary term of the Ohnesorge number (Z) in inkjet printing has a major role in determining the capability of ink droplet formation. It is a dimensionless number and is related to the surface tension as well as the viscosity of the printing inks:

$$Oh = \frac{\mu}{\sqrt{\rho \sigma L}} = \frac{\sqrt{\mathrm{We}}}{\mathfrak{R}}$$

where t μ represents the viscosity, σ the surface tension of the ink, and ρ r the density. L, We, and Reynold number are the diameter of the droplets, the Weber number, and the Reynolds number, respectively.

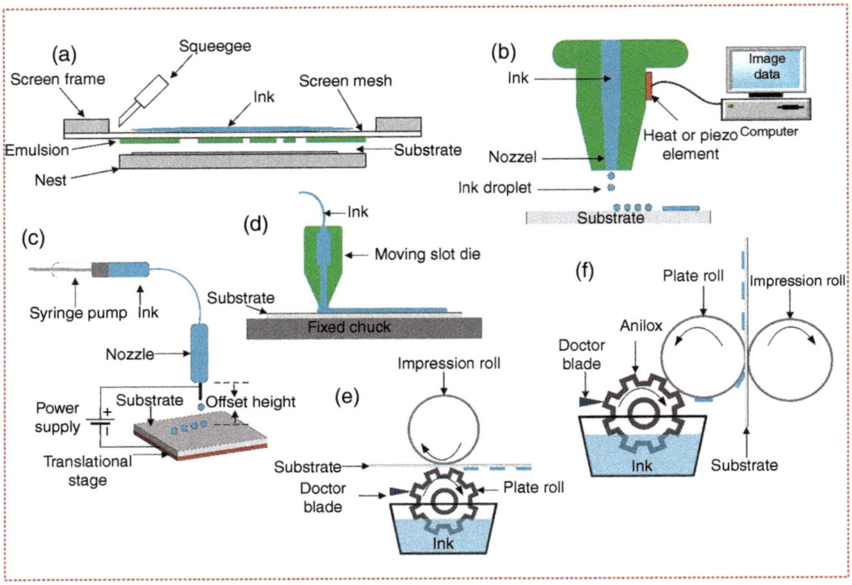

Figure 7.3 Different printing techniques: (a) screen printing, (b) inkjet printing, (c) electrohydrodynamic, (d) slot die coating, (e) gravure printing, (f) flexographic printing.

7.2.2 Screen Printing

Screen printing is beneficial to print inks on a variety of substrates with flexible or rigid structure materials. The entire process is remarkably simple, flexible, and economical. During printing, the ink passes through a stencil made of a porous material and deposits onto the substrate material. A delicate mesh made of nylon textile is used as whole screen and stretched over a wooden or aluminum frame with a fitness of 150–400 fibers per inch. Sometimes, metal screens are also chosen, with a fitness of 850 fibers per inch. The areas in the pattern zones open, thereby permitting penetration of the printable ink. The toughened photosensitive material present in other areas blocks the penetration, thereby creating a screen plate at the top. As shown in Figure 7.3a, inks are passed over the screen plate to fill all the mesh openings. The mesh is moved down using a squeegee, and the squeegee is pushed onto the back of the screen. The amount of ink to be pumped in the mesh openings is controlled by the capillary effect. The tension on the screen at the rear side decreases during the forward movement of the squeegee, thereby forming meshes from the substrate where the ink is deposited. Thus, on the substrate, a design is created, which can be used in a later stage. During printing, inks undergo four steps: filling, contact, adhesion, and release.

7.2.3 Slot Die Coating

Deposition of uniform, homogeneous wet films in cross-directionality through a non-contact method for large-scale production is done using slot die coating.

The process can handle different viscosities ranging from <1 mPa s to a few thousand Pa s. The coating head mounted near the web has a slot inside, which aids in pumping the ink as shown in Figure 7.3d. After the meniscus formation, the moving substrate along with the web leads to a homogeneous layer deposition, which results in a continuous coating on the substrate with even thickness. The film thickness is given by

$$d = \frac{f}{Sw} \cdot \frac{c}{\rho}$$

where c is the ink concentration, f is the rate of flow, S is the coating speed, ρ is the final film material density, and w is the width of coating.

7.2.4 Electrohydrodynamic (EHD) Printing

Electrohydrodynamic (EHD) printing is a high-resolution printing technique that uses an electric field to produce an essential fluid flow for ink delivery onto a substrate . It produces two to five smaller droplets compared to the nozzle size. The positive impact on the shape regulation of nanoparticles in the droplets developed for printing is due to the electric field existing between the substrate and the nozzle. It is also preferred in additive manufacturing of 3D structures. As shown in Figure 7.3c, some of the significant elements in EHD printing are pressure regulator, nozzle tip, translational stage, ink chamber, and substrate. The conditions of printing are influenced by factors such as backpressure, applied voltage (V), offset height (h). The relationship for the jetting frequency (f) with V and h is given by

$$f = K\left(\frac{V}{h}\right)^{3/2}$$

7.2.5 Gravure Printing

Functional inks are directly transferred through imprinted structures by physical contact with the substrate in the gravure printing technique. A high rate of speed is required for processing the functional materials in the roll-to-roll (R2R) pattern. Figure 7.3e presents the operating principle of this technique. The printing roller is prepared with imprinted cells designed on the surface, which moves against the doctor blade and eliminates the excessive ink from the roller surface. The ink is moved to the substrate surface between the impression and printing roller. Gravure printing is also referred to as "inverse direct gravure printing," in which with the help of the impression roll in a flat plate the pattern is transferred onto the substrate, which is illustrated in Figure 7.3e. The morphology and fidelity of the pattern printed are determined by the rheological behavior of the ink, characterized as Ca number.

$$\mathrm{Ca} = \frac{\eta U}{\gamma}$$

where U is the rate of speed of printing, γ is the surface tension, and η is the ink viscosity.

At low Ca, the ink is dragged out from the cells, thereby deteriorating the pattern fidelity, whereas at high Ca, the ink gets deposited in the non-patterned region due to inefficient doctoring. Optimal printing can be achieved at Ca ≈ 1 by regulating the ink parameters and the printing speed.

7.2.6 Flexographic Printing

Flexible relief plates are used in flexographic printing. Ink is pulled out using an imprinted cell on an anilox cylinder, and any excess materials are removed using a doctor blade. The raised-up image on the printing flexible relief plates are transferred onto the substrate once the ink in the cells are moved to the relief plate. A schematic design of this process is shown in Figure 7.3f. Different types of inks on porous and nonporous substrate materials are used in flexographic printing.

7.2.7 Roll-to-roll (R2R) Printing

High processing speed and addictive nature are the merits of electronics bulk manufacturing, which are utilized in R2R printing. R2R printing is mostly preferred for bulk development of electronics due to its addictive nature and swiftness. Inkjet printing is the only non-contact printing technique that is widely comparable to the R2R process.

Large-scale inkjet printers outfitted with the industrial print heads can realize the R2R printing process by combining with a flash sintering unit based on the photonic effect. Conductive patterns can be delivered at a rate of 10 m min^{-1}.

7.2.8 Printable Nanomaterials Requirements

Figure 7.4 compares the competency for minimum dot sizes among a variety of printing methodologies. Before setting the research projects' suggestions and production plans, the important factor to be considered is the choice of the printing method. New and advanced inkjet printers retain features down to 1 μm but are not into the production scale. Minimum printing area of <1 μm has drawn great attention to EHD printing. Various printing approaches require different rheological ink properties to match the viscosity of inks to be used before print. The desired PE inks must retain outstanding printability, thus satisfying further requirements. As an initial stage, ink preparation must be lucid with high output. Next, the range of surface tension and viscosity of the printable inks should be considered to assort with various printing technologies. Durable stability of ink at room temperature is another consideration and is expected to give high conductivity. For lower temperatures below 150 °C, the reproduced patterns with conductive inks are expected to give bulk conductivity on sintering. Metal ink oxidation is prevented by insulator inks, which possess good insulation properties. Further improvements are required in the field of PE research and modern material chemistry regarding inorganic conductive ink production. Till now numerous materials, for example, carbon, conductive polymers, and metallic nanoparticles, are synthesized and find their application

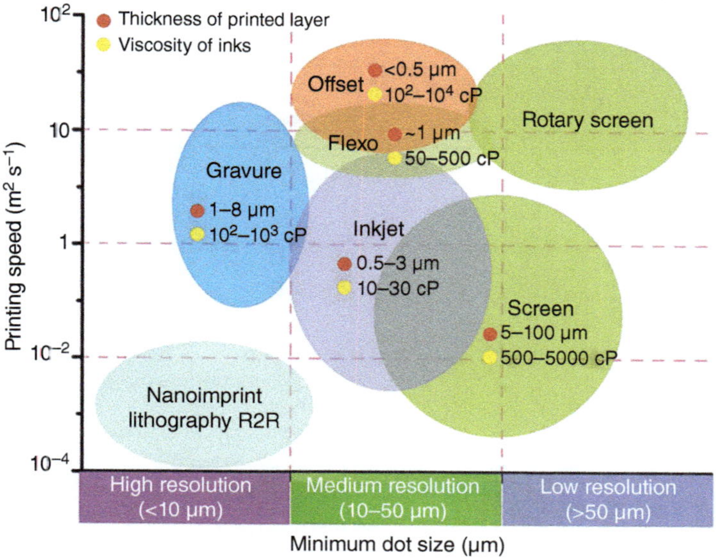

Figure 7.4 Resolution and throughput for different printing technologies. Source: Reproduced with permission [1]. Copyright Royal Society of Chemistry.

as functional inks for designing and manufacturing plastics, PE devices based on paper substrates and textiles, which have paved a way for the growth of research in printable electronics.

7.3 Synthesis and Preparation of Nanomaterial-based Inks

Wiring is an important technology for all types of electronics. PE technology uses several kinds of electrical conducting materials such as organic molecules (PEDOT/PSS [poly 3,4-ethylenedioxy-thiophene] poly [styrene sulfonate]), metallic materials (Ag, Au, Cu), and ceramics (C nanomaterials and oxides). Electrical conductivity is the primary characteristic of a wiring material. Electrical conductivities of different materials are compared in Table 7.1. The bulk material properties hinder certain carbon materials such as graphene and carbon nanotubes (CNTs) from direct comparison of their electrical conductivity with other materials. Metallic materials such as silver and copper have the highest value among other materials. In addition, ceramics and organic materials retain certain properties.

Aqueous or organic solvents generally used for dissolving cannot be used here for metals. A traditional approach to create printable metals is performed by mixing metal particles with a polymer of high viscosity and a paste is formed. The paste has poor electrical conductivity (10^{-3}–10^{-4} Ω cm) although it gives better reliability and good mechanical interconnection. Bulk materials possess electrical conductivity in the range of 10^{-6} Ω cm. Based on recent progress of research in the field of

Table 7.1 Comparison of electrical conductivities of conductive nanomaterials for printed electronics.

Materials		Electrical conductivity (siemens cm^{-1})	Remarks	Ref
Metals	Ag	6.2×10^5	Bulk properties	[2]
	Cu	5.9×10^5		
	Au	4.4×10^5		
	Pt	1.0×10^5		
	Ni	1.4×10^5		
Organic ceramics	PEDOT/PSS	$1–10^3$	Depends on doping/composition, oxygen defects, and crystalline, High-temperature treatment	
	ITO	$10^3–10^4$		
1D, 2D	CNT	1×10^5 cm^2 V^{-1} s^{-1}	Theoretical values of single-walled (SW) CNT electron mobility	
	Graphene	2×10^5 cm^2 V^{-1} s^{-1}	One- and two-dimensional materials	

surface modification and nanotechnology, with a minimum quantity of surfactants and without resins, organic solvents or water can create a dispersion of metallic nanoparticles (ink). Once the ink is printed and the solvent dried, a self-supported uniform film is formed with closely contacted metal nanoparticles with minimum lubricant residues. These films thus formed exhibit better electrical properties compared to metallic pastes but poor mechanical characteristics.

In addition to ink formulation design for a particular printing technique, synthesis of conductive nanomaterials should also be considered as a major concern during the application of various printing methods for fabrication of functional devices. The inks should provide good electrical conductivity for patterns printed during the process. The ink should be stable without any aggregation and must have a high concentration for conductive nanomaterials. High printing resolution and precision are achieved based on the physical properties of the ink, which include surface tension and viscosity. The essential materials for PE shown in Figure 7.5 include a suitable filler material that satisfies the characteristic feature of the device, solvent to enhance ink flowability, binding material to enhance stretchability, and further additives.

7.3.1 Metallic-based Inks

Top-down and bottom-up are the two main methods used for preparation of metal nanoparticles. In the top-down approach, processes such as pyrolysis, grinding, milling, and etching are used to break bulk metals to smaller particles to form

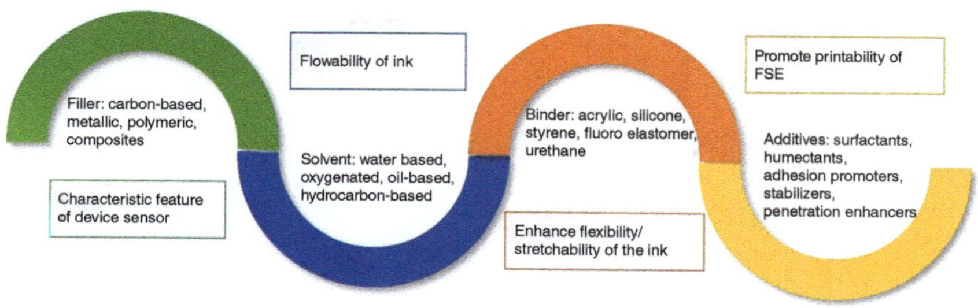

Figure 7.5 Essential materials for printed electronics.

nanoparticles. In the bottom-up method, atomic-scale precursors are used to synthesize nanosized materials. Among these two approaches, the bottom-up is largely popular due to its control on the size of the nanoparticles. Wet chemistry process is preferred among all bottom-up approaches to produce metallic nanoparticles (mainly Cu NPs and Ag NPs) from ionic precursors (CH_3COOAg, $Cu(NO_3)_2$, $AgNO_3$, $CuCl_2$) by the reduction reaction. The physical approach for Ag NP synthesis can be done using various techniques, which come under the top-down approach such as thermal decomposition, evaporation–condensation, pulsed wire discharge, laser ablation, mechano-chemical synthesis, spray pyrolysis. The merits of the physical approach include solvent contaminant restriction, homogeneous particle growth, and highly pure nanoparticles with less agglomeration. There are also some demerits: prolonged NP synthesis time, expensive equipment utilization, and high energy consumption.

7.3.1.1 Silver Nanoparticles

Magdassi et al. prepared concentrated citrate-reduced silver nanocolloids as ink for inkjet printing on surfaces such as glass and paper. The material was prepared by redispersion following lyophilization of newly prepared colloids. Carboxymethyl cellulose (CMC), a polymer stabilizer, was used as a binder and stabilizer. The material was characterized by its high electrical conductivity, reactivity, and optical properties [3].

Shankar et al. synthesized Ag NPs by the reduction method using tin acetate as the reducing agent with a passivating shell of dodecylamine (DDA). This process was carried out in the toluene medium. The aim was to formulate nonaqueous conductive ink to print circuits for antennas and RFID tags on surfaces such as Epson photo paper through aerosol jet printing. The characterization revealed their low resistivity and 10% high electrical conductivity (Figure 7.6) than the bulk silver material [4].

Zhang et al. synthesized highly focused monodisperse Ag NPs by the reduction method using dextrose and adipoyl hydrazide. Conductive inks were prepared using the synthesized nanoparticle using solvents such as ethanol, water, ethylene glycol. The formulated ink was used to print electronics circuits such as electrodes and sensor devices on the paper substrate through inkjet printing. This approach was found to be cost-effective while improving the conductivity due to the appropriate ratio of Ag-PVP/$AgNO_3$ [5].

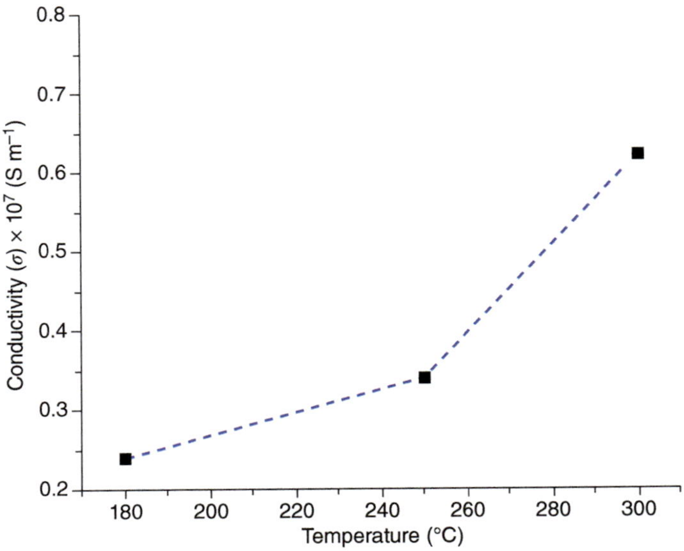

Figure 7.6 Temperature vs. conductivity for printed silver tracks on polyimide. Source: Reproduced with permission Shankar et al. [4]. Copyright Royal Society of Chemistry.

Walker and others created reactive silver inks by modified Tollens' process to inkjet print larger devices such as photovoltaics, transistors, displays, batteries, antennas, sensors with high conductivity. The reactive silver ink was prepared by using a combination of silver acetate in aqueous ammonium hydroxide (NH_4OH) titrated with the formic acid. The conductivity studies performed on different types of silver inks revealed that the created reactive silver inks possessed the highest conductivity as shown in Figure 7.7 [6].

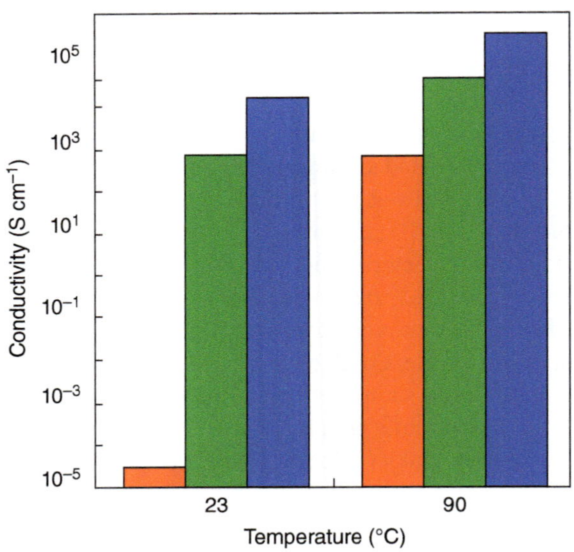

Figure 7.7 Electrical conductivity vs. temperature for the developed Ag NPs (red), the silver particle ink developed for pen-on-paper printing (green), and the reactive silver ink (blue). Source: Reproduced with permission Walker and Lewis [6]. Copyright ACS Publications.

Lai et al. presented a review article on synthesis of silver NP and inks. The authors reported that the preparation of Ag NPs followed by the formulation of silver ink is the most crucial step. Also, control of chemical concentration is especially important to maintain the size and morphology of the nanoparticles. The most prominently suggested method to synthesize nanoparticles is chemical reduction. It requires a reducing agent to decrease the ion solution to the physical nanoparticle form. Some of the common reducing agents are sodium citrate, dimethylformamide (DMF), ascorbate, Tollens reagent, and ethylene glycol. The authors concluded that most printing metal inks on the market are silver based owing to their good electrical properties [7].

Huang et al. synthesized Ag NPs by reducing silver nitrate with triethanolamine in the presence of polyacrylic acid as shown in Figure 7.8. The conductive ink was prepared by dispersion of Ag NPs into a mixture of water and ethylene glycol. The prepared ink was used to print silver patterns on paper through the inkjet technology. The conductivity of the printed patterns increased, which had proven its feasibility to print flexible electronic devices [8].

Zhou et al. performed a study on DDA-protected Ag NPs by modifying a small amount of dodecanethiol as a co-protective agent to formulate conductive silver inks for inkjet printing. The study revealed that the electrical properties can be enhanced by heat-treating at 90 °C quickly, thereby proving its suitability to print electronic circuits [9].

The same team also synthesized water-dispersible Ag NPs using polyacrylic acid as capping agents to inkjet print on paper substrates. After treating the Ag NPs with the poly(diallyldimethylammonium chloride) (PDAC) solution at room temperature, the nanoparticles coalesced together following the detachment of polyacrylic acid (PAA). The films showed alterations in the morphology and electrical properties. This identification is significant for printed electronic devices [10].

Wu et al. synthesized and characterized carbon–core/Ag–shell NPs Figure 7.9 at different thicknesses by successive reduction of silver ammonia. Conductive ink was formulated to print RFID antennas on the paper substrate. It was produced at a large scale and replaced the conventional silver paste as a conductive ink. Low cost and

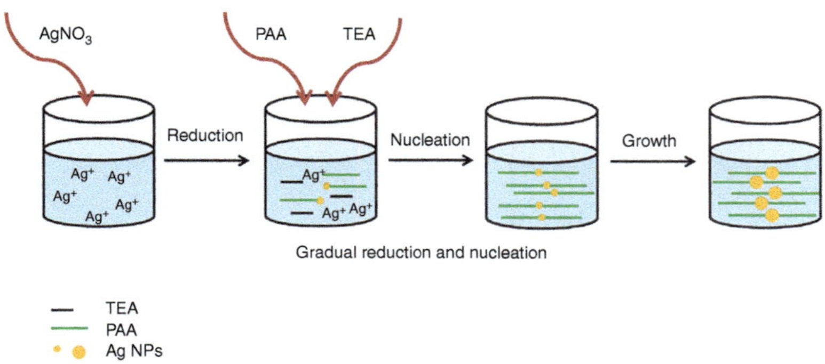

Figure 7.8 Schematic representation of the preparation of Ag NPs. Source: Reproduced with permission Huang et al. [8]. Copyright Elsevier.

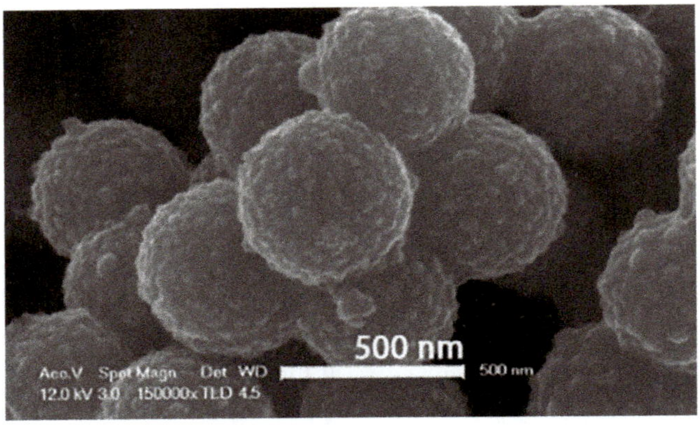

Figure 7.9 SEM images of Ag core–shell. Source: Reproduced with permission Wu et al. [11], Elsevier.

high conductivity enable this nanoparticle for the replacement of Ag paste to form a conductive paste [11].

Titkov et al. developed Ag NPs with organic silver salt and polymer additives to formulate inkjet printing silver inks. Silver nitrate of reagent grade, nitric acid, sodium hydroxide, caprylic acid, neodecanoic acid, ethylene glycol, toluene, isopropanol, and butyl glycol acetate were used in the experiment. Caprylic acid was used as a stabilizer. Toluene or xylene was used as the solvent for the inks based on Ag NPs. It was found that the presence of the polymeric surfactant hinders the conductive layer formation [12].

Vaseem et al. developed particle-free silver-organo-complex (SOC) ink to print flexible electronics with high conductivity as shown in Figure 7.10. Silver ink was inkjet printed to create inductors on different types of substrates with the same surface morphology, which served as proof of concept [13].

Zhang et al. developed tannic acid (TA)-stabilized Ag NPs for inkjet printing with enhanced conductivity for flexible electronics as represented by the scheme in Figure 7.11. TA was used as a capping and reducing agent and also in an aqueous solution. A conductive pattern was attained by inkjet printing by means of a common color. The prints showed good adhesion. It has proven to effectively print conductive LED device circuits [14].

Cai et al. developed a conductive pen made of Ag NPs by a microwave method. The obtained Ag NPs possessed uniform particle size and could be evenly scattered

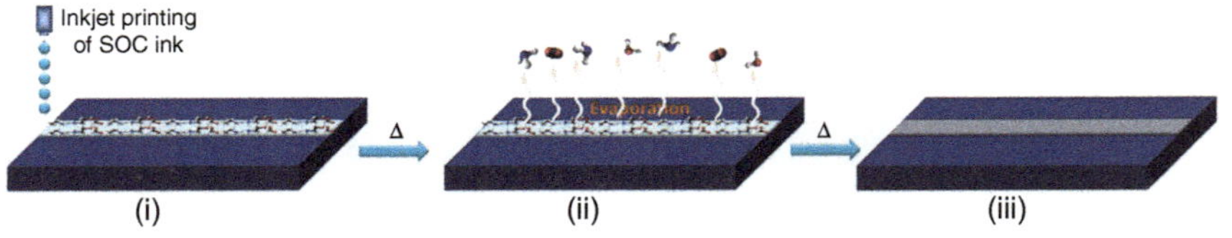

Figure 7.10 Schematic presentation of thermal reduction process. Source: Reproduced with permission Vaseem et al. [13]. Copyright ACS Publications.

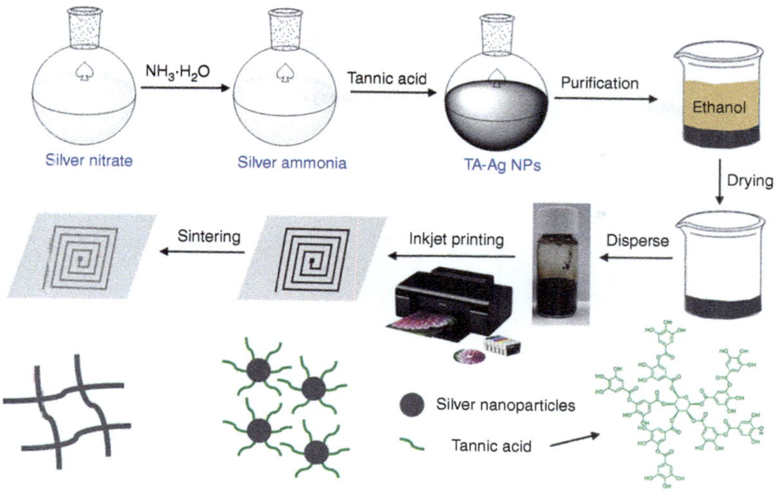

Figure 7.11 Schematic on the formulation of TA-Ag NP inks and their function in the field of conductive flexible electronics. Source: Reproduced with permission Zhang et al. [14], Royal Society of Chemistry.

in deionized water. The ink was injected into a blank mark pen to produce the conductive pen to draw on a photo paper. Studies prove that the fabricated conductive pen and ink find their major application in the field of flexible electronics [15].

Zhang et al. performed a synthesis of Ag NPs for paper-based electronics through inkjet printing. Conductivity of the printed silver tracks was enhanced by a multiple-step room-temperature post-processing method. Electrical properties, mechanical flexibility, and strain sensitivity have proven to be effective for the material to serve the future of 3D electronics [16].

Kastner et al. and Shahariar developed Ag NPs for inkjet printing of conductive lines on textiles. The reactive silver ink contained silver in the complexed form and no dispersions were found. The ink was tested on various substrates as shown in Figure 7.12. The inkjet printer electronic circuit paths on the fabric were found to withstand washing tests, moisture, and heat [17, 18].

Chang et al. performed a study involving water-in-oil type emulsion for H_2O-based Ag NPs synthesized using diethanolamine, polyacrylic acid, and silver nitrate. Ag NP morphology after emulsification was analyzed with electron microscopy to explicate the engulfment of particles. The Ag NP encapsulation revealed a low deposition rate due to the decrease in density of Ag NPs/water droplets encapsulation, preventing aggregation and good ink stability. The ink was utilized to print good conductive tracks through inkjet printing as shown in Figure 7.13 [19].

Hassan et al. printed transparent, flexible, and mesh-type electrodes based on Ag NPs on polyethylene terephthalate (PET) substrate through inkjet printing. The sheet resistance was found to decrease with the decrease in transparency due to the hole size. The electrodes developed were superior to the CNT electrodes. The *V-I* characteristics indicate that the current decreases with an increase in the hole size as illustrated in Figure 7.14 [20].

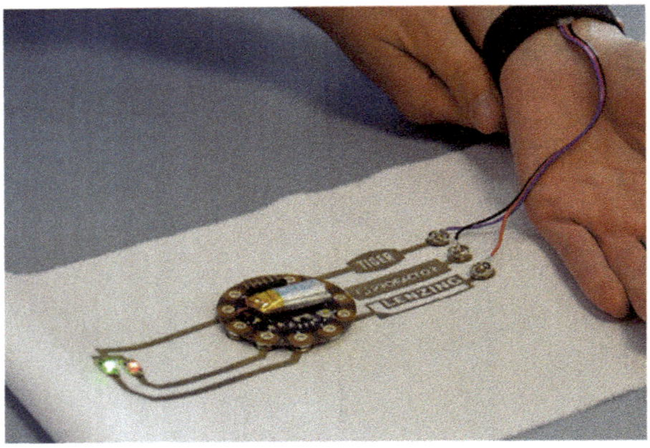

Figure 7.12 Homemade conductive ink printed onto the textile. Source: Reproduced with permission Kastner et al. [17], Elsevier.

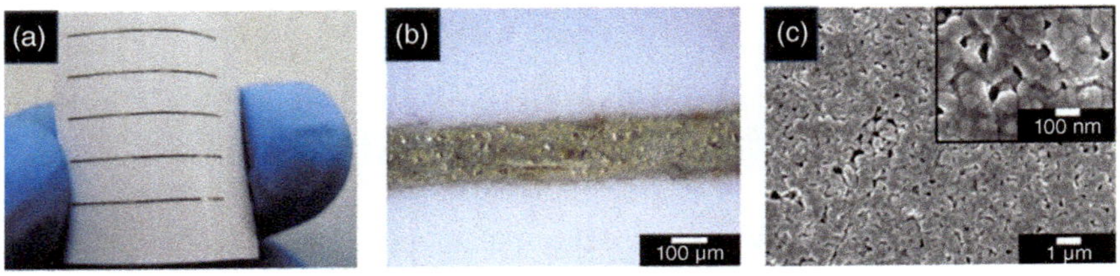

Figure 7.13 SEM image of conductive tracks printed on paper substrate with Ag NP-encapsulated nanoemulsion: (a) appearance, (b) microscopic image, and (c) top-view. Source: Reproduced with permission Chang et al. [19], Elsevier.

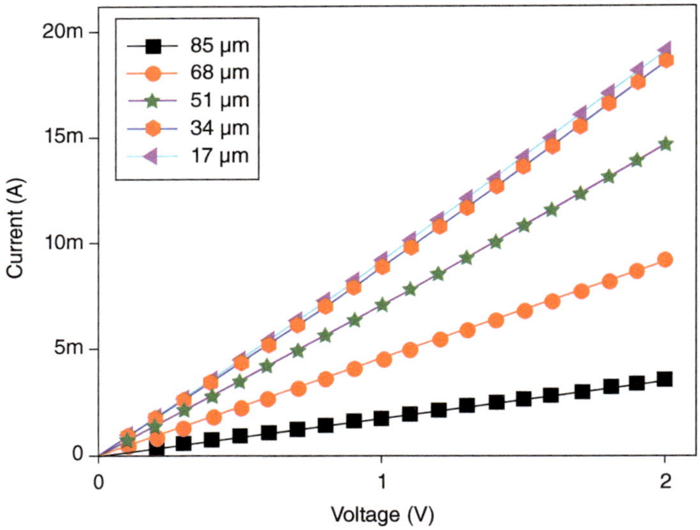

Figure 7.14 *I*–*V* graph and resistance vs. at different hole size. Source: Reproduced with permission Hassan et al. [20]. Copyright Springer.

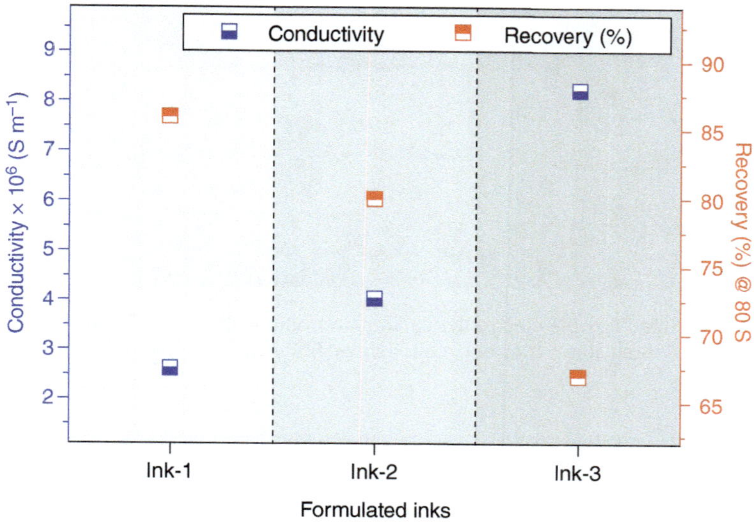

Figure 7.15 Electrical conductivity of three different silver ink combinations for screen-printed films. Source: [21] / with permission of Springer Nature.

Manjunath et al. developed a screen printable Ag ink, which finds its application in screen printing technology for fabricating highly conductive electrodes. The particle-free ink was formed using citric acid and Ag nitrate with the help of the sodium-carboxymethylcellulose binder. Three loading concentrations in the Ag ink were considered for optimizing the printing process: 14, 18, and 22%. Results showed a linear variation between the silver nitrate concentration in the ink and conductivity as illustrated in Figure 7.15. The printed silver films have found application in thin-film transistors (TFTs) [21].

Mo et al. presented a review article on ink based on Ag NPs, which can be applied in the field of printed and flexible electronics, and summarized upon Ag NP–based inks. The ink was prepared with moderate sintering. The authors highlighted various strategies to tailor the protective agents. Various sintering methods such as photonic sintering, plasma sintering, electrical sintering, microwave sintering are also discussed. The key applications discussed in the review article are RFID antenna, conductive film, biosensor, TFTs, and stretchable electronics [22].

Hao et al. prepared Ag NPs using humic acid (HA). HA chemically reduced the silver salts into the liquid phase, thus acting as a stabilizer. The thus prepared conductive ink was used to print Ag conductive patterns on a photo paper using a domestic inkjet printer. The patterns possessed the capability to be incorporated into an alarm or LED device, signifying their practicability in flexible printing electronics as illustrated in Figure 7.16 [23].

Fernandes et al. synthesized and characterized Ag NP conductive inks. The process included silver nitrate chemical reduction and formulated seven different

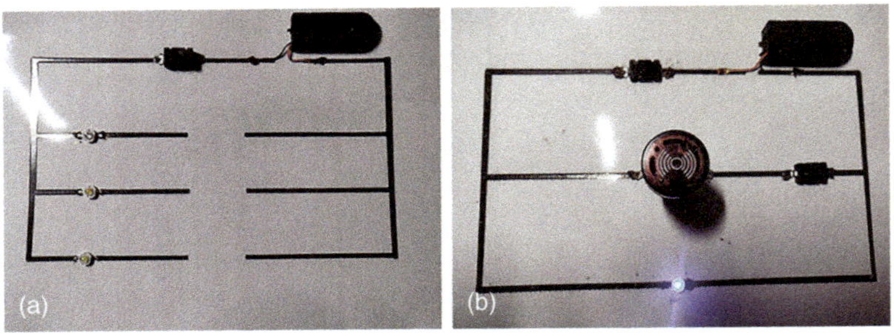

Figure 7.16 Inkjet printed flexible electronics circuits over photopapers (a) Printed LED electronics circuit (b) Printed alarm control circuit. Source: [23] / with permission of Springer Nature.

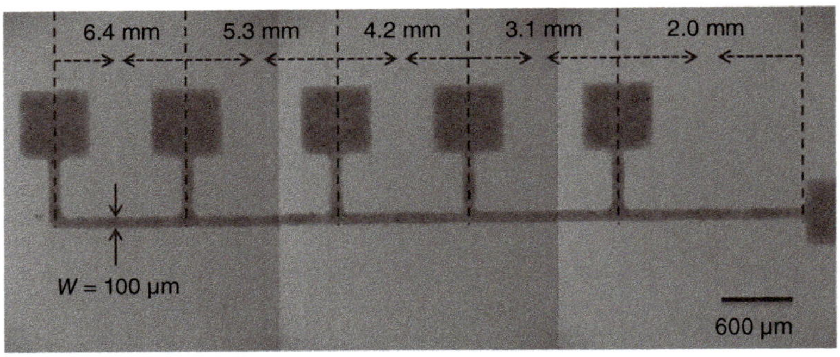

Figure 7.17 Optical microscopy image of a screen-printed Ag NW pattern. Source: [25] / with permission of John Wiley & Sons.

conductive inks. Investigations were carried out on viscosity, resistivity, inkjet printability, surface tension, and adhesion. Studies have proved the suitability of the inks for utilization in flexible electronics especially for biological sensing [24].

7.3.1.2 Silver Nanowires

Liang et al. developed water-based Ag nanowires (Ag NWs) to fabricate wearable TFTs and stretchable conductors through screen printing on PET sheets. New water-based Ag NW ink was prepared using a mixture of (hydroxypropyl) methylcellulose, fluorosurfactant; deionized water was used as the solvent and an antifoaming agent as an organic binder. Investigations were carried out to connect the ink rheological properties with its compositions. The conductivity of the ink was measured to be high for various widths of the printed lines as shown in Figures 7.17 and 7.18 [25].

Chen et al. discussed a systematic procedure for Ag NW ink. The authors commented that Ag NW inks made by researchers aware immature due to unclear research procedure to develop qualified inks. In their study, the authors discussed the electric double-layer model, bubble theory, Young equation, percolation theory, Kelvin equation, and Rhodes equation [26].

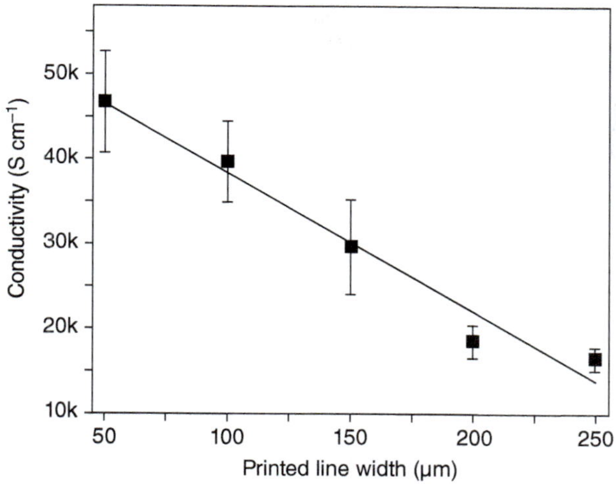

Figure 7.18 Conductivity of the screen-printed Ag NW lines with various line widths. Source: Reproduced with permission Liang et al. [25]. Copyright Wiley.

Huang et al. developed water-based Ag NW ink, which has its application in the field of flexible electronics. Gravure printing was done though it is limited by resolution and electrical properties such as conductivity. The substrate used was plastic. The major concern during the development of highly conductive Ag NW patterns on the PET sheet for flexible electronics applications were printing conditions, ink properties, post-printing treatment, and so on as shown in Figure 7.19 [27].

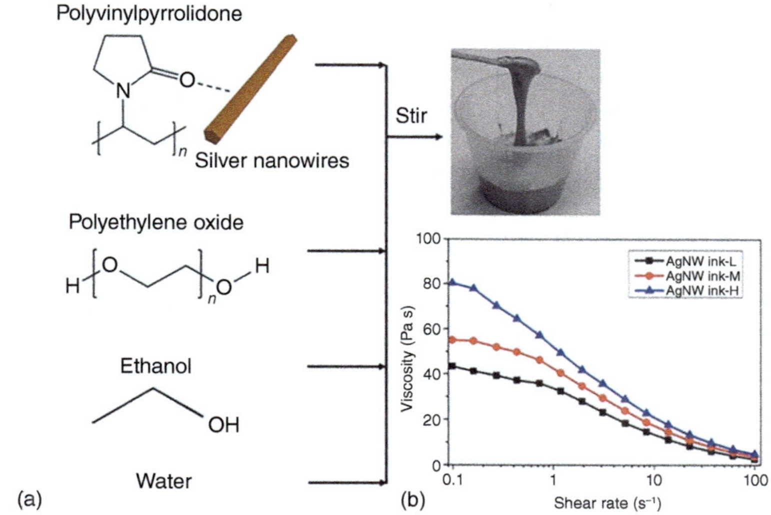

Figure 7.19 (a) Ag NW ink fabrication process and its digital image of 5 wt% Ag NW ink. Source: Reproduced with permission Huang et al. [27], Springer Nature. (b) Viscosity vs. shear rate for various AG NW ink concentrations. Source: [27] Huang et al. / Springer Nature / CC BY 4.0.

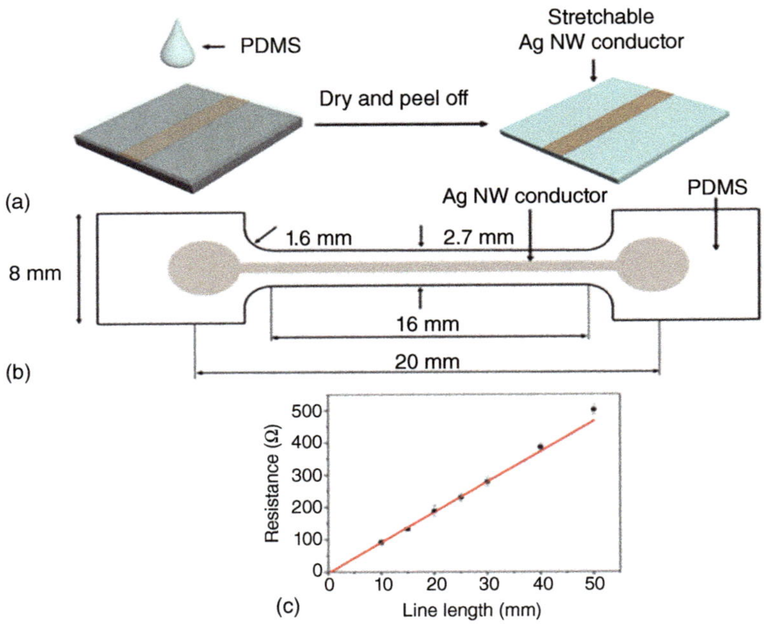

Figure 7.20 (a) Schematic diagram of a stretchable Ag NW conductor fabrication. (b) Shape and dimensions of the device. (c) Measured resistance with different line lengths. Source: Reproduced with permission [28]. Copyright ACS Publications.

Huang et al. performed a study on inkjet printing of high-concentration Ag NW ink on flexible PET substrates to form heater as shown in Figure 7.20. All the characteristic studies have proven the tremendous potential of Ag NW inkjet printing in the mass production of flexible electronics [28].

Li et al. performed their study in screen printing of Ag NWs by focusing on the following study aspects: the balance of transparency and conductivity of the screen-printed stretchable Ag nanofilm ink formulation, post-treatment, and optimization of parameters as shown in Figure 7.21. The screen-printed patterns exhibited tremendous electrical stability and mechanical repeatability [29].

Williams et al. developed high-aspect-ratio water-based Ag NW ink. The developed ink is used for direct writing of electronic tattoos. Hydroxypropyl methylcellulose (HPMC) was used as a viscosifier in a minimum quantity for suspension of synthesized Ag NWs in water. Printing was done on non-planar materials such as fruits and a human finger as shown in Figure 7.22 [30].

Patil et al. performed a study on inkjet printing of Ag NWs synthesized by the polyol method. The study focused on conductivity improvement printed pattern stability on flexible surfaces and techniques followed. The optimization was done by varying the concentration of isopropyl alcohol and synthesized ethylene glycol where Ag NWs mixture was suspended. The patterns obtained by printing using the developed ink demonstrated high conductivity and strength. The flowchart of the printing procedure and electrical characteristics is shown in Figure 7.23 [31].

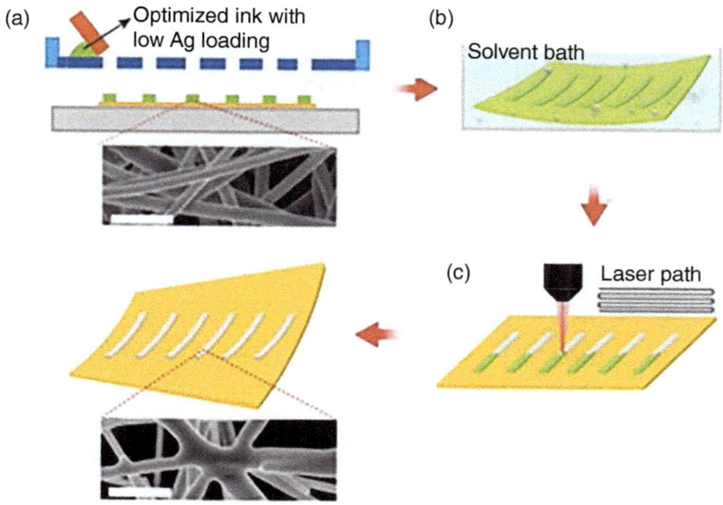

Figure 7.21 (a) Optimization of ink formulation and printing parameters. (b) Immersion of the printed films in a solvent bath. (c) Treatment with a strong laser. Source: Reproduced with permission Li et al. [29], Springer Nature.

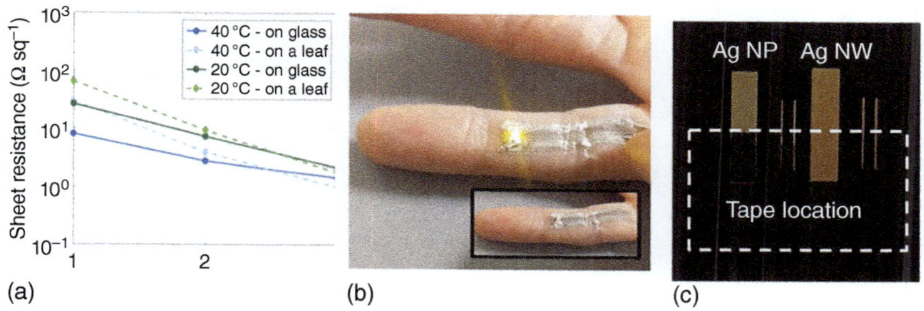

Figure 7.22 (a) Printed trace sheet resistance on biological and inorganic substrates. (b) Conductive traces of print on the finger. Source: Reproduced with permission Williams et al. [30]. Royal Society of Chemistry. (c) Scotch tape adhesion test comparison between a printed film of Ag NPs and Ag NWs.

7.3.2 1D and 2D Material-based Inks

7.3.2.1 CNT Ink

Carbon nanomaterials and graphene-based carbon nanomaterials consist of the honeycomb structure with single-layer single-walled carbon nanotubes (SWCNTs) and multi-walled carbon nanotubes (MWCNTs) consisting of multiple shells, whose structures are shown in Figure 7.24. Each structure possesses its own optical, electronic, mechanical, and thermal properties. These materials serve as a good choice in the field of PE compared to metal-based nanomaterials.

CNTs are manufactured by three techniques: laser ablation, chemical vapor deposition, and electric arc discharge. The widely preferred methodology to manufacture

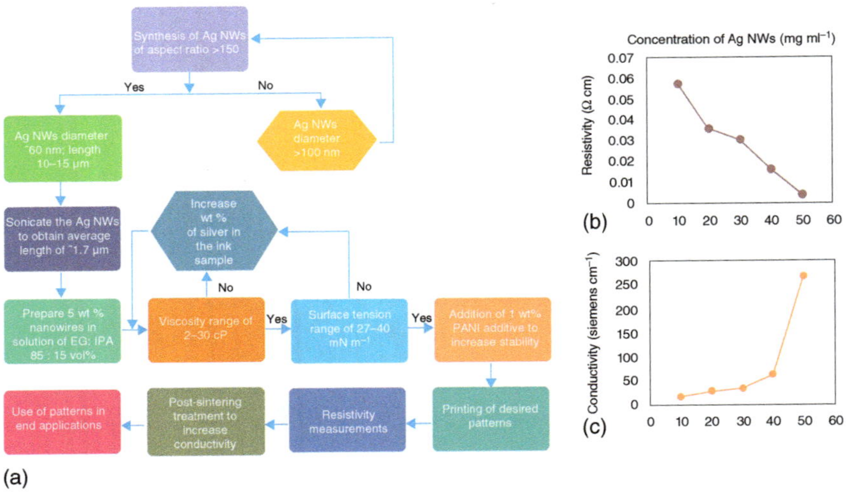

Figure 7.23 (a) Ag NW conducting ink printing procedure in the form of a flowchart and the Ag NW concentration effect on (b) resistivity and (c) conductivity. Source: Reproduced with permission Patil et al. [31]. Copyright Royal Society of Chemistry.

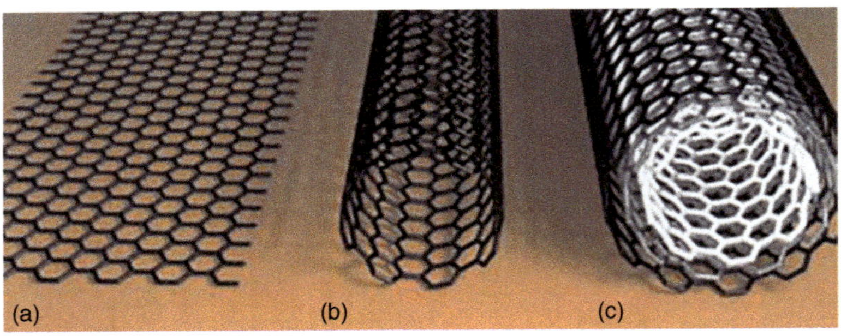

Figure 7.24 Structures of different graphene-based nanomaterials. (a) The basic forms of planar graphene sheet (b) single-walled carbon nanotubes (c) multi-walled carbon nanotubes. Source: Reproduced with permission Kamyshny and Magdassi [32], Royal Society of Chemistry.

SWCNTs and MWCNTs is the electric discharge method in which a DC discharge is applied between graphite electrodes which are heated at the graphite melting point in a chamber full of inert gas at subatmospheric pressure. MWCNTs are produced in the absence of catalyst, while SWCNTs are produced in the presence of metal catalysts such as iron, nickel, cobalt, molybdenum. Laser ablation of pure graphite and hydrothermal or plasma-enhanced catalytic decomposition are some effective methods to produce high-quality SWCNTs. Bottom-up synthesis methods are used for the production of long CNTs. MWCNTs are widely produced with hydrothermal processing by the electrolytic process of water dilution polyethylene or ethylene glycol with the nickel catalyst at 700–800 °C and a pressure of ≈60–100 MPa. Partition of SWCNTs with diverse chirality is attained by surfactant adsorption, and later selective extraction is performed to split up the composite surfactant/SWCNT complexes. The major problem in preparing CNT inks is to obtain evenly dispersed, low-viscous non-aggregated nanotubes. Hydrophobic CNTs when dispersed in aqueous medium or organic solvents become a printable dispersion. Some of the typical organic solvents used are DMF, *N*-methyl-2-pyrrolidone, isopropyl alcohol, and g-butyrolactone. Preparation of printable aqueous inks necessitates the utilization of dispersing agents, namely, Triton X-100, SDS, dodecylbenzene sulfonate, sodium cholate, Tween 20 and Tween CTAB, and dodecyl trimethylammonium bromide.

Fan et al. fabricated and characterized MWCNT-based ink by catalytic decomposition of propylene on the iron catalyst in a nano-agglomerated fluidized bed reactor. The prepared ink was purified and centrifuged as shown in Figure 7.25. The ink was inkjet printed on a paper and characterized. With the increasing concentration of MWCNTs, the electrical resistance decreased exponentially and exhibited good adhesion with the paper, proving its suitability for PE application [33].

Song et al. performed a study in inkjet printing of SWCNTs and performed electrical characterization of the printed pattern and investigated substrate heating and surface hydrophilicity. SWCNTs were prepared by the electric arc discharge method and lines were printed with the number of overwrites as shown in Figure 7.26. The authors concluded that the inkjet printing method of SWCNTs will help in the fabrication of electronic devices for industrial purposes [34].

Denneulin et al. performed a study on impact of ink formulation on CNT network organization within inkjet printer conductive films. The SWCNTs were dispersed in an aqueous solvent by sonication. Inkjet printing was done on the aqueous suspension of carboxylic acid–functionalized SWCNTs. The CNT network arrangement

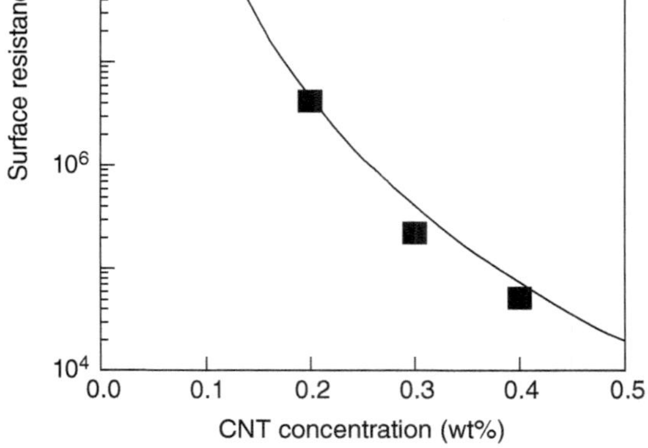

Figure 7.25 The relationship between the concentration of MWCNTs and the printed strip surface resistance on paper substrate. Source: Reproduced with permission Fan et al. [33]. Copyright Springer.

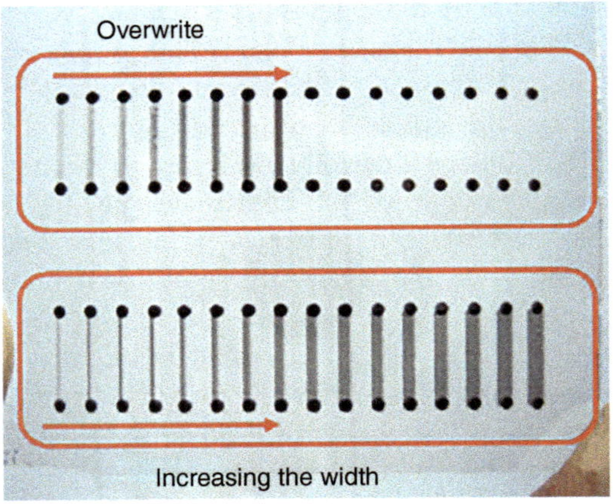

Figure 7.26 Line images fabricated on a glass substrate between Pt dot electrodes. Source: [34] Song et al. / with permission of IOP Publishing.

within the printed layer was visualized. Electrical characterization was carried out by four-probe measurements of sheet resistance, and linearity of the measurement was verified [35].

Niu presented an article on carbon nanotube transparent conducting films. A brief insight was given into the manufacturing methods, thin-film deposition methods, electrical characteristics, optical characteristics, and microstructural characteristics in their study, a schematic representation of which is given in Figure 7.27. The scopes and advantages of replacement of thin-film CNTs on ITO (tin-doped indium oxide) were elucidated. The future perspectives and applications of the thin-film CNTs were highlighted and consolidated in the field of flexible electronics [36].

Zhao et al. performed their study on selective dopamine determination in human serum with inkjet-printed MWCNT chips. The CNT ink was examined with homogeneous double layers to increase the efficiency of dopamine detection. This method successfully resulted in the demonstration of the samples. The low-cost approach carried out in the formulation of ink paves a way for futuristic applications in the field of printed sensors [37].

Guo et al. synthesized SWCNT and graphene oxide composite ink (GO-SWCNT ink) for supercapacitors as shown in Figure 7.28. The fabricated paper electrodes exhibited different capacitance values for different ink concentrations and 85% of capacitance retention were identified, which support the application of ink for energy storage devices [38].

Menon performed a study on curable conducting ink using MWCNTs for PE applications. The ink was used to print a patch antenna on various flexible substrates like Mylar, photo paper, cellulose acetate sheets, and silicone rubber as shown in Figure 7.29. Characterization studies revealed good electrical properties of the printed substrates with less substrate roughness [39].

7.3.2.2 Graphene Ink

Graphene is obtained by bottom-up and top-down approaches. The bottom-up method depends on the growth of carbon atoms into 2D carbon layers by the process called chemical vapor deposition, leading to superior quality sheets of graphene

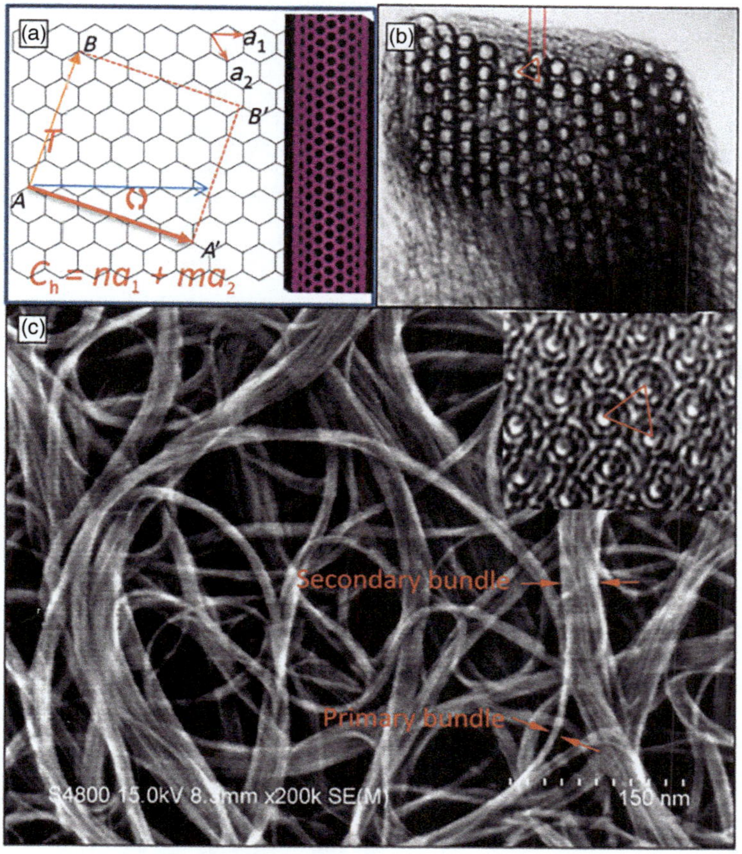

Figure 7.27 (a) Schematic image of SWCNTs, (b) high-resolution transmission electron microscopy (HRTEM) image of the end of an SWCNT bundle, and (c) SEM image of a high-purity and high-conductivity double-walled CNT sample. (a) Source: Niu [36]/Springer Nature. (b–c) Source: Reproduced with permission Niu [36], Materials Research Society.

on metal substrates. But this approach has limitations such as cost and process complexities. Otherwise, the top-down approaches involve graphene production from available sources of bulk carbon by methods such as reduction of graphene oxide, exfoliation of graphite, and carbonization of relevant materials. This approach is widely preferred owing to its elevated yield, low cost, and product quality. Graphene printing is mostly based on the liquid-phase ink approach, consisting of graphene fillers in stabilized solutions, which are made feasible through a top-down approach in formulating conductive inks. The formulated graphene inks must possess stability against precipitation to result in smooth conductive patterns. This can be achieved by optimization of graphene ink composition such as solvents, fillers, and additive materials. The ink needs good fluidic properties, with appropriate surface tension and viscosity for ease of printing. Incompatible ink would always fail the printing device with poor conductive patterns. The ink must possess good

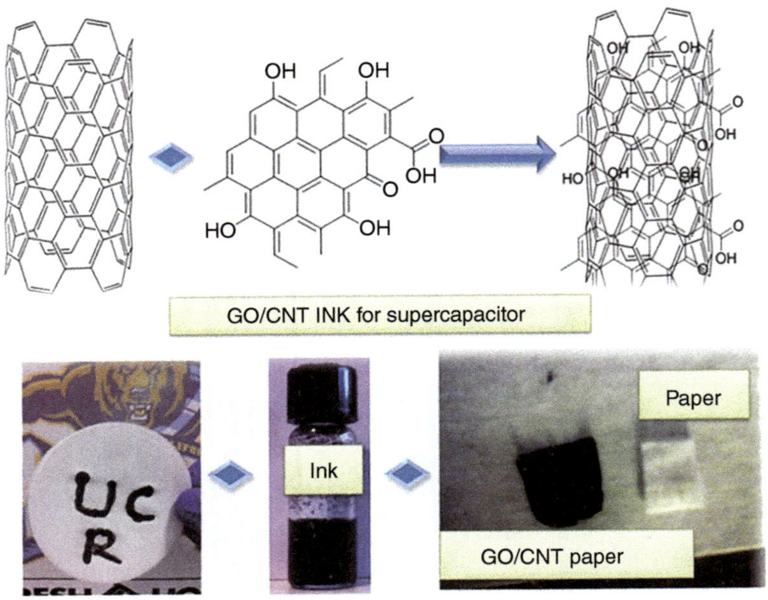

Figure 7.28 Non-covalent interactions of GO-SWCNT and preparation of supercapacitor electrodes using GO/SWCNT ink on paper substrate. Source: Reproduced with permission Guo et al. [38], Materials Research Society.

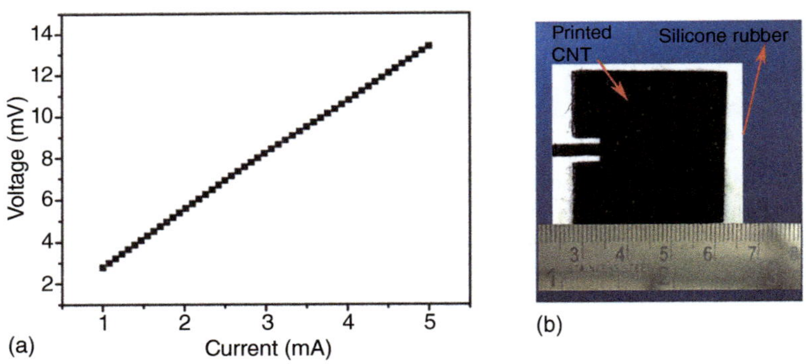

Figure 7.29 (a) Printed CNT pattern *V-I* characteristics. (b) Printed CNT patch antenna pattern on Si rubber substrate. (a) Source: Menon [39]. (b) Source: Reproduced with permission [39], Royal Society of Chemistry.

adhesion to the substrate on which it is being printed. An economical production with high yield is preferred to produce graphene inks for real-time applications.

Combined with mechanical exfoliation and chemical vapor deposition approach, the liquid-phase exfoliation of graphite is an efficient approach for the production of superior quality pristine graphene. The parallel layers of graphene in bulk graphite are stacked one over the other due to the van der Waals force of attraction, and this can be overcome to attain the single sheets of graphene. Even though there

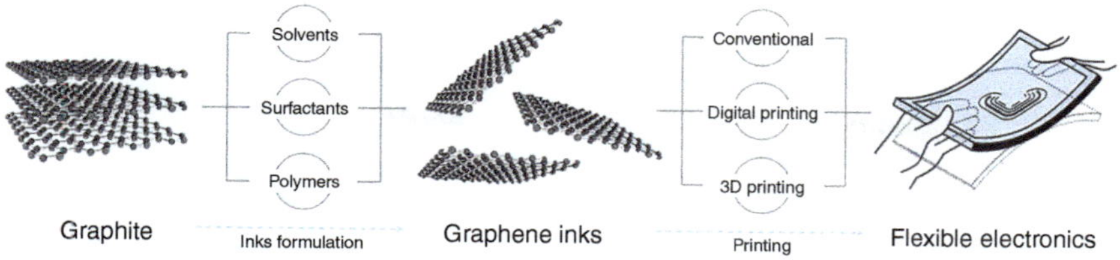

Figure 7.30 Dispersion of graphene. Source: Reproduced with permission Tran et al. [40]. Copyright Elsevier.

is weak van der Waals force between the parallel layers leading to layer sliding, full exfoliation into individual sheets of graphene poses a challenge. Hence, the liquid-phase exfoliation concept using liquid immersion reduces the intermolecular force of attraction between the graphene layers. The adjacent layers on the bulk graphite get peeled off due to high energy fluctuations, thus causing exfoliation during sonification. The individual sheets of graphene obtained after exfoliation are stabilized using solvents to lessen the intermolecular attraction of the dispersed graphene, which is presented in Figure 7.30. Substantial research studies focus on direct exfoliation of graphite into the elevated quality of graphene inks, in which raw graphite is directly exfoliated into graphene using liquid media sonication. This method is cheap, with good processability and good conductivity.

Li et al. performed their study on efficient inkjet printing of graphene. They proposed a method to effectively inkjet print graphene with regained electrical performance. Graphene was exfoliated from the flakes of graphite in DMF. By distillation, terpineol replaced the DMF. Graphene was significantly concentrated with terpineol in much lower volume than DMF. A little amount of ethyl cellulose polymer was added into the collected graphene and DMF dispersion to avoid agglomeration of the graphene flakes. The formulated ink was printed using piezoelectric Dimatix Materials Printer to print resistors, TFTs, micro-supercapacitors, which possess good electrical properties as shown in Figure 7.31 [41].

Gao et al. performed their study on printed patterns using inkjet printing. Here, ethyl cellulose-stabilized pristine graphene on PET substrate as well as on flexible PE substrates was used, which is highly conductive. This ink was formulated by graphite exfoliation using supercritical carbon dioxide assisted by ultrasound.

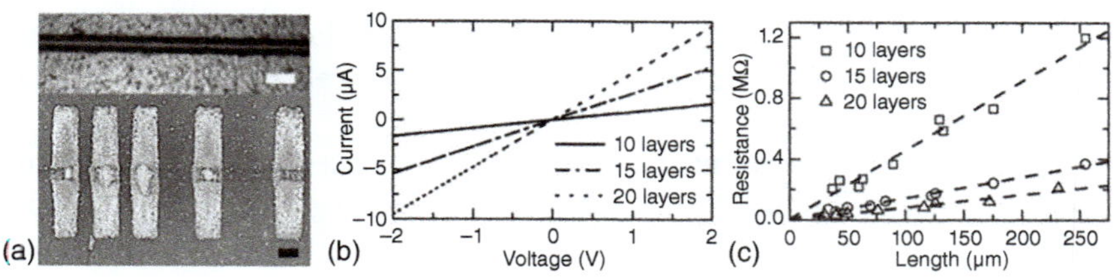

Figure 7.31 (a) Optical images of resistors. (b) Current–voltage curves of the printed line resistors. (c) Variation of resistance in accordance with the device length for resistors with different printing layers. Source: Reproduced with permission [41], John Wiley & Sons.

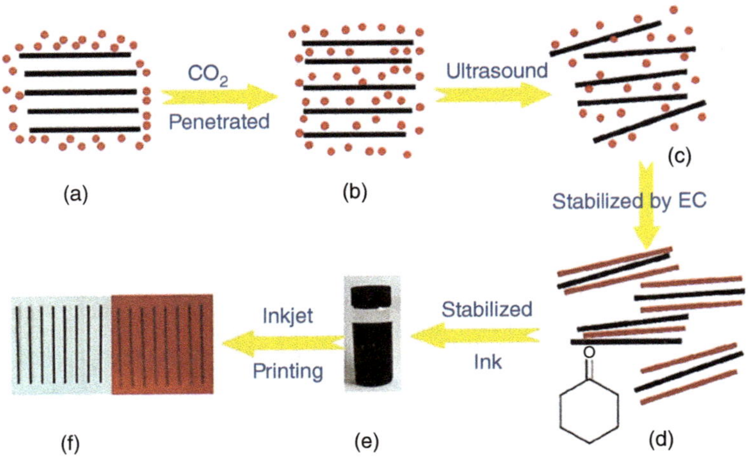

Figure 7.32 Schematic design of pristine graphene ink preparation. (a) Supercritical carbon dioxide immersed layered graphite. (b) Penetration and intercalation of gas molecules in the graphite interlayer. (c) Formation of single-layer or few-layer-thick graphene sheets. (d) Cyclohexanone-stabilized graphene sheets, and (e) prepared stabilized graphene ink. (f) Printed graphene electrodes on PET and polyimide (PI) substrates. (e) Source: Reproduced with permission Gao et al. [42], American Chemical Society.

The inkjet printing patterns were uniform and continuous. The study exhibited good electrical properties and proved to be an efficient method to formulate ink and fabricate conductive electrodes, providing high potential for graphene-based flexible electronic devices as shown in Figure 7.32 [42].

Secor et al. performed a study on graphene gravure printing for flexible electronics applications. Graphene-EC powder dispersion in ethanol and terpineol was performed by bath sonication. The graphene inks were printed onto DuPont Kapton FPC films using a direct gravure printing system. The electrical characteristic studies have proven its suitability for flexible electronic applications [43].

Yang and others synthesized graphene-based conductive inks for flexible electronics devices. The authors have commented that the random combination of ink formulation as well as the pattern deposition methods will lead to the future progress in the field of graphene and its inks for industrial flexible electronics applications [44].

Arapov et al. performed a study on gelation of graphene/polymer dispersions triggered by mild heating. A colloidally stabilized and highly concentrated graphene pastes were prepared from the graphene gel, and the printing was done on industrially related paper and plastic substrates and exhibited excellent performance and good electrical properties [45].

He et al. studied inkjet-printed graphene for ion selective electrodes with post-print thermal annealing. The printed devices exhibited good sensitivity toward human fluids including plasma, serum, and sweat [46].

Secor et al. performed a study on combustion-assisted photonic annealing of printable graphene inks by using nitrocellulose exothermic binders. This resulted in

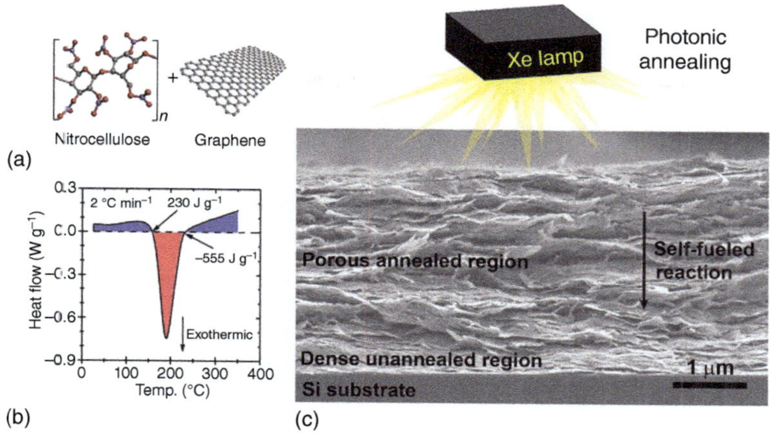

Figure 7.33 General idea of the process and concept proof: (a) nitrocellulose (left) and graphene chemical structure (right); (b) degree of difference in scanning calorimetry of graphene/nitrocellulose displaying a large exothermic reaction of nitrocellulose at ~200 °C. (c) Schematic diagram of photonic annealing. Cross-sectional SEM image reveals the resulting porous graphene microstructure even on a thermally conductive Si substrate. (c) Source: Reproduced with permission [47], American Chemical Society.

outstanding electrical properties that can be matched to extended thermal annealing as shown in Figure 7.33 [47].

Karagiannidis et al. performed a study on microfluidization of graphite and graphene-based conductive ink formulation as shown in Figure 7.34. The material was stabilized without centrifugation at concentrations using sodium salt to prepare printable conductive inks. The blade-coated film sheet resistance was found to be below 2 Ω. The conductive inks have proven their application for wide-area printing in flexible electronics [48].

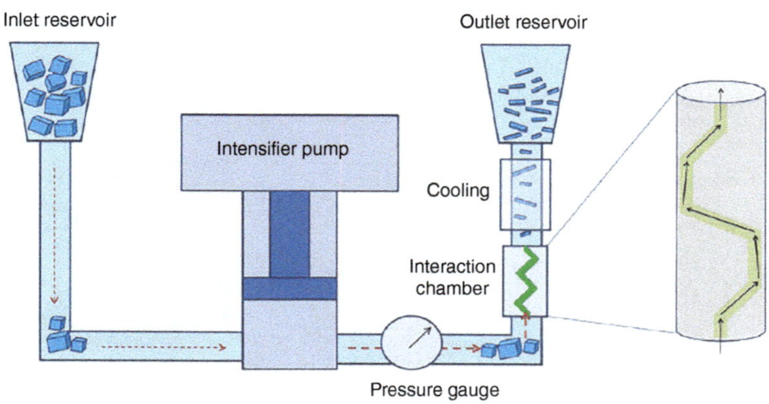

Figure 7.34 Schematic of the microfluidization process. Source: Karagiannidis et al. [48].

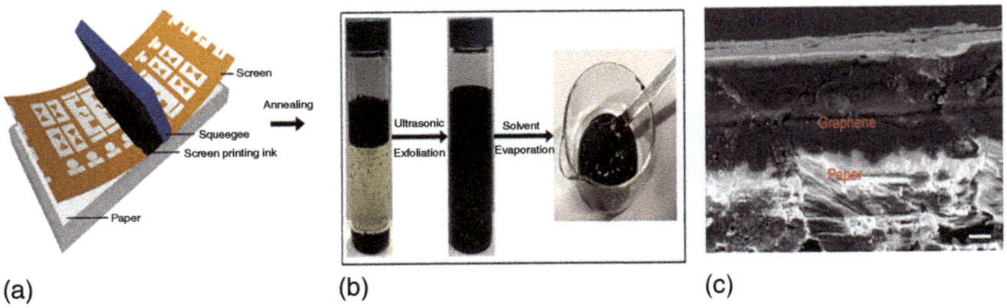

Figure 7.35 Screen-printed graphene antenna. (a) Graphene patterning on the screen using ink and squeegee. (b) Cyrene-based highly concentrated graphene ink. (c) Cross-sectional image of the printed graphene antenna: SEM image. (a) Source: Pan et al. [49]/Springer nature/CC BY 4.0. (b–c). Source: [49] Pan et al. / with permission of Springer Nature.

Tran et al. presented a review article on printable graphene inks for flexible electronics application. The printed techniques discussed were gravure printing, screen printing, and inkjet printing [40].

Pan et al. discussed the fabrication of highly conductive multilayer graphene ink. The formed ink has found its application in IoT applications and wireless connectivity as shown in Figure 7.35. The authors have commented that the inks based on graphene will dominate this technology due to their cost effectiveness and direct application on materials like textile and paper due to the use of nontoxic solvent dihydrolevoglucosenone (Cyrene). The printing exhibited good conductivity and was further extended for wireless connectivity antenna applications [49].

He et al. worked on highly conductive screen-printing graphene nanoplatelets ink in a cost-effective and large-scale manufacturing for flexible PE. The substrates used for printing the formulated ink included paper and plastic as shown in Figure 7.36. Combined with post-printing treatments such as compression rolling and thermal annealing, the printed sample displayed high conductivity. The graphene printed patterns can also function as current collectors in supercapacitors [50].

Secor et al. worked on highly conductive inkjet printing using graphene patterns for flexible electronic applications. The ink was prepared by solution-phase exfoliation of graphene using ethanol, and ethyl cellulose as a stabilizing polymer [51].

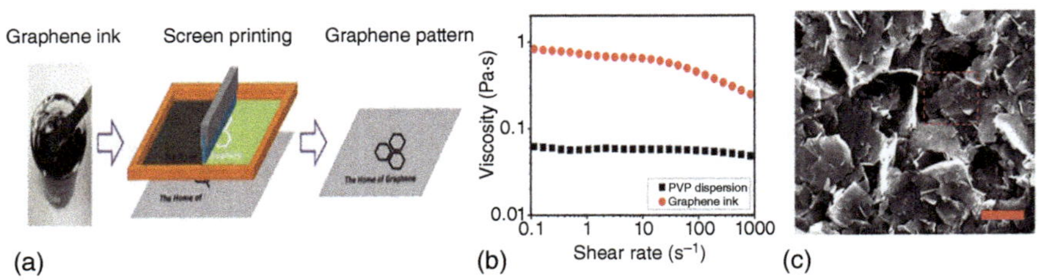

Figure 7.36 (a) Pictorial diagram of the screen-printing process. (b) Measured viscosity vs. shear rate of graphene nanoparticles (GNP)-ink and polymer solution. (c) Dried GNP-ink SEM image with a scale bar of 10 μm. (a and c) Source: Reproduced with permission [50], American Chemical Society.

7.4 Outlooks and Perspectives

The field of PE is diversifying in all fields with technological advancements leading to effective and efficient performance characteristics of the manufactured devices compared with the same conventional devices. In this chapter, fundamentals, features, and progress in various printing technologies for PE were discussed with substantial literature illustrations. The chapter also provided an overview of different nanomaterial inks such as metallic, organic, ceramics followed by the synthesis procedures of specific nanomaterials such as silver-based nanomaterials, 1D and 2D carbon-based nanomaterials. The various research reports elucidated the utilization of these respective inks for their specific PE applications. Observations made from various literature reports revealed the technological developments and utilization of functional inks. Also, it is understood that the performance of PE devices is highly influenced by the optimal combinations of printable nanomaterials. It is to be noted that still researchers attempt to formulate finest conductive inks for PE and endeavor to identify those materials to make them the best among those available. Further research studies are ongoing to find printable inorganic nanomaterials, recognize their functioning mechanisms and their respective real-time applications while retaining economic perspectives by considering techno-economical aspects. To conclude, it is to accept the fact that PE aids in making everyday life smart. With sufficient research activities, a high demand exists in future for the possibilities for PE device applications in diversified fields.

References

1 Wu, W. (2017). Inorganic nanomaterials for printed electronics: a review. *Nanoscale* 9 (22): 7342–7372. https://doi.org/10.1039/c7nr01604b.

2 Pavlovic, M. and Balint, B. (2012). *Introduction to Printed Electronics*. Springer.

3 Magdassi, S., Bassa, A., Vinetsky, Y., and Kamyshny, A. (2003). Silver nanoparticles as pigments for water-based ink-jet inks. *Chem. Mater.* 15 (11): 2208–2217. https://doi.org/10.1021/cm021804b.

4 Shankar, R., Groven, L., Amert, A. et al. (2011). Non-aqueous synthesis of silver nanoparticles using tin acetate as a reducing agent for the conductive ink formulation in printed electronics. *J. Mater. Chem.* 21 (29): 10871–10877. https://doi.org/10.1039/c0jm04521g.

5 Zhang, Z., Zhang, X., Xin, Z. et al. (2011). Synthesis of monodisperse silver nanoparticles for ink-jet printed flexible electronics. *Nanotechnology* 22 (42): 0–7. https://doi.org/10.1088/0957-4484/22/42/425601.

6 Walker, S.B. and Lewis, J.A. (2012). Reactive silver inks for patterning high-conductivity features at mild temperatures. *J. Am. Chem. Soc.* 134 (3): 1419–1421. https://doi.org/10.1021/ja209267c.

7 Lai, C.Y., Cheong, C.F., Mandeep, J.S. et al. (2014). Synthesis and characterization of silver nanoparticles and silver inks: review on the past and recent

technology roadmaps. *J. Mater. Eng. Perform.* 23 (10): 3541–3550. https://doi.org/10.1007/s11665-014-1166-6.

8 Huang, Q., Shen, W., Xu, Q. et al. (2014). Properties of polyacrylic acid-coated silver nanoparticle ink for inkjet printing conductive tracks on paper with high conductivity. *Mater. Chem. Phys.* 147 (3): 550–556. https://doi.org/10.1016/j.matchemphys.2014.05.030.

9 Zhou, X., Li, W., Wu, M. et al. (2014). Enhanced dispersibility and dispersion stability of dodecylamine-protected silver nanoparticles by dodecanethiol for ink-jet conductive inks. *Appl. Surf. Sci.* 292: 537–543. https://doi.org/10.1016/j.apsusc.2013.12.006.

10 Huang, Q., Shen, W., Xu, Q. et al. (2014). Room-temperature sintering of conductive Ag films on paper. *Mater. Lett.* 123: 124–127. https://doi.org/10.1016/j.matlet.2014.02.101.

11 Wu, W., Yang, S., Zhang, S. et al. (2014). Fabrication, characterization and screen printing of conductive ink based on carbon@Ag core-shell nanoparticles. *J. Colloid Interface Sci.* 427: 15–19. https://doi.org/10.1016/j.jcis.2013.10.064.

12 Titkov, A.I., Bukhanets, O.G., Gadirov, R.M. et al. (2015). Conductive inks for inkjet printing based on composition of nanoparticles and organic silver salt. *Inorg. Mater. Appl. Res.* 6 (4): 375–381. https://doi.org/10.1134/S2075113315040243.

13 Vaseem, M., McKerricher, G., and Shamim, A. (2016). Robust design of a particle-free silver-organo-complex ink with high conductivity and inkjet stability for flexible electronics. *ACS Appl. Mater. Interfaces* 8 (1): 177–186. https://doi.org/10.1021/acsami.5b08125.

14 Zhang, N., Luo, J., Liu, R., and Liu, X. (2016). Tannic acid stabilized silver nanoparticles for inkjet printing of conductive flexible electronics. *RSC Adv.* 6 (87): 83720–83729. https://doi.org/10.1039/c6ra19800g.

15 Cai, Y., Piao, X., Gao, W. et al. (2017). Large-scale and facile synthesis of silver nanoparticles: via a microwave method for a conductive pen. *RSC Adv.* 7 (54): 34041–34048. https://doi.org/10.1039/c7ra05125e.

16 Zhang, X.Y., Xu, J.-J., Wu, J.-Y. et al. (2017). Seeds triggered massive synthesis and multi-step room temperature post-processing of silver nanoink – application for paper electronics. *RSC Adv.* 7 (1): 8–19. https://doi.org/10.1039/c6ra27163d.

17 Kastner, J. et al. (2017). Silver-based reactive ink for inkjet-printing of conductive lines on textiles. *Microelectron. Eng.* 176: 84–88. https://doi.org/10.1016/j.mee.2017.02.004.

18 Shahariar, H., Kim, I., Soewardiman, H., and Jur, J.S. (2019). Inkjet printing of reactive silver ink on textiles. *ACS Appl. Mater. Interfaces* https://doi.org/10.1021/acsami.8b18231.

19 Chang, C.W., Cheng, T.Y., and Liao, Y.C. (2018). Encapsulated silver nanoparticles in water/oil emulsion for conductive inks. *J. Taiwan Inst. Chem. Eng.* 92: 8–14. https://doi.org/10.1016/j.jtice.2018.01.046.

20 Hassan, G., Bae, J., and Lee, C.H. (2018). Ink-jet printed transparent and flexible electrodes based on silver nanoparticles. *J. Mater. Sci. Mater. Electron.* 29 (1): 49–55. https://doi.org/10.1007/s10854-017-7886-2.

21 Manjunath, G., Pujar, P., Gupta, B. et al. (2019). Low-temperature reducible particle-free screen-printable silver ink for the fabrication of high conductive electrodes. *J. Mater. Sci. Mater. Electron.* 30 (20): 18647–18658. https://doi.org/10.1007/s10854-019-02217-9.

22 Mo, L. et al. (2019). Silver nanoparticles based ink with moderate sintering in flexible and printed electronics. *Int. J. Mol. Sci.* 20 (9): https://doi.org/10.3390/ijms20092124.

23 Hao, Y., Xu, Z., Gao, J. et al. (2019). Humic acid assisted chemical synthesis of silver nanoparticles for inkjet printing of flexible circuits. *J. Mater. Sci. Mater. Electron.* 30 (23): 20400–20409. https://doi.org/10.1007/s10854-019-02372-z.

24 Fernandes, I.J. et al. (2020). Silver nanoparticle conductive inks: synthesis, characterization, and fabrication of inkjet-printed flexible electrodes. *Sci. Rep.* 10 (1): 1–11. https://doi.org/10.1038/s41598-020-65698-3.

25 Liang, J., Tong, K., and Pei, Q. (2016). A water-based silver-nanowire screen-print ink for the fabrication of stretchable conductors and wearable thin-film transistors. *Adv. Mater.* 28 (28): 5986–5996. https://doi.org/10.1002/adma.201600772.

26 Chen, S. et al. (2017). A systematic and effective research procedure for silver nanowire ink. *J. Alloys Compd.* 706: 164–175. https://doi.org/10.1016/j.jallcom.2017.02.214.

27 Huang, Q. and Zhu, Y. (2018). Gravure printing of water-based silver nanowire ink on plastic substrate for flexible electronics. *Sci. Rep.* 8 (1): 1–10. https://doi.org/10.1038/s41598-018-33494-9.

28 Huang, Q., Al-Milaji, K.N., and Zhao, H. (2018). Inkjet printing of silver nanowires for stretchable heaters. *ACS Appl. Nano Mater.* 1 (9): 4528–4536. https://doi.org/10.1021/acsanm.8b00830.

29 Li, W., Yang, S., and Shamim, A. (2019). Screen printing of silver nanowires: balancing conductivity with transparency while maintaining flexibility and stretchability. *Flex. Electron.* 3 (1): 10.1038/s41528-019-0057-1.

30 Williams, N.X., Noyce, S., Cardenas, J.A. et al. (2019). Silver nanowire inks for direct-write electronic tattoo applications. *Nanoscale* 11 (30): 14294–14302. https://doi.org/10.1039/c9nr03378e.

31 Patil, P., Patil, S., Kate, P., and Kulkarni, A.A. (2021). Inkjet printing of silver nanowires on flexible surfaces and methodologies to improve the conductivity and stability of the printed patterns. *Nanoscale Adv.* https://doi.org/10.1039/d0na00684j.

32 Kamyshny, A. and Magdassi, S. (2019). Conductive nanomaterials for 2D and 3D printed flexible electronics. *Chem. Soc. Rev.* 48 (6): 1712–1740. https://doi.org/10.1039/c8cs00738a.

33 Fan, Z., Wei, T., Luo, G., and Wei, F. (2005). Fabrication and characterization of multi-walled carbon nanotubes-based ink. *J. Mater. Sci.* 40 (18): 5075–5077. https://doi.org/10.1007/s10853-005-1622-7.

34 Song, J.W., Kim, J., Yoon, Y.H. et al. (2008). Inkjet printing of single-walled carbon nanotubes and electrical characterization of the line pattern. *Nanotechnology* 19 (9): https://doi.org/10.1088/0957-4484/19/9/095702.

35 Denneulin, A., Bras, J., Carcone, F. et al. (2011). Impact of ink formulation on carbon nanotube network organization within inkjet printed conductive films. *Carbon N. Y.* 49 (8): 2603–2614. https://doi.org/10.1016/j.carbon.2011.02.012.

36 Niu, C. (2011). Carbon nanotube transparent conducting films. *MRS Bull.* 36 (10): 766–773. https://doi.org/10.1557/mrs.2011.213.

37 Zhao, J. et al. (2013). Sensitive and selective dopamine determination in human serum with inkjet printed Nafion/MWCNT chips. *Electrochem. Commun.* 37: 32–35. https://doi.org/10.1016/j.elecom.2013.10.007.

38 Guo, S., Wang, W., Ozkan, C.S., and Ozkan, M. (2013). Assembled graphene oxide and single-walled carbon nanotube ink for stable supercapacitors. *J. Mater. Res.* 28 (7): 918–926. https://doi.org/10.1557/jmr.2012.421.

39 Menon, H., Aiswarya, R., and Surendran, K.P. (2017). Screen printable MWCNT inks for printed electronics. *RSC. Adv.* 7 (70): 44076–44081. https://doi.org/10.1039/c7ra06260e.

40 Tran, T.S., Dutta, N.K., and Choudhury, N.R. (2018). Graphene inks for printed flexible electronics: graphene dispersions, ink formulations, printing techniques and applications. *Adv. Colloid Interf. Sci.* 261: 41–61. https://doi.org/10.1016/j.cis.2018.09.003.

41 Li, J., Ye, F., Vaziri, S. et al. (2013). Efficient inkjet printing of graphene. *Adv. Mater.* 25 (29): 3985–3992. https://doi.org/10.1002/adma.201300361.

42 Gao, Y., Shi, W., Wang, W. et al. (2014). Inkjet printing patterns of highly conductive pristine graphene on flexible substrates. *Ind. Eng. Chem. Res.* 53 (43): 16777–16784. https://doi.org/10.1021/ie502675z.

43 Secor, E.B., Lim, S., Zhang, H. et al. (2014). Gravure printing of graphene for large-area flexible electronics. *Adv. Mater.* 26 (26): 4533–4538. https://doi.org/10.1002/adma.201401052.

44 Yang, W. and Wang, C. (2016). Graphene and the related conductive inks for flexible electronics. *J. Mater. Chem. C* 4 (30): 7193–7207. https://doi.org/10.1039/c6tc01625a.

45 Arapov, K., Rubingh, E., Abbel, R. et al. (2016). Conductive screen printing inks by gelation of graphene dispersions. *Adv. Funct. Mater.* 26 (4): 586–593. https://doi.org/10.1002/adfm.201504030.

46 He, Q. et al. (2017). Enabling inkjet printed graphene for ion selective electrodes with postprint thermal annealing. *ACS Appl. Mater. Interfaces* 9 (14): 12719–12727. https://doi.org/10.1021/acsami.7b00092.

47 Secor, E.B., Gao, T.Z., Dos Santos, M.H. et al. (2017). Combustion-assisted photonic annealing of printable graphene inks via exothermic binders. *ACS Appl. Mater. Interfaces* 9 (35): 29418–29423. https://doi.org/10.1021/acsami.7b07189.

48 Karagiannidis, P.G. et al. (2017). Microfluidization of graphite and formulation of graphene-based conductive inks. *ACS Nano* 11 (3): 2742–2755. https://doi.org/10.1021/acsnano.6b07735.

49 Pan, K. et al. (2018). Sustainable production of highly conductive multilayer graphene ink for wireless connectivity and IoT applications. *Nat. Commun.* 9 (1): https://doi.org/10.1038/s41467-018-07632-w.

50 He, P. et al. (2019). Screen-printing of a highly conductive graphene ink for flexible printed electronics. *ACS Appl. Mater Interfaces* 11 (35): 32225–32234. https://doi.org/10.1021/acsami.9b04589.

51 Secor, E.B., Prabhumirashi, P.L., Puntambekar, K. et al. (2013). Inkjet printing of high conductivity, flexible graphene patterns. *J. Phys. Chem. Lett.* 4 (8): 1347–1351. https://doi.org/10.1021/jz400644c.

8

Metal-oxide Semiconductors for Noninvasive Diagnosis of Breast Cancer

Mohan Velumani[1], Ivneet Banga[2], Anirban Paul[2], Asokan Prasanth[1], Samir Ranjan Meher[3], Elizabeth Rufus[1], Sriram Muthukumar[2], Shalini Prasad[2], and Zachariah C. Alex[1]

[1] *Vellore Institute of Technology, School of Electronics Engineering, Department of Sensor and Biomedical Technology, Tiruvalam Road, Katpadi, Vellore, 632014, India*
[2] *University of Texas at Dallas, Department of Bioengineering, 800 W Campbell Road, Richardson, TX 75080, USA*
[3] *Vellore Institute of Technology, School of Advanced Sciences, Department of Physics, Tiruvalam Road, Katpadi, Vellore, 632014, India*

8.1 Introduction

Cancer is the most dreaded disease and the leading cause of death worldwide. Prostate, lung, breast, and colon cancers are the most common types. Its incidence trend has been growing in most developed countries such as the United States, Canada, the United Kingdom and in developing countries, for example, India. In the United States, an estimated 0.6 million people died due to cancer, and 1.8 million new cancer cases were diagnosed in 2020. Also, according to a recent report published by the National Cancer Registry Programme (NCRP), India, by 2025, there will be a 20% increase in the cancer patients. Breast cancer has the second highest mortality rate in both developing and developed nations [1]. The high mortality due to cancer is a serious concern for the scientific community; hence, continuous efforts have been made for several years to ensure its early diagnosis and treatment. One of the major breakthroughs achieved in the field of cancer prognosis is its early detection. Cancer causes several metabolic changes in the human body at an early phase, which result in an imbalance (elevation/ depression) of measurable metabolite concentrations in the body. These metabolites are called biomarkers. Biomarkers can be solid, liquid, or volatile. Figure 8.1 represents classes and subclasses of the origin of cancer biomarkers [2].

In the past few years, research in biomedical engineering has enabled clinicians achieve early medical diagnosis with advanced technologies. Although these conventional diagnostic techniques help in identification of cancer, they have certain drawbacks: cost, painful to patients, and may not always give accurate results, especially in the early stage of the disease. For example, the detection of cancer and its

Nanotechnology in Electronics: Materials, Properties, Devices, First Edition.
Edited by Visakh P.M., Artem Semkin, Raneesh Balakrishnan, and Saša Lazović.

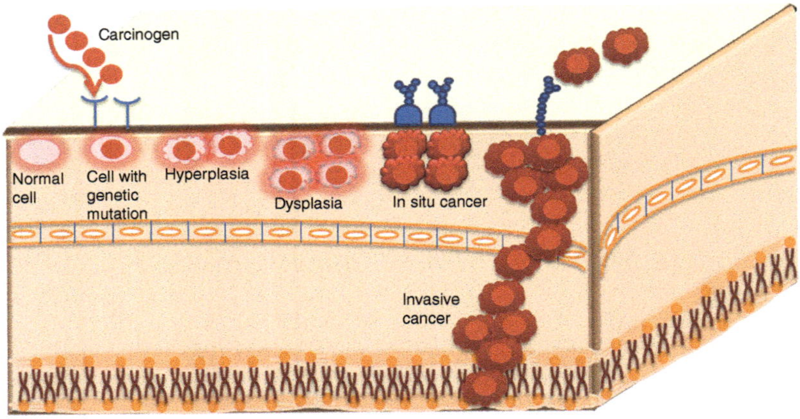

Figure 8.1 Schematic representation of breast cancer tumorigenesis.

treatment significantly affect many people around the world. Treatment and maintenance costs are also considerably high. In 2014, of cancer treatment and diagnosis cost in the United States was estimated to be $87.8 billion [3]. Recent research has shown that the overall cost of cancer diagnosis and treatment is indeed decreasing for patients treated in the early stage, and, hence, the prognostic management of cancer remains a topic of key interest in the scientific community, from fabrication of device to even machine learning [4]. Over the past few decades, advances in molecular biology and biotechnology have achieved significant milestones to identify cellular changes at the molecular level, which may be a promising approach for early disease detection. Moreover, present technologies can quantify cellular changes at an early phase of disease development. This may help in identifying particular pathophysiological processes and underlying conditions to build more accurate devices [5–8]. Although constant efforts are being made by researchers toward achieving this goal the need of the hour is to develop noninvasive diagnostic platforms that can offer prognostic analysis for each patient and should be sensitive, specific, robust, and cost-effective [9–12]. One of the key focus areas for researchers is noninvasive techniques. Fundamentally, a noninvasive technique involves no puncturing, cut, or break in the skin for sample collection. It only involves human samples that are released by the body such as sweat, saliva, urine [13–15]. The common example of noninvasive diagnostic technique is glucose sensing from sweat or saliva. The requirement of glucose monitoring in the human body is extremely critical for the diagnosis of diabetes [16–18]. The currently available methods are painful and may cause infections. Hence, noninvasive glucose monitoring is highly desirable. Generally, noninvasive measurement is associated without need to test blood and hence, does away with the need to puncture the skin. For this purpose, sweat and saliva have proved to be an excellent choice to monitor glucose levels for diabetes. Our group has reported several examples of noninvasive glucose level detection techniques [19–22]. The concept of noninvasive disease diagnosis is also valid for other chronic diseases, for example, myocardial infraction and even cancer [23, 24].

Table 8.1 Sources of biomarkers, its categories, subcategories, and classes.

Sources	Categories		Subcategories	Classes
Blood Saliva	Protein		Peptides, globular proteins, fibrous proteins, membrane proteins	Antecedent biomarkers
Urine	Genomic	DNA	Genes, mutations, polymorphisms, epigenetics, DNA quantity	Screening biomarkers
Feces		RNA	MicroRNA, messenger RNA, gene expression	Diagnostic biomarker
Cerebrospinal fluid	Carbohydrate		Glucose, sucrose, glycan	
Amniotic fluid	Lipid		Cholesterol, acyl glycerol, phospholipids, prostaglandins	Staging biomarker
Tissue cells	Metabolites		Reactive oxygen species, dicarboxylic acid inorganic ions	Prognostic biomarker
Breath	Organic compounds		Volatile organic compounds	

Regarding the detection of biomarkers associated with diseases, in this case, cancer, there are various classifications of cancer biomarkers, which are listed in Table 8.1. The potential biomarkers can range from proteins, enzymes in the body fluid to even volatile organic compounds (VOCs) in breath [25]. The early diagnosis of cancer is crucial for patient survival; therefore, sensitive and specific methods are required for early cancer diagnosis. These methods aid clinicians in successfully diagnosing and treating a patient, thereby, increasing their chances of survival [26]. The conventional diagnostic methods for cancer detection involve histological characterization such as staining based on cell morphology and analyzing the slides using microscopic techniques. These are invasive techniques that require human sample collection by taking a biopsy sample followed by sample processing techniques that involve cell fixation and is further analyzed using microscope to identify and detect cancer cells [27–29]. These tests, however, can sometimes be inconclusive as cancer cells may not be detected in biopsy samples at the early stage of the disease and further testing and characterization is required to identify the malignancy. Recently, techniques such as immunoassays (e.g. ELISA) are used to detect cancer biomarkers, and these are carried out in hospital laboratories. However, in most cases, only one biomarker is targeted and analyzed as a disease indicator. Immunoassays are very sensitive and selective but can be time consuming and expensive [30–34]. Furthermore, some ELISA test kits are not highly sensitive for the detection of low-level concentrations of target biomarkers. Molecular tools mainly based on genomic and proteomics studies to understand molecular levels and pathways of disease progression are used recently. These techniques aid in understanding of disease progression but are highly complex and time consuming. Hence, the need of the hour is to develop diagnostic platforms that can help in achieving rapid diagnosis requirements in cancer detection. Rapid diagnosis helps reduce the mortality rate and increase chances of disease detection at an early stage. Currently used methods for rapid diagnosis involve dipstick-based platforms that use the direct monitoring of antibody–antigen interactions but are partially sensitive. The design and development of label-free affinity-based target capture probes is the most researched field, in both academic and industry, to help establish alternative methods to the existing ELISA-based immunoassays.

8.2 Sensing Material and Techniques

A sensor is a device that detects events or changes in its environment and provides a proportional output. A typical sensor consists of an analyte, a transducer, and a data processing system. Sensors can be classified based on the stimulus they can sense and signal they produce [35]. Materials used in sensors play an important role in noninvasive disease diagnosis, and current technology advancements demand not only advanced but smart materials integrated with complex electronics [36]. In general, noninvasive techniques involve sampling from sweat, saliva, interstitial fluid, etc., and there is a unique challenge lies on sensitivity of sensors as in most cases the target analyte present in such bodily fluids is low in concentration. Hence, signal amplification with a reliable mathematical model is extremely critical. There have been multiple reports of various types of materials used for noninvasive sensing, with varying size, chemical, physical, spectral, and thermal properties [37]. Wearable technologies have profound impact on the health sector as they provide many diagnostic parameters that help people to stay healthy. Tracking such parameters includes heart rate monitor, spO_2 monitor, and even stress monitor also. All these trackers work based on different fundamentals and principles integrated with smart and sensitive materials, which provide valuable information of human physiology [38]. Semiconductor materials are always a prime choice for scientific community for the development of noninvasive sensor platforms as they can be reformed to nanomaterials along with their semiconducting property. A huge number of nano-sized semiconducting materials have been used for noninvasive diagnostic management [39]. Breath analysis involves easy sample collection and analysis compared to blood or tissue analysis [40–43]. Its application is being explored in clinical diagnosis as the technique is less time consuming and can provide easy, on-site, and point-of-care applicability. However, some of the currently used sensors and analytical methods are not fast enough in terms of sensor response to allow breath-to-breath resolved sampling. Therefore, additional modifications and developments are necessary to provide well-defined and reproducible breath sample measures to produce reliable systems. Some substances present in exhaled breath are blood borne and originate from the metabolism of the endogenous pathways- alveoli, others, such as NO and CO, are generated in the airways. Hence, thoroughly controlled breath sampling is a key requirement for reliable analysis and quantification of endogenously produced breath biomarkers [44, 45].

To integrate sensor technology into clinical breath analysis, sensor devices that are automatically controlled and allow adjustable sampling done as per the patient's requirements are needed. These devices should be readily available to the concerned population, economically friendly, and require low sample volume for analysis. For standardization of the sensing device, data optimization such as sampling flow rate, total volume of the sample, or CO_2 partial pressures present in the individual breath during sampling must be recorded continuously and should be set into baseline parameters for the device. [46, 47]. This is especially important in the dynamically changing state of critically ill patients as their body metabolism varies largely from others. Despite a number of promising results revealing interesting diagnostic properties of different biomarkers, analysis of VOCs in the breath of the critically ill has not yet been introduced into clinical practice and needs further optimization and characterization. Therefore, considerable effort has been put, and research is being evolved continuously in establishing reproducible and reliable analytical methods for breath analysis and in understanding how various compound levels vary in the breath. Three important questions govern the research involved in studying volatile markers in breath of critically ill patients. (1) How should breath samples be collected and what is the volume of exhaled breath that is required? (2) Given the low concentrations of VOCs in breath, how can they be detected and quantified so as to have a sensitive and specific detection platform? And finally, (3) what is the correlation between levels of endogenously produced volatile substances and particular diseases?

Breast cancer is a result of interplay of various factors that lead to genetic mutations. Hereditary and environmental factors play a major role in causing these mutations, thereby inactivating DNA repair genes in the breast cells [48]. Figure 8.2 indicates the different stages of breastcancer [49]. The main biomarkers associated with targeted therapy of breast cancer are estrogen receptors (ERs) or progesterone receptors (PgRs). Growth factors such as epidermal growth factors, mainly ErbB-1 (EGFR) and ErbB-2 (ERBB2, HER2/neu, HER2), are also regulated in breast cancer tumorigenesis [49, 50]. EGFR overexpression is correlated to overexpression of HER2, levels of which are used for targeted therapy using the drug trastuzumab [51]. However, classification of breast cancer type is based on gene expression profiling, which requires a large amount of tumor tissue for analysis and is a highly expensive and difficult procedure. Moreover, gene expression in an individual changes due to post-transcriptional modifications and may lead to upregulation or downregulation of certain genes [52]. Use of protein levels for disease diagnosis and understanding biochemical modifications involved in the occurrence of breast cancer remain a difficult task as there is a risk of sample degradation arising from lack of protein stability, post-translational modifications [53].

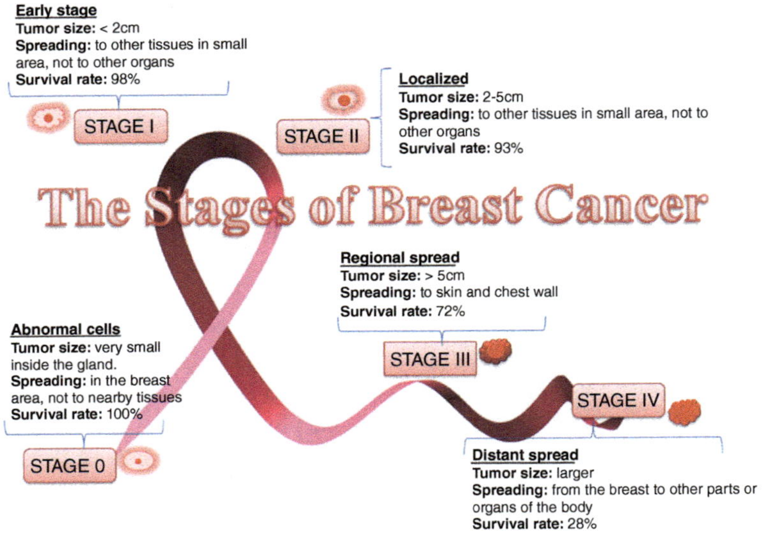

Figure 8.2 Different stages of breast cancer and its associated characteristics.

8.3 Biomarkers for Noninvasive Diagnosis of Breast Cancer

Breast cell–related diseases can be associated either with the regulation of hormone levels linked to the reproductive system or can be altered due to other underlying disease conditions. The diagnosis of breast cell–related diseases requires extensive biopsy and sampling for confirmation. These invasive diagnostic techniques cause distress until the diagnosis is confirmed through biopsy, and sometimes even after treatment there is a high chance of relapse; therefore, the patient should get biopsy done every few months. Some of the benign tumors, such as premalignant hyperplasia, when left untreated can also lead to invasive breast cancer (IBC), which is one of the most common malignancies diagnosed in women. According to the multistep model for breast cancer progression, premalignant hyperplastic may affect breast regions severely, and this causes in situ carcinoma, which ultimately will result in IBC. Most common imaging techniques, such as mammography, ultrasound, magnetic resonance imaging (MRI), in combination with conventional biopsy techniques are employed for disease diagnosis. However, these methods are less accurate in the diagnosis of women under the age of 50 for benign breast conditions. The benign breast condition is prevalent in women under 50 years. Mammography is an imaging technique that heavily relies on the breast density and hence is less accurate for women under 50 in the premenopause stage. It is considered as a gold standard for overaged (50+) women in postmenopause or women under the age of 50 having suitable sampling condition due to heredity. The interesting fact is that the lesions diagnosed via mammography for ductal carcinoma in situ (DCIS) are 11 and 20%. The fact is also observed when examining the carcinoma in vivo upon complete excision. On the other hand, techniques like ultrasound are not able to recognize breast tumors clearly but are frequently used as a complementary technique along with mammography in many cases. MRI is one of the high-tech imaging techniques often employed to visualize in situ breast carcinoma. Although it is a very popular imaging tool, it shows significantly lower specificity than mammography and false-positive results for IBC, which make healthy subjects bear unnecessary burden of additional diagnosis like biopsy and follow-up tests. Due to such disadvantages of the conventional techniques, scientific fraternity has started exploring alternative reliable approaches that are not only complementary but also rigorous and cost-effective. Monitoring the patterns of VOCs in exhaled breath is a new way to visualize disease diagnosis as it has the ability to not only identify the disease progression but also it helps the medical community to understand the severity of the disease at an early stage. The release of VOCs in blood is associated with inflammation occurring in breast cells that leads to secretion of VOCs, dissolved in blood via the oxidative stress pathway. It is also observed that in some genes at benign lesions, the DNA methylation has been promoted by the blood cell along with volatile biomarkers at a very early stage of cancer diagnosis. One part of this blood cell along with volatile biomarkers are directly transmitted to exhaled breath.

By gas chromatography–mass spectrometry (GC-MS), a combination of specific volatile biomarkers in breath samples can be used to identify breast cancer. The test accuracy has also improved for identification of monomethylated alkanes present in exhaled breath that are associated with breast cancer, and the biochemical background of cancer-specific volatile biomarkers in exhaled breath has been investigated. However, GC-MS is time consuming, requires expensive equipment and experienced operators, and suitable pre-concentration methods. Researchers have achieved a technological breakthrough: a breath test based on a nanoscale artificial nose that distinguishes patients with cancer (breast, lung, colon, and prostate) and healthy controls, irrespective of age, gender, lifestyle, smoking habits, and other confounding factors.

8.3.1 Body Metabolism for VOC Generation

A collection of VOCs making the volatilome of the healthy human body listing 1849 VOCs was recently published by Costello et al. [55]. The report highlights that 874

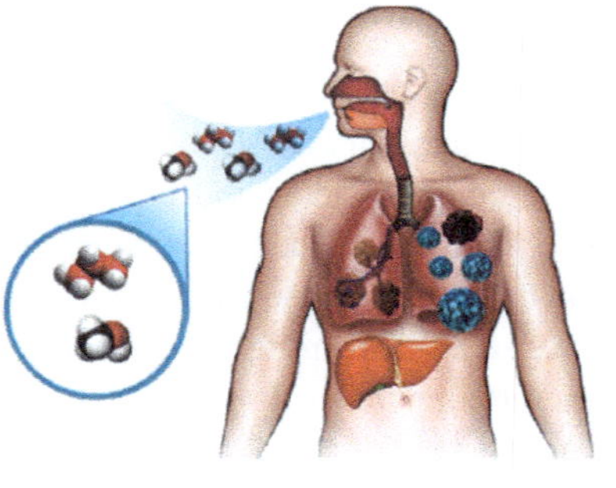

Figure 8.3 Exhaled biomarkers. Source: [54] \ with permission from American Chemical Society.

VOCs are present in breath, 279 in urine, 353 in saliva, 130 in blood, 504 in skin emanations or lesions, and 381 in feces. Some VOCS are present in more than one body fluids, such as the VOCs present in breath overlap with blood, saliva, and feces [56–59].

The bulk matrix of exhaled breath is made up of a mixture of gases such as oxygen, carbon dioxide, water and VOCs as well as non-VOCs shown in Figure 8.3 [54]. VOCs are present in trace amounts on the order of nmol l^{-1} (ppbv) or pmol l^{-1} (pptv). To monitor the changes in the metabolic processes in the body that lead to cancer and other diseases, concentrations of VOCs need to be monitored. Some common VOCs such as ethane, pentane, acetone, isoprene are used to study different biochemical metabolic pathways, and concentration levels of these compounds can help in differentiating between healthy and diseased human state. Moreover, the study of VOCs is advantageous over the abundant nonvolatile compounds and gases that are present in the human body as nonvolatile compounds have limited stability and no clear relationship can be established between the airway concentration and analyte concentration at the site of production [60]. VOCs, on the other hand, can be easily studied because their solubilities can be correlated back to the kinetics at the site of production and they are highly stable. Till date, more than 500 different VOCs have been identified in the human breath both endogenous and exogenous in origin [61]. Some examples of endogenous markers used for disease diagnosis are as follows:

1. Alcohols or oxygen-containing compounds – methanol, ethanol, acetone, acetaldehyde, 2-propanol, etc.
2. Some sulfur-containing compounds – dimethyl sulfide, carbon disulfide, mercaptans (methyl, ethyl), etc.
3. Nitrogen-containing compounds – ammonia, dimethylamine, trimethylamine, etc.
4. Hydrocarbons – ethane, pentane, isoprene, etc.

The clinical significance of these compounds and their levels inside the body can be understood only when we are able to understand the biochemical pathways associated with them and how they are generated in the body. Some metabolic pathways involve in VOC generation based on the chemical structure of the compounds, which is explained in what follows. Oxidative stress is closely linked to cancer development. It causes activation of oxidase enzymes, mainly cytochrome P450 (CYP 450). Oxidative stress is defined in terms of the equilibrium between the formed and deactivated reactive oxygen species (ROS) as well as the free radicals [62]. Cellular processes and metabolic pathways in the mitochondria release ROS in the cell. Accumulation of these compounds in the cell causes the ROS to attack the polyunsaturated fatty acids and proteins. Thus, during the state of oxidative stress, these ROS and free radicals cause excessive cellular damage and lead to tumorigenesis. On the other hand, saturated hydrocarbons are produced from omega-3 and omega-6 fatty acids that are fundamental components of cell membranes [63, 64]. Pentane and ethane are expressed in cell cultures exposed to ROS and can be correlated to lipid peroxidation [65].

Moreover, aldehydes such as acetaldehyde are released as a result of amino acid and carbohydrate catabolism mechanisms [66]. Ketones are produced due to stress-related conditions that cause an increase in the activity of fatty acid oxidation. Acetone concentrations are found to vary in the body as a result of food consumption habits. Furthermore, isoprene is commonly produced by the muscle and liver cells [67–69]. On the other hand, sulfur-containing compounds are produced in the body via the transamination pathway when incomplete metabolism of methionine occurs and levels of the target sulfur compounds are altered [63]. Sulfur-containing compounds are used as biomarkers to detect liver failure. Compounds such as dimethyl sulfide, ethyl mercaptan are present in breath of patients with liver cirrhosis [70, 71]. Liver cirrhosis causes alteration in the levels of these compounds, and, thus, the breath of patients with liver disease smells like rotten cabbage [72]. Additionally, nitrogen-containing compounds, mainly ammonia, is released as a result of protein breakdown. Conversion of ammonia to urea occurs in the liver, and failure to cause this conversion can lead to increased levels of ammonia in urine [73]. Concentration of ammonia in breath is altered by smoking activity [74].

8.3.2 Various Components of Human Breath and Its Related Diseases

The study of human exhaled breath samples, called breath profile, reveals the occurrence of more than 1000 VOCs in exhaled breath, which indicates health status of the patients [75]. Thus, exhaled breath profiling can be used to monitor and diagnose various diseases and certain other characteristics of patients such as diet, phenotype, exercise. Broza et al. compiled the disease-specific functional groups from different reports [76]. Table 8.2 represents the diseases, corresponding VOCs, and their respective sources.

Table 8.2 Various disease-related VOCs and their sources.

S. no	Disease	Disease-related VOCs	Sample sources	References
1	Diabetes	Acetone, ethanol, methyl nitrate	Breath	[77][78]
2	Lung cancer	Pentane, heptane, octane, decane, 2-methylpentane, 2,3,3-trimethylpentane, 2,3,5-trimethyl hexane, 2,4-dimethyl-1-heptane, 4-methyloctane, 2,4-dimethyl-1-heptane, isoprene, 1-propanol, acetone, butyl acetate, 2-pentane, benzene, toluene, styrene.	Breath, blood, urine	[54]
3	Breast cancer	3,3-dimethylpentane, 2-amino-5-isopropyl-8-methyl-1-azulenecarbonitrile, 5-(2-methylpropylnonane, 2,3,4-trimethyldecane	Breath	[43][79]
4	Halitosis	H_2S, methane thiol	Breath	[80]
5	Kidney diseases	Ammonia	Breath	[81][82]

8.3.3 Breast Cancer-related VOCs

Mammography and ultrasound are the common tools for detection of breast cancer–related abnormalities even though they have low sensitivity and specificity. Misdiagnosis using these techniques is highly common and can cause failure to treat cancerous lesions. Moreover, currently available personalized cancer treatments as a result of the development of personal genomics are based on complex genome profiling and involve time-consuming analysis of genetic parameters [83]. All these aspects urge the need for development of new diagnostic tools that can allow for faster cancer diagnosis with high sensitivity and specificity. Breathomics was first explored for noninvasive disease diagnosis in the early 1990s. It was found that women with breast cancer had increased levels of pentane in exhaled breath [84]. Exhaled breath is now being used as a noninvasive disease detection method as it allows easy point-of-care diagnostics. Human exhaled breath is a combination of N_2, CO_2, O_2, water vapor, some inert gases as well as VOCs and nonvolatile compounds [85, 86]. Exhaled breath is analyzed for the presence of VOCs as elevated levels of these compounds can serve as disease biomarkers. However, endogenously produced VOCs are used as biomarkers for disease detection [87]. Some common VOCs assessed in clinical practice for detection of various diseases are ethanol detection, urea breath testing for detection of H. pylori, and asthma detection using the nitric oxide level in breath [88–91].

Table 8.3 Classification of common volatile organic compounds associated with breast cancer depending on the type of functional group present.

S.no	Chemical group	Volatile organic compound	Reference
1	Aliphatic hydrocarbons	Nonane, 2-methyl-propane, 2-methyl-octane, 5-methyl-tridecane, 3-methyl-undecane, 6-methyl-pentadecane, 3-methyl-nonadecane, 4-methyl-dodecane, 2-methyl-1-pentene	[50]
		Undecane, Tridecane, Tetradecane	[79]
		Dodecane	[92]
		3,3-Dimethyl-pentane, 5-(2-methylpropyl)-nonane	[43]
2	Primary and secondary alcohols	2-Propanol	[93]
3	Aldehydes	Heptanal	[93]
4	Aromatic compounds	1-Phenyl-ethanone	[93]
5	Nitriles	2-Amino-5-isopropyl-8-methyl-1-azulenecarbonitrile	[43] [93]
6	Others	Ammonium acetate Isopropyl myristate	[43] [93]

Breath analysis has been widely explored as a noninvasive diagnostic test for detection of breast cancer [94]. Increased oxidative stress is a key factor that is regulated in breast cancer, which causes damage to the cellular DNA, lipids, and proteins. Also, cytochrome P450 enzyme, expressed in breast cancer, causes lipid peroxidation of polyunsaturated fatty acids and leads to overexpression of VOCs such as alkanes and alkane derivatives in breath [62, 95, 96]. Apart from lipid peroxidation, free radical–induced oxidation of certain amino acids, proteins, and peptides has resulted in the production of hydrocarbons and monoaldehydes [97, 98]. Some of the VOCs associated with breast cancer are listed in Table 8.3, and they are classified based on the functional group present.

8.4 Sensing Elements

8.4.1 Various Materials for VOC Sensing

Various materials are widely used for VOC sensing, which are classified as follows:

1. Semiconducting metal oxides
2. Graphene-based materials
3. Carbon nanotube-based materials
4. Polymer-based materials
5. Silicon-based materials

This classification can be further subclassified into doped, functionalized, nanocomposites, and hybrid materials to improvise the sensitivity and selectivity toward various VOCs. Semiconductor metal oxides (SMOs) ZnO, SnO_2, CuO, ZrO_2, WO_3, etc. are important VOC-sensitive materials due to their high sensitivity, low cost, thermal stability, ease of fabrication, and ability to tune their properties [99]. Among the carbon-based materials graphene and carbon nanotubes (CNTs) are significantly incorporated for various VOC sensing applications. CNTs can be classified as single-walled (SWCNTs) and multi-walled carbon nanotubes (MWCNTs) according to their structure. CNTs are widely used for VOC sensing because of their large surface-area-to-volume ratio, electrical properties, mechanical stability, size, and compatibility with other nanomaterials toward VOC sensing [100]. Graphene is another potential material for VOC sensing due to its attractive properties and exhibit excellent thermal, electrical, and mechanical properties. By tuning the graphene synthesis methods, its sensing performance toward VOC can be enhanced [101]. Polymers have huge significance over other materials; polymer-based VOC sensors are mechanically flexible, cheap and consumes low energy [102]. Silicon-based materials have an edge over other materials because of their properties such as large carrier mobility, chemical functionalization, compatibility with complementary metal-oxide semiconductor (CMOS) and very large-scale integration (VLSI) technologies [103]. We have explored the applicability of room-temperature ionic liquids for VOC sensing using electrochemical methods of transduction [104–110].

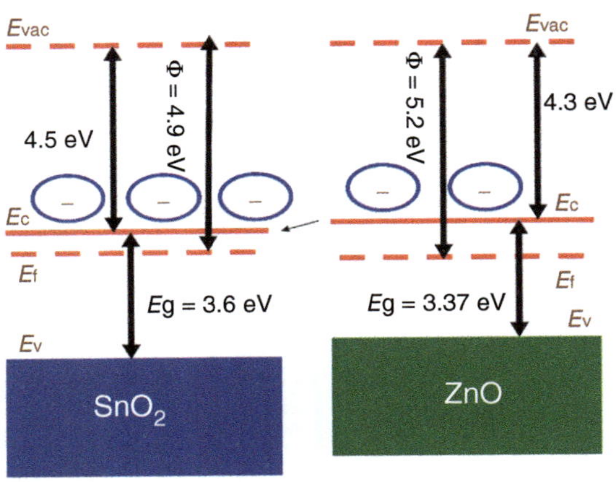

Figure 8.4 Composite semiconducting metal oxides in band diagram representation.

8.4.2 Metal Oxides for VOC Sensing

SMO-based materials are widely investigated as sensing elements in the field of gas sensing research. They are increasingly used because of their interesting properties such as tunable size, ability to operate at higher operating temperatures, biocompatibility [111]. Also, they possess unique electrical, optical, mechanical, and magnetic properties, enabling the users to choose from a wide range of properties based on their application. Koo et al. investigated the pristine WO_3 nanotubes for exhaled breath analysis and observed that pure nanotubes exhibited a better response to NO gas than toluene. But Pt and Pd–WO_3 exhibited opposite response at 400 °C. The results indicated that Pt–WO_3 nanotubes show better sensitivity and selectivity to toluene compared to NO [112]. Metal oxides have different activation and reaction temperature towards various gases. Also, a larger surface area enhances the probability of target molecule interaction at the sensing surface. Also the balance between the operating temperature and surface area improves the sensing properties of gas sensors [113].

8.4.3 Significance of Composite Metal Oxides

To tune sensitivity, reduce hysteresis, and enhance the performance characteristics, composite metal oxides have been proposed, the energy band representation of the composite semiconductors is shown in Figure 8.4. Composite metal oxides exhibit the properties of both materials, which in turn show synergistic effects on gas sensing properties. Velumani et al. studied impedimetric humidity sensing characteristics of pure and composite sensors and observed better sensitivity to various RH% [114].

8.5 Fabrication Methods

8.5.1 Thin Film Deposition

Thin films are best described by their birth process rather than thickness. A thin film is a material created ab initio by random nucleation and growth processes of individually condensing or reacting atomic/ionic/molecular species on a substrate.

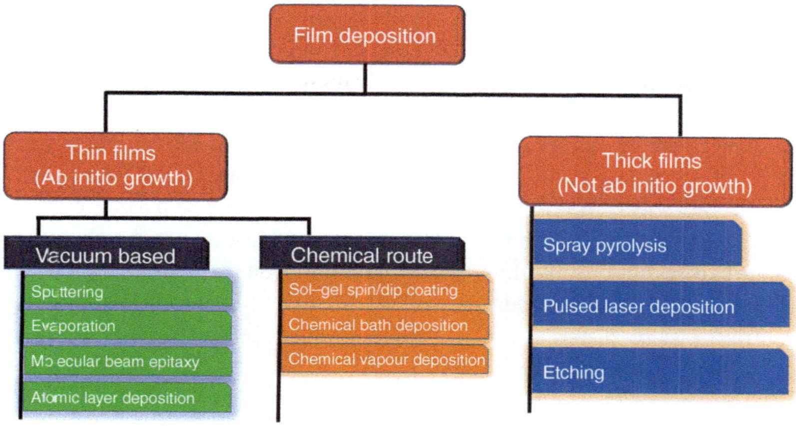

Figure 8.5 Different methods for the preparation of thin and thick films.

The thickness of thin films ranges between few nm to 10's of μm. Thin films exhibit exotic properties that can be tuned by changing the deposition parameters in a controlled manner. One of the major advantages of thin films is that they possess a relaxed phase diagram and hence alloying or doping is easier without the formation of secondary phases. There is a group of other materials known as thick films, which are mostly obtained by thinning the materials through a variety of techniques. The growth process of thick films is not ab initio, and, therefore, the corresponding microstructure, which significantly affects its VOC sensing ability, is different from that of thin films. Different methods for the preparation of thin and thick films are shown in Figure 8.5.

The thin film growth mechanism can be explained in the following steps: (i) In the first step, nucleation takes place where the adsorbed nuclei are statistically distributed over the substrate surface. (ii) In the second step, the growth of the nuclei takes place, and beyond the critical radius, the Gibbs free energy change increases, leading to stable nuclei. (iii) The stable nuclei coalesce with each other and form larger islands and tend to cover the entire substrate surface. (iv) Finally, the empty channels are filled and a continuous film is obtained. There are different growth modes possible depending on the cohesive and adhesive forces between the adatoms and substrate atoms. In the island growth (Volmer–Weber), three-dimensional islands are formed, leading to nanocrystalline/polycrystalline films. In this case, the adatoms are more strongly bound to each other than to the substrate atoms. In the layer-by-layer (Frank-van der Merwe) growth process, the adhesive force between the ad-atoms and the substrate atoms is very strong, forming continuous films and hence resulting in high crystallinity. The layer-by-layer growth mechanism is mostly observed in epitaxial films. In the intermediate growth process, which is known as the Stranski–Krstanov growth, initially the film is formed layer by layer and then the two-dimensional island formation takes place. Viable thin film deposition techniques for VOC sensor fabrication can broadly be classified as

1. Vacuum-based techniques
2. Chemical routes

8.5.1.1 Vacuum-based Techniques

Vacuum-based techniques provide a very precise control over the deposition parameters and are industrially feasible. Vacuum leads to an increased mean free path for sputtered as well as evaporated atoms. The film contamination or the impurity level also decreases with the decrease in pressure. For the evaporation-based deposition, the boiling point or the sublimation point decreases in accordance with the vapor pressure curve. Moreover, the breakdown voltage for striking the plasma during sputtering also decreases remarkably at lower pressures. According to the chamber pressure, there can be different gas flow regimes depending on the Knudsen number, which is defined as the ratio between the mean free path (λ_{MFP}) and the chamber dimension (D):

$$K_{\text{n}} = \frac{\lambda_{\text{MFP}}}{D} \tag{8.1}$$

For $K_{\text{n}} < 0.01$, the flow is termed as the viscous regime and mechanical pumps such as rotary and root pumps are used to create rough vacuum. For intermediate flow ($1 > K_{\text{n}} > 0.01$) and molecular flow regimes ($K_{\text{n}} > 1$), high vacuum pumps such as diffusion, turbomolecular, cryogenic pumps are used. For the preparation of thin film–based sensors, major industrially practical vacuum-based techniques are magnetron sputtering and electron beam evaporation. In sputtering, ions are generated in the form of plasma and are directed toward the target. The target atoms are ejected out due to the momentum transfer and are transported to the substrate for condensation. Radiofrequency (RF) magnetron sputtering is an ideal technique for the development of metal-oxide and chalcogenide-based thin film sensors. The RF voltages can be coupled capacitively through the insulating targets of metal oxides or chalcogenides to the plasma. The RF is high enough to sustain the plasma discharge due to the self-biasing of the target. In the electron beam evaporation, a focused beam of electrons is used to locally heat the evaporant material in a crucible. The local heating minimizes any contamination from the crucible. Electron beam evaporation is one of the ideal techniques for reactive evaporation of metal oxides, metal nitrides, and chalcogenides, which can be used as thin film–based sensors.

8.5.1.2 Chemical Routes

The chemical routes for preparing thin film–based sensors are mainly the sol–gel-based spin/dip coating and chemical vapor deposition (CVD). CVD is the process of chemically reacting a volatile compound of a material to be deposited with other gases to produce a nonvolatile solid that deposits atomistically on a substrate. CVD can be categorized as atmospheric pressure CVD, low-pressure CVD, plasma-enhanced CVD, metal–organic CVD, etc. On the other hand, the sol–gel-based spin/dip coating is less expensive and requires considerably less equipment compared to conventional thin film growth techniques like CVD, sputtering, or evaporation. However, the most important advantage of sol–gel processing over conventional coating methods is the ability to precisely control the microstructure of the deposited films with atomic-scale mixing. Prior to gelation, the fluid sol or solution is ideal for preparing thin films by common processes like spinning or dipping.

8.5.2 Thick Film Deposition

Thick films are obtained by collective deposition of materials on a substrate surface, and they do not involve ab initio nucleation and growth. Generally, the thickness of thick films is of the order of few micrometers. The major thick film growth techniques employed for the fabrication of sensors are spray pyrolysis (SP) and pulsed laser deposition (PLD). In the SP process, the solution is sprayed on a heated surface, upon which the constituents react to form a chemical compound. SP is a simple and cost-effective technique without the need of any clean environment and can be used for the deposition of large area films. The substrate temperature is a crucial parameter that determines the film morphology and hence the sensing ability. In PLD, high-power laser pulses are used to evaporate the materials from the target. The laser ablation produces a transient, highly luminescent plasma plume, and the ablated material condenses on the substrate surface. PLD ensures the proper stoichiometric transfer from the target to the film and thus is an ideal technique for the growth of highly crystalline and stoichiometric films. However, the major limitation of PLD is that it cannot be used for large area coatings but is ideal for the fabrication of VOC sensors.

A few representative examples of thin-film and thick-film-based sensors fabricated by the aforementioned techniques for VOC sensing are given in Table 8.4.

8.5.3 Growth of Nanomaterials

The key elements of nanotechnology are nanomaterials whose size is 10 nm, which is one billionth of a meter. Compared to bulk materials, nanoscale materials exhibit high surface-to-volume ratio, which tends to increase the reactivity at the molecular

Table 8.4 Thick-film and thin-film-based VOC sensors and their fabrication techniques.

Sensing material	Preparation technique	Target VOC	Reference
ZnO	RF magnetron sputtering	Acetone, ethanol, toluene, isopropanol	[115]
Al : NiO	RF magnetron sputtering	Hydrogen, methane	[116]
Al : ZnO	RF magnetron sputtering	Formaldehyde	[117]
WO_3	E-beam evaporation	Cyclohexane	[118]
$In_2O_3 + SnO_2$	Thermal evaporation	Ethanol	[119]
$WO_3 + Cr_2O_3$	Sol–gel dip coating	Acetone	[120]
SiOCH	Filament-assisted CVD	Hydrocarbons	[121]
SnO_2	Aerosol-assisted CVD	Ethanol, acetone, hexane	[122]
In_2S_3	Spray pyrolysis	Methanol, ethanol, acetone	[123]
SnO_2	Spray pyrolysis	Methanol, ethanol, acetone	[124]
Gd : CeO_2	PLD	Acetone	[125]
Pr : CeO_2	PLD	Toluene	[126]

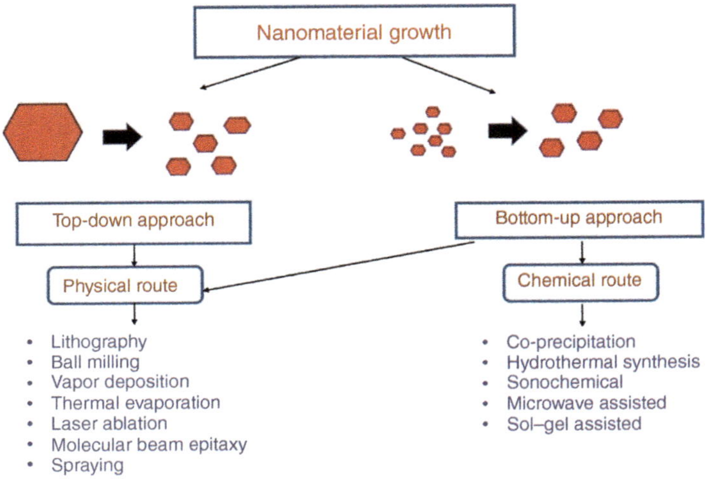

Figure 8.6 Nanomaterial growth techniques.

level. Also electrical, mechanical, optical, and chemical properties are enhanced at the nanoscale. Based on their dimensionality, nanomaterials can be classified into nanosheets [127], nanorods [128], and nanoparticles [129]. Electronic properties of a material are based on its electronic band structure wherein the energy levels are altered based on the particle size of the grown nanomaterials.

Nanomaterials can be grown by top-down and bottom-up approaches. These approaches are further classified into physical and chemical methods as shown in Figure 8.6.

8.5.3.1 Physical Methods

The growth of nanomaterials through physical methods are categorized as "top-down" and "bottom-up" approaches. In the former approach, larger materials are grounded into small particles by mechanical milling. A significant drawback of this approach is very low possibility of getting the desired size and shape. The latter approach is based on condensation of gas- or liquid-phase nanoparticles to form larger materials through chemical reactions. In magnetron sputtering technique, the target material is bombarded with a high-energy Ar^+ ion source to remove the surface atoms. The removed surface atoms are directed toward the substrate surface, where they agglomerate and react to form thin films whose particles are nanometers in size [130]. In the laser evaporation method, a high-energy laser beam is used as the source to evaporate the target material. The evaporated material is formed as nanoparticles outside the evaporation zone because of the difference in the temperature gradient. Size of the nanoparticles can be altered by varying the laser power [126]. In the thermal evaporation method, the material to be evaporated is placed in the boats and rods and is evaporated by passing the electric current or focusing the electron beam to heat the sample. This breaks chemical bonds in compounds and reduces them into smaller ones, which is an endothermic reaction [126].

8.5.3.2 Chemical Methods

The chemical route for the preparation of nanomaterials is the bottom-up approach through which pure, doped, and nanocomposite materials are synthesized. In the chemical route, different techniques are used for the preparation of nanomaterials. Size, type of nanomaterials, properties of nanocomposites, and ease of synthesis are the deciding factors for choosing an appropriate method of preparation.

Co-precipitation is one of the simplest and most widely used synthesis methods to prepare pure, doped, and mixed metal oxides. In this method, precursor materials such as nitrates, acetates, chlorides are dissolved in an aqueous solution, and then the prepared hydroxides are precipitated by adding a base such as sodium hydroxide. The precipitated material is then washed and calcined at a higher temperature to remove the organic contents and improve crystallinity. The size of the nanoparticles is uniform because the precipitation rate is not controlled and it can be obtained by keeping pH constant during the synthesis [131]. Sol–gel is one of the wet-chemical-based synthesis techniques. It involves the following processes: the metal alkoxide solution undergoes hydrolysis with water or alcohols followed by polycondensation. Then, the water molecules are condensed, where the gel phase is converted into the powder phase. Upon heating, crystallinity of the powder is improved. This method is suitable for the preparation of oxides, composites, and doped materials. The main advantage is its simple procedure, high surface area, and stable surface [124]. In hydrothermal techniques, the prepared solution is subjected to high temperature and pressure. To obtain a homogeneous solution, the organic solvent along with precursors is stirred. The prepared solution is then placed in a sealed stainless-steel autoclave. By selecting suitable solvents and factors such as pressure, pH, temperature, aging time of the reaction, desired size and morphology of specific nanoparticles are obtained. Compared to co-precipitation, yield is less, but crystallinity is improved at relatively lower temperatures [132]. The sonochemical method is one of the novel techniques for the synthesis of nanomaterials using high-intensity ultrasound. For the preparation of nanomaterials, this approach does not require high temperature, long reaction time, and high pressure. Ultrasonic irradiation collapses the cavities inside the liquid medium, which chemically excites the surrounding matter. In the microwave synthesis technique, the material prepared is heated by applying microwave radiation. Smaller particles are prepared within a shorter duration by applying microwave radiation in the frequency range of 1–2.5 GHz to produce a temperature of 100–200 °C [133, 134].

8.6 Noninvasive Techniques for Breast Cancer Diagnosis

8.6.1 Selected Ion Flow Tube Mass Spectrometry (SIFT-MS)

Selected ion flow tube mass spectrometry was first developed in 1976 for studying ion neutral reactions [135]. This technique involves the use of a mass filter for selection of one ionic species at a time, which is then passed into the tube at lower energy than the neutral gas. A schematic representation of the setup is shown in Figure 8.7. The selected ion is injected at a lower energy to avoid collisional fragmentation of

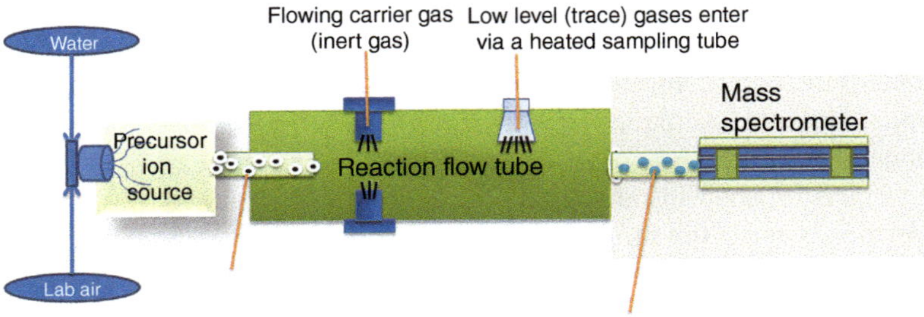

Figure 8.7 Schematic representation of SIFT setup involving the precursor ion source that selects the ion that flows along with an inert gas into the reaction flow tube and they are finally detected by the mass spectrometer present at the end of the tube.

specific ions. Therefore, this technique is often called mass-selected ion injection. It is widely used to study ionic reactions and understand the associated kinematics [136–141]. SIFT allows real-time quantification of target analytes in trace mixtures. SIFT can be operated typically at two modes: full-scan mode and multiple ion monitoring mode. The sampling procedures use ambient air, exhaled breath, and headspace above liquids, solids, and in vitro cell cultures.

In a study by Kato et al. [142], cell lysates were used to measure formaldehyde levels in treated and untreated cells. Detection of formaldehyde levels was calibrated by using a standard stock solution of formalin. The results thus obtained helped in studying the mechanism of cytotoxicity against two important anticancer drugs, doxorubicin and daunorubicin.

8.6.2 Proton Transfer Reaction Mass Spectrometry (PTR-MS)

Proton transfer reaction mass spectrometry has been widely used for the detection of VOCs with high sensitivity and specificity. The technique was first developed at the University of Innsbruck in Austria by Lindinger and his team in the late 1990s [143]. The technique is based on the H_3O^+ ions that allow easy proton transfer in non-dissociative manner such that the positively charged ions interact only with the target trace organic analytes but not with the components present in clean air. VOCs that have higher affinity for proton exchange than for water will participate in the reaction. A quadruple mass spectrometer is used at the end of the reactor tube that collects the positive ions. Hence, the signal at the mass spectrometer is directly proportional to the concentration of VOCs in the target sample. However, a disadvantage of this technique is that VOCs with same mass cannot be identified separately. It has advantages over other spectrometric methods in the case of having higher measurement frequency that is, it allows one measurement per second as compared to GC-MS that only allows one measurement per thirty second [144].

8.6.3 Gas Chromatography–Mass Spectrometry (GC-MS)

The application of exhaled breath as a source of VOCs to allow for therapeutic monitoring and early disease detection is widely being explored. Volatile biomarkers for

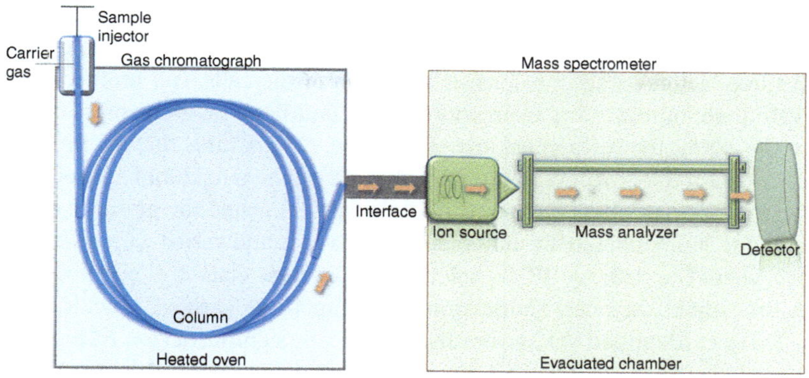

Figure 8.8 Schematic representation of the gas chromatography–mass spectrometry setup.

breast cancer diagnosis can be detected using GC-MS. The method is based on the sampling of concentrated human breath containing VOCs, which is then injected into a gas GC column. The GC column separates different VOCs present based on their boiling point, polarity, and other physicochemical properties. These individually separated VOCs are then injected into a mass spectrometer where they are analyzed based on their m/z ratios, such that each VOC is broken down into fragments and analyzed by a detector [145]. A schematic representation of the setup is shown in Figure 8.8. GC-MS gives us a mass spectrum of fingerprints, which can be used to identify unknown compounds and can serve as a fingerprint for identification. The results so obtained can be compared with a computer-based spectral library, and unknown VOCs can be identified. GC-MS is used as a dual analytical technique that allows for the separation and identification of individual compounds present as VOCs in breath.

Linus Pauling, the Nobel laureate awarded with two Nobel Prizes, was able to quantify 250 VOCs present in breath samples using GC [146]. Li et al. [147] investigated the role of four potential biomarkers in exhaled breath for the detection of breast cancer. They used Bio-VOC for sample collection and the assay was done using GC–MS. The results were obtained using the Kruskal–Wallis one-way analysis of variance test along with binary logistic regression curves. The data was analyzed using the receiver operating characteristics curve. The study used the elevated levels of breath aldehyde biomarkers for the early stage, noninvasive diagnosis of breast cancer. Three subject groups were selected and studied in this research work. A total of 24 healthy controls were selected, along with 22 patients with breast cancer and 17 patients with benign breast cancer. The patients were selected randomly. Mainly, C5–C10 monofunctional aldehydes that are end results of lipid peroxidation were investigated as cancer-related breath biomarkers. The role of hexanal, heptanal, nonanal, and octanal was studied, and their area under the curves was analyzed for sensitivity and specificity. It was observed that combined use of these four biomarkers as a diagnostic panel is more efficient than detection of a specific compound.

Lavra et al. [148] investigated the role of different VOCs and correlated their presence with characteristics of breast cancer cells. The biomarkers were analyzed based on three factors – the transformed condition of the cells, cell proliferation time or cell doubling time, and expression of ERs and PgRs or overexpression of HER2. In this study, the researchers analyzed the VOC signatures that are present in the headspace of the cancer cell culture media and can help to identify the occurrence of specific breast cancer biomarkers. The results indicated the presence of 13 VOCs that were susceptible to the aforementioned factors and varied with different cancer cell lines. The detected VOCs belonged to different classes of compounds, mainly hydrocarbons, ketones, amine aromatic compounds, carboxylic acids, and alcohols. These 13 identified VOCs were also statistically significant ($p = 7.7 \times 10^{-3}$) to differentiate between the tumor-derived and normal cell lines. Hydrocarbons, mainly, 2,4-dimethyl-1-heptene, 2-xylene, 2,3-dimethylhexane, 2,2-dimethylbutane were found in higher concentrations. Alcohols such as cyclohexanol and ketones such as 2-dodecanone were also found to be higher. However, a decrease in ketone concentrations of (2-nonanone and 4-methyl-2-heptanone) was observed. These results helped in identification and characterization of specific VOC biomarkers present in the headspace of cell lines for breast cancer–derived tumors.

In another study by Phillips et al. [50], a combined case control as well as a cross-sectional study, samples were collected from women who had abnormal mammograms. GC-MS analysis was used to derive the breath methylated alkane contour (BMAC). The researchers performed a three-dimensional analysis study of the alveolar gradient of specific alkane length hydrocarbons (C4–C20) and monomethylated alkanes. The BMAC levels were studied to contrast between women with and without breast cancer. The data set involved 201 breath samples from women who showed abnormal signs in the mammogram and had biopsies done. The results were read by two different pathologists to increase the confidence level. The breath test showed a sensitivity of 94.1% and a specificity of 62.7% between women with breast cancer and healthy women. Moreover, the negative predictive value (NPV) was found to be greater for the breath test compared to screening mammogram, and a higher positive predictive value was found for screening mammogram than the screening breath test. Thus, the breath test used for identification of oxidative stress biomarkers showed superior NPV and promising results to be employed as a diagnostic tool for noninvasive disease diagnosis.

In a systemic study performed by Wang et al. [149] for the analysis of cancer, VOCs present in breath were identified by studying the gas profiles for the presence of different metabolites. Exhaled breath was collected from 85 different subjects who was diagnosed positive for breast cancer by histological testing and was compared to 45 healthy subjects. The data obtained using GC-MS was analyzed by the principal component analysis and the method of partial least square discriminant analysis for final data processing from different subjects. The experimental results ($p < 0.05$) revealed that three potential biomarkers 1,4-dimethoxy-2,3-butanediol, cyclohexanone, and 2,5,6-trimethyloctane could be correlated to the presence of breast cancer specifically. The presence of these biomarkers in exhaled breath could serve as a novel platform for non-IBC diagnosis.

Clinical testing was done by Mandy et al. [150] for the detection of typical discriminatory VOC markers in breath. The study involved 10 patients with breast cancer and 10 healthy women. The breath samples were taken and analyzed using GC-MS to detect the presence of different compounds. This pilot study revealed the presence of mainly five biomarkers (methylhexane, decene, caryophyllene, naphthalene, and trichlorethylene) that can be associated with the presence of breast cancer. The altered levels of these compounds can be correlated to the change in the metabolic activity of the cytochrome P450 enzyme as well as accumulation of other tumor metabolites.

Using GC, Shibamoto et al. [151] quantitatively analyzed volatile carbonyl compounds in the expired air collected from tumor-bearing transgenic mice and humans. Quantitative analysis was done for three biomarkers, formaldehyde, acetone, and acetaldehyde. Increased levels of formaldehyde were observed in the tumor-bearing mice as compared to the control mice per unit metabolic size. The exhaled breath was analyzed for formaldehyde levels in three healthy women and three women with breast cancer. The results indicated a high likelihood of using these carbonyl compounds as biomarkers for breast cancer diagnosis.

8.6.4 Differential Mobility Spectrometer (DMS)

Differential mobility spectrometry (DMS) is employed for identification and characterization of gas-phase ions, also known as field ion spectrometry (FIS) [152] and field asymmetric waveform ion mobility spectrometry (FAIMS) [153]. The technique is based on the principle of the change in mobility of ions in the presence of different strengths of electric fields. It is a powerful technique that is selective in nature and is easy to use. The apparatus for DMS can be operated in two geometries – planar and coaxial. It is modeled using computational [154] and mathematical models [155]. Mathematical modeling is used to understand general principles involved, whereas computational modeling helps in solving quantitative calculations [156]. This is a new and rapid detection technique used for non-invasive diagnosis of breast cancer.

8.6.5 Chemiresistive Sensing Mechanism

Chemiresistive VOC sensors work based on the chemical interaction between the material surface and target molecules shown in Figure 8.9 [157]. These sensors require activation energy for the chemical adsorption of oxygen molecules present in air. At temperatures ranging from 100 to 400 °C, the oxygen molecules are ionized as different receptor states (O_2^-, O^-, O^{2-}). The electrons in the surface of the semiconductor are captured, resulting in an increase in the depletion layer width [113]. Based on the charge carriers present, the semiconducting materials can be classified into n- and p-type materials. Also, the target VOC species can be differentiated into two reducing gas (electron donors) and oxidizing gas (electron acceptors). The carrier concentration of the charges of the SMO can be altered by the chemical reaction between the sensor surface and target gases. Thus, changes in the sensor resistance depend on the type of SMO and the target VOC molecules. The electrons from the

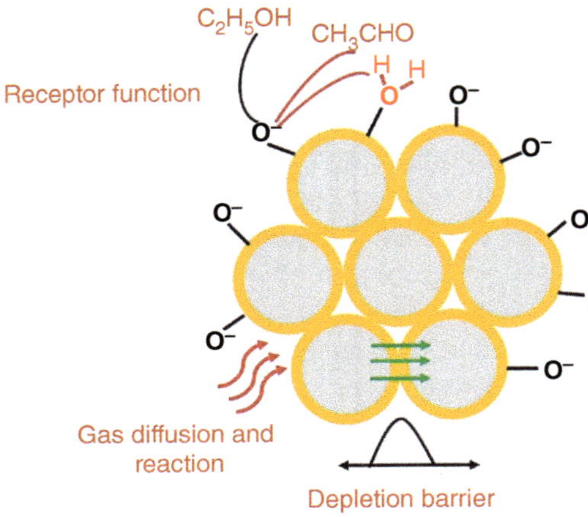

Figure 8.9 Three basic factors for chemiresistive gas sensors.

conduction band of the semiconductor material are pulled by the adsorbed oxygen ions, which results in an increase in the resistance of the sensing material. When the sensor surface is exposed to the reducing gas, the chemical reaction releases the electrons back to the conduction band of the semiconductor, which results in a decrease in the sensor resistance. The inverse mechanism is for the oxidizing gases. The response of the SMO VOC sensor is determined by the surface and bulk properties of the material as well as the concentration of the target gases [113].

The SMO gas sensing materials can be classified into two groups based on their activation temperature and material properties: (i) materials that are mainly based on the surface conductance effect and (ii) on bulk conductance. The former group of SMOs operates at a lower temperature range (400– 600 °C) and the latter group at a higher temperature range (>700 °C) [158]. The SMOs such as ZnO, SnO_2, CuO are the examples of surface conducting oxides. At lower operating temperatures, the effect of bulk conductance change is slow due to surface adsorption of oxygen species. Bulk conductance materials such as TiO_2, CeO_2, Nb_2O_5 respond to conductivity changes in oxygen partial pressure at higher operating temperatures. The conductivity of the SMO and oxygen partial pressure is explained by the following relation:

$$S = D \exp\left(-E_A/kT\right) po\, 1/N \tag{8.2}$$

where S is electrical conductivity, D is constant, E_A is activation energy, T is temperature, k is the equilibrium constant, and N is the constant depends on bulk defects.

8.6.6 Fiber-optic Sensors

Recently, implementation of optical fibers in sensor applications are rapidly increasing due to ease of fabrication, low cost, and immunity to electromagnetic interferences. Especially, in VOC gas sensors, optical fibers are widely used because of their fast response and sensitivity toward gas molecules. Different types of optical fibers have been reported in sensor applications such as plastic optical fibers, silica optical fibers, fiber Bragg grating optical fibers [159], photonic crystal optical fibers (PCFs) [160]. Other than the geometries and types of optical fibers, various sensing mechanisms are reported including absorption-based sensors, evanescent mode sensors [161], plasmonic mode optical sensors such as surface plasmon resonance (SPR) [162], localized SPR [163], and lossy mode resonance (LMR) [164].

Optical fibers contain three main parts: core, clad, and out jacket; in most sensor applications, a portion of clad is removed and replaced with nanomaterial layers called the sensing layer. The mechanism of sensors purely depends up on the type of the nanomaterial coated over the core layer. The optical property of nanomaterials decides the type of sensor mechanism, where metal oxides such as zinc oxide (ZnO), tin oxide (SnO_2), indium tin oxide (ITO) are widely used in evanescent mode sensing mechanism due to its leaky mode condition at a wide spectral range. Evanescent wave fiber-optic sensors are discussed in detail next.

8.6.6.1 Evanescent Wave Fiber-optic VOC Sensors

Based on the principle of total internal reflection, light propagation takes place inside the core, and, thus, transmission of optical signals occurs. In evanescent wave optical sensors, a portion of the clad is replaced with absorbing materials. When an analyte interacts with the absorbing material, evanescent decay gets violated and its reflected intensity varies. This is the mechanism behind evanescent wave optical fibers represented in Figure 8.10 [161]. The change in the analyte concentration varies the intensity of the reflected light. The variation of intensity depends on the property of the nanomaterial and the nature of the gas.

8.6.6.2 Lossy Mode Resonance Fiber-optic VOC Sensors

LMR is the coupling phenomenon of the core and lossy modes of thin films at a specific thickness. Due to the phase match between the core and lossy modes of thin films, resonance takes place and results in loss of light intensity. In the angular interrogation technique, at a specific angle, resonance occurs, called resonance angle, whereas in the wavelength interrogation technique at a specific wavelength, resonance take place, called resonance wavelength. The resonance wavelength is highly sensitive to the refractive index of the surrounding medium, permittivity of the material, thickness and type of the material and its analyte. The change in the

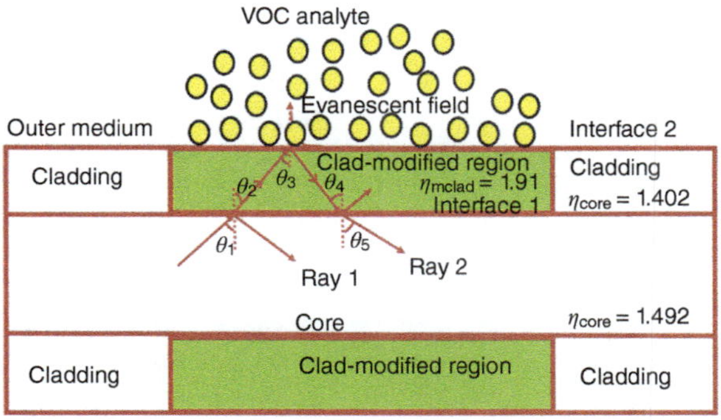

Figure 8.10 Schematic diagram of an evanescent wave optical fiber gas sensor.

thickness of the thin film induces the number of lossy modes, whereas the increase in the thin film thickness increases the number of resonance peaks. This property causes the LMR-based optical sensors to be used in various applications. Tiwari et al. [165] developed an ammonia sensor based on LMR using tapered optical fibers with porphyrin-incorporated titanium dioxide.

8.6.7 Surface Plasmon Resonance Sensor

SPR sensing principle has received much attention after Liedberg et al. demonstrated its potential for chemical sensing. SPR is a phenomenon that occurs due to the interaction of light with mobile charge carriers in a metal film (plasma surface) at the interface of media with different refractive indices (e.g. a metal/dielectric layer). The surface plasmons are transverse magnetic (TM) waves, which can be excited by polarized light. Resonance occurs when the wave vector and frequency of incident light coincide with those of surface plasmon [166]. The wave vector K_{sp} of the surface plasmon on the boundary between the metal and the sample at frequency w is

$$K_{sp} = w/c\sqrt{(\varepsilon m\varepsilon s/\varepsilon m + \varepsilon s)} \tag{8.3}$$

where εm and εs are the real parts of the dielectric constants of the metal and the sample, c is the velocity of light in vacuum.

Kretschmann proposed an SPR technique to excite and detect collective oscillations of free electrons (known as surface plasmons) by an evanescent wave from a high refractive index prism – coated with a thin metal film touching the sample (Figure 8.11) in which light is focused onto a metal film through a glass prism and subsequent reflection is detected [167].

At a definite incident angle, absorption of light takes place owing to which energy of the reflected light reduces which is due to the resonation of plasmons with incident light. This is observed as a sharp dip in the R–θ curve in Figure 8.12. The resonance angle is very sensitive to variations in the refractive index (or the dielectric constant) of the sample medium.

SPR components can be miniaturized using optical fibers. A fiber-optic SPR sensor eliminating the coupling prism is shown in Figure 8.13.

The sensing element in the fiber is prepared by removing the cladding and depositing a metal film on the core using electron beam evaporation. To make it a biosensor,

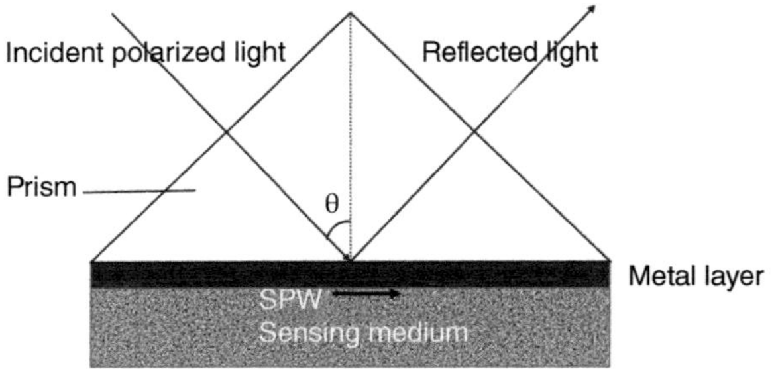

Figure 8.11 Kretschmann configuration of a prism-based SPR sensor.

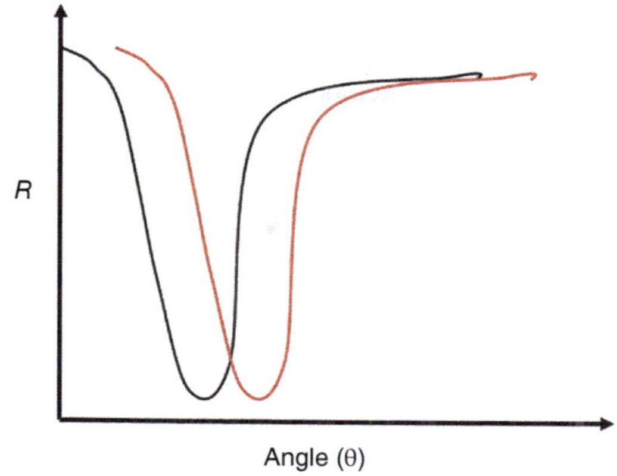

Figure 8.12 Reflectance as a function of angle of incidence at the prism–metal interface.

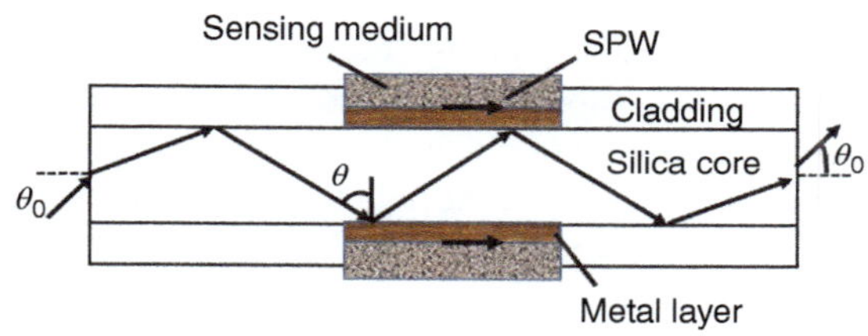

Figure 8.13 Fiber-optic SPR sensor probe.

antibodies are immobilized on this metallic film. Probe configuration of a typical biosensor is shown in Figure 8.14.

8.6.8 Calorimetric Sensors

Calorimetric-based sensing is one of the conventional techniques used for biosensing applications. As the name suggests, the sensor changes its color in the presence of VOCs, and the color change ratio depends on the concentration of the proposed analyte and its flow rate over the sensor surface [168]. For the detection of volatile analytes, a typical gas–solid reaction model can be proposed. The sensing probe is prepared on a solid/flexible substrate, wherein target analytes are diffused into the sensor surface and react with the proposed sensing material. The target analyte and the sensing material interaction process is as follows:

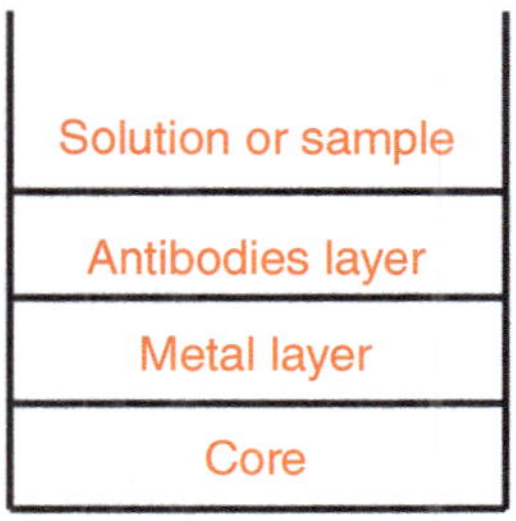

Figure 8.14 Probe configuration for a typical biosensor.

1. Adsorption of the target VOCs on the sensor surface
2. Diffusion of the analyte molecules
3. Color change based on the concentration of VOCs

To improve cross reactivity and selectivity, two different sensing platforms sensor array and lock-and-key models are used widely. In the lock-and-key technique, the concentration of the target VOCs can be analyzed using optical signals [169].

The sensor array model mimics the mammalian olfactory system, wherein the target VOCs are identified by the pattern recognition technique. In a typical calorimetric sensor array, chemo-responsive indicators are metal salts – redox reaction; pH analytes – Bronsted acid/base; metal ion dye – Lewis basicity; nucleophilic indicators – electrophilic analytes. These sensor arrays were proposed to identify various breath-based diseases, in which VOCs were the target analytes. Mazonne et al. [170] proposed that using the calorimetric sensor array various types of lung cancers were detected with good accuracy. Alagirisamy et al. detected 0.05 $\mu g\,l^{-1}$ hydrogen sulfide, a target biomarker for halitosis, using the iodine solution and starch in a calorimetric sensor array [171].

8.7 Conclusion

Early disease detection is crucial to ensure timely treatment and decrease the mortality rate, especially for diseases such as cancer. The development of devices for early detection of diseases can play a significant role in achieving this goal. The latest advancements in the field of biosensors are making it possible to a large extent to detect many types of cancer at an early stage. They are not able to detect biomarkers, which are good indicators of a disease, even in very low concentrations in body fluids, thus offering a reliable, noninvasive or minimally invasive, and cost-effective detection methodology. There is scope that these biosensors can, to a large extent, replace complex techniques such as invasive biopsies, expensive labelling processes, and time-consuming analyses. However, there is a need to introduce the sensor technology into clinical breath analysis and develop easy-to-handle cost-effective and adaptable sensors. A lot of research has focused on developing sensor platforms using different materials as electrode materials and architecture play crucial roles in the performance of sensors. This chapter outlined breast cancer–related VOCs and the detection mechanism currently used for diagnosis. Researchers are also focused on developing sensors that can measure multiple biomarkers simultaneously. The development of multi-analyte immunosensors, as they are called, is in early stage. While we have seen the development of a large number of sensors that are able to meet sensitivity and selectivity requirements, they are still far from meeting the commercialization needs. Research has shown that biosensors are going to lead future disease diagnosis and health monitoring, but more efforts, including large-scale sampling, are needed to ensure that they are able to reach the patient's bedside.

Acknowledgements

The authors express their sincere thanks to MHRD Scheme for promotion of academic and research collaboration (SPARC)-IIT KGP, India, for providing financial support to promote academic and research collaboration through the project [SPARC/2018–2019/P461/SL]. The authors would like to thank Vellore Institute of Technology (VIT), India, and the University of Texas at Dallas (UTD) for their support in academic and research collaboration. The author Mohan Velumani would like to thank the CSIR–SRF grant [09/844(0047)/2018-EMR-I] for financial support.

References

1 Siegel, R.L., Miller, K.D., and Jemal, A. (2020). Cancer statistics, 2020. *CA Cancer J. Clin.* 70 (1): 7–30.

2 Goossens, N., Nakagawa, S., Sun, X., and Hoshida, Y. (2015). Cancer biomarker discovery and validation. *Transl. Cancer Res.* 4 (3): 256–269.

3 Yabroff, K.R., Lund, J., Kepka, D., and Mariotto, A. (2011). Economic burden of cancer in the United States: estimates, projections, and future research. *Cancer Epidemiol. Biomarkers Prev.* 20 (10): 2006–2014.

4 Kourou, K., Exarchos, T.P., Exarchos, K.P. et al. (2015). Machine learning applications in cancer prognosis and prediction. *Comput. Struct. Biotechnol. J.* 13: 8–17.

5 Solassol, J., Marin, P., Demettre, E. et al. (2005). Proteomic detection of prostate-specific antigen using a serum fractionation procedure: potential implication for new low-abundance cancer biomarkers detection. *Anal. Biochem.* 338 (1): 26–31.

6 Singh, S., Gill, A.A.S., Nlooto, M., and Karpoormath, R. (2019). Prostate cancer biomarkers detection using nanoparticles based electrochemical biosensors. *Biosens. Bioelectron.* 137: 213–221.

7 Singh, P., Pandey, S.K., Singh, J. et al. (2016). Biomedical perspective of electrochemical nanobiosensor. *Nano-Micro Lett.* 8 (3): 193–203.

8 Wei, F., Patel, P., Liao, W. et al. (2009). Electrochemical sensor for multiplex biomarkers detection. *Clin. Cancer Res.* 15 (13): 4446–4452.

9 Bohunicky, B. and Mousa, S.A. (2010). Biosensors: the new wave in cancer diagnosis. *Nanotechnol. Sci. Appl.* 4: 1–10.

10 Cheng, N., Du, D., Wang, X. et al. (2019). Recent advances in biosensors for detecting cancer-derived exosomes. *Trends Biotechnol.* 37 (11): 1236–1254.

11 Khanmohammadi, A., Aghaie, A., Vahedi, E. et al. (2020). Electrochemical biosensors for the detection of lung cancer biomarkers: a review. *Talanta* 206: 120251.

12 Tothill, I.E. (2009). Biosensors for cancer markers diagnosis. *Semin. Cell Dev. Biol.* 20 (1): 55–62.

13 Fitzgerald, J. and Fenniri, H. (2017). Cutting edge methods for non-invasive disease diagnosis using E-tongue and E-nose devices. *Biosensors* 7 (4).

14 Huang, S., Li, H., Ding, X., and Xiong, C. (2009). Presence and characterization of cell-free seminal RNA in healthy individuals: implications for noninvasive disease diagnosis and gene expression studies of the male reproductive system. *Clin. Chem.* 55 (11): 1967–1976.

15 Pereira, J., Porto-Figueira, P., Cavaco, C. et al. (2015). Breath analysis as a potential and non-invasive frontier in disease diagnosis: an overview. *Metabolites* 5 (1): 3–55.

16 Paul, A., Vyas, G., Paul, P., and Srivastava, D.N. (2018). Gold-nanoparticle-encapsulated zif-8 for a mediator-free enzymatic glucose sensor by amperometry. *ACS Appl. Nano Mater.* 1 (7): 3600–3607.

17 Paul, A. and Srivastava, D.N. (2018). Amperometric glucose sensing at nanomolar level using MOF-encapsulated TiO_2 platform. *ACS Omega* 3 (11): 14634–14640.

18 Paul, A., Chiriacò, M.S., Primiceri, E. et al. (2019). Picomolar detection of retinol binding protein 4 for early management of type II diabetes. *Biosens. Bioelectron.* 128 (2018): 122–128.

19 Munje, R.D., Muthukumar, S., and Prasad, S. (2017). Lancet-free and label-free diagnostics of glucose in sweat using zinc oxide based flexible bioelectronics. *Sensors Actuators B Chem.* 238: 482–490.

20 Munje, R.D., Muthukumar, S., Jagannath, B., and Prasad, S. (2017). A new paradigm in sweat based wearable diagnostics biosensors using room temperature ionic liquids (RTILs). *Sci. Rep.* 7 (1): 1950.

21 Bhide, A., Muthukumar, S., Saini, A., and Prasad, S. (2018). Simultaneous lancet-free monitoring of alcohol and glucose from low-volumes of perspired human sweat. *Sci. Rep.* 8 (1): 6507.

22 Bhide, A., Muthukumar, S., and Prasad, S. (2018). Continuous lifestyle awareness through sweat platform (CLASP): a novel sensor for simultaneous detection of alcohol and glucose from passive perspired sweat. *Biosens. Bioelectron.* 117: 537–545.

23 Albers, M.J., Bok, R., Chen, A.P. et al. (2008). Hyperpolarized ^{13}C lactate, pyruvate, and alanine: noninvasive biomarkers for prostate cancer detection and grading. *Cancer Res.* 68 (20): 8607–8615.

24 Lu, G., Halig, L., Wang, D. et al. (2014). Spectral-spatial classification for non-invasive cancer detection using hyperspectral imaging. *J. Biomed. Opt.* 19 (10): 106004.

25 Bhatt, A., Mathur, R., Farooque, A. et al. (2010). Cancer biomarkers – current perspectives. *Indian J. Med. Res.* 132: 129–149.

26 Tripathi, P.P., Arami, H., Banga, I. et al. (2018). Cell penetrating peptides in preclinical and clinical cancer diagnosis and therapy. *Oncotarget* 9 (98): 37252–37267.

27 Alturkistani, H.A., Tashkandi, F.M., and Mohammedsaleh, Z.M. (2015). Histological stains: a literature review and case study. *Glob. J. Health Sci.* 8 (3): 72–79.

28 Zeng, C., Vangveravong, S., Xu, J. et al. (2007). Subcellular localization of sigma-2 receptors in breast cancer cells using two-photon and confocal microscopy. *Cancer Res.* 67 (14): 6708–6716.

29 Kaufmann, R., Müller, P., Hildenbrand, G. et al. (2011). Analysis of Her2/neu membrane protein clusters in different types of breast cancer cells using localization microscopy. *J. Microsc.* 242 (1): 46–54.

30 Xuan, Z., Li, M., Rong, P. et al. (2016). Plasmonic ELISA based on the controlled growth of silver nanoparticles. *Nanoscale* 8 (39): 17271–17277.

31 Vaisocherová, H., Faca, V.M., Taylor, A.D. et al. (2009). Comparative study of SPR and ELISA methods based on analysis of CD166/ALCAM levels in cancer and control human sera. *Biosens. Bioelectron.* 24 (7): 2143–2148.

32 Ambrosi, A., Airò, F., and Merkoçi, A. (2010). Enhanced gold nanoparticle based ELISA for a breast cancer biomarker. *Anal. Chem.* 82 (3): 1151–1156.

33 Scholler, N., Crawford, M., Sato, A. et al. (2006). Bead-based ELISA for validation of ovarian cancer early detection markers. *Clin. Cancer Res.* 12 (7 Pt 1): 2117–2124.

34 Arya, S.K. and Estrela, P. (2018). Recent advances in enhancement strategies for electrochemical ELISA-based immunoassays for cancer biomarker detection. *Sensors* 18 (7).

35 Middelhoek, S. and Hoogerwerf, A.C. (1985). Smart sensors: when and where? *Sens. Actuators* 8 (1): 39–48.

36 Tyagi, D., Wang, H., Huang, W. et al. (2020). Recent advances in two-dimensional-material-based sensing technology toward health and environmental monitoring applications. *Nanoscale* 12 (6): 3535–3559.

37 Morgan, H., Rout, C.S., and Late, D.J. (2019). Chapter 14 – Future prospects of 2D materials for sensing applications. In: *Woodhead Publishing Series in Electronic and Optical Materials* (ed. M. Hywel, C.S. Rout and D.J. Late), 481–482. Woodhead Publishing.

38 Darwish, A. and Hassanien, A.E. (2011). Wearable and implantable wireless sensor network solutions for healthcare monitoring. *Sensors* 11 (6): 5561–5595.

39 Anichini, C., Czepa, W., Pakulski, D. et al. (2018). Chemical sensing with 2D materials. *Chem. Soc. Rev.* 47 (13): 4860–4908.

40 Konvalina, G. and Haick, H. (2014). Sensors for breath testing: from nanomaterials to comprehensive disease detection. *Acc. Chem. Res.* 47 (1): 66–76.

41 Phillips, M., Cataneo, R.N., Cummin, A.R.C. et al. (2003). Detection of lung cancer with volatile markers in the breath. *Chest* 123 (6): 2115–2123.

42 Phillips, M., Altorki, N., Austin, J.H.M. et al. (2008). Detection of lung cancer using weighted digital analysis of breath biomarkers. *Clin. Chim. Acta* 393 (2): 76–84.

43 Peng, G., Hakim, M., Broza, Y.Y. et al. (2010). Detection of lung, breast, colorectal, and prostate cancers from exhaled breath using a single array of nanosensors. *Br. J. Cancer* 103 (4): 542–551.

44 Zhao, D., Miller, D., Xian, X. et al. (2014). A novel real-time carbon dioxide analyzer for health and environmental applications. *Sens. Actuators B Chem.* 195: 171–176.

45 Cristescu, S.M., Mandon, J., Harren, F.J.M. et al. (2013). Methods of NO detection in exhaled breath. *J. Breath Res.* 7 (1): 17104.

46 Brinkman, P., Zee, A.H.M.d., and Wagener, A.H. (2019). Breathomics and treatable traits for chronic airway diseases. *Curr. Opin. Pulm. Med.* 25 (1): 94–100.

47 van der Schee, M.P., Paff, T., Brinkman, P. et al. (2015). Breathomics in lung disease. *Chest* 147 (1): 224–231.

48 DeVita, V.T., Lawrence, T.S., and Rosenberg, S.A. (2008). *DeVita, Hellman, and Rosenberg's Cancer: Principles & Practice of Oncology*. Philadelphia, PA: Wolters Kluwer/Lippincott Williams & Wilkins.

49 Cleator, S., Heller, W., and Coombes, R.C. (2007). Triple-negative breast cancer: therapeutic options. *Lancet Oncol.* 8 (3): 235–244.

50 Phillips, M., Cataneo, R.N., Ditkoff, B.A. et al. (2003). Volatile markers of breast cancer in the breath. *Breast J.* 9 (3): 184–191.

51 Dean-Colomb, W. and Esteva, F.J. (2008). Her2-positive breast cancer: herceptin and beyond. *Eur. J. Cancer* 44 (18): 2806–2812.

52 Kodadek, T. (2001). Protein microarrays: prospects and problems. *Chem. Biol.* 8 (2): 105–115.

53 Tyers, M. and Mann, M. (2003). From genomics to proteomics. *Nature* 422 (6928): 193–197.

54 Hakim, M., Broza, Y.Y., Barash, O. et al. (2012). Volatile organic compounds of lung cancer and possible biochemical pathways. *Chem. Rev.* 112 (11): 5949–5966.

55 De Lacy Costello, B., Amann, A., Al-Kateb, H. et al. (2014). A review of the volatiles from the healthy human body. *J. Breath Res.* 8 (1).

56 Mochalski, P., King, J., Klieber, M. et al. (2013). Blood and breath levels of selected volatile organic compounds in healthy volunteers. *Analyst* 138 (7): 2134–2145.

57 Miekisch, W., Schubert, J.K., Vagts, D.A., and Geiger, K. (2001). Analysis of volatile disease markers in blood. *Clin. Chem.* 47 (6): 1053–1060.

58 Al-Kateb, H., de Lacy Costello, B., and Ratcliffe, N. (2013). An investigation of volatile organic compounds from the saliva of healthy individuals using headspace-trap/GC-MS. *J. Breath Res.* 7 (3): 36004.

59 Amann, A. and Smith, D. (2013). *Volatile Biomarkers: Non-Invasive Diagnosis in Physiology and Medicine*. Amsterdam, Boston, MA: Elsevier.

60 Effros, R.M., Hoagland, K.W., Bosbous, M. et al. (2002). Dilution of respiratory solutes in exhaled condensates. *Am. J. Respir. Crit. Care Med.* 165 (5): 663–669.

61 Boots, A.W., van Berkel, J.J.B.N., Dallinga, J.W. et al. (2012). The versatile use of exhaled volatile organic compounds in human health and disease. *J. Breath Res.* 6 (2): 27108.

62 Ambrosone, C.B. (2000). Oxidants and antioxidants in breast cancer. *Antioxid. Redox Signal.* 2 (4): 903–917.

63 Miekisch, W., Schubert, J.K., and Noeldge-Schomburg, G.F.E. (2004). Diagnostic potential of breath analysis--focus on volatile organic compounds. *Clin. Chim. Acta* 347 (1–2): 25–39.

64 Mazzatenta, A., Di Giulio, C., and Pokorski, M. (2013). Pathologies currently identified by exhaled biomarkers. *Respir. Physiol. Neurobiol.* 187 (1): 128–134.

65 Aghdassi, E. and Allard, J.P. (2000). Breath alkanes as a marker of oxidative stress in different clinical conditions. *Free Radic. Biol. Med.* 28 (6): 880–886.

66 Haick, H., Broza, Y.Y., Mochalski, P. et al. (2014). Assessment, origin, and implementation of breath volatile cancer markers. *Chem. Soc. Rev.* 43 (5): 1423–1449.

67 Kushch, I., Arendacká, B., Stolc, S. et al. (2008). Breath isoprene–aspects of normal physiology related to age, gender and cholesterol profile as determined in a proton transfer reaction mass spectrometry study. *Clin. Chem. Lab. Med.* 46 (7): 1011–1018.

68 Foster, W.M., Jiang, L., Stetkiewicz, P.T., and Risby, T.H. (1996). Breath isoprene: temporal changes in respiratory output after exposure to ozone. *J. Appl. Physiol.* 80 (2): 706–710.

69 King, J., Koc, H., Unterkofler, K. et al. (2010). Physiological modeling of isoprene dynamics in exhaled breath. *J. Theor. Biol.* 267 (4): 626–637.

70 Scislowski, P.W. and Pickard, K. (1994). The regulation of transaminative flux of methionine in rat liver mitochondria. *Arch. Biochem. Biophys.* 314 (2): 412–416.

71 Tangerman, A., Meuwese-Arends, M.T., and van Tongeren, J.H. (1983). A new sensitive assay for measuring volatile sulphur compounds in human breath by Tenax trapping and gas chromatography and its application in liver cirrhosis. *Clin. Chim. Acta* 130 (1): 103–110.

72 Van den Velde, S., Nevens, F., Van Hee, P. et al. (2008). GC–MS analysis of breath odor compounds in liver patients. *J. Chromatogr. B Anal. Technol. Biomed. Life Sci.* 875 (2): 344–348.

73 Turner, C., Spanel, P., and Smith, D. (2006). A longitudinal study of ammonia, acetone and propanol in the exhaled breath of 30 subjects using selected ion flow tube mass spectrometry, SIFT-MS. *Physiol. Meas.* 27 (4): 321–337.

74 Davies, S., Spanel, P., and Smith, D. (1997). Quantitative analysis of ammonia on the breath of patients in end-stage renal failure. *Kidney Int.* 52 (1): 223–228.

75 Broza, Y.Y. and Haick, H. (2013). Nanomaterial-based sensors for detection of disease by volatile organic compounds. *Nanomedicine* 8 (5): 785–806.

76 Vishinkin, R. and Haick, H. (2015). Nanoscale sensor technologies for disease detection via volatolomics. *Small* 11 (46): 6142–6164.

77 Novak, B.J., Blake, D.R., Meinardi, S. et al. (2007). Exhaled methyl nitrate as a noninvasive marker of hyperglycemia in type 1 diabetes. *Proc. Natl. Acad. Sci. U.S.A.* 104 (40): 15613–15618.

78 Galassetti, P.R., Novak, B., Nemet, D. et al. (2005). Breath ethanol and acetone as indicators of serum glucose levels: an initial report. *Diabetes Technol. Ther.* 7 (1): 115–123.

79 Phillips, M., Cataneo, R.N., Saunders, C. et al. (2010). Volatile biomarkers in the breath of women with breast cancer. *J. Breath Res.* 4 (2).

80 Choi, K.-I., Kim, H.-J., Kang, Y.C., and Lee, J.-H. (2014). Ultraselective and ultrasensitive detection of H_2S in highly humid atmosphere using CuO-loaded SnO_2 hollow spheres for real-time diagnosis of halitosis. *Sens. Actuators B Chem.* 194: 371–376.

81 Krishnan, S.T., Devadhasan, J.P., and Kim, S. (2017). Recent analytical approaches to detect exhaled breath ammonia with special reference to renal patients. *Anal. Bioanal. Chem.* 409 (1): 21–31.

82 Obermeier, J., Trefz, P., Happ, J. et al. (2017). Exhaled volatile substances mirror clinical conditions in pediatric chronic kidney disease. *PLoS One* 12 (6): e0178745.

83 Senkus, E., Kyriakides, S., Penault-Llorca, F. et al. (2013). Primary breast cancer: ESMO clinical practice guidelines for diagnosis, treatment and follow-up. *Ann. Oncol. Off. J. Eur. Soc. Med. Oncol.* 24 Suppl 6: vi7–vi23.

84 Barash, O., Zhang, W., Halpern, J.M. et al. (2015). Differentiation between genetic mutations of breast cancer by breath volatolomics. *Oncotarget* 6 (42): 44864–44876.

85 Amann, A., Costello, B.D.L., Miekisch, W. et al. (2014). The human volatilome: volatile organic compounds (VOCs) in exhaled breath, skin emanations, urine, feces and saliva. *J. Breath Res.* 8 (3): 034001.

86 Phillips, M. (1992). Breath tests in medicine. *Sci. Am.* 267 (1): 74–79.

87 Anh, D.T.V., Olthuis, W., and Bergveld, P. (2005). A hydrogen peroxide sensor for exhaled breath measurement. *Sens. Actuators B Chem.* 111: 494–499.

88 Harding, P. and Field, P.H. (1987). Breathalyzer accuracy in actual law enforcement practice: a comparison of blood- and breath-alcohol results in Wisconsin drivers. *J. Forensic Sci.* 32 (5): 1235–1240.

89 Atreja, A., Fu, A.Z., Sanaka, M.R., and Vargo, J.J. (2010). Non-invasive testing for Helicobacter pylori in patients hospitalized with peptic ulcer hemorrhage: a cost-effectiveness analysis. *Dig. Dis. Sci.* 55 (5): 1356–1363.

90 Petrone, P., Sarkisyan, G., Fernández, M. et al. (2011). Small intestinal bacterial overgrowth in patients with lower gastrointestinal symptoms and a history of previous abdominal surgery. *Arch. Surg.* 146 (4): 444–447.

91 Robroeks, C.M., van Berkel, J.J., Jöbsis, Q. et al. (2013). Exhaled volatile organic compounds predict exacerbations of childhood asthma in a 1-year prospective study. *Eur. Respir. J.* 42 (1): 98–106.

92 Gaspar, E.M., Lucena, A.F., Duro da Costa, J., and Chaves das Neves, H. (2009). Organic metabolites in exhaled human breath – a multivariate approach for identification of biomarkers in lung disorders. *J. Chromatogr. A* 1216 (14): 2749–2756.

93 Phillips, M., Cataneo, R.N., Ditkoff, B.A. et al. (2006). Prediction of breast cancer using volatile biomarkers in the breath. *Breast Cancer Res. Treat.* 99 (1): 19–21.

94 Shuster, G., Gallimidi, Z., Reiss, A.H. et al. (2011). Classification of breast cancer precursors through exhaled breath. *Breast Cancer Res. Treat.* 126 (3): 791–796.

95 Kasapović, J., Pejić, S., Stojiljković, V. et al. (2010). Antioxidant status and lipid peroxidation in the blood of breast cancer patients of different ages after chemotherapy with 5-fluorouracil, doxorubicin and cyclophosphamide. *Clin. Biochem.* 43 (16–17): 1287–1293.

96 Mense, S.M., Remotti, F., Bhan, A. et al. (2008). Estrogen-induced breast cancer: alterations in breast morphology and oxidative stress as a function of estrogen exposure. *Toxicol. Appl. Pharmacol.* 232 (1): 78–85.

97 Kessler, W. and Remmer, H. (1990). Generation of volatile hydrocarbons from amino acids and proteins by an iron/ascorbate/GSH system. *Biochem. Pharmacol.* 39 (8): 1347–1351.

98 Clemens, M.R., Remmer, H., and Waller, H.D. (1984). Phenylhydrazine-induced lipid peroxidation of red blood cells in vitro and in vivo: monitoring by the production of volatile hydrocarbons. *Biochem. Pharmacol.* 33 (11): 1715–1718.

99 Katwal, G., Paulose, M., Rusakova, I.A. et al. (2016). Rapid growth of zinc oxide nanotube-nanowire hybrid architectures and their use in breast cancer-related volatile organics detection. *Nano Lett.* 16 (5): 3014–3021.

100 Kim, S.-J., Choi, S.-J., Jang, J.-S. et al. (2017). Innovative nanosensor for disease diagnosis. *Acc. Chem. Res.* 50 (7): 1587–1596.

101 Tripathi, K.M., Kim, T.Y., Losic, D., and Tung, T.T. (2016). Recent advances in engineered graphene and composites for detection of volatile organic compounds (VOCs) and non-invasive diseases diagnosis. *Carbon N.Y.* 110 (July 2018): 97–129.

102 Lv, F., Qiu, T., Liu, L. et al. (2016). Recent advances in conjugated polymer materials for disease diagnosis. *Small* 12 (6): 696–705.

103 Zhang, G.-J. and Ning, Y. (2012). Silicon nanowire biosensor and its applications in disease diagnostics: a review. *Anal. Chim. Acta* 749: 1–15.

104 Upasham, S., Banga, I.K., Jagannath, B. et al. (2020). Electrochemical impedimetric biosensors, featuring the use of room temperature ionic liquids (RTILs): special focus on non-faradaic sensing. *Biosens. Bioelectron.* 177: 112940.

105 Bhide, A., Jagannath, B., Graef, E. et al. (2018). Versatile duplex electrochemical sensor for the detection of CO_2 and relative humidity using room temperature ionic liquid. *ECS Trans.* 85: 751–765.

106 Graef, E.W., Munje, R.D., and Prasad, S. (2017). A robust electrochemical CO_2 sensor utilizing room temperature ionic liquids. *IEEE Trans. Nanotechnol.* 16 (5): 826–831.

107 Bhide, A., Jagannath, B., Tanak, A. et al. (2020). CLIP: carbon dioxide testing suitable for low power microelectronics and IOT interfaces using room temperature ionic liquid platform. *Sci. Rep.* 10 (1): 2557.

108 Banga, I., Paul, A., Muthukumar, S., and Prasad, S. (2020). Characterization of room-temperature ionic liquids to study the electrochemical activity of nitro compounds. *Sensors* 20 (4): 1124.

109 Banga, I., Paul, A., Sardesai, A.U. et al. (2021). MATH: Methanol vapor analytics through handheld sensing platform. *Electrochim. Acta* 368: 137624.

110 Paul, A., Muthukumar, S., and Prasad, S. (2020). Review–room-temperature ionic liquids for electrochemical application with special focus on gas sensors. *J. Electrochem. Soc.* 167: 37511.

111 Liu, X., Cheng, S., Liu, H. et al. (2012). A survey on gas sensing technology. *Sensors* 12 (7): 9635–9665.

112 Koo, W.-T., Choi, S.-J., Kim, N.-H. et al. (2016). Catalyst-decorated hollow WO_3 nanotubes using layer-by-layer self-assembly on polymeric nanofiber templates and their application in exhaled breath sensor. *Sens. Actuators B Chem.* 223: 301–310.

113 Wang, C., Yin, L., Zhang, L. et al. (2010). Metal oxide gas sensors: sensitivity and influencing factors. *Sensors* 10 (3): 2088–2106.

114 Velumani, M., Meher, S.R., and Alex, Z.C. (2019). Composite metal oxide thin film based impedometric humidity sensors. *Sens. Actuators B Chem.* 301 (May): 127084.

115 Al-Salman, H.S. and Abdullah, M.J. (2015). Preparation of ZnO nanostructures by RF-magnetron sputtering on thermally oxidized porous silicon substrate for VOC sensing application. *Measurement* 59: 248–257.

116 Gagaoudakis, (Manolis), E., Michail, G., Katerinopoulou, D. et al. (2020). Transparent p-type NiO: Al thin films as room temperature hydrogen and methane gas sensors. *Mater. Sci. Semicond. Process.* 109: 104922.

117 Chi, C.-Y., Chen, H.-I., Chen, W.-C. et al. (2018). Formaldehyde sensing characteristics of an aluminum-doped zinc oxide (AZO) thin-film-based sensor. *Sens. Actuators B Chem.* 255: 3017–3024.

118 Khojier, K., Zolghadr, S., Teimoori, F., and Goudarzi, S. (2020). Fabrication and characterization of porous WO_3 thin film as a high accuracy cyclohexene sensor. *Mater. Sci. Semicond. Process.* 118 (March): 105220.

119 Vaishnav, V.S., Patel, P.D., and Patel, N.G. (2005). Indium tin oxide thin film gas sensors for detection of ethanol vapours. *Thin Solid Films* 490 (1): 94–100.

120 Gao, P., Ji, H., Zhou, Y., and Li, X. (2012). Selective acetone gas sensors using porous WO_3–Cr_2O_3 thin films prepared by sol-gel method. *Thin Solid Films* 520 (7): 3100–3106.

121 Jousseaume, V., Yeromonahos, C., El Sabahy, J. et al. (2018). Filament assisted chemical vapor deposited organosilicate as chemical layer for nanometric hydrocarbon gas sensors. *Sens. Actuators B Chem.* 271: 271–279.

122 Haddad, K., Abokifa, A., Kavadiya, S. et al. (2018). SnO_2 nanostructured thin films for room-temperature gas sensing of volatile organic compounds. *ACS Appl. Mater. Interfaces* 10 (35): 29972–29981.

123 Souissi, R., Bouguila, N., Bouricha, B. et al. (2020). Thickness effect on VOC sensing properties of sprayed In_2S_3 films. *RSC Adv.* 10 (32): 18841–18852.

124 Touidjen, N.H., Bendahmane, B., Zeggar, M.L. et al. (2016). SnO_2 thin film synthesis for organic vapors sensing at ambient temperature. *Sens. Bio-Sensing Res.* 11: 52–57.

125 Nagaraju, P., Vijayakumar, Y., and Reddy, M.V. (2017). Preparation and characterization of nanostructured Gd doped cerium oxide thin films by pulsed laser deposition for acetone sensor application. *Mater. Sci. Eng. B* 226: 99–106.

126 Ueda, T., Defferriere, T., Hyodo, T. et al. (2020). Nanostructured Pr-doped Ceria (PCO) thin films as sensing electrodes in solid-electrolyte type gas sensors with enhanced toluene sensitivity. *Sens. Actuators B Chem.* 317: 128037.

127 Li, G., Cheng, Z., Xiang, Q. et al. (2019). Bimetal PdAu decorated SnO_2 nanosheets based gas sensor with temperature-dependent dual selectivity for detecting formaldehyde and acetone. *Sens. Actuators B Chem.* 283: 590–601.

128 Nguyen, H. and El-Safty, S.A. (2011). Meso- and macroporous Co_3O_4 nanorods for effective VOC gas sensors. *J. Phys. Chem. C* 115 (17): 8466–8474.

129 Segev-Bar, M. and Haick, H. (2013). Flexible sensors based on nanoparticles. *ACS Nano* 7 (10): 8366–8378.

130 Velumani, M., Meher, S.R., and Alex, Z.C. (2018). Impedometric humidity sensing characteristics of SnO_2 thin films and SnO_2–ZnO composite thin films grown by magnetron sputtering. *J. Mater. Sci. Mater. Electron.* 29 (5): 3999–4010.

131 Rahimi, R., Maleki, A., Maleki, S. et al. (2014). Synthesis and characterization of magnetic dichromate hybrid nanomaterials with triphenylphosphine surface modified iron oxide nanoparticles (Fe_3O_4@SiO_2@PPh_3@Cr_2O_{72}−). *Solid State Sci.* 28: 9–13.

132 Tang, W. and Wang, J. (2015). Mechanism for toluene detection of flower-like ZnO sensors prepared by hydrothermal approach: charge transfer. *Sens. Actuators B Chem.* 207 (Part A): 66–73.

133 Tompsett, G.A., Conner, W.C., and Yngvesson, K.S. (2006). Microwave synthesis of nanoporous materials. *ChemPhysChem* 7 (2): 296–319.

134 Hangxun, X., Zeiger, B.W., and Suslick, K.S. (2013). Sonochemical synthesis of nanomaterials. *Chem. Soc. Rev.* 42 (7): 2555–2567.

135 Adams, N.G. and Smith, D. (1976). The selected ion flow tube (SIFT); a technique for studying ion-neutral reactions. *Int. J. Mass Spectrom.* 21 (3): 349–359.

136 Smith, D., Spanel, P., Thompson, J.M. et al. (1998). The selected ion flow tube method for workplace analyses of trace gases in air and breath: its scope, validation, and applications. *Appl. Occup. Environ. Hyg.* 13 (12): 817–823.

137 Smith, D. and Adams, N.G. (1980). Elementary plasma reactions of environmental interest. *Plasma Chem.* I: 1–43.

138 Smith, D. and Spanel, P. (1996). The novel selected-ion flow tube approach to trace gas analysis of air and breath. *Rapid Commun. Mass Spectrom.* 10 (10): 1183–1198.

139 Smith, D. and Španěl, P. (1996). Application of ion chemistry and the SIFT technique to the quantitative analysis of trace gases in air and on breath. *Int. Rev. Phys. Chem.* 15 (1): 231–271.

140 Smith, D., Wang, T., and Španěl, P. (2003). Analysis of ketones by selected ion flow tube mass spectrometry. *Rapid Commun. Mass Spectrom.* 17 (23): 2655–2660.

141 Senthilmohan, S.T., McEwan, M.J., Wilson, P.F. et al. (2001). Real time analysis of breath volatiles using SIFT-MS in cigarette smoking. *Redox Rep.* 6 (3): 185–187.

142 Kato, S., Burke, P.J., Koch, T.H., and Bierbaum, V.M. (2001). Formaldehyde in human cancer cells: detection by preconcentration-chemical ionization mass spectrometry. *Anal. Chem.* 73 (13): 2992–2997.

143 Lindinger, W., Hansel, A., and Jordan, A. (1998). On-line monitoring of volatile organic compounds at pptv levels by means of proton-transfer-reaction mass spectrometry (PTR-MS) medical applications, food control and environmental research. *Int. J. Mass Spectrom. Ion Process.* 173 (3): 191–241.

144 Steeghs, M., Bais, H.P., de Gouw, J. et al. (2004). Proton-transfer-reaction mass spectrometry as a new tool for real time analysis of root-secreted volatile organic compounds in *Arabidopsis*. *Plant Physiol.* 135 (1): 47–58.

145 Wu, Z., Huang, Z., Lehmann, R. et al. (2009). The application of chromatography-mass spectrometry: methods to metabonomics. *Chromatographia* 69 (1): 23–32.

146 Pauling, L., Robinson, A.B., Teranishi, R., and Cary, P. (1971). Quantitative analysis of urine vapor and breath by gas-liquid partition chromatography. *Proc. Natl. Acad. Sci. U.S.A.* 68 (10): 2374–2376.

147 Li, J., Peng, Y., Liu, Y. et al. (2014). Investigation of potential breath biomarkers for the early diagnosis of breast cancer using gas chromatography–mass spectrometry. *Clin. Chim. Acta* 436: 59–67.

148 Lavra, L., Catini, A., Ulivieri, A. et al. (2015). Investigation of VOCs associated with different characteristics of breast cancer cells. *Sci. Rep.* 5: 13246.

149 Wang, C., Sun, B., Guo, L. et al. (2014). Volatile organic metabolites identify patients with breast cancer, cyclomastopathy and mammary gland fibroma. *Sci. Rep.* 4 (1): 5383.

150 Mangler, M., Freitag, C., Lanowska, M. et al. (2012). Volatile organic compounds (VOCs) in exhaled breath of patients with breast cancer in a clinical setting. *Ginekol. Pol.* 83 (10): 730–736.

151 Ebeler, S.E., Clifford, A.J., and Shibamoto, T. (1997). Quantitative analysis by gas chromatography of volatile carbonyl compounds in expired air from mice and human. *J. Chromatogr. B. Biomed. Sci. Appl.* 702 (1–2): 211–215.

152 Son, I.-S., Yoon, S.-M., Kim, H.-S. et al. (2014). A study on characteristics of sampling flow and pressure conditions for chemical detection optimization. *J. Korea Inst. Mil. Sci. Technol.* 17: 258–264.

153 Guevremont, R. and Purves, R.W. (1999). Atmospheric pressure ion focusing in a high-field asymmetric waveform ion mobility spectrometer. *Rev. Sci. Instrum.* 70 (2): 1370–1383.

154 Krylov, E.V. (2003). Comparison of the planar and coaxial field asymmetrical waveform ion mobility spectrometer (FAIMS). *Int. J. Mass Spectrom.* 225 (1): 39–51.

155 Buryakov, I.A. (2006). Mathematical analysis of ion motion in a gas subjected to an alternating-sign periodic asymmetric-waveform electric field. *Tech. Phys.* 51 (9): 1121–1126.

156 Krylov, E.V., Nazarov, E.G., and Miller, R.A. (2007). Differential mobility spectrometer: model of operation. *Int. J. Mass Spectrom.* 266 (1): 76–85.

157 Joshi, N., Hayasaka, T., Liu, Y. et al. (2018). A review on chemiresistive room temperature gas sensors based on metal oxide nanostructures, graphene and 2D transition metal dichalcogenides. *Mikrochim. Acta* 185 (4): 213.

158 Bârsan, N., Schweizer-Berberich, M., and Goepel, W. (1999). Fundamental and practical aspects in the design of nanoscaled SnO_2 gas sensors: a status report. *Fresenius J. Anal. Chem.* 365: 287–304.

159 Jiang, T., Ren, L., Jia, Z. et al. (2017). Application of FBG based sensor in pipeline safety monitoring. *Appl. Sci.* 7 (6): 540.

160 Lopez-Torres, D., Elosua, C., and Arregui, F.J. (2020). Optical fiber sensors based on microstructured optical fibers to detect gases and volatile organic compounds – a review. *Sensors* 20 (9).

161 Narasimman, S., Balakrishnan, L., Meher, S.R. et al. (2017). Influence of surface functionalization on the gas sensing characteristics of ZnO nanorhombuses. *J. Alloys Compd.* 706: 186–197.

162 Lee, K.H., Zhang, Y.-Z., Jiang, Q. et al. (2020). Ultrasound-driven two-dimensional Ti_3C_2Tx MXene hydrogel generator. *ACS Nano* 14 (3): 3199–3207.

163 Baldini, F., Brenci, M., Chiavaioli, F. et al. (2012). Optical fibre gratings as tools for chemical and biochemical sensing. *Anal. Bioanal. Chem.* 402 (1): 109–116.

164 Wang, Q. and Zhao, W.-M. (2018). A comprehensive review of lossy mode resonance-based fiber optic sensors. *Opt. Lasers Eng.* 100: 47–60.

165 Tiwari, D., Mullaney, K., Korposh, S. et al. (2017). An ammonia sensor based on Lossy Mode Resonances on a tapered optical fibre coated with porphyrin-incorporated titanium dioxide. *Sens. Actuators B Chem.* 242: 645–652.

166 Liedberg, B., Nylander, C., and Lunström, I. (1983). Surface plasmon resonance for gas detection and biosensing. *Sens. Actuators* 4 (C): 299–304.

167 Kretschmann, E. and Raether, H. (1968). radiative decay of nonradiative surface plasmons excited by light. *Z. Naturforsch. A.* 23: 2135–2136.

168 Park, N.H., Akamatsu, T., Itoh, T. et al. (2014). Calorimetric thermoelectric gas sensor for the detection of hydrogen, methane and mixed gases. *Sensors* 14 (5): 8350–8362.

169 Hurot, C., Scaramozzino, N., Buhot, A., and Hou, Y. (2020). Bio-inspired strategies for improving the selectivity and sensitivity of artificial noses: a review. *Sensors* 20 (6): 1–28.

170 Mazzone, P.J., Wang, X.F., Xu, Y. et al. (2012). Exhaled breath analysis with a colorimetric sensor array for the identification and characterization of lung cancer. *J. Thorac. Oncol.* 7 (1): 137–142.

171 Alagirisamy, N., Hardas, S.S., and Jayaraman, S. (2010). Novel colorimetric sensor for oral malodour. *Anal. Chim. Acta* 661 (1): 97–102.

9

Down-conversion Photoluminescence Properties of $ZrO_2 : Ln^{3+}$ (Ln = Eu, Sm, Er, Tb, Ho, Tm, Pr, Gd, Dy) Films Formed by Plasma Electrolytic Oxidation

Aleksandar Ćirić[1] *and Stevan Stojadinović*[2]

[1] *Radiation Chemistry and Physics, Optical Materials and Spectroscopy Group, University of Belgrade, Vinča Institute of Nuclear Sciences – National Institute of the Republic of Serbia, P.O. Box 522, Belgrade 11001, Serbia*
[2] *Department of Applied Physics and Metrology, Faculty of Physics, University of Belgrade, Studentski trg 12-16, Belgrade, 11000, Serbia*

9.1 Introduction

Luminescence from lanthanide ions (Ln^{3+}) doped materials has attracted significant scientific and technological attention due to their potential applications in optoelectronic devices, solid-state lighting, luminescence thermometry, etc. [1–5]. Unique optical and luminescence properties of Ln^{3+}-doped materials originate from a ladder-like energy-level structures and intra-configurational 4f–4f transitions, which are barely sensitive to the ions' surroundings due to the shielding effect of outer 5s and 5p shell electrons [6, 7]. Luminescence from Ln^{3+} ions can span a wide spectral window, from the ultraviolet to near-infrared regions.

The characteristic sharp emission from Ln^{3+} ions are broadly divided into two processes: down- and up-conversion photoluminescence (PL). Down-conversion PL is a phenomenon in which the absorption of a single high-energy photon is followed by the emission of lower-energy photons [3]. In most cases, emission of only one low-energy photon occurs. This process obeys the Stokes law of wavelength change known as the Stokes shift. Up-conversion PL is a nonlinear process in which the absorption of two or more low-energy photons leads to the emission of a higher-energy photon, i.e. the wavelength of the emitted radiation is shorter than that of the excitation radiation [8].

Metal oxides are eminent host matrices for Ln^{3+} due to their outstanding physicochemical characteristics such as good thermal, chemical, and physical stability, wide bandgap, broad optical transparency, low phonon frequency. Among them, zirconium oxide (zirconia, ZrO_2) is an excellent host material because its low phonon energy of about 470 cm^{-1} can increase the probability of radiative transition occurrence and reduce non-radiative multiphonon relaxation of excited Ln^{3+} throughout the vibrational bands of the ZrO_2 host lattice, thus foreseeing a high yield of intra-4f transitions from the Ln^{3+} [9–11].

Nanotechnology in Electronics: Materials, Properties, Devices, First Edition.
Edited by Visakh P.M., Artem Semkin, Raneesh Balakrishnan, and Saša Lazović.

In recent years, we have shown that plasma electrolytic oxidation (PEO) is a very convenient process for formation Ln^{3+}-doped metal-oxide semiconductor films such as ZrO_2 [12–20], Al_2O_3 [21–27], TiO_2 [28, 29], Zn [30], Nb_2O_5 [31], HfO_2 [32], Y_2O_3 [33], Gd_2O_3 [34] by PEO of pure Zr, Al, Ti, Zn, Nb, Hf, Y, and Gd metals, respectively, in electrolytes containing suitable lanthanide oxide (Ln_2O_3) particles. In this chapter, we review the results of investigations of down-conversion PL properties of $ZrO_2 : Ln^{3+}$ (Ln = Eu, Sm, Er, Tb, Ho, Tm, Pr, Gd, Dy) films formed by PEO.

The formation mechanism of the PEO film is complex, as it involves electro-, thermo-, and plasma-chemical reactions at the electrolyte and metal–electrolyte interface, due to locally high temperature (10^3–10^4 K) and pressure (~10^2 MPa) at the micro-discharge sites [35]. During the PEO process of zirconium in an alkaline electrolyte containing Ln_2O_3 particles, ZrO_2 films are formed by PEO as a result of the oxidation of zirconium substrate via a sequence of localized micro-discharging events. Zr^{4+} ions formed at the zirconium/oxide interface ($Zr \rightarrow Zr^{4+} + 4e^-$) enter the micro-discharge channels and react with O^{2-}/OH^- ions from the electrolyte. Reaction products are ejected from the micro-discharge channels onto the film surface where they rapidly solidify in contact with the low-temperature electrolyte-forming ZrO_2 structures. The Ln_2O_3 particles take part in the PEO process through the electrophoretic and micro-discharging mechanisms [36]. Ln_2O_3 particles have a negative zeta potential in alkaline media; thus, its particles are negatively charged. Due to the electrophoretic effect, the Ln_2O_3 particles move in the direction of the zirconium anode. Melting point of Ln_2O_3 particles (~2400 °C) is much lower than locally high temperatures (~7000 °C) induced at the micro-discharging sites [37], allowing the molten Ln_2O_3 particles to react with ZrO_2 and form Ln^{3+}-doped ZrO_2 films.

9.2 Experiment

Zirconium foils of 99.95% purity (purchased from Alfa Aesar) with dimensions 25 × 10 × 0.2 mm were used as anodes and a substrate for film growth during the PEO process. Preprocessing included degreasing in acetone, ethanol than distilled water, by ultrasonic cleaner and subsequent drying by hot air. The insulation resin was applied to one part of the foil, leaving the active surface of 15 × 10 mm accessible to the electrolyte. The experimental setup used for PEO is described in Ref. [38]. Foils were immersed into the electrolyte inside a 500 ml Pyrex beaker. Basic electrolytes used were a water solution of 10 g l^{-1} $Na_3PO_4{\cdot}10H_2O$. Ln_2O_3 particles (Ln = Eu, Sm, Er, Ho, Tm, Pr, Gd, Dy, and Tb) were of 99.9% min. purity, purchased from Pangea International, were added to the basic electrolyte. Platinum electrodes were used as a cathode. The DC power supply used was Consort EV261 (0–600 V, 0–1000 mA) with which the current density was kept at constant 150 mA cm^{-2}. The electrolyte temperature was kept at around 20 °C by a temperature-controlled bath. A magnetic stirrer ensured the equal distribution of particles added to the electrolyte. After 10 minute run of PEO, the samples were rinsed in distilled water to prevent any additional deposition of the electrolyte constituents.

Scanning electron microscope (SEM) JEOL 840A equipped with energy-dispersive X-ray spectroscopy (EDS) was used to characterize the morphology and chemical composition of the films formed. Crystallinity was investigated by X-ray diffraction (XRD, Rigaku Ultima IV diffractometer) in Bragg–Brentano geometry with Ni-filtered CuKα radiation (λ = 1.54178 Å). Diffraction data were recorded from 20° to 80° 2θ scattering angle, and the diffraction peaks were identified by Rigaku PDXL 2 software and COD database.

PL spectra were measured at room temperature using Horiba Jobin Yvon Fluorolog FL3-22 spectrofluorometer, with a 450 W xenon (Xe) lamp as an excitation light source, coupled to a double-excitation grating monochromator in a wavelength range from 220 to 600 nm, a double-grating emission monochromator in a wavelength range from 290 to 850 nm, and a side-on photomultiplier (Hamamatsu, model 928 P). Spectra obtained were corrected for the spectral response of the measuring system and spectral distribution of the Xe lamp.

Judd–Ofelt analysis of $ZrO_2 : Eu^{3+}$ was performed with the JOES application software and the CIE chromaticity diagram [39]. Temperature stability of PL was investigated on a custom-built hot/cold stage [40].

9.3 Results and Discussion

9.3.1 Morphology, Chemical, and Phase Composition

Typical top view and cross-sectional SEM images of PEO films formed for 10 minutes in the basic electrolyte with addition of Ln_2O_3 particles are shown in Figure 9.1. The surface morphology and thickness of the films are not significantly influenced by changing the concentration of the Ln_2O_3 particles up to 4 g l^{-1} in the basic electrolyte [12, 14–20], since numerous micro-discharge channels and regions resulting from the rapid cooling of molten material decorate the film surface. Thickness of the films is around 8.4 ± 0.3 μm.

Results of the EDS analyses revealed that the main elements of the films are Zr from substrate and O and Ln from the electrolyte, and all elements are rather

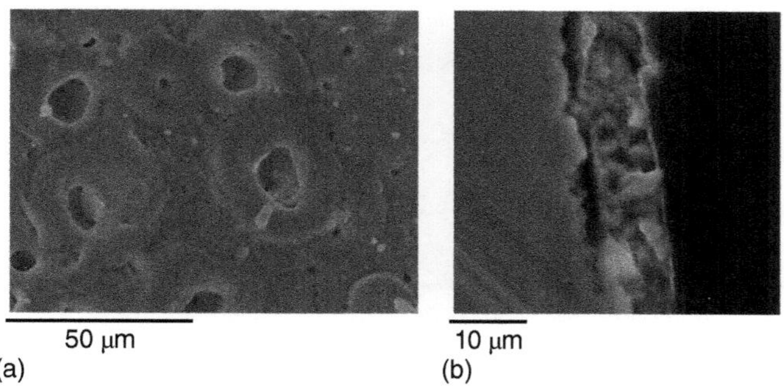

Figure 9.1 Typical (a) top view; (b) cross-sectional SEM images of the films formed by PEO of zirconium.

Table 9.1 Concentrations of rare-earth oxide powders added to the supporting electrolyte during the 10 minutes PEO process and concentrations of incorporated lanthanides within the film. The optimum concentrations yielding the highest intensities are selected.

	Eu	Sm	Er	Tb	Ho	Tm	Pr	Gd	Dy
C [gl^{-1}]	2	2	4	4	0.5	4	1	4	4
C [at%]	0.44	0.67	0.22	0.60	0.61	3.42	0.28	1.50	0.63

uniformly distributed [12, 14–20]. The incorporated lanthanide concentration is proportional to the concentration of Ln_2O_3 particles added to the basic electrolyte. The optimal concentrations that resulted in the highest PL are given in Table 9.1.

The XRD patterns of the films are shown in Figure 9.2. $ZrO_2 : Ln^{3+}$ films consist of monoclinic and tetragonal crystal phases of ZrO_2, in various ratios. In ZrO_2 films doped with Dy, Ho, Tb, Er, Sm, and Eu, the dominant crystal phase is the monoclinic

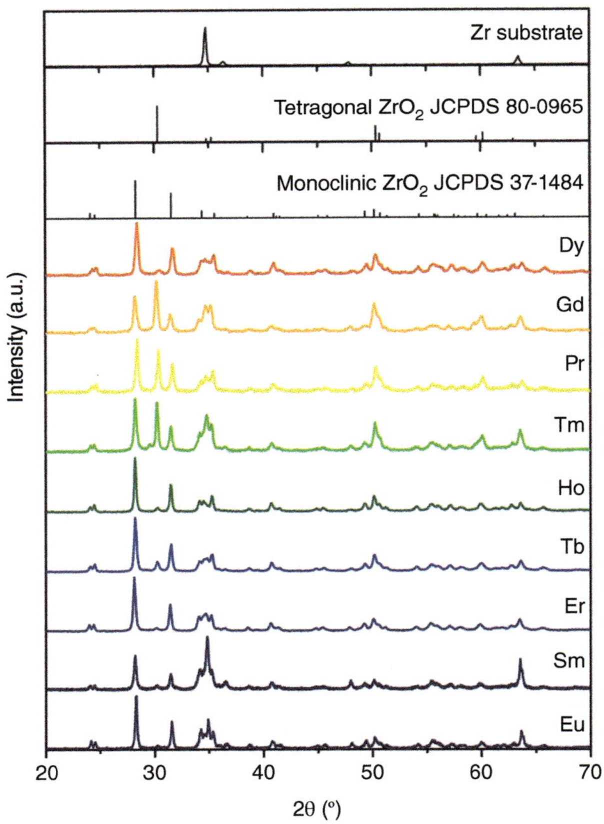

Figure 9.2 XRD patterns of ZrO_2:Ln^{3+}(Ln = Eu, Sm, Er, Tb, Ho, Tm, Pr, Gd, Dy) created by 10 minutes PEO process.

phase. In Gd-, Pr-, and Tm-doped samples, there is a mixture of approximately equal amount of tetragonal and monoclinic phases, meaning that the dopant was partially successful in tetragonal phase stabilization.

9.3.2 PL of ZrO_2 Films

9.3.2.1 PL of Undoped ZrO_2

Figure 9.3 shows PL excitation and emission spectra of undoped ZrO_2 film. From 375 to 700 nm observed is the wide PL band originating from optical transitions in PL centers. The PL centers are the defect centers related to defects and oxygen vacancies [41]. The equivalent of the emission maximum at ca. 490 nm in the PL excitation spectrum is the peak at ca. 275 nm, suggesting that the wide PL emission band originates from the broad excitation in the UV region (Figure 9.3a). Under UV excitation, the total PL will be the sum of PL originating from the ZrO_2 matrix and 4f–4f transitions of Ln^{3+} dopant ions.

9.3.2.2 PL of $ZrO_2 : Eu^{3+}$

Excitation spectrum of $ZrO_2 : Eu^{3+}$ monitored at the most intense emission transition $^5D_0 \rightarrow {}^7F_2$ at 613 nm is presented in Figure 9.4. The broad and intense band in the middle UV region is attributed to electron transfer between the completely filled 2p orbital of oxygen ions and partially filled 4f orbital of europium ions [42], referred to as the charge transfer band (CTB). The sharp peaks originate from the intra-configurational 4f transitions of Eu^{3+}.

Emission spectra under two excitations, into the CTB and directly into the most intense 4f–4f excitation, are given in Figure 9.5. Under middle UV excitation, both emissions from the host matrix and 4f levels of Eu^{3+} were observed. The most intense excitation and emission intra-configurational transitions of Eu^{3+} in the ZrO_2 host matrix were identified [43] (Table 9.2).

Judd–Ofelt intensity parameters were estimated from the emission spectrum [44]:

$$\Omega_\lambda = \frac{9.6 \cdot 10^{-42}\ \mathrm{esu^2cm^2}}{e^2 U^\lambda} \left(\frac{\nu_{MD}}{\nu_\lambda}\right)^3 \frac{\chi_{MD}}{\chi_\lambda} \frac{I_\lambda}{I_{MD}} \tag{9.1}$$

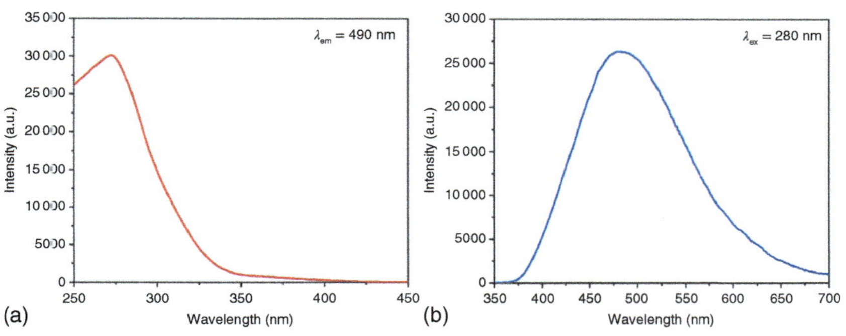

Figure 9.3 (a) Excitation and (b) emission spectrum of undoped ZrO_2 films created by PEO.

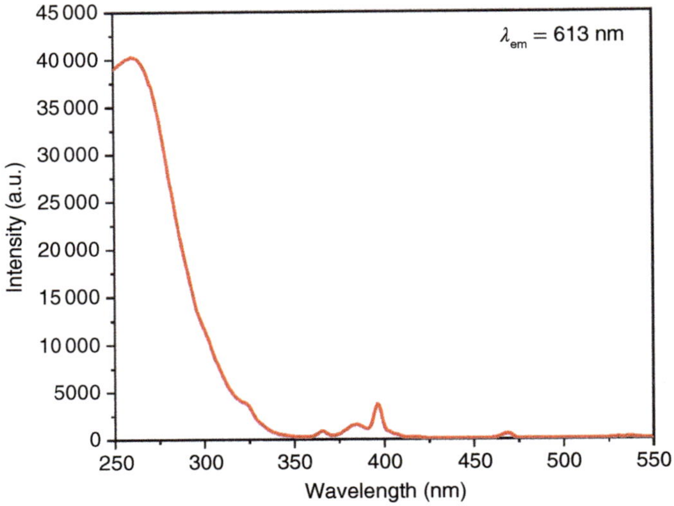

Figure 9.4 Excitation spectrum of ZrO_2:Eu^{3+} monitored at 613 nm.

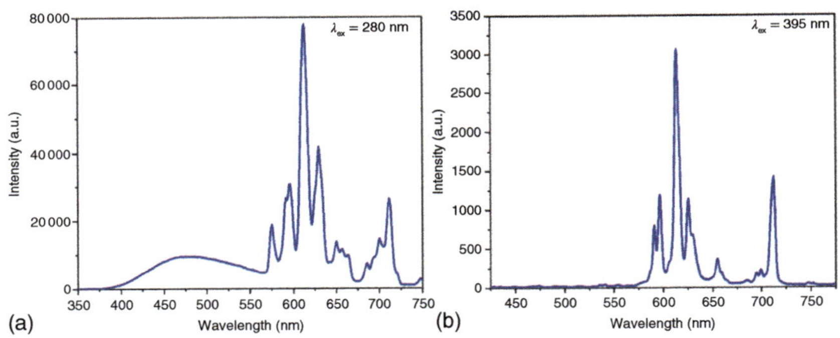

Figure 9.5 Emission spectra of ZrO_2:Eu^{3+} by (a) 280 nm, (b) 395 nm excitation.

where $e = 4.803 \cdot 10^{-10}$ esu cm is the elementary charge, ν is the emission barycenter in cm^{-1}, I are intensities in counts (for conversion of counts into irradiance, see Ref. [45]): $I_{2,4}$ of emissions $^5D_0 \rightarrow {}^7F_{2,4}$, respectively, and I_{MD} of $^5D_0 \rightarrow {}^7F_1$ magnetic dipole transition. U^λ are the squared reduced matrix elements that are equal to $U^2 = 0.0032$, $U^4 = 0.0023$ [46, 47]. χ are the local field corrections for emission given by [48–50]

$$\chi_\lambda = \frac{n(n^2+2)^2}{9}, \quad \chi_{MD} = n^3 \tag{9.2}$$

where n is the refractive index at the wavelength of the emission barycenter. For our sample, $\Omega_2 = 4.836 \cdot 10^{-20}$ cm^2 and $\Omega_4 = 1.306 \cdot 10^{-20}$ cm^2. From the Judd–Ofelt parameters, other derived quantities were estimated using the equations given in Ref. [43]. The most important ones are the radiative transition probabilities of emissions, which are equal to 144, 499, and 65 $seconds^{-1}$ for $^5D_0 \rightarrow {}^7F_{1,2,4}$ emissions,

Table 9.2 PL intra-configurational transitions between 4f levels of ZrO_2:Eu^{3+}.

Excitation			Emission		
Initial	**Final**	**λ (nm)**	**Initial**	**Final**	**λ (nm)**
$^7F_0$	$^5D_4$	363	$^5D_0$	$^7F_0$	577
$^7F_0$	$^5L_7$	383		$^7F_1$	596
$^7F_0$	$^5L_6$	395		$^7F_2$	613
$^7F_0$	$^5D_3$	415		$^7F_3$	655
$^7F_0$	$^5D_2$	466		$^7F_4$	712
$^7F_0$	$^5D_1$	528			
$^7F_1$	$^5D_1$	535			

giving the branching ratios of 20, 71, and 9%, and the radiative lifetime of $^5D_0$ level is estimated to be 1.413 ms.

Temperature stability of emission was tested from room temperature to 272 °C (Figure 9.6). In a span of over 250 °C, the PL intensity of emissions from $^5D_0$ level dropped to almost 50% (Figure 9.6a inset). By observing the peak at ca. 545 nm attributed to emission from $^5D_1$ level, it is evident that it is less affected by the temperature quenching mechanisms, due to the energy level population by the thermal energy from the $^5D_0$ level. The intensity ratio of these two emissions follows the Boltzmann distribution (Figure 9.6b) [51, 52].

$$\mathrm{LIR} = B\exp\left(\frac{-\Delta E}{kT}\right) \tag{9.3}$$

where ΔE is the energy gap between $^5D_{0,1}$ levels, and B is the temperature invariant constant that depends on the host material properties. The estimated gap of 1795 cm^{-1} closely matches the reported value for Eu^{3+} ions [47, 53]. The relative change of luminescence intensity ratio (LIR) with temperature is given by the relative sensitivity [54]: $S_r = \Delta E/kT^2$, and it is equal to 3.07% °C^{-1} at 17 °C.

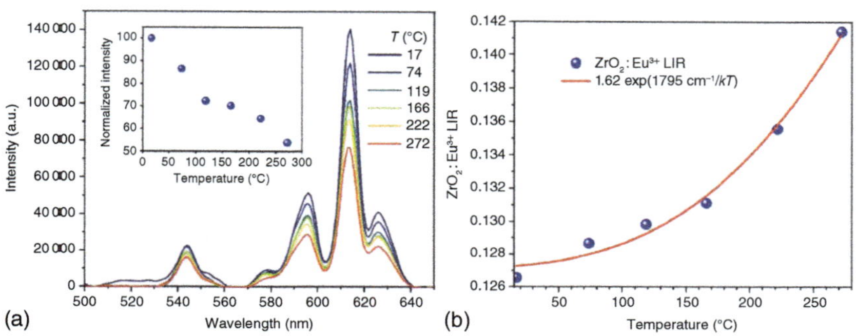

Figure 9.6 (a) Temperature dependent spectra of ZrO_2:Eu^{3+}; inset intensity of emissions from $^5D_0$ normalized to maximum at 17 °C. (b) LIR of $^5D_1$ and $^5D_0$ emissions.

9.3.2.3 PL of $ZrO_2 : Sm^{3+}$

The excitation spectrum of $ZrO_2 : Sm^{3+}$ monitored at the most intense 4f–4f emission is given in Figure 9.7a, featuring a broad excitation band and a series of weak sharp peaks attributed to the direct excitations into 4f levels of Sm^{3+}. The emission spectrum was recorded under irradiation, corresponding to the maximum of the most intense peak in the excitation spectrum. The emission spectrum consists of a broad PL band originating from the defects and oxygen vacancies in the zirconia host matrix and sharp peaks due to 4f–4f transitions of Sm^{3+}. The excitation and emission 4f–4f transitions are identified and given in Table 9.3 [55].

9.3.2.4 PL of $ZrO_2 : Er^{3+}$

The excitation spectrum of $ZrO_2 : Er^{3+}$ is given in Figure 9.8, featuring a broad and intense CTB in the middle UV region [56], and a series of sharp peaks due to the excitations into higher 4f levels of Er^{3+}.

PL emission spectra of $ZrO_2 : Er^{3+}$ under middle UV excitation and into the most intense 4f–4f excitation are presented in Figure 9.9. In the former, the total PL results from the emissions from the ZrO_2 matrix and Er^{3+} ions. In the latter, no host-matrix emissions are observed, and the emission spectra consist of solely the 4f–4f transitions of Er^{3+}. The most intense 4f–4f excitations and emissions of Er^{3+} are given in Table 9.4 [57, 58].

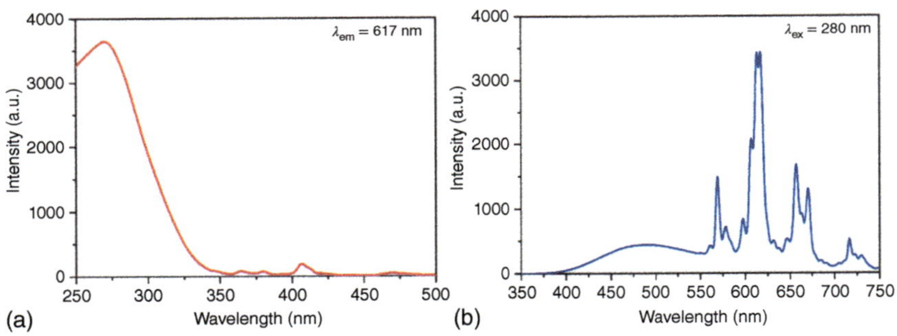

Figure 9.7 (a) Excitation monitored at 617 nm, (b) emission spectra under 280 nm irradiation of $ZrO_2:Sm^{3+}$.

Table 9.3 PL intra-configurational transitions between 4f levels of $ZrO_2:Sm^{3+}$.

Excitation			Emission		
Initial	**Final**	**λ (nm)**	**Initial**	**Final**	**λ (nm)**
$^6H_{5/2}$	$^4L_{17/2}$	364	$^4G_{5/2}$	$^6H_{5/2}$	570
$^6H_{5/2}$	$^6P_{5/2}$	380	$^4G_{5/2}$	$^6H_{7/2}$	617
$^6H_{5/2}$	$^4F_{7/2}$	406	$^4G_{5/2}$	$^6H_{9/2}$	660
$^6H_{5/2}$	$^4P_{5/2}$	421	$^4G_{5/2}$	$^6H_{11/2}$	720
$^6H_{5/2}$	$^4I_{13/2}$	466			
$^6H_{5/2}$	$^4I_{11/2}$	476			

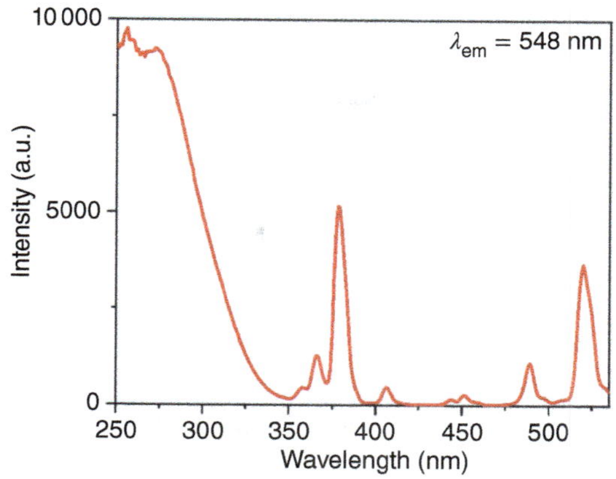

Figure 9.8 Excitation spectra of ZrO_2:Er^{3+} monitored at 548 nm.

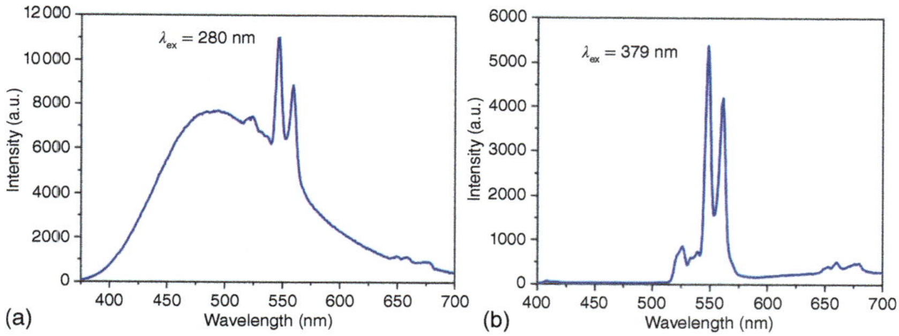

Figure 9.9 Emission spectra of ZrO_2:Er^{3+} under (a) 280 nm, (b) 379 nm irradiation.

Table 9.4 PL intra-configurational transitions between 4f levels of ZrO_2:Er^{3+}.

Excitation			Emission		
Initial	**Final**	**λ (nm)**	**Initial**	**Final**	**λ (nm)**
$^4I_{15/2}$	$^4G_{9/2}$	368	$^2H_{9/2}$	$^4I_{15/2}$	408
$^4I_{15/2}$	$^4G_{11/2}$	379	$^2H_{11/2}$	$^4I_{15/2}$	526
$^4I_{15/2}$	$^2H_{9/2}$	407	$^4S_{3/2}$	$^4I_{15/2}$	548
$^4I_{15/2}$	$^4F_{3/2}$	443	$^4F_{9/2}$	$^4I_{15/2}$	660
$^4I_{15/2}$	$^4F_{5/2}$	452			
$^4I_{15/2}$	$^4F_{7/2}$	490			
$^4I_{15/2}$	$^2H_{11/2}$	523			

9.3.2.5 PL of ZrO_2 : Tb^{3+}

The excitation spectrum of ZrO_2 : Tb^{3+} by monitoring the most intense emission at 544 nm is presented in Figure 9.10a. The broad band in middle UV dominates the spectrum and is attributed to the inter-configurational excitation from 4f to 5d levels [59]. As the 4f–5d transitions are parity allowed, the electric dipole is an

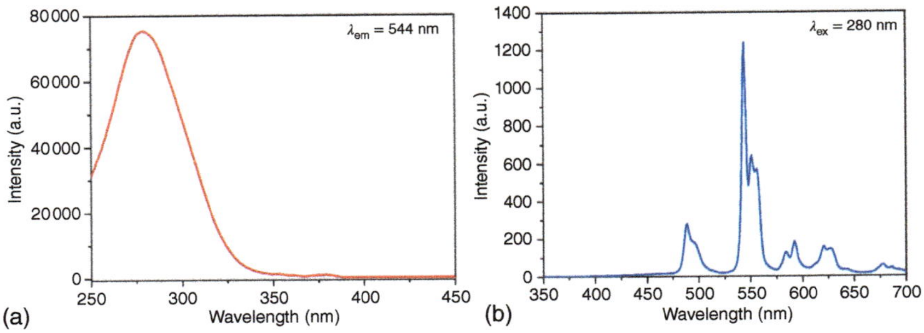

Figure 9.10 (a) Excitation spectrum of ZrO_2:Tb^{3+} recorded by monitoring emission at 544 nm; (b) emission under 280 nm excitation.

Table 9.5 PL emissions of ZrO_2:Tb^{3+}.

Initial	Final	λ (nm)
$^5D_4$	$^7F_6$	488
$^5D_4$	$^7F_5$	543
$^5D_4$	$^7F_4$	592
$^5D_4$	$^7F_3$	620
$^5D_4$	$^7F_2$	677

allowed transition and the intensities are much higher than the parity-forbidden intra-configuration 4f transitions [49, 60]. Under 4f–5d excitation by a 280 nm beam, the emission spectrum is recorded (Figure 9.10b), featuring very weak host-matrix PL. The emission peak positions and identification are given in Table 9.5.

9.3.2.6 PL of ZrO_2 : Ho^{3+}

The broad band in the excitation spectrum of ZrO_2 : Ho^{3+} is attributed to the CTB caused by electron transfer between Ho^{3+} and O^{2-} ions [61]. The sharp peaks originate from 4f–4f transitions of Ho^{3+} (Figure 9.11).

Under 280 nm irradiation, the ZrO_2 : Ho^{3+} emission spectrum is featured by broad PL from the host matrix overlapped with the sharp PL peaks from Ho^{3+} (Figure 9.12a). By direct excitation into excited 4f levels, only Ho^{3+} emissions are observed (Figure 9.12b). All major 4f–4f emissions and excitations are identified in Table 9.6 [62].

9.3.2.7 PL of ZrO_2 : Tm^{3+}

The excitation spectrum of ZrO_2 : Tm^{3+} film is presented in Figure 9.13, which is obtained by monitoring the 456 nm emission. The broad band is due to excitation into the CTB, while the intense sharp peak is attributed to $^3H_6 \rightarrow {}^1D_2$ transition of Tm^{3+}.

The emission spectrum under 280 nm irradiation is recorded and displayed in Figure 9.14a, featuring a broad band due to the host-matrix emission related

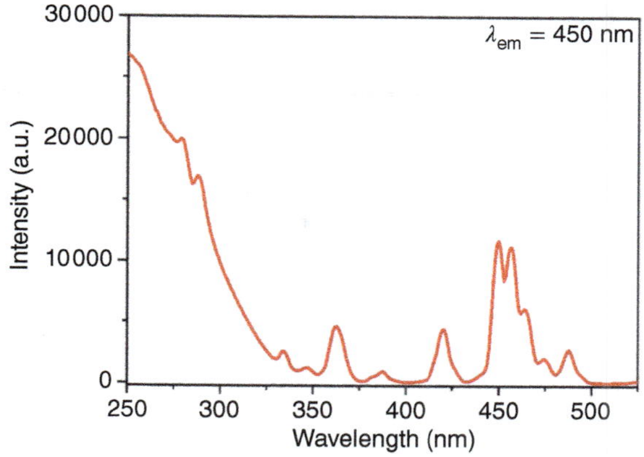

Figure 9.11 Excitation spectrum of $ZrO_2:Ho^{3+}$ monitored at 450 nm.

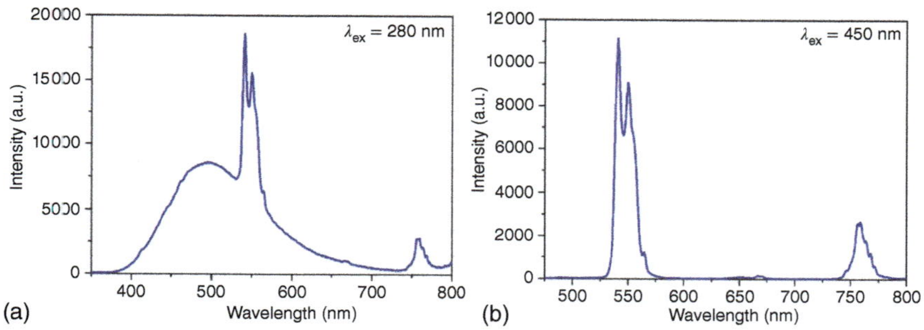

Figure 9.12 Emission spectra of $ZrO_2:Ho^{3+}$ under (a) 280 nm, (b) 450 nm irradiation.

Table 9.6 PL intra-configurational transitions between 4f levels of $ZrO_2:Ho^{3+}$.

Excitation			Emission		
Initial	**Final**	**λ (nm)**	**Initial**	**Final**	**λ (nm)**
$^5I_8$	$^3F_4$	335	$^5F_4$	$^5I_8$	540
$^5I_8$	$^5G_3$	374	$^5S_2$	$^5I_8$	550
$^5I_8$	$^3H_6$	363	$^5F_5$	$^5I_8$	670
$^5I_8$	$^5G_4$	387	$^5F_4$	$^5I_7$	755
$^5I_8$	$^5G_5$	420	$^5S_2$	$^5I_7$	765
$^5I_8$	$^5G_6$	450			
$^5I_8$	$^5F_1$	575			
$^5I_8$	$^3K_8$	464			
$^5I_8$	$^5F_2$	475			
$^5I_8$	$^5F_3$	488			

Source: Based on Carnall et al. [62]/John Wiley & Sons.

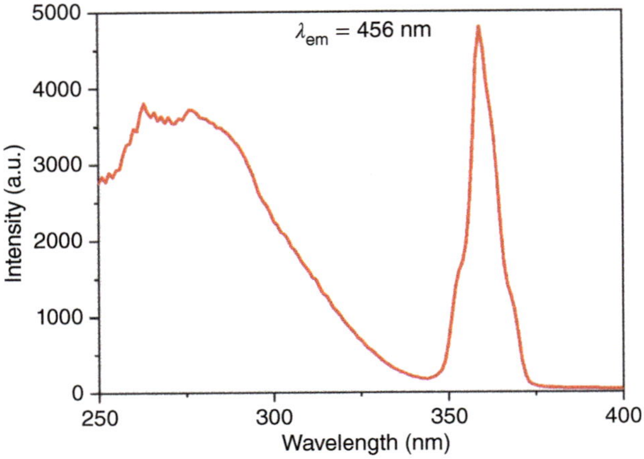

Figure 9.13 Excitation spectrum of ZrO_2:Tm^{3+} monitored at 456 nm.

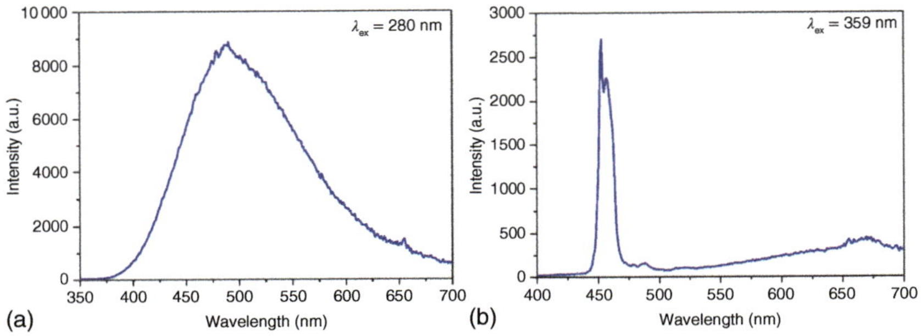

Figure 9.14 Emission spectra of ZrO_2:Tm^{3+} under excitation by (a) 280 nm, (b) 359 nm beam.

to oxygen vacancies and defect sites in the zirconia matrix, dominating over 4f–4f emissions of Tm^{3+}. Figure 9.14b displays an emission spectrum by 359 nm excitation, featuring no host-matrix PL. The most intense is the emission at 456 nm attributed to $^1D_2 \rightarrow {}^3F_4$-induced electric dipole transition. All major observed excitations and emissions are given in Table 9.7 [34].

9.3.2.8 PL of $ZrO_2 : Pr^{3+}$

As the Pr^{3+} has low-lying 5d levels, the excitation broad band has its origin in the middle UV region due to the inter-configurational 4f–5d excitation [63, 64] or charge transfer (CT) transitions [26] (Figure 9.15).

The emission spectra when the sample is irradiated by a 280 nm beam display emissions from the zirconia host matrix overlapping with 4f–4f transitions of Pr^{3+} (Figure 9.16a). By direct excitation into $^3P_2$ level of Pr^{3+}, no host-matrix emissions are observed (Figure 9.16b). However, the spectra of $ZrO_2 : Pr^{3+}$ are complex as they present overlapping emissions from $^3P_{0,1}$ and $^1D_2$ excited levels (see Table 9.8) [47].

Table 9.7 PL intra-configurational transitions between 4f levels of ZrO_2:Tm^{3+}.

Excitation			Emission		
Initial	**Final**	**λ (nm)**	**Initial**	**Final**	**λ (nm)**
$^3H_6$	$^1D_2$	359	$^1D_2$	$^3F_4$	456
$^3H_6$	$^1I_6+^3P_0$	282	$^1D_2$	$^3H_5$	517
$^3H_6$	$^3P_1$	276	$^1G_4$	$^3F_4$	654
$^3H_6$	$^3P_2$	263	$^1G_4$	$^3H_6$	485
			$^3F_2$	$^3H_6$	669
			$^3F_3$	$^3H_6$	694

Source: Based on Carnall et al. [62]/John Wiley & Sons.

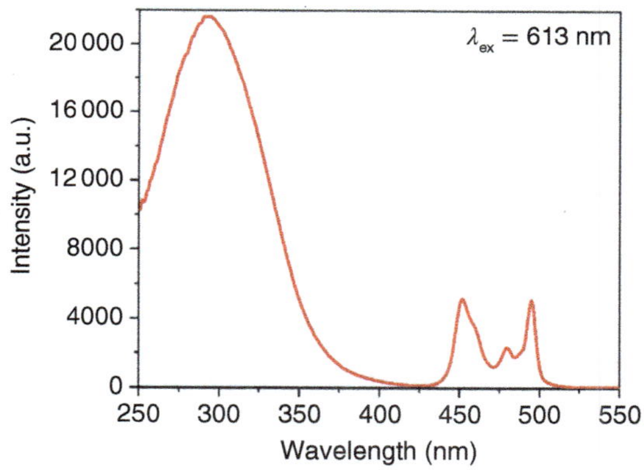

Figure 9.15 Excitation spectrum of ZrO_2:Pr^{3+}, monitored at 613 nm.

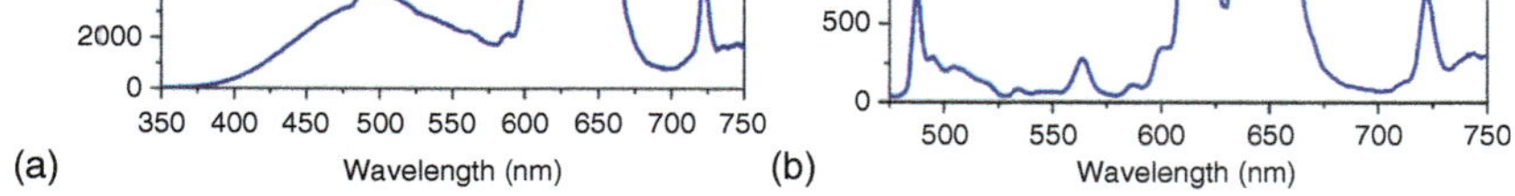

Figure 9.16 Emission spectra of ZrO_2:Pr^{3+} under (a) 280 nm, (b) 452 nm irradiation.

Table 9.8 PL intra-configurational transitions between 4f levels of $ZrO_2:Pr^{3+}$.

Excitation			Emission		
Initial	Final	λ (nm)	Initial	Final	λ (nm)
$^3H_4$	$^3P_2$	452	$^3P_1$	$^3H_4$	476
$^3H_4$	$^1I_6$	480	$^3P_0$	$^3H_4$	490
$^3H_4$	$^3P_1$	487	$^3P_1$	$^3H_5$	520
$^3H_4$	$^3P_0$	495	$^3P_0$	$^3H_5$	533
$^3H_4$	$^1D_2$	593	$^3P_1$	$^3H_6$	564
			$^3P_0$	$^3H_6$	608
			$^3P_0$	$^3F_2$	636
			$^3P_0$	$^3F_4$	722
			$^1D_2$	$^3H_4$	595
			$^1D_2$	$^3H_5$	670

Source: Based on Carnall et al. [47].

9.3.2.9 PL of $ZrO_2 : Gd^{3+}$

Figure 9.17a presents the excitation spectrum of $ZrO_2 : Gd^{3+}$ by observing the most intense Gd^{3+} emission at 313 nm, corresponding to $^6P_{7/2} \rightarrow {}^8S_{7/2}$ transition. The emission spectrum in Figure 9.17b features a weak peak at 307 nm due to the $^6P_{5/2} \rightarrow {}^8S_{7/2}$, an intense peak at 313 nm due to the $^6P_{7/2} \rightarrow {}^8S_{7/2}$ of Gd^{3+} (Table 9.9) [65], and a broad band from 350 to 700 nm attributed to the host-matrix emissions.

9.3.2.10 PL of $ZrO_2 : Dy^{3+}$

$ZrO_2 : Dy^{3+}$ excitation spectrum is presented in Figure 9.18, featuring broad excitations due to the CTB between O^{2-} and Dy^{3+} ions and a series of sharp peaks due

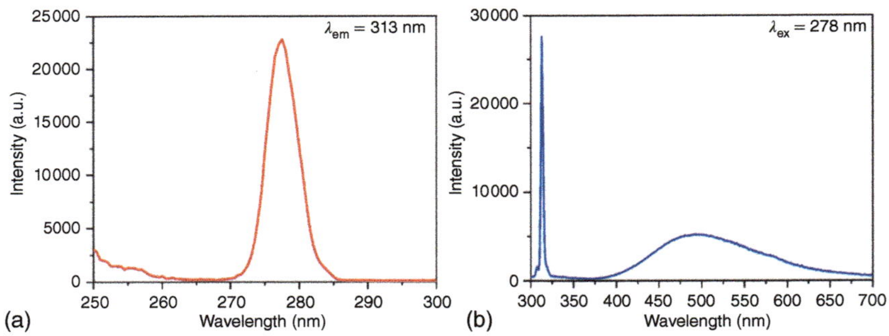

Figure 9.17 (a) Excitation spectrum of $ZrO_2:Gd^{3+}$ monitored at 313 nm, (b) emission observed by 278 nm irradiation.

Table 9.9 PL intra-configurational transitions between 4f levels of $ZrO_2:Gd^{3+}$.

Excitation			Emission		
Initial	**Final**	**λ (nm)**	**Initial**	**Final**	**λ (nm)**
$^8S_{7/2}$	^6D_J	250–260	$^6P_{5/2}$	$^8S_{7/2}$	307
$^8S_{7/2}$	^6I_J	278	$^6P_{7/2}$	$^8S_{7/2}$	313

Source: Based on Carnall et al. [65]/John Wiley & Sons.

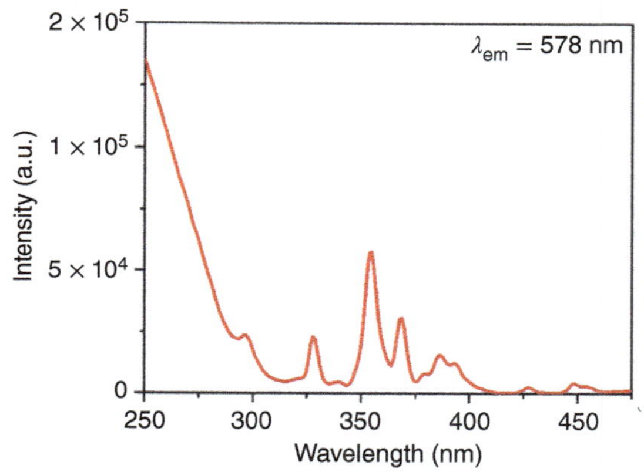

Figure 9.18 Excitation spectrum of $ZrO_2:Dy^{3+}$ monitored at 578 nm.

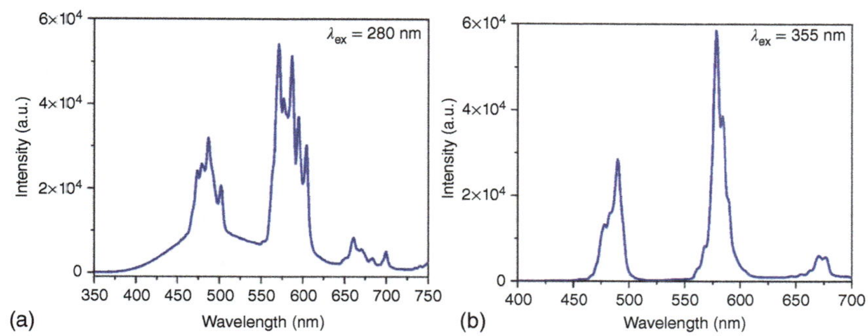

Figure 9.19 Emission spectra of $ZrO_2:Dy^{3+}$ by (a) 280 nm, (b) 355 nm irradiation.

to the intra-configurational 4f transitions of Dy^{3+}. Under 280 nm irradiation, the resulting emission spectrum consists of broad PL from host-matrix overlapping with sharp peaks attributed to the 4f–4f transitions of Dy^{3+} (Figure 9.19a). When excited directly into the higher 4f levels, the host matrix does not emit light (Figure 9.19b). The emission and excitation of 4f–4f transitions of Dy^{3+} are identified and presented in Table 9.10 [47].

Table 9.10 PL intra-configurational transitions between 4f levels of ZrO_2:Dy^{3+}.

Excitation			Emission		
Initial	**Final**	**λ (nm)**	**Initial**	**Final**	**λ (nm)**
$^6H_{15/2}$	$^2H_J + ^4L_J + ^4K_{11/2} + ^4G_{5/2} + ^4P_J + ^4F_{3/2}$	280	$^4F_{9/2}$	$^6H_{15/2}$	487
$^6H_{15/2}$	$^4D_{7/2}$	297	$^4F_{9/2}$	$^6H_{13/2}$	578
$^6H_{15/2}$	$^4G_{9/2} + ^4M_{17/2}$	328	$^4F_{9/2}$	$^6H_{11/2}$	670
$^6H_{15/2}$	$^4M_{15/2} + ^6P_{7/2}$	354			
$^6H_{15/2}$	$^6P_{5/2} + ^6P_{3/2}$	369			
$^6H_{15/2}$	$^4M_{19/2}$	380			
$^6H_{15/2}$	$^4I_{13/2} + ^4F_{7/2} + ^4K_{17/2}$	387 + 393			
$^6H_{15/2}$	$^4M_{21/2}$	400			
$^6H_{15/2}$	$^4G_{11/2}$	428			
$^6H_{15/2}$	$^4I_{15/2}$	449 + 455			
$^6H_{15/2}$	$^4F_{9/2}$	468 + 474			

Source: Based on Carnall et al. [47]/John Wiley & Sons.

Table 9.11 CIE 1931 chromaticity coordinates, correlated color temperatures (CCT), dominant wavelengths (λ_{dom}), and color purities (CP) of ZrO_2:Ln^{3+}, irradiated by 280 nm UV light or by excitation directly into 4f levels of Ln^{3+}.

Dopant	Irradiation into	*x*	*y*	CCT (K)	λ_{dom} (nm)	CP (%)
Eu	UV	0.548	0.350	1491	603.5	69.5
Eu	4f	0.632	0.347	1084	606.1	93.7
/	UV	0.236	0.336	11 746	492.1	32.8
Sm	UV	0.467	0.357	2180	597.7	47.3
Er	4f	0.359	0.617	N/A	558.2	93.6
Er	UV	0.245	0.370	N/A	497.1	27.7
Tb	UV	0.354	0.563	N/A	558.2	75.8
Ho	UV	0.256	0.408	N/A	504.8	23.6
Ho	4f	0.295	0.691	N/A	549	97.8
Tm	UV	0.245	0.350	10 250	494.2	28.8
Tm	4f	0.280	0.174	N/A	N/A	N/A
Pr	UV	0.443	0.343	2368	603	35.6
Pr	4f	0.661	0.330	1000	611.3	97.5
Gd	UV/4f	0.262	0.356	8982	495.7	22.9
Dy	UV	0.384	0.398	4058	575	34.8
Dy	4f	0.431	0.421	3246	579.3	55.7

Source: Based on Ćirić et al. [39]/John Wiley & Sons.

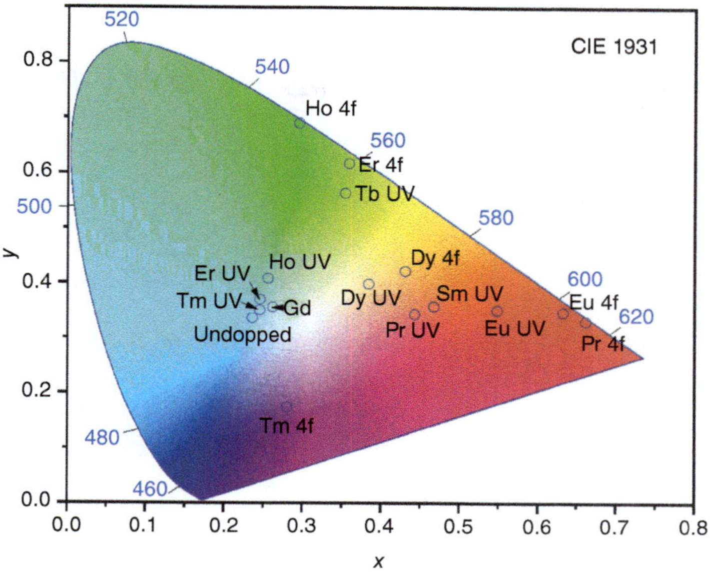

Figure 9.20 CIE 1931 chromaticity diagram of ZrO_2:Ln^{3+}, excited by 280 nm (UV) or directly into 4f levels.

9.4 CIE Chromaticity of $ZrO_2 : Ln^{3+}$

Chromaticity coordinates of $ZrO_2 : Ln^{3+}$ under UV irradiation or into excited 4f levels of Ln^{3+} are estimated by JOES application software [39], and the results are presented in Table 9.11. Correlated color temperatures, dominant wavelengths, and color purities are estimated using the OSRAM Color Calculator software [66]. The emission colors are given in the CIE 1931 chromaticity diagram in Figure 9.20. Under 280 nm UV radiation, all samples emit with reduced color purity, i.e. their color drifts toward the cool white. $ZrO_2 : Ho^{3+}$ under excitation into its 4f levels emits pure green color. Pr^{3+} and Eu^{3+} samples show high purities for red emissions.

9.5 Conclusion

In this chapter, we showed that Ln^{3+} (Ln = Eu, Sm, Er, Tb, Ho, Tm, Pr, Gd, Dy) doped ZrO_2 films can be formed by PEO of zirconium in alkaline electrolytes containing Ln_2O_3 particles. The following conclusions were drawn:

- PL emission and excitation spectra of formed films exhibit well-pronounced bands inherent to ZrO_2 and Ln^{3+} ions incorporated into ZrO_2 films.
- PL emission spectra of Ln^{3+} feature sharp bands related to 4f–4f transitions. The most intense 4f–4f transitions of Ln^{3+} are: in the orange region $^5D_0 \rightarrow {}^7F_2$ of Eu^{3+}, $^4G_{5/2} \rightarrow {}^6H_{7/2}$ of Sm^{3+}, and $^3P_0 \rightarrow {}^3H_6$ of Pr^{3+}; in the yellow region $^4F_{9/2} \rightarrow {}^6H_{13/2}$ of Dy^{3+}; in the green region $^4S_{3/2} \rightarrow {}^4I_{15/2}$ of Er^{3+}, $^5D_4 \rightarrow {}^7F_5$ of Tb^{3+}, and $^5F_4 \rightarrow {}^5I_8$

of Ho^{3+}; in the blue region $^1D_2 \rightarrow {}^3F_4$ of Tm^{3+};and in the UV region $^6P_{7/2} \rightarrow {}^8S_{7/2}$ of Gd^{3+}.

- Ln^{3+} ions in ZrO_2 host can be directly excited from ground state to higher 4f levels. For Eu^{3+}, more efficient excitation is through electron transfer transition from 2p orbital of O^{2-} to 4f orbital of Eu^{3+}, while for Tb^{3+} through the inter-configurational excitation from 4f to 5d levels.

Acknowledgments

This research was funded by the Ministry of Education, Science and Technological Development of the Republic of Serbia.

References

1 Kenyon, A.J. (2002). Recent developments in rare-earth doped materials for optoelectronics. *Prog. Quantum Electron.* 26: 225–284.

2 Suta, M. and Wickleder, C. (2019). Synthesis, spectroscopic properties and applications of divalent lanthanides apart from Eu^{2+}. *J. Lumin.* 210: 210–238.

3 Wang, L., Xie, R.-J., Suehiro, T. et al. (2018). Down-conversion nitride materials for solid state lighting: recent advances and perspectives. *Chem. Rev.* 118: 1951–2009.

4 Sarkar, D., Ganguli, S., Samanta, T., and Mahalingam, V. (2019). Design of lanthanide-doped colloidal nanocrystals: applications as phosphors, sensors, and photocatalysts. *Langmuir* 35: 6211–6230.

5 Dramićanin, M.D. (2016). Sensing temperature via downshifting emissions of lanthanide doped metal oxides and salts – a review. *Methods Appl. Fluoresc.* 4: 042001.

6 Dhoble, S.J., Pawade, V.B., Swart, H.C., and Chopra, V. (ed.) (2020). *Spectroscopy of Lanthanide-Doped Oxide Materials*, Series in Electronic and Optical Materials. Oxford: Woodhead Publishing.

7 Liu, G. and Jacquier, B. (ed.) (2005). *Spectroscopic Properties of Rare Earths in Optical Materials*, Springer Series in Materials Science. Berlin Heidelberg: Springer-Verlag.

8 Zhou, J., Liu, Q., Feng, W. et al. (2015). Upconversion luminescent materials: advances and applications. *Chem. Rev.* 115: 395–465.

9 Li, J., Meng, S., Niu, J., and Lu, H. (2017). Electronic structures and optical properties of monoclinic ZrO_2 studied by first-principles local density approximation + U approach. *J. Adv. Ceram.* 6: 43–49.

10 Cabello, G., Lillo, L., Caro, C. et al. (2008). Structure and optical characterization of photochemically prepared ZrO_2 thin films doped with erbium and europium. *J. Non-Cryst. Solids* 354: 3919–3928.

11 Martínez-Olmos, R.C., Guzmán-Mendoza, J., Báez-Rodríguez, A. et al. (2015). Synthesis, characterization and luminescence studies in $ZrO_2 : Dy^{3+}$ and

$ZrO_2 : Dy^{3+}$, Gd^{3+} films deposited by the Pyrosol method. *Opt. Mater.* 46: 168–174.

12 Ćirić, A. and Stojadinović, S. (2019). Photoluminescence properties of Pr^{3+} doped ZrO_2 formed by plasma electrolytic oxidation. *J. Alloys Compd.* 803: 126–134.

13 Stojadinović, S. and Vasilić, R. (2016). Formation and photoluminescence of Eu^{3+} doped zirconia coatings formed by plasma electrolytic oxidation. *J. Lumin.* 176: 25–31.

14 Stojadinović, S., Tadić, N., and Vasilić, R. (2016). Photoluminescence of Sm^{3+} doped ZrO_2 coatings Formed by plasma electrolytic oxidation of zirconium. *Mater. Lett.* 164: 329–332.

15 Stojadinović, S., Tadić, N., and Vasilić, R. (2019). Down- and up-conversion photoluminescence of $ZrO_2 : Ho^{3+}$ and $ZrO_2 : Ho^{3+}/Yb^{3+}$ coatings formed by plasma electrolytic oxidation. *J. Alloys Compd.* 785: 1222–1232.

16 Ćirić, A. and Stojadinović, S. (2019). Photoluminescence studies of $ZrO_2 : Tm^{3+}/Yb^{3+}$ coatings formed by plasma electrolytic oxidation. *J. Lumin.* 214: 116568.

17 Stojadinović, S., Tadić, N., and Vasilić, R. (2018). Down-conversion photoluminescence of $ZrO_2 : Er^{3+}$ coatings formed by plasma electrolytic oxidation. *Mater. Lett.* 219: 251–255.

18 Stojadinović, S., Tadić, N., and Vasilić, R. (2018). Structural and photoluminescent properties of $ZrO_2 : Tb^{3+}$ coatings formed by plasma electrolytic oxidation. *J. Lumin.* 197: 83–89.

19 Ćirić, A. and Stojadinović, S. (2020). Photoluminescence of $ZrO_2 : Gd^{3+}$ and $ZrO_2 : Dy^{3+}$ coatings formed by the plasma electrolytic oxidation. *J. Alloys Compd.* 832: 154907.

20 Stojadinović, S., Tadić, N., and Vasilić, R. (2019). Photoluminescence properties of Er^{3+}/Yb^{3+} doped ZrO_2 coatings formed by plasma electrolytic oxidation. *J. Lumin.* 208: 296–301.

21 Stojadinović, S. and Vasilić, R. (2018). Eu^{2+} photoluminescence in Al_2O_3 coatings obtained by plasma electrolytic oxidation. *J. Lumin.* 199: 240–244.

22 Stojadinović, S., Tadić, N., and Vasilić, R. (2017). Photoluminescence of Sm^{2+}/Sm^{3+} doped Al_2O_3 coatings formed by plasma electrolytic oxidation of aluminum. *J. Lumin.* 192: 110–116.

23 Stojadinović, S. and Vasilić, R. (2018). Photoluminescence of $Ce3^{+}$ and Ce^{3+}/Tb^{3+} ions in Al_2O_3 host formed by plasma electrolytic oxidation. *J. Lumin.* 203: 576–581.

24 Stojadinović, S. and Ćirić, A. (2019). Ce^{3+}/Eu^{2+} doped Al_2O_3 coatings formed by plasma electrolytic oxidation of aluminum: photoluminescence enhancement by $Ce^{3+} \rightarrow Eu^{2+}$ energy transfer. *Coatings* 9: 819.

25 Stojadinović, S. and Ćirić, A. (2020). Dy^{3+} and Dy^{3+}/Ce^{3+} doped Al_2O_3 coatings obtained by plasma electrolytic oxidation: photoluminescence and energy transfer from Ce^{3+} to Dy^{3+}. *J. Lumin.* 226: 117403.

26 Stojadinović, S. and Ćirić, A. (2020). Highly-increased photoluminescence of Pr^{3+} due to the $Eu^{2+} \rightarrow Pr^{3+}$ energy transfer in Al_2O_3 coatings formed by plasma electrolytic oxidation. *J. Lumin.* 226: 117407.

27 Stojadinović, S. and Ćirić, A. (2021). Sensitizing effect of Ce^{3+} on Tm^{3+} blue emission in $Al_2O_3 : Tm^{3+}/Ce^{3+}$ coatings formed by plasma electrolytic oxidation. *J. Mater. Sci. Mater. Electron.* 32: 7841–7847.

28 Stojadinović, S., Radić, N., Grbić, B. et al. (2016). Structural, photoluminescent and photocatalytic properties of $TiO_2 : Eu^{3+}$ coatings formed by plasma electrolytic oxidation. *Appl. Surf. Sci.* 370: 218–228.

29 Stojadinović, S., Tadić, N., Radić, N. et al. (2018). Effect of Tb^{3+} doping on the photocatalytic activity of TiO_2 coatings formed by plasma electrolytic oxidation of titanium. *Surf. Coating. Technol.* 337: 279–289.

30 Stojadinović, S. and Ćirić, A. (2021). Photoluminescence of $ZnO : Eu^{3+}$ and $ZnO : Tb^{3+}$ coatings formed by plasma electrolytic oxidation of pure zinc substrate. *J. Lumin.* 235: 118022.

31 Stojadinović, S. and Vasilić, R. (2016). Orange–red photoluminescence of $Nb_2O_5 : Eu^{3+}, Sm^{3+}$ coatings formed by plasma electrolytic oxidation of niobium. *J. Alloys Compd.* 685: 881–889.

32 Stojadinović, S., Tadić, N., Ćirić, A., and Vasilić, R. (2018). Photoluminescence properties of Eu^{3+} doped HfO_2 coatings formed by plasma electrolytic oxidation of hafnium. *Opt. Mater.* 77: 19–24.

33 Ćirić, A. and Stojadinović, S. (2020). Structural and photoluminescence properties of Y_2O_3 and $Y_2O_3 : Ln^{3+}$ (Ln = Eu, Er, Ho) films synthesized by plasma electrolytic oxidation of yttrium substrate. *J. Lumin.* 217: 116762.

34 Ćirić, A. and Stojadinović, S. (2019). Photoluminescence of Gd_2O_3 and $Gd_2O_3 : Ln^{3+}$ (Ln = Eu, Er, Ho) formed by plasma electrolytic oxidation of pure gadolinium substrate. *Opt. Mater.* 99: 109546.

35 Kaseem, M., Fatimah, S., Nashrah, N., and Ko, Y.G. (2021). Recent progress in surface modification of metals coated by plasma electrolytic oxidation: principle, structure, and performance. *Prog. Mater. Sci.* 117: 100735.

36 Lu, X., Mohedano, M., Blawert, C. et al. (2016). Plasma electrolytic oxidation coatings with particle additions – a review. *Surf. Coat. Technol.* 307: 1165–1182.

37 Stojadinović, S., Radić-Perić, J., Vasilić, R., and Perić, M. (2014). Spectroscopic investigation of direct current (DC) plasma electrolytic oxidation of zirconium in citric acid. *Appl. Spectrosc.* 68: 101–112.

38 Ćirić, A., Stojadinović, S., Ristić, Z. et al. (2021). Supersensitive Sm^{2+}-activated Al_2O_3 thermometric coatings for high-resolution multiple temperature read-outs from luminescence. *Adv. Mater. Technol.* 6: 2001201.

39 Ćirić, A., Stojadinović, S., Sekulić, M., and Dramićanin, M.D. (2019). JOES: an application software for Judd-Ofelt analysis from Eu^{3+} emission spectra. *J. Lumin.* 205: 351–356.

40 Ćirić, A., Stojadinović, S., and Dramićanin, M.D. (2019). Custom-built thermometry apparatus and luminescence intensity ratio thermometry of $ZrO_2 : Eu^{3+}$ and $Nb_2O_5 : Eu^{3+}$. *Meas. Sci. Technol.* 30: 045001.

41 Trivinho-Strixino, F., Guimarães, F.E.G., and Pereira, E.C. (2008). Zirconium oxide anodic films: optical and structural properties. *Chem. Phys. Lett.* 461: 82–86.

42 Hui, Y., Zou, B., Liu, S. et al. (2015). Effects of Eu^{3+}-doping and annealing on structure and fluorescence of zirconia phosphors. *Ceram. Int.* 41: 2760–2769.

43 Binnemans, K. (2015). Interpretation of europium(III) spectra. *Coord. Chem. Rev.* 295: 1–45.

44 Ćirić, A., Stojadinović, S., Brik, M.G., and Dramićanin, M.D. (2020). Judd-Ofelt parametrization from emission spectra: the case study of the Eu^{3+} $5D_1$ emitting level. *Chem. Phys.* 528: 110513.

45 Suta, M. and Meijerink, A. (2020). A theoretical framework for ratiometric single ion luminescent thermometers – thermodynamic and kinetic guidelines for optimized performance. *Adv. Theory Simul.* 3: 2000176.

46 Ćirić, A., Stojadinović, S., and Dramićanin, M.D. (2019). An extension of the Judd-Ofelt theory to the field of lanthanide thermometry. *J. Lumin.* 216: 116749.

47 Carnall, W.T., Fields, P.R., and Rajnak, K. (1968). Electronic energy levels in the trivalent lanthanide aquo ions. I. Pr^{3+}, Nd^{3+}, Pm^{3+}, Sm^{3+}, Dy^{3+}, Ho^{3+}, Er^{3+}, and Tm^{3+}. *J. Chem. Phys.* 49: 4424–4442.

48 Hehlen, M.P., Brik, M.G., and Krämer, K.W. (2013). 50th anniversary of the Judd–Ofelt theory: an experimentalist's view of the formalism and its application. *J. Lumin.* 136: 221–239.

49 Görller-Walrand, C. and Binnemans, K. (1998). Chapter 167: Spectral intensities of f–f transitions. In: *Handbook on the Physics and Chemistry of Rare Earths*, vol. 25 (ed. K.A. Gschneidner Jr. and L. Eyring), 101–264. Elsevier.

50 Ćirić, A., Zeković, I., Medić, M. et al. (2020). Judd-Ofelt modelling of the dual-excited single band ratiometric luminescence thermometry. *J. Lumin.* 225: 117369.

51 Dramićanin, M.D. (2020). Trends in luminescence thermometry. *J. Appl. Phys.* 128: 040902.

52 Wade, S.A., Collins, S.F., and Baxter, G.W. (2003). Fluorescence intensity ratio technique for optical fiber point temperature sensing. *J. Appl. Phys.* 94: 4743.

53 Nikolić, M.G., Antić, Ž., Ćulubrk, S. et al. (2014). Temperature sensing with Eu^{3+} doped TiO_2 nanoparticles. *Sensors Actuators B Chem.* 201: 46–50.

54 Dramićanin, M.D. (2018). *Luminescence Thermometry, Methods, Materials and Applications*. San Diego, CA: Elsevier Science & Technology, Woodhead Publishing.

55 Lakshminarayana, G., Yang, R., Mao, M. et al. (2009). Photoluminescence of Sm^{3+}, $Dy3^{+}$, and Tm^{3+}-doped transparent glass ceramics containing CaF_2 nanocrystals. *J. Non-Cryst. Solids* 355: 2668–2673.

56 Soares, M.R.N., Holz, T., Oliveira, F. et al. (2015). Tunable green to red ZrO_2 : Er nanophosphors. *RSC Adv.* 5: 20138–20147.

57 Ćirić, A., Gavrilović, T., and Dramićanin, M.D. (2021). Luminescence intensity ratio thermometry with Er^{3+}: performance overview. *Crystals.* 11: 189.

58 Yan, S.-A., Wang, J.-W., Chang, Y.-S. et al. (2011). Synthesis and luminescence properties of Ln^{3+} (Ln^{3+} = Er^{3+}, Sm^{3+})-doped barium lanthanum tungstate $BaLa_2WO_7$ phosphors. *Opt. Mater.* 34: 147–151.

59 Reisfeld, R., Zelner, M., and Patra, A. (2000). Fluorescence study of zirconia films doped by Eu^{3+}, Tb^{3+} and Sm^{3+} and their comparison with silica films. *J. Alloys Compd.* 300–301: 147–151.

60 Yuan, J.-L., Zeng, X.-Y., Zhao, J.-T. et al. (2008). Energy transfer mechanisms in Tb^{3+}, Yb^{3+} codoped Y_2O_3 downconversion phosphor. *J. Phys. D. Appl. Phys.* 41: 105406.

61 You, J., Ma, L., Qu, Y. et al. (2016). UC/DC luminescence of Ho^{3+} doped pyrochlore structured $La_{2(1-x)}Yb_{2x}TiO_5$ phosphor synthesized by sol-gel method. *J. Rare Earths* 34: 235–239.

62 Carnall, W.T., Crosswhite, H., and Crosswhite, H.M. (1978). *Energy Level Structure and Transition Probabilities in the Spectra of the Trivalent Lanthanides in LaF_3*. Argonne, IL: Argonne National Laboratory.

63 Fidelus, J.D., Yatsunenko, S., Godlewski, M. et al. (2009). Relation between structural properties of Pr^{3+}-doped yttria-stabilized zirconia nanopowders and their luminescence efficiency. *Scr. Mater.* 61: 415–418.

64 Srivastava, A.M. (2013). On the Pr^{3+} interconfigurational optical transitions in glasses and an comparative study of the photon cascade emission in crystals and glasses. *Opt. Mater.* 35: 1194–1199.

65 Carnall, W.T., Fields, P.R., and Rajnak, K. (1968). Electronic energy levels of the trivalent lanthanide aquo ions. II. Gd^{3+}. *J. Chem. Phys.* 49: 4443–4446.

66 OSRAM Sylvania, LED color calculator, (2019). https://www.osram.us/cb/tools-and-resources/applications/led-colorcalculator/index.jsp (accessed 1 September 2020).

10

Multiferroics for Spintronic Applications

Jini K. Jose and Raneesh Balakrishnan

Department of Physics, Catholicate College, Pathanamthitta, Kerala, 689645, India

10.1 Magnetoelectric Multiferroic Materials

Materials that can react to various external stimuli and demonstrate powerful cross-coupling effects have received attention of researchers worldwide in the past few decades. In the past, just one single property (such as electric, magnetic, or elastic) was used in the fabrication of devices. Single-phase materials that may possess fundamentally new properties through high cross-coupling effects are currently in high demand. Magnetoelectric , elastodielectric, piezomagnetic, and magnetoelastic effects are considered cross-coupling effects. This would also improve the performance of existing devices, make it difficult to introduce new functionalities in the past, or overcome external restrictions that hinder performance enhancement through insuperable physical barriers.

Multiferroics is a special class of materials that simultaneously possess two or more primary ferroic order parameters such as ferroelectricity, ferromagnetism, ferroelasticity, ferrotoroidicity in the same phase [1]. In the past few decades, a lot of research works have been carried out in the field of multiferroics worldwide. A material that simultaneously possesses multiferroic orders is not sufficient. There should be strong coupling interaction between different orders to achieve potential applications. Magnetoelectric coupling refers to linear magnetoelectric effects [2]. One can manipulate electric polarization by a magnetic field and magnetization by an electric field [3], which makes them a potential candidate in device fabrication for different applications – memory devices [4–6], spintronics, sensors [7, 8], etc. The ultimate goal for device functionality would be a single-phase multiferroic that exhibits strong magnetoelectric effects at room temperature. Single-phase multiferroic materials are defined as homogeneous and chemically isotropic compounds in which electric and magnetic states coexist. Strict symmetric conditions are required for the coexistence of magnetic and electric order states in the same phase. To overcome the difficulty in obtaining

Nanotechnology in Electronics: Materials, Properties, Devices, First Edition.
Edited by Visakh P.M., Artem Semkin, Raneesh Balakrishnan, and Saša Lazović.

single-phase multiferroic materials with sufficient room temperature magnetoelectric coupling, researchers focus their attention on multiferroic composite materials. Unlike single-phase multiferroic materials, the composite materials the order phases are physically separated from each other within the material. The magneto–electric effects measured in typical composite multiferroics are a few orders of magnitude larger than those of single-phase multiferroics. It has been reported that most composite multiferroics operate at or above room temperature. Nowadays, different efficient techniques are available for the fabrication of novel multiferroic composite structures at the nanoscale. This brings new aspects to the field of multiferroics, such as nanostructure and thin film fabrication, which could potentially facilitate their integration with electronic circuits and spintronic devices.

10.2 Spintronics

10.2.1 Fundamental Aspects of Spintronics

Spintronics, spin-based electronics, is a multidisciplinary and fast-developing research area that is currently emerging. Spintronics research is closely related to physics, chemistry, materials science, electrical engineering and computer science. In spintronic devices, electron spin is employed in addition to electron charge, whereas in conventional electronic devices, electron charge is used to transfer and store information. The intrinsic properties such as nonvolatility, smaller size, high speed, ultralow power consumption make spin-based devices a potential candidate in the microelectronics industry.

In spintronics, the spin of an electron and its electronic charge are used for electronic applications. The giant magnetoresistance (GMR) effect is the basic concept of spintronics, in which the resistance of a magnetic multilayer varies with an applied magnetic field. The significance of spintronics lies in the ability to inject, manipulate, and detect spins in solid-state systems.

Spintronics refers to the study of the role played by the spin of electrons in solid-state physics and in the fabrication of possible devices that explicitly exploit spin properties in addition to the charge degree of freedom. For example, spin relaxation and spin transport in metals and semiconductors are of fundamental research interest because these phenomena have potential applications in electronics technology.

The read head and memory storage cell, the prototype spintronic devices used in industry is the giant magnetoresistive sandwich structure consisting of alternating ferromagnetic and nonmagnetic metal layers. The device resistance changes from low (parallel magnetization) to high (antiparallel magnetization) depending on the relative orientation of magnetization in the magnetic layers. This change in resistance, called magnetoresistance, is used to sense a change in a magnetic field. Recent advancements in GMR technology have led to the invention of magnetic tunnel junction (MJT) devices in which the tunneling current relies on the spin orientations of electrodes. The two major approaches involved in the design and fabrication of spintronic devices are (i) perfecting the existing GMR-based technology and (ii) finding novel ways for the generation and utilization of spin-polarized currents.

10.2.2 Giant Magnetoresistance

The GMR is a quantum mechanical effect observed in multilayered thin-film structures formed by alternating ferromagnetic and nonmagnetic layers. Some of the materials used for the GMR structure are Fe/Cr and Co/Cu magnetic multilayers. When the magnetic moments of the ferromagnetic layers are parallel, spin-dependent scattering of carriers is minimized and the material is in its lowest resistance. When the ferromagnetic layers are anti-aligned, spin-dependent scattering of carriers is maximized, and the material has its highest resistance. The directions of the magnetic moments are manipulated by external magnetic fields applied to materials [9]. Due to this reason, the effect is called GMR, a giant change in electrical resistance in the presence of a magnetic field [10, 11]. Baibich et al. reported the GMR of Fe/Cr magnetic superlattices by varying the magnetic field, thickness of the nonmagnetic(Cr) layer, and the number of superlattice structures [10]. They reported a large change in resistance by applying a small magnetic field, which has wide applications in designing MRAM memories and magnetic read heads, biosensors, MEMS devices, etc.

A lot of experiments have been carried out using polarized neutron reflectometry (PNR), which clearly illustrates Fe/Cr superlattices that led to the GMR effect [12], but PNR has some limitations also [13]. In 1994, Daughton et al. reported the design of a magnetic sensor to detect very weak magnetic fields, which is considered the first GMR-based device [14]. The major achievement of GMR in reading heads for magnetic hard-disk drives was done by IBM in 1997 [15]. The innovation of GMR-based devices has made a big impact on the storage capacity of hard disks from 1 to 20 gigabytes. GMR made it possible for the magnetic read heads of hard disks to be much sensitive to changes in the magnetic field, which boosts the storage capacity by allowing data to be stored in much smaller regions on the disk surface. Another important milestone in the field of spintronics is the discovery of magnetic random-access memory (MRAM). The notable fact about MRAM is that, instead of an electrical charge to store the data, it uses magnetic storage elements. Therefore, MRAM devices have specific features to retain data even after the power is removed from a system unlike in regular random-access memory. MRAM devices had a long storage lifetime, and the first GMR-based RAM chips were produced by Honeywell in 1997. Nowadays, GMR-based read heads are commonly used in many portable devices such as laptops/computers, iPods, and CD/DVD players. The best design for GMR read head is the one in which small changes in the magnetic field can give a maximum rate of change in resistance or a high GMR value. In GMR read heads, changes in resistance 1% per oersted were observed [15]. The GMR ratio can be calculated as follows:

$$\text{GMR ratio} = \frac{\text{Antiparallel resistance} - \text{Resistance}}{\text{Resistance}} \times 100$$

10.2.3 Tunneling Magnetoresistance

An MTJ consists of two layers of a magnetic material separated by an ultrathin layer of insulator with a thickness of about 1 nm. If a bias voltage is applied between the two metal electrodes, electrons can tunnel through the barrier as the insulating layer is so thin. The tunneling current in MTJs depends on the relative orientation of magnetization of two ferromagnetic layers, which can be changed by an applied magnetic field [16–18]. This phenomenon is called tunneling magnetoresistance (TMR), a consequence of spin-dependent tunneling. Tunneling conductance is high when magnetization of ferromagnets is parallel, which allows the best matching of densities of spin states on the two sides of the barrier. In contrast, lowest conductance is observed when magnetization is antiparallel. MTJs differ from the GMR by the fact that the electrodes are magnetically uncoupled and must, therefore, have different reversal fields to achieve parallel and antiparallel configurations by an external magnetic field. The first successful fabrication of a high-quality MTJ showing TMR was reported in 1995 by Moodera et al. [17].

According to Julliere's model, TMR can be written as follows:

$$\mathrm{TMR} = \frac{R_{\mathrm{ap}} - R_{\mathrm{p}}}{R_{\mathrm{p}}} = \frac{2P_1P_2}{1 - P_1P_2}$$

where P_1, P_2 are the spin polarization values of the two electrodes as measured by the Meservey–Tedrow technique [19] and R_{ap} and R_{p} are the junction resistances with antiparallel and parallel magnetization orientations, respectively, for the ferromagnetic electrodes.

Spin-dependent tunneling materials have been attracting huge attention since 1995 when a sizable TMR value of 10% was achieved at room temperature [17], which showed the potential for new device applications. In the early stage of the development of MTJs, the tunneling barrier used in the MTJ structure was typically an amorphous insulating layer; in particular, Al_2O_3 was widely adapted because of the well-characterized fabrication technique. Researchers successfully achieved a TMR value of about 70% [20] at room temperature with Al_2O_3 as the barrier. Later, crystal MgO thin layers were applied in MTJ fabrication, and a TMR value of about 500% [21] was obtained at room temperature.

It was initially thought that spin polarization of the tunneling electrons and the TMR were simply reflecting spin polarization of electronic density of states at the Fermi level in ferromagnetic electrodes. It now turns out that the TMR also depends on the insulating material of the tunnel barrier and, more specifically, on the details of the electronic structure at the electrode/barrier interface.

One of the promising applications of MTJs is MRAM [22]. The basic concept of MRAM is to use the magnetization direction in MTJs for information storage "0" and "1" corresponding to parallel and antiparallel magnetization in an MTJ. The information bits can be written by passing a current through the MTJ, and they can be read out by measuring the change in resistance. The MRAM potentially combines key advantages such as nonvolatility, infinite endurance, and fast random access that make it an ideal candidate to become the "universal memory," one of the chief aims of nanoelectronics [23].

10.3 Spintronic Devices

Nowadays, all conventional electronic devices are based on the flow of electrons, which is considered a binary system (on/off state). If electrons flow in a circuit, it is the on state, and no electron flow means the off state. Researchers all around the world are focusing on the fabrication of new spintronic devices that can offer an increase in the power of a device by up to ten thousand times compared to conventional electronic devices. The magnetic materials required for fabrication of spintronics devices should have a stable structure, good electrical property, and the ability to be applied in a thin film form. Taz et al. reported that a room-temperature amorphous ferromagnetic oxide semiconductor (replacing conventional crystalline oxide metals) could substantially reduce the cost and complexity associated with the use of crystalline materials for spintronic devices. They reported a novel material $(Fe_{0.66}Dy_{0.24}Tb_{0.1})_3O_{7-x}$ (FDTO) with semiconducting behavior, suitable electrical conductivity, and high magnetic moment, all of which could be controlled by altering the oxygen concentration during deposition. Magnetoelectric devices were made by integrating ultrathin FDTO with multiferroic $BiFeO_3$ (Figure 10.1). The magnetism in FDTO coupled to the $BiFeO_3$ at room temperature resulted in magnetoelectric switching and it was probed through the fabrication of a simple spin valve structure [24].

The performance expected is that, incorporating multiferroic materials in the subject of spintronics ought to allow, one to take the privilege of pair functionalities (for example, ferromagnetism and ferroelectricity); however, it also needs coupling between the two orders. The control of magnetization by an electric field or control polarization by magnetic field is a result of the magnetoelectric coupling between magnetism and ferroelectricity. This coupling presents fascinating viewpoints for the outline of ferroelectric memories with a nondestructive reading or MRAM along with an electric writing procedure (MERAM) [25, 26] and allows the realization of a four-state logic in a single device [27]. This paves a new path to a finer combination for these memories. Multiferroic materials are among the scarce materials and Most of them are weak ferromagnets or antiferromagnets [4]. The possibility of

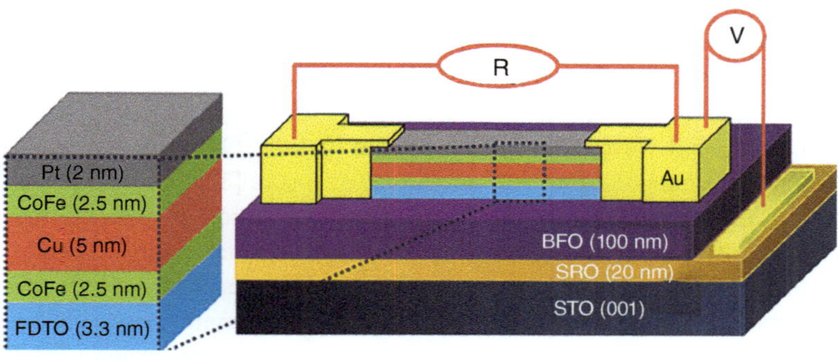

Figure 10.1 Schematic of the GMR heterostructure used to study magnetoelectric switching capabilities of amorphous FDTO. Source: [24] / Springer Nature/ CC BY 4.0.

broadening the varieties of multiferroic systems will produce great results. They can be designed by artificial multiferroics merging with ferromagnetic materials with a huge amount of magnetostriction coefficients and ferroelectric or piezoelectric materials [28]. Though the stain effect the indirect magnetoelectric coupling is achieved, and precisely these are the result of such artificial compounds. There could be more complex implementation through the strain effect [29].

To modify the behavior of magnetic skyrmion states in the presence of an electric field, Xu et al. proposed a new mechanism EPDQ, where the electric field (E) switching of the magnetic topological charge (Q) was possible in a controllable and reversible manner via electric polarization (P) and Dzyaloshinskii–Moriya interaction (D) [30]. They demonstrated the EPDQ mechanism using multiferroic VOI_2 monolayers. Earlier studies have predicted VOI_2 monolayers exhibit both ferroelectric and ferromagnetic behavior [31]. Their study provided a new perspective about the most preferred electric field control of magnetism and proved VOI_2 was an appropriate candidate for fabrication of spintronic devices.

Recently, a work was reported on double perovskite Pr_2FeCrO_6 (PFCO). Studies revealed that PFCO could be considered a promising multiferroic material for spintronic applications. The field dependence of magnetoresistance (MR) for PFCO showed MR% of 44% with a clear distinction of resistance states in parallel and antiparallel states between 0 and ±5 KOe observed [32]. The considerable change in magnetoresistance with field revealed PFCO was a potential candidate for MTJs in the field of spintronics using multiferroic materials.

10.3.1 Spin Valve

A spin valve is a layered structure of magnetic and nonmagnetic (spacer) materials whose electrical resistance depends on the spin state of electrons passing through the device and so can be controlled by an external magnetic field. The discovery of GMR and TMR in metallic spin valves have revolutionized applications such as magnetic recording and memory. Spin valves are also considered the heart of addressable data storage devices, such as MRAM.

Spin valve devices are largely used to study and measure the properties associated with spin transport in paramagnetic metals or semiconductors. Spin diffusion length L_s is the important parameter that can be measured from the spin valve device and it can be represented as $L_s = \sqrt{(D_s \tau_s)}$, where D_s is the spin diffusion coefficient and τ_s spin relaxation time [33].

The fabrication of spin valves was made possible in 1991 [34]. Later on, spin valves made a drastic change in the field of magnetic sensors [35, 36], read heads of magnetic sensors [37, 38], and MRAM. The spin valve structure in either the current-in-plane (CIP) or current-perpendicular-to-plane (CPP) form can be used to analyze the GMR effect. A spin valve structure generally consists of four layers: the topmost layer (free layer) is the first sensing layer of a ferromagnetic material. The second layer (spacer layer) – a nonmagnetic material separating the free and fixed layer; the third layer (pinned or fixed layer) – a magnetic material; the fourth layer (exchange layer) – antiferromagnetic materials that fix the magnetic orientation of the pinned layer. If sufficient bias voltage is applied between the electrodes, the electrons can tunnel through the nonmagnetic region. TMR effect is a direct consequence of spin-dependent tunneling. The tunneling current in a spin valve depends on the relative orientation of magnetization of two magnetic layers, which can be changed by applying a magnetic field.

The sensitivity of spin valves to the generation and detection of spin currents in the electric field is very crucial during construction. The spin current can be generated and detected using ferromagnetic–nonmagnetic interfaces. The efficiency of the interface is determined by interfacial spin polarizations P_S, which manifest in magnetoresistance. The sign reversal is a significant change in P_S. The reversal of electrical current leads to the reversal of the P_S sign and the corresponding change in the MR sign can be viewed as a spin-rectification effect for the ferromagnetic–nonmagnetic interface, which could be useful for tunable spin filtering [39].

Due to low production cost and flexibility, organic spin valves are ideal for spintronic applications [40]. A conventional trilayer base structure consists of an organic layer sandwiched between two magnetic layers serving as magnetic electrodes. The necessity to attain high magnetoresistance is the critical issue associated with the evolution of spin valve devices. The initial work was carried out by Xiong et al. using tris-(8-hydroxyquinoline) aluminum (Alq_3) as the spacer layer between ferromagnetic $La_{0.67}Sr_{0.33}MnO_3$ (LSMO) and cobalt electrodes and a giant negative magnetoresistance of 40% was measured at 11 K [41]. Later, extensive research work was done for studying various types of organic spin valves, aiming at both enhancing magnetoresistance and understanding the underlying mechanism. Recently, a paper was published on the spin-rectification effect in an organic spin valve structure consisting of ferroelectric croconic acid ($C_5H_2O_5$) sandwiched between ferromagnetic electrodes $La_{0.7}Sr_{0.3}MnO_3$ and Co, which can be switched between a high-resistance and a low-resistance state by a poling voltage [39]. The spin-polarized transport through the organic layer can be verified by combining inelastic electron tunneling spectroscopy (IETS) and MR measurements. Galbiati et al. reported that only the combination of IETS and MR measurements should provide a complete proof of spin-polarized transport through the organic layer. Only from MR measurements, it was hard to distinguish the electric transport through the organic layer from the electric transport through the organic layer, which was short-circuited in FM/organic/Tunnel Barrier/FM devices [42].

10.3.2 Multiferroic Tunnel Junctions

Multiferroic tunnel junction (MFTJ) nanopillars are one of the important devices of spintronics used for logic and memory applications, which combine magnetism and electronics. MFTJs, consisting of two ferromagnetic electrodes separated by a thin ferroelectric insulating layer of thickness in nanometers, have received great interest for potential multifunctional applications in spintronics, due to their switchable electric polarization with high tenability. They exhibit fast read/write speed, infinite endurance, nonvolatility, etc. [43]. Instead of a nonmagnetic layer used in the spin valve structure, a ferroelectric thin film is used as a barrier in an MJT, which gives them an additional level of functionality. Using a ferroelectric barrier in an MTJ makes it multiferroic. In such an MFTJ, tunneling electroresistance (TER) and TMR effects coexist [44], which could lead to new applications such as multilevel data storage [45]. They show the performance of four resistance states [46]. To attain promising applications in spintronics, MFTJs should have a large TMR ratio. Dai reported that an interfacial electronic reconstruction driven by in situ screening of ferroelectric polarization could result in large TMR and TER effects. The features of magnetic- and ferroelectric-dependent nonvolatile multi-states can be used to create spintronic-based artificial intelligence devices such as synapse [47]. For practical applications, one of the layers is pinned (reference layer), while other magnetic layers store a binary state based on the relative orientation of two ferromagnetic layers. Ikeda reported a TMR ratio greater than 604% by using the MgO oxide barrier, and this makes MTJs an excellent candidate for sensor design [48]. Today, there are different ways to improve the performance of MTJs such as perfect material choice, scalability, energy efficiency, reliability, power consumption, area. Different approaches such as thermally assisted switching (TAS) and spin-transfer torque (STT) are investigated, but both of them suffer power or reliability issues.

Electrically controlled intermediate ferroelectric polarization states in MFTJs can be used for multistate memory applications [49]. Recently, Fang reported that in the presence of gradually reversed ferroelectric domains a $La_{0.67}Sr_{0.33}MnO_3$(LSMO)/$PbZr_{0.2}Ti_{0.8}O_3$(PZT)/Co structured MFTJ device could show multilevel resistance states via TER and TMR, respectively [50]. The bi-ferroic memristive properties of MFTJs were detailed by Luo et al. using Co/$PbTiO_3$(PTO)/$La_{0.3}Sr_{0.7}MnO_3$(LSMO) MFTJs. They achieved a broad range of TMR (between 3 and 30%) and TER (1000%) values [51].

Significant research efforts are focused on using organic materials for spintronic applications [52]. Organic materials are composed of light elements that show very weak spin–orbit coupling (SOC), allowing the electron state to be sustained over a long period [53]. Spin relaxation times in organic materials is several orders of magnitude greater than that in inorganic materials, and thus spin diffusion lengths in organic materials are of the order of hundreds of nanometers. These make organic materials a better candidate for barriers in organic MTJs. These results lead to the conclusion that the understanding of the transport properties is very crucial for designing high-performance OMFTJs for memristor and spintronic applications.

Currently, many reports have been published on organic multiferroic tunnel junctions (OMFTJs) based on most typical ferroelectric polymer poly(vinylidene fluoride) (PVDF) barriers. Xiao et al. could explain the temperature dependence of transport mechanisms in OMFTJs. They reported a systematic study of the temperature-dependent transport behavior in $La_{0.6}Sr_{0.4}MnO_3$/PVDF/Co OMFTJs [54]. Subedi et al. reported that organic magnetoelectric-based information storage with four-state bits is feasible at room temperature [55]. They constructed OMFTJs using organic ferroelectric P(VDF-TrFE) copolymers (poly(vinylidene fluoride-trifluoroethylene) as a barrier sandwiched between $La_{2/3}Sr_{1/3}MnO_3$ (LSMO) and Co ferromagnetic electrodes and showed the TMR significantly

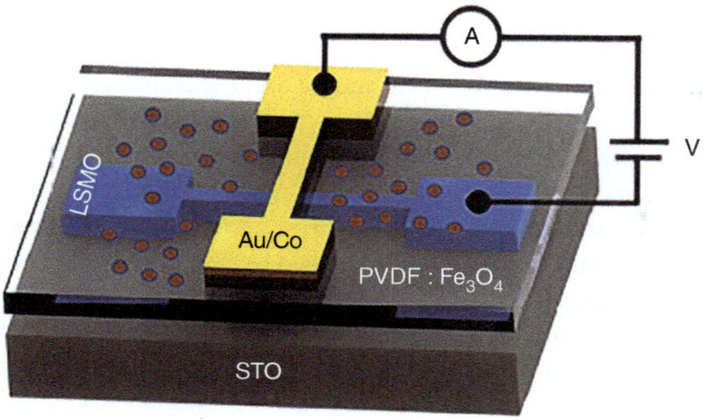

Figure 10.2 Schematics of the structure and measurement setup of the LSMO/PVDF : Fe_3O_4/Co device. Source: [57] / with the permission of AIP Publishing.

quenches in the unpolarized MFTJ from 21% at 20 K to 0.7% at 296 K. The report detailed that TER could reach around 1000% and the TMR about 30% at 200 K. In 2018, Liang et al. studied OMFTJs based on $La_{0.6}Sr_{0.4}MnO_3$/P(VDF-TrFE)/Co structures with different thicknesses of P(VDF-TrFE) and obtained the evidence of ferroelectric "ailing-channel" in the organic barrier [56].

The improvement of ferroelectric properties of OMFTJ devices is urgently required to promote potential memristor and spintronic applications.

Gao et al. reported on OMFTJ based on an organic barrier of the poly(vinylidene fluoride) (PVDF) : Fe_3O_4 nanocomposite. They fabricated $La_{0.6}Sr_{0.4}MnO_3$/PVDF : Fe_3O_4 nanocomposite/Co OMFTJs and found that the ferroelectric properties of the OMFTJ were considerably improved compared with a pure PVDF barrier by adding Fe_3O_4 nanoparticles into the PVDF barrier (Figure 10.2). A drastic change in the TER value of OMFTJ was also reported and a TMR of about 450% at 10 K and 100% at room temperature (RT), which were much higher than those of the pure PVDF-based device (70% at 10 K and 7% at RT) [57].

10.3.3 Spin FET

In 1990, Datta and Das proposed the first spin-polarized field-effect transistor (spin–FET) using the concept of spin transistor [58]. The spin–FET was proposed as an electronic analog of electro-optic modulators. It is considered one of the most advanced applications of spintronics technology. Datta and Das provided the theoretical background for the construction of spin–FET, but it is not yet fabricated completely. The design of the spin transistor is analogous to that of conventional MOSFETs, which consists mainly of three terminals: source, drain, and channel. The channel between the source and drain is controlled by the gate. In a spin–FET, the channel is made of a nonmagnetic layer, which is used for transmitting spin-polarized electrons from a ferromagnetic source to another ferromagnetic electrode (drain). The operation of the device is based on the Rashba SOC effect [59], where the spin-polarized current from the source to drain is controlled by an electric field through the gate. A large current can be obtained when the spins are parallel to magnetization in the source and drain. Otherwise, when the spins are antiparallel to magnetization in the ferromagnets, the spins are scattered and the transistor is off. Thus, there are oscillations in the output electric current as a function of the gate voltage. The modulation is identical to the intensity modulation in electro-optic modulators. After the first spin–FET was proposed, a wide range of spin transistors based on various working principles were reported [60, 61]. In 2009, the first working device made of InAs heterostructure was presented by Koo et al. [62].

The major problem associated with the fabrication of spin–FET is the short diffusion length. Zhang et al. reported that at room temperature the spin diffusion length in graphene could not surpass several micrometers [63]. A feasible solution is to use collective spin phenomena (e.g. spin waves). It was reported that spin waves were able to maintain coherence over much greater distances than single electrons. Covington et al. reported that in conducting ferromagnetic materials such as $Ni_{81}Fe_{19}$, the spin coherence length exceeds tens of micrometers [64]. In nonconducting ferrites such as $Y_3Fe_2(FeO_4)_3$, the spin coherence length exceeds millimeters at room temperature [65].

Balinskiy et al. reported their work on using synthetic multiferroics for voltage-controlled spin-wave modulation, which may use much lower electric field for operation [66]. They described a spin-wave modulator based on the stress-mediated coupling of piezoelectric and magnetostrictive materials (Figure 10.3). The switching between the spin-wave conducting and nonconducting states was achieved by applying an electric field of $\pm 0.3\,MV\,m^{-1}$. They reported a modulation depth of over 300% at a distance of 80 μm away from the excitation port at room temperature. The manifestation of the spin-wave modulator revealed a potential solution to the issues associated with a short spin diffusion length.

In the last few years, spin-based memories have attained great research interest due to their high durability and nonvolatility. However, compared to semiconductor logic transistors, which are capable of large resistance changes, spin-based memories provide small changes in resistance. Recent developments in the availability of multiferroic magnetoelectric materials offer an opportunity to directly combine the change in the spin states of a magnetic memory with a change in the charge of a semiconductor transistor.

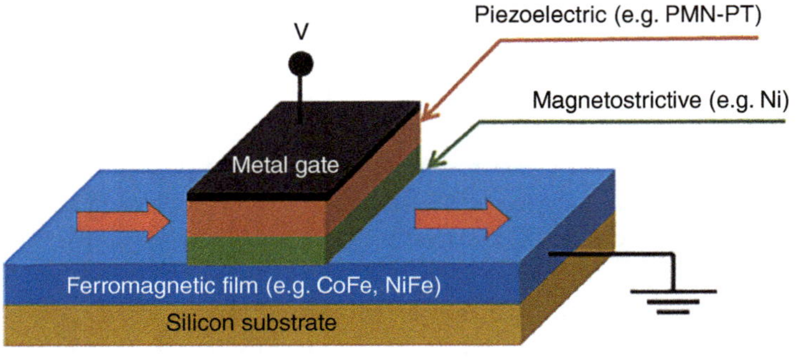

Figure 10.3 Schematic of a spin-wave modulator. It consists of a semiconductor substrate. Source: [66] \ Springer Nature\ CC BY 4.0.

In 2020 Li et al. proposed the spin–orbit torque field-effect transistor (SOTFET), a new magnetoelectric memory device using CoFe/$BiFeO_3$ gate stack [67]. The goal of the fabrication of magnetoelectric memory devices is to combine the benefits of magnetic memories with large changes in resistance of FETs, providing both memory and logic functionality. A change in magnetization of a spin–orbit/ferromagnetic layer is used in SOTFETs to monitor the semiconductor channel using a multiferroic magnetoelectric layer, resulting in readouts with several orders of magnitude change in resistance. In the spin FET design, the SOC of the channel material regulates the feasibility of managing the spin of electrons via an external electric field. Due to their high SOC and long spin lifetime, traditional III–V semiconductors were originally suggested as channel materials. However, they produce weak magnetoresistance signals as the mean velocity of the carriers is very small and the associated spin diffusion length is relatively short. Carbon-based nanomaterials are exclusively capable of making channel materials in spintronic devices with long (up to ~100 μm) spin diffusion lengths [68–70]. But the small SOC of carbon-based nanomaterials limits the possibility of manipulating electrons via an externally applied field and becomes necessary to combine it with a material with high SOC. In 2016, Zanolli reported the first-principles investigation of the electronic and magnetic properties of the graphene–magnetoelectric multiferroic material ($BaMnO_3$) interface [71]. It was reported that a strong interaction between graphene π electrons and the magnetic Mn d states induced significant spin polarization in graphene and suggested a series of possible applications in spintronics (e.g. spin filters, spin injectors) for hybrid organic–multiferroic materials.

10.3.4 Spin LED

A spin-polarized light-emitting diode (LED) is an important device to determine the effectiveness of injection of spin-polarized current into semiconductor heterostructures. In a usual LED, recombination of electrons and holes produces unpolarized light, while in spin LED, recombination of spin-polarized carriers results in circularly polarized light [72]. The spin injection in a spin LED has been successfully carried out by a few groups [72–74]. Later, organic light-emitting diodes (OLEDs) and OLED displays were introduced to the scientific community. The report mentioned is only referring to the spin LED; however, using multiferroic materials has not been recorded due to limited research work.

10.4 Summary

The multidisciplinary aspect of spintronics has revolutionized the research in electronics and materials science in recent decades. It is clear that multiferroic spintronics devices are capable of making further significant breakthroughs. In this chapter, the background of spintronics and the recent works published in the field of multiferroic heterostructure-based spintronics were reported. Organic materials for spintronic applications also gained greatest research interest nowadays, and a huge number of research studies are being conducted on using organic materials for spintronic applications such as organic spin valves, organic MJTs.

References

1 Schmid, H. (1994). Multi-ferroic magnetoelectrics. *Ferroelectrics* 162 (1): 317–338.

2 Fiebig, M. (2005). Revival of the magnetoelectric effect. *J. Phys. D. Appl. Phys.* 38 (8).

3 Ramesh, R. and Spaldin, N.A. (2009). Multiferroics: progress and prospects in thin films. *Nat. Mater.* 6 (1): 21–29.

4 Hill, N.A. (2000). Why are there so few magnetic ferroelectrics? *J. Phys. Chem.* 104 (29): 6694–6709.

5 Israel, C., Mathur, N.D., and Scott, J.F. (2008). A one-cent room-temperature magnetoelectric sensor. *Nat. Mater.* 7 (2): 93–94.

6 Veev, J.P., Duan, C.G., Burton, J.D. et al. (2009). Magnetic tunnel junctions with ferroelectric barriers: prediction of four resistance states from first principles. *Nano Lett.* 9 (1): 427–432.

7 Jahns, R., Piorra, A., Lage, E. et al. (2013). Giant magnetoelectric effect in thin-film composites. *J. Am. Ceram. Soc.* 96 (6): 1673–1681.

8 Jahns, R., Greve, H., Woltermann, E. et al. (2012). Sensitivity enhancement of magnetoelectric sensors through frequency-conversion. *Sensors Actuators A Phys.* 183: 16–21. https://doi.org/10.1016/j.sna.2012.05.049.

9 Fert, A. (2008). Nobel lecture: origin, development, and future of spintronics. *Rev. Mod. Phys.* 80 (4): 1517–1530.

10 Baibich, M.N., Broto, J.M., Fert, A. et al. (1988). Giant magnetoresistance of (001)Fe/(001)Cr magnetic superlattices. *Phys. Rev. Lett.* 61 (21): 2472–2475.

11 Binasch, G., Grünberg, P., Saurenbach, F., and Zinn, W. (1989). Enhanced magnetoresistance in layered magnetic structures with antiferromagnetic interlayer exchange. *Phys. Rev. B* 39 (7): 4828–4830.

12 Moyerman, S., Eckert, J.C., Borchers, J.A. et al. (2006). Magnetic structure variations during giant magnetoresistance training in spin valves with picoscale antiferromagnetic layers. *J. Appl. Phys.* 99 (8): 8–11.

13 Li, M., Cui, W., Yu, J. et al. (2015). Magnetic proximity effect and interlayer exchange coupling of ferromagnetic/topological insulator/ferromagnetic trilayer. *Phys. Rev. B – Condens. Matter Mater. Phys.* 91 (1): 1–7.

14 Daughton, J., Brown, J., Chen, E. et al. (1994). Magnetic field sensors using GMR multilayer. *IEEE Trans. Magn.* 30 (6): 4608–4610.

15 Prinz, G.A. (1998). Magnetoelectronics. *Science.* 282 (5394): 1660–1663.

16 Miyazaki, T. and Tezuka, N. (1995). Giant magnetic tunneling effect in Fe/Al_2O_3/Fe junction. *J. Magn. Magn. Mater.* 139 (3): 94–97.

17 Moodera, J.S., Kinder, L.R., Wong, T.M., and Meservey, R. (1995). Large magnetoresistance at room temperature in ferromagnetic thin film tunnel junctions. *Phys. Rev. Lett.* 74 (16): 3273–3276.

18 Miyazaki, T. and Tezuka, N. (1995). Spin polarized tunneling in ferromagnet/insulator/ferromagnet junctions. *J. Magn. Magn. Mater.* 151 (3): 403–410.

19 Meservey, R., Tedrow, P.M., and Fulde, P. (1970). Magnetic field splitting of the quasiparticle states in superconducting aluminum films. *Phys. Rev. Lett.* 25 (18): 1270–1272.

20 Wang, D., Nordman, C., Daughton, J.M. et al. (2004). 70% TMR at room temperature for SDT sandwich junctions with CoFeB as free and reference layers. *IEEE Trans. Magn.* 40 (4 II): 2269–2271.

21 Lee, Y.M., Hayakawa, J., Ikeda, S. et al. (2007). Effect of electrode composition on the tunnel magnetoresistance of pseudo-spin-valve magnetic tunnel junction with a MgO tunnel barrier. *Appl. Phys. Lett.* 90 (21): 8–11.

22 Daughton, J.M. (1997). Magnetic tunneling applied to memory (invited). *J. Appl. Phys.* 81 (8 PART 2A): 3758–3763.

23 Chappert, C., Fert, A., and Van Dau, F.N. (2009). The emergence of spin electronics in data storage. *Nat. Mater.* 6: 813–823.

24 Taz, H., Prasad, B., Huang, Y.L. et al. (2020). Integration of amorphous ferromagnetic oxides with multiferroic materials for room temperature magnetoelectric spintronics. *Sci. Rep.* 10 (1): 1–10.

25 Bibes, M. and Barthélémy, A. (2008). Multiferroics: towards a magnetoelectric memory. *Nat. Mater.* 7 (6): 425–426.

26 Chen, X., Hochstrat, A., Borisov, P., and Kleemann, W. (2006). Magnetoelectric exchange bias systems in spintronics. *Appl. Phys. Lett.* 89 (20): 1–4.

27 Gajek, M., Bibes, M., Fusil, S. et al. (2007). Tunnel junctions with multiferroic barriers. *Nat. Mater.* 6 (4): 296–302.

28 Eerenstein, W., Wiora, M., Prieto, J.L. et al. (2007). Giant sharp and persistent converse magnetoelectric effects in multiferroic epitaxial heterostructures. *Nat. Mater.* 6 (5): 348–351.

29 Duan, C.G., Jaswal, S.S., and Tsymbal, E.Y. (2006). Predicted magnetoelectric effect in Fe/$BaTiO_3$ multilayers: ferroelectric control of magnetism. *Phys. Rev. Lett.* 97 (4): 13–16.

30 Xu, C., Chen, P., Tan, H. et al. (2020). Electric-field switching of magnetic topological charge in type-I multiferroics. *Phys. Rev. Lett.* 125 (3): 37203. https://doi.org/10.1103/PhysRevLett.125.037203.

31 Tan, H., Li, M., Liu, H. et al. (2019). Two-dimensional ferromagnetic-ferroelectric multiferroics in violation of the d0 rule. *Phys. Rev. B* 99 (19): 195434. https://doi.org/10.1103/PhysRevB.99.195434.

32 Ravi, S. (2020). High Curie temperature and room temperature magnetoresistance in Pr_2FeCrO_6 material for spintronics applications. *Mater Lett [Internet].* 278: 128448. https://doi.org/10.1016/j.matlet.2020.128448.

33 Bandyopadhyay, S. and Cahay, M. (2008). *Introduction to Spintronics*. New York: CRC Press.

34 Dieny, B., Speriosu, V.S., Parkin, S.S.P. et al. (1991). Giant magnetoresistive in soft ferromagnetic multilayers. *Phys. Rev. B* 43 (1): 1297–1300.

35 Heim, D.E., Tsang, J., Speriosu, V.S. et al. (1994). Design and operation of spin valve sensors. *IEEE Trans. Magn.* 30 (2): 316–321.

36 Freitas, P.P., Leal, J.L., Melo, L.V. et al. (1994). Spin-valve sensors exchange-biased by ultrathin TbCo films. *Appl. Phys. Lett.* 65 (4): 493–495.

37 Leal, L., Oliveira, N.J., Rodrigues, L.M. et al. (1994). Unshielded spin-valve sensors exchange-biased by thin TbCo layers. *IEEE Trans. Magn.* 30 (6): 3831–3833.

38 Tsang, C., Fontana, R.E., Lin, T. et al. (1994). Design, fabrication & testing of spin-valve read heads for high density recording. *IEEE Trans. Magn.* 30 (6): 3801–3806.

39 Yin, Y., Yin, Y., Jiang, X. et al. (2020). Spin rectification and electrically controlled spin transport in molecular-ferroelectrics-based spin valves. *Phys. Rev. Appl.* 13 (6): https://doi.org/10.1103/PhysRevApplied.13.064011.

40 Sun, D., Yin, L., Sun, C. et al. (2010). Giant magnetoresistance in organic spin valves. *Phys. Rev. Lett.* 104 (23): 1–4.

41 Xiong, Z.H., Wu, D., Vardeny, Z.V., and Shi, J. (2004). Giant magnetoresistance in organic spin-valves. *Nature* 427 (6977): 821–824.

42 Galbiati, M., Tatay, S., Delprat, S. et al. (2015). Is spin transport through molecules really occurring in organic spin valves? a combined magnetoresistance and inelastic electron tunnelling spectroscopy study. *Appl. Phys. Lett.* 106 (8): 10–14.

43 Zhao, W., Belhaire, E., Chappert, C., and Mazoyer, P. (2009). Spin transfer torque (STT)-MRAM – based runtime reconfiguration FPGA circuit. *Trans. Embed. Comput. Syst.* 9 (2): 14.

44 Zhuravlev, M.Y., Jaswal, S.S., Tsymbal, E.Y., and Sabirianov, R.F. (2005). Ferroelectric switch for spin injection. *Appl. Phys. Lett.* 87 (22): 1–3.

45 Garcia, V. and Bibes, M. (2014). Ferroelectric tunnel junctions for information storage and processing. *Nat. Commun.* 5: 4289.

46 Liang, S., Yang, H., Yang, H. et al. (2016). Ferroelectric control of organic/ferromagnetic spinterface. *Adv. Mater.* 28 (46): 10204–10210.

47 Dai, J.Q. (2016). Tunneling magnetoresistance and electroresistance in Fe/$PbTiO_3$/Fe multiferroic tunnel junctions. *J. Appl. Phys.* 120 (7): https://doi.org/10.1063/1.4960695.

48 Ikeda, S., Hayakawa, J., Ashizawa, Y. et al. (2008). Tunnel magnetoresistance of 604% at 300 K by suppression of Ta diffusion in CoFeBMgOCoFeB pseudo-spin-valves annealed at high temperature. *Appl. Phys. Lett.* 93 (8): 67–70.

49 Matsunaga, S., Hayakawa, J., Ikeda, S. et al. (2008). Fabrication of a nonvolatile full adder based on logic-in-memory architecture using magnetic tunnel junctions. *Appl. Phys. Express* 1 (9): 0913011–0913013.

50 Fang, M., Zhang, S., Zhang, W. et al. (2019). Nonvolatile multilevel states in multiferroic tunnel junctions. *Phys. Rev. Appl.* 12 (4): https://doi.org/10.1103/PhysRevApplied.12.044049.

51 Luo, Z.D., Apachitei, G., Yang, M.M. et al. (2018). Bi-ferroic memristive properties of multiferroic tunnel junctions. *Appl. Phys. Lett.* 112 (10): https://doi.org/10.1063/1.5023877.

52 Szulczewski, G., Sanvito, S., and Coey, M. (2009). A spin of their own. *Nat. Mater.* 8 (9): 693–695. https://doi.org/10.1038/nmat2518.

53 Drew, A.J., Hoppler, J., Schulz, L. et al. (2009). Direct measurement of the electronic spin diffusion length in a fully functional organic spin valve by low-energy muon spin rotation. *Nat. Mater.* 8 (2): 109–114. https://doi.org/10.1038/nmat2333.

54 Xiao, C., Sun, H., Cheng, L. et al. (2020 Jun 3). Temperature dependence of transport mechanisms in organic multiferroic tunnel junctions. *J. Phys. D. Appl. Phys.* 53 (32): 325301. https://doi.org/10.1088/1361-6463/ab841b.

55 Subedi, R.C., Geng, R., Luong, H.M. et al. (2017). Large magnetoelectric effect in organic ferroelectric copolymer-based multiferroic tunnel junctions. *Appl. Phys. Lett.* 110 (5): 053302.

56 Liang, S., Yu, Z., Devaux, X. et al. (2018). Quenching of spin polarization switching in organic multiferroic tunnel junctions by ferroelectric "ailing-channel" in organic barrier. *ACS Appl. Mater. Interfaces* 10 (36): 30614–30622.

57 Gao, X., Liang, S., Ferri, A. et al. (2020). Enhancement of ferroelectric performance in PVDF : Fe_3O_4 nanocomposite based organic multiferroic tunnel junctions. *Appl. Phys. Lett.* 116 (15): 152905.

58 Datta, S. and Das, B. (1990). Electronic analog of the electro-optic modulator. *Appl. Phys. Lett.* 56 (7): 665–667.

59 Bychkov, Y.A. and Rashba, E.I. (1984). Oscillatory effects and the magnetic susceptibility of carriers in inversion layers. *J. Phys. C Solid State Phys.* 17 (33): 6039–6045.

60 Sugahara, S. and Tanaka, M. (2004). A spin metal-oxide-semiconductor field-effect transistor using half-metallic-ferromagnet contacts for the source and drain. *Appl. Phys. Lett.* 84 (13): 2307–2309.

61 Sugahara, S. and Tanaka, M. (2006). Spin MOSFETs as a basis for spintronics. *ACM Trans Storage.* 2 (2): 197–219.

62 Koo, H.C., Kwon, J.H., Eom, J. et al. (2009). Control of spin precession in a spin-injected field effect transistor. *Science.* 325 (5947): 1515–1518.

63 Zhang, P. and Wu, M.W. (2011). Electron spin diffusion and transport in graphene. *Phys. Rev. B – Condens. Matter Mater. Phys.* 84 (4): 1–12.

64 Covington, M., Crawford, T.M., and Parker, G.J. (2002). Time-resolved measurement of propagating spin waves in ferromagnetic thin films. *Phys. Rev. Lett.* 89 (23): 1–4.

65 Serga, A.A., Chumak, A.V., and Hillebrands, B. (2010). YIG magnonics. *J. Phys. D. Appl. Phys.* 43 (26).

66 Balinskiy, M., Chavez, A.C., Barra, A. et al. (2018). Magnetoelectric spin wave modulator based on synthetic multiferroic structure. *Sci. Rep.* 8 (1): 1–10.

67 Li, X., Casamento, J., Dang, P. et al. (2020). Spin-orbit torque field-effect transistor (SOTFET): proposal for a magnetoelectric memory. *Appl. Phys. Lett.* 116 (24): 242405.

68 Hueso, L.E., Pruneda, J.M., Ferrari, V. et al. (2007). Transformation of spin information into large electrical signals using carbon nanotubes. *Nature* 445 (7126): 410–413.

69 Han, W., Kawakami, R.K., Gmitra, M., and Fabian, J. (2014). Graphene spintronics. *Nat. Nanotechnol.* 9 (10): 794–807. https://doi.org/10.1038/nnano.2014.214.

70 Tombros, N., Jozsa, C., Popinciuc, M. et al. (2007). Electronic spin transport and spin precession in single graphene layers at room temperature. *Nature* 448 (7153): 571–574.

71 Zanolli, Z. (2016). Graphene-multiferroic interfaces for spintronics applications. *Sci. Rep.* 6 (July): 1–6. https://doi.org/10.1038/srep31346.

72 Flederling, R., Kelm, M., Reuscher, G. et al. (1999). Injection and detection of a spin-polarized current in a light-emitting diode. *Nature* 402 (6763): 787–790.

73 Jonker, B.T., Park, Y.D., Bennett, B.R. et al. (2000). Robust electrical spin injection into a semiconductor heterostructure. *Phys. Rev. B – Condens. Matter Mater. Phys.* 62 (12): 8180–8183.

74 Ohno, Y., Arata, I., Matsukura, F. et al. (2001). Electrical spin injection in ferromagnetic/nonmagnetic semiconductor heterostructures. *Phys. E Low-Dimens. Syst. Nanostruct.* 10 (1–3): 489–492.

11

Quartz Tuning Fork–Based Nanosensors

S. Abraham Sampson[1] and Suwarna S. Datar[2]

[1] *VIT Vellore, Department of Sensors and Biomedical Engineering, School of Electronics, Vellore, Tamilnadu 632014, India*
[2] *Defence Institute of Advanced Technology, Department of Applied Physics, Girinagar, Pune, Maharashtra, 411025, India*

11.1 Introduction

The modern world is rapidly progressing towards total and complete automation of machines and systems. A very important and integral part of this development is the rapid pace at which sensors are being developed and deployed. Any modern commonplace system or machine, with even a very basic level of automation, requires the services of and inputs from many sensors for its unhindered operation. Various chemical reactions, electrical changes, mechanical and structural variations, magnetic disruptions, and biological processes form stimuli for different sensors. Today, hundreds of thousands of sensors are being employed in everyday life, which allow precise real-time monitoring and control, thus facilitating increased productivity and stability.

Modern technologies have made us acutely aware of dangerous chemicals and vapors in our everyday environment. Sensors play a very important role in determining the safety standards, thus safeguarding human life and health. Equally important is the role of sensors in manufacturing industries, medical centers, and food processing units as they all rely on accuracy and control they have on their environments. There is thus an insatiable need for better and smaller sensors, which has led to an explosion of research on various chemical and gas sensors using different microelectromechanical systems (MEMS) and nanoelectromechanical system (NEMS) technologies [1, 2].

As the amazing developments in the field of nanotechnology continue unabated, nanostructured materials have also gained considerable importance in sensor technologies due to their exciting potential for application in various situations. Sensitivity and selectivity of microsensors based on these nanostructured materials such as nanoparticles, nanowires, nanorods, nanofibers, nanotubes and nanobelts of conducting polymers and carbon-based structures [1–3] are greatly

Nanotechnology in Electronics: Materials, Properties, Devices, First Edition.
Edited by Visakh P.M., Artem Semkin, Raneesh Balakrishnan, and Saša Lazović.

increased [4]. Sensors based on the principle of adsorption make full utilization of their application potential, as their performance is directly related to granularity, porosity, and the surface-area-to-volume ratio of the sensing materials [5–7].

Such path-breaking discoveries have led to the development of a wide range of novel physical, chemical, and biological sensors [8–11]. These sensors can provide highly sensitive and low-cost portable tools for in-situ chemical and biological analyses in pharmaceutical, food, biomedical, environmental, security, industrial safety, clinical, indoor monitoring applications, etc. with very little power consumption [12–14]. Detection of volatile organic compounds (VOCs), which are used in industrial and domestic applications, are another topic of interest due to their serious environmental and human health impacts [15, 16]. This has impelled the development of a new era of chemical sensor technologies [17, 18].

11.2 Chemical Sensors

Chemical sensors and gas sensors, like all other sensors, are largely varied in terms of the materials used, physical principles of sensing, selectivity, sensitivity, applications, and various other criteria. There can be a myriad of variables such as environmental factors, economic considerations and sensor performances that need to be considered when developing a sensor for a particular application. It is not possible for a single kind of sensor to cater to all requirements and specifications of all the possible use cases. A particular sensor that is best suited in a particular situation might be the least preferable one in another situation. Thus, the quest for better sensors is a never-ending process, with each iteration taking us closer to perfection, but never attaining it.

The general working principle of gas and chemical sensors is based on the physical or chemical interaction between the analyte and sensing material, and this interaction is converted into some form of electrical energy, which can be quantified to identify the extent of the interaction between the two. Broadly speaking, chemical sensors can be classified into direct and indirect sensors. In direct sensors, analytes cause changes in directly measurable electrical parameters such as resistance, capacitance, voltage, or current. Examples are metal-oxide sensors, whose resistances vary due to the interaction with the analytes [19], and electrochemical sensors, which may further be classified as amperometric, potentiometric, conductometric, etc. [20].

The main disadvantage of these is that they require the sensor to be conducting or semi-conducting in nature. However, because many of the sensing materials are non-conductive or conditionally conductive [21], measuring changes in their electrical properties becomes extremely cumbersome, as in the case of many polymer-based sensors. Resistive sensors, especially metal-oxide sensors, have shown great sensitivity at higher temperatures when the resistance approaches measurable values, but suffer from poor selectivity. Some polymer-based sensors have displayed good sensitivity and selectivity but have much shorter lifetimes as they degrade easily.

Complex chemical sensors (indirect sensors) require transducers, which will convert chemical or physical interactions into measurable entities such as changes in light modulation, mass, stiffness, heat/temperature. Second-stage transducers would then convert these to electrical signals, which can then be mapped to the underlying analyte interactions. Optical variations, which manifest as chemiluminescence or as changes in intensity, polarization, and velocity of light in that medium, can be detected by various spectroscopic techniques as well as by using fiber-optic and LED-based sensors [20]. These, however, are susceptible to optical noises and require highly sophisticated and costly equipment in controlled environments for accurate measurements.

Compared to these, measuring the mechanical properties is much easier and more practical. There are mainly two types of mechanical sensors: those that convert the mechanical changes directly to voltage, such as bulk acoustic wave (BAW) and surface acoustic wave (SAW) devices, and those that convert these mechanical changes to frequency, which is then mapped. Sensors that are based on the measurement of frequency have been shown to be more sensitive than acoustic wave–based sensors because the availability of a wider range of crystals with oscillating frequencies across the spectrum offers more sensor material choices and allows more customizations.

Microcantilever and quartz crystal microbalance (QCM) based sensors, where subtle changes in the mass of the sensing material are detected by monitoring the changes in resonant frequencies of the vibrating cantilever or crystal, are preferred as they are highly sensitive to mass changes [22–24]. Their surfaces are usually modified with polymers and other surfactants to improve sensitivity and achieve better selectivity. These, however, have a major disadvantage in that both QCM- and microcantilever-based sensors can detect changes in mass only and not any other mechanical property such as changes in density or stiffness of the sensing material, which is present on the sensor surface. This means that the sensors need controlled environments to avoid inadvertent mass loading due to contaminant materials or accidental mass variations due to the movement of air. Also, polymer-based mass sensors have been reported to have slow response time since it is governed by the diffusion of gas molecules into the polymer films [25]. These can be overcome if the sensor is able to detect changes in stiffness as well as mass of the sensing material, as is possible with quartz tuning fork (QTF)-based sensors discussed next.

11.3 Quartz Tuning Forks (QTFs)

QTFs, shown in Figure 11.1, are extremely inexpensive, single crystal quartz mechanical oscillators which have been used in wristwatches and other high precision timing circuits for decades. The unique property of the piezoelectric material quartz, which enables the conversion of electrical signals into mechanical vibrations and vice versa, allow the QTFs to be excited electrically and their resonant frequencies to be read out electrically. These QTFs can be considered mechanical transducers since changes in the mechanical properties of the system can be

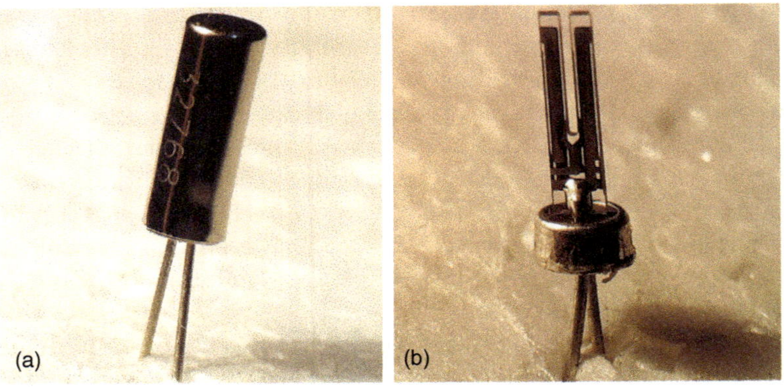

Figure 11.1 Commercially available QTFs. (a) Sealed QTF. (b) Open QTF.

monitored by measuring the changes in resonant frequency of the system, which is measured in the form of an electrical signal. Their high precision resonance frequency, high quality (Q) factor, high mechanical stability, near-zero thermal deflection, and low power consumption make them extremely good candidates for use in timing circuits, probe microscopy techniques, gyroscopes, micro-balances, and sensors [26, 27]. Gyroscopes and micro-balances rarely use commercial QTFs, instead preferring specially micro-machined QTFs that are double ended or are H shaped.

$$f = \frac{1}{2\pi}\sqrt{\frac{k}{m}} \tag{11.1}$$

QTFs can be oscillated in numerous modes, but the most commonly used, especially in commercial purposes, is the fundamental mode of operation, where the tines oscillate in antiphase to each other with a resonant frequency of 32.768 kHz. The tuning fork is essentially two cantilevers joined together at the base such that their oscillations are coupled to each other. When operated in the fundamental mode, the resonant frequency f of the QTF is governed by Eq. (1.1), where k is the effective spring constant and m is the effective mass of the QTF. This effectively means that the resonant frequency of the QTF can either be increased or decreased by suitably modifying the QTFs rather than changing the material properties or the dimensions of the QTFs. This presents us with an opportunity that is missing in the case of QCMs and microcantilevers, both of which can detect only changes in mass of the sensing material: detecting changes in stiffness of the sensing material. The resonant frequency of the QTF is affected by changes in the mass as well as the spring constant of the sensing material if the sensing material is modified onto the QTF in a suitable manner, making it a more sensitive and desirable frequency-based sensor as will be explained in detail in the forthcoming pages.

The distinct properties of QTFs allow it to be used as a gas/chemical sensor in three unique ways with very different principles of operation. The first technique is where it is used as a force sensor in force microscopic techniques and also near-field optical microscopy to study the sample properties, which may or may not be affected by the presence of analytes. The second method is where the QTF is used in photoacoustic and photothermal spectroscopy where the QTF is used as a transducer to convert optical variations into electrical signals and thereby to detect trace gases that are interacting with the laser light. The third and most direct way is where the QTF itself becomes a gas/chemical sensor, and variations in resonant frequency due to either mass changes or spring constant changes are measured to detect the effect of analytes.

In all these three techniques, the two prongs of the QTF are exposed by removing the hermetic seals so that the prongs might be subjected to some form of force or energy: attractive, repulsive, acoustic, thermal, electrical, etc. In force microscopy and the two spectroscopy techniques, bare QTFs are generally used with minimal or no modifications of the prongs with any nanomaterials or structures. Very rarely are modified QTFs used in these cases as the QTF is not the primary detector, but rather a second-level transducer. For these two use cases, the QTF is used primarily as an oscillator crystal while taking advantage of quartz's piezoelectric properties, and, hence, the resonant frequency peak is not the most important factor.

For example, in quartz-enhanced photoacoustic spectroscopy (QEPAS), a modulated optical output of a light source (laser or LED) is passed through the trace gas in a collimator where the gas molecules produce characteristic acoustic waves (proportional to the concentration of the target gas), which impinge on the QTF (which acts as an acoustic wave transducer), which converts those waves into electrical signals, as shown in Figure 11.2. And in quartz-enhanced photothermal spectroscopy (QEPTS), the light, after passing through the trace gas molecules, is directed at the bare quartz of the QTF, which reflects, absorbs, and partially transmits the light. This causes a photothermal–elastic conversion inducing mechanical vibrations proportional to the light falling on the QTF, which is then converted into electrical output, as shown in Figure 11.3 [28].

In the case of the QTF itself being transformed into a direct sensor, which is the least explored of all these techniques (based on the number of publications), it is imperative that the QTF be modified with other materials or structures that would provide sensitivity and selectivity to the sensor as it is impossible for the bare QTF to show sensitivity or selectivity toward multiple analytes. Thus, various types of polymers and nanomaterials and nanostructures have been used to modify the QTFs to convert them into sensors, which have proven to be both sensitive and selective to various different types of analytes. In this case, the resonant frequency or amplitude of oscillation is one of the most important factors as any and all changes in the properties of the modified materials result in a change in the QTF–polymer system's resonant frequency, and, therefore, amplitude at the oscillating frequency would need to be constantly measured and logged.

The resonant frequency of the QTF sensor can either be increased or decreased depending on how the QTF is modified with the functionalizing material. There are two possible ways for modifying the QTF. The first is by modifying the surface of the QTF using surface coatings or strategic depositions of materials, which would react with the analyte, much like the surface modifications done on QCMs and cantilevers. In this case, there is an addition of mass, and the resonant frequency of the

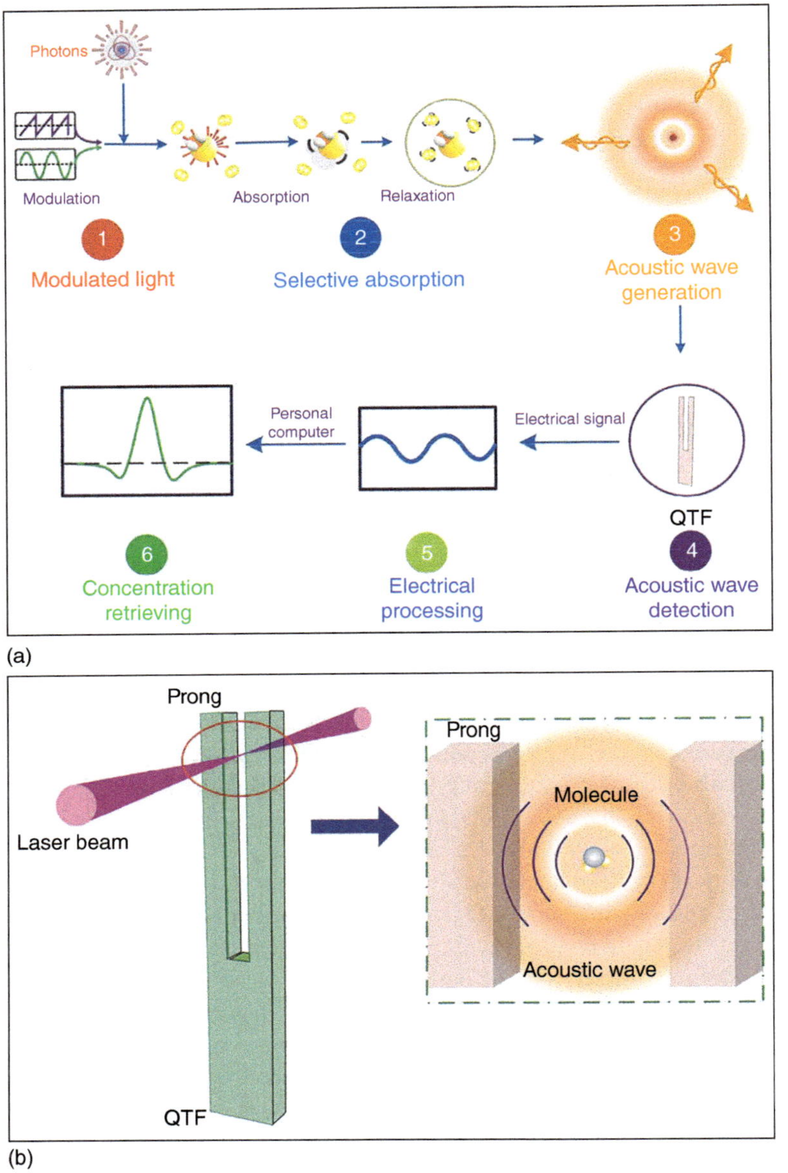

(b)

Figure 11.2 QTF used as an acoustic sensor in quartz-enhanced photo-acoustic spectroscopy (QEPAS). (a) Schematic diagram of QEPAS. (b) Schematic of interaction of acoustic wave and QTF. Source: Ma [28]/Frontiers Media/CC BY 4.0.

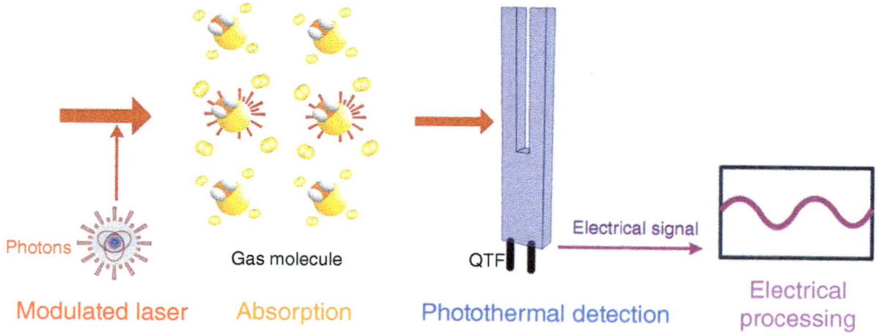

Figure 11.3 QTF used as a thermal sensor in quartz-enhanced photothermal spectroscopy (QEPTS). Source: Ma [28]/Frontiers Media/CC BY 4.0.

QTF system will decrease proportionally to the increase in mass. The other method is to bridge the two prongs of the QTF using polymer wires or films, which will then act as a sensing medium for the analyte. This is the mechanical equivalent of adding a spring between the two prongs of the QTF and offers the unique possibility of the resonant frequency either increasing or decreasing.

QTF–polymer systems can be modified with the polymer bridges in two ways:

1. The polymer is loosely held on top of two prongs and therefore does not exert any strain, which is equivalent to QTFs with coatings or depositions on them (Figure 11.4a). Thus, resonant frequency reduces after polymer modification due

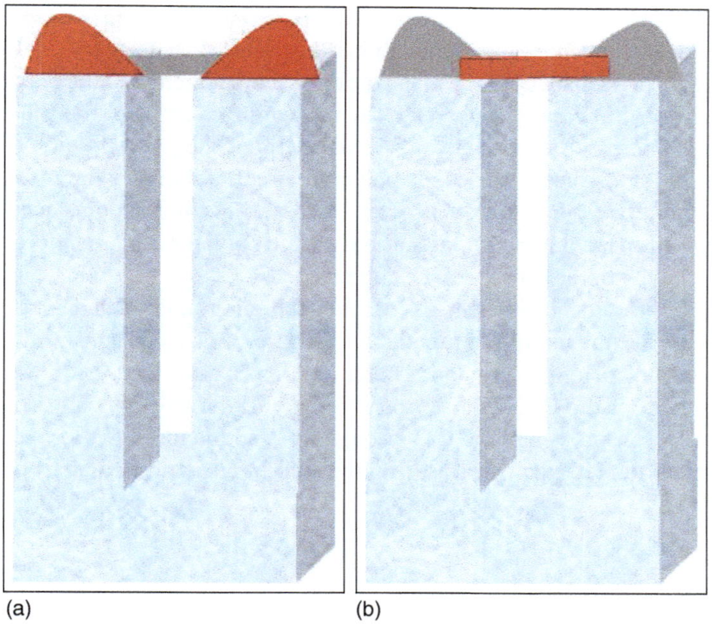

Figure 11.4 Illustration of QTF-modified polymer bridges. (a) Loose polymer bridge. (b) Tight polymer bridge.

to the small mass attached, and such QTF systems are therefore referred to as mass-loaded QTFs.

2. The polymer attaches tightly on two prongs, pulling them in when oscillated in the fundamental mode of operation (Figure 11.4b). Therefore, effective force constant of the QTF-polymer system increases and resonant frequency increases; such QTF systems are called spring-loaded QTFs.

   The equations governing the resonant frequency of the QTF sensors are as follows:

$$f = \frac{1}{2\pi}\sqrt{\frac{k}{m + \Delta m}} \tag{11.2}$$

$$f = \frac{1}{2\pi}\sqrt{\frac{k + \Delta k}{m + \Delta m}} \tag{11.3}$$

where Δk and Δm are the change in the spring constant and the change in mass, respectively, due to the addition of the polymer on the QTF. Eq. (1.2) corresponds to mass-loaded QTFs, whereas Eq. (1.3) corresponds to spring-loaded QTFs.

The resonant frequency of the tuning fork after being spring loaded with the polymer, shifts based on Young's modulus of the polymer wire, as given by the following equation:

$$E = \frac{2lk_{fork}}{Af_0}\Delta f_0 \tag{11.4}$$

where l is the length and A is the cross-sectional area of the polymer wire/film, k_{fork} is effective spring constant, f_0 is the resonance frequency of the bare fork, and Δf_0 is the shift in the resonance frequency due to polymer modification. Here, mass change of the QTF system due to polymer functionalization is neglected since it is very small compared to the mass of the bare QTF (by many orders). The interaction of the polymer with the analyte can increase or decrease the resonant frequency of the QTF based on whether it changes the mass or the force constant of the polymer wire. If there is a change in the force constant of the polymer wire, it dominates the mass change and, therefore, the resonant frequency will increase. This interaction can be tuned by choosing a proper combination of the analyte and polymer, and, therefore, sensitivity and selectivity of the sensor can be enhanced.

Since the broad scope of this book is the use of nanotechnology and related material properties in devices, the use of nonfunctionalized bare QTFs as sensors is outside the purview of the current discussion. Therefore, this chapter focuses on the historical development of the QTF as a sensor and the current research status of the third type of QTF sensors where the QTF prongs are functionalized with materials and the analytes are detected by measuring the shift in resonant frequency of QTF–polymer systems.

11.4 Early QTF Development

To appreciate the development of functionalized QTF–polymer systems as a chemical sensor, it is imperative that we go back a few decades and see how the QTF, which was designed to be a simple timing crystal for wristwatches, became a very useful scientific tool with multiple applications. The invention of the wristwatch, in the 1960s, in its current ubiquitous form is of particular significance. The use of quartz as timing crystals in clocks and watches was not new. They were used in various forms such as bars, balance wheels, and tuning forks, with resonant frequencies varying from hundreds of Hz to MHz, with different oscillation modes. But with electronic wristwatches, due to power and size limitations, a drastically different kind of quartz timing crystal was needed in order to meet the requirements for timing circuits: small in size, low power consumption, and resistant to shocks and vibrations, among other things.

The development of a bar-shaped quartz crystal by Max P. Forrer of Centre Electronique Horloger (CEH), Neuchâtel, Switzerland, for timing applications was an initial step toward this direction and precipitated the development of QTFs. He presented his findings entitled "A flexure-mode quartz for an electronic wrist-watch" at the 23rd Annual Symposium on Frequency Control in 1969. The quartz crystal bar was expressly designed for timing applications in electronic wristwatches, but was little bulky, power hungry, and less resistant to vibrations and shocks [29]. As the market for quartz-driven electronic wristwatches picked up, selling nearly 1.2 million units worldwide in 1969, the race for perfecting the crystal unit heated up companies in Switzerland, Japan, the United States, and France, spearheading the research on quartz timing crystals [30].

In 1972, at the 26th session of the same symposium, Hiroshi Yoda and his team from Nihon Dempa Kogyo Co. Ltd., Tokyo, presented the properties of QTFs, which have become the most commonly used commercial QTFs today. They demonstrated how the MU-type QTF with a resonant frequency of 32.768 kHz was best suited for use in commercial electronic wristwatches by establishing that their QTF form required very low power consumption, was sufficiently resistant to vibrations and shocks, was mostly insensitive to temperature, and had excellent frequency aging characteristics [30]. This analysis has stood the test of time, and their design remains largely unchanged to this day. As mentioned in the previous section, the resonant peak of 32.768 kHz was obtained in the fundamental mode of operation. It is worth mentioning that the QTF has several other modes of operation, each of which exhibits markedly different resonant frequencies over a wide range, with their own advantages and disadvantages [31, 32].

11.5 QTF as a Sensor

Though the commercially available QTF was first designed as a timing crystal for wristwatches in the early 1970s, it held, and still holds, much promise as a scientific tool. However, it would be more than a decade before the QTF's potential as a sensor was explored. In 1982, Dinger of Asulab SA, Switzerland, used the torsional mode of oscillation of the QTF as a temperature sensor for the first time. He used wristwatch packages of 1.5 mm diameter and 5.3 mm length, with a linear temperature coefficient of 35 ppm/°C and a resonant frequency of around 262 kHz. He concluded that the QTF as a temperature sensor offered better linearity and long-term stability but could be used only for medium and low precision applications as its temperature sensitivity depended on the cut angle and the dimensional ratios, leading to large variations in thermal properties [33].

Many scientists and engineers started focusing on various capabilities of the QTF, most of which involved the prongs being exposed to the atmosphere by removing the hermetic seal. This warranted a study of the properties of the QTF without the protective vacuum seal. In 1983, Michel Christen, a colleague of Dinger, reported a study of the damping effects of air and gas on the prongs of a QTF whose seal was broken, for using QTF as a simple pressure sensing element. Under normal atmospheric conditions, at room temperature, the resonant frequency decreased and the Q factor reduced by a factor of 10 [34].

Christen analyzed the vibration amplitude damping, change in Q factor, and variation in resonant frequency as a function of gas pressure and attributed these changes to radiation, mass addition, and viscosity. He discovered that for pressures greater than 100 mbar, up to a few bars when the Q value drops below a set threshold, there is a linear change in resonant frequency as pressure increases and drops by nearly 500 ppm bar^{-1} as shown in Figure 11.5. Therefore, he concluded, the QTF, from a practical vantage point, could be used as a simple pressure sensor [34].

In 1989, another scientific application of the QTF as in the field of force microscopy was uncovered by Gnther et al. They reported the use of 32.768 kHz tuning forks as a force sensor in the scanning near-field acoustic microscope. This was a two-part study where the first part of the study was to determine how the amplitude of oscillation and the resonant frequency changed when the QTF, oscillating perpendicular to the reference surface, was in proximity to an object, in this case a metal ball.

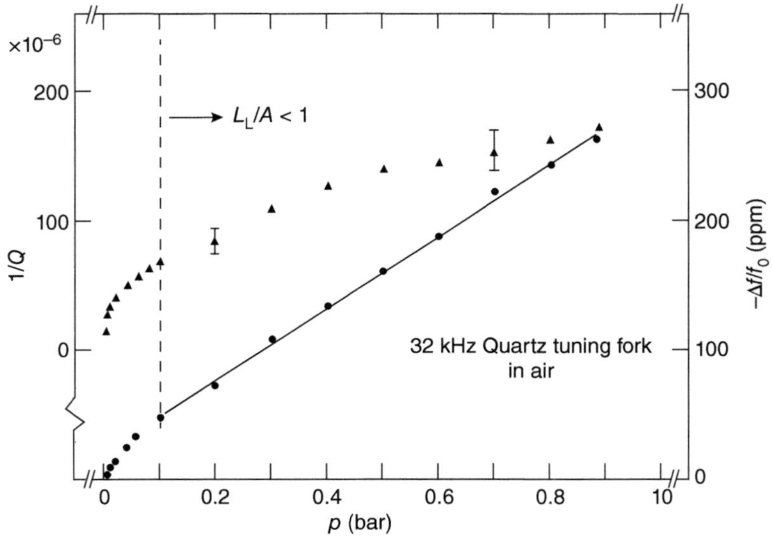

Figure 11.5 Typical damping and relative frequency shift of the QTF as a function of air pressure. Source: Christen [34]/With permission of Elsevier.

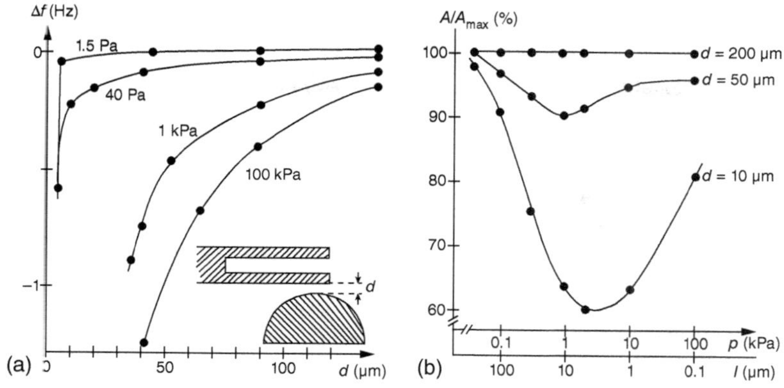

Figure 11.6 Change in resonant frequency and vibration amplitude when QTF prong end-face is oscillated parallel to the metal ball surface. (a) Resonant frequency as a function of distance *d*. (b) Vibration amplitude as a function of air pressure; *l* is mean free path. Source: Republished with permission of Springer Nature, from Günther et al. [35]. Permission conveyed through Copyright Clearance Center, Inc.

Their study showed that both amplitude and frequency started decreasing when the distance between the object and the end-face of the prong closest to the object was less than 200 μm, as shown in Figure 11.6. However, the quantum of decrease was also a function of gas pressure, very similar to what Christen had demonstrated a few years earlier. If the QTF was resonated parallel to the reference surface, frequency was observed over a range of 200 μm [35].

In the same work, a tip with a radius of curvature of only a few micrometers was attached to the QTF and was used for interaction with the reference surface instead of the QTF prong face; the range of interaction could be reduced to 200 μm and deviations in surface morphology as small as 5 nm could be detected. In this case also, there was a strong dependence of the QTF performance on gas pressure in the measurement chamber. In essence, the QTF was used as a resonator to oscillate the tip close to the sample surface whose resonance frequency was a direct function of surface micro-variations and was thus a very important tool in morphological studies done with a force microscope [35].

In 1995, Khaled Karra of Technische Universität München, Germany, and Robert D Grober of Yale picked up on the idea and reported studies using tuning forks to measure and control the distance between the optical fiber and the sample surface in a scanning near-field optical microscope. They glued an optical fiber with an aluminized tapered tip to one of the prongs. Here also, the QTF was used as a force sensor whose signals were used to detect the change in the distance between the QTF and the sample surface, which indirectly measured the sample surface deformations. The QTF was driven mechanically by an oscillator, and the electrical output was measured to keep track of the oscillations, which would decrease when the tip was too close to the sample surface. In this case, the QTF and the tip, which acted as a shear force pickup, were oscillated in parallel to the sample surface [36, 37].

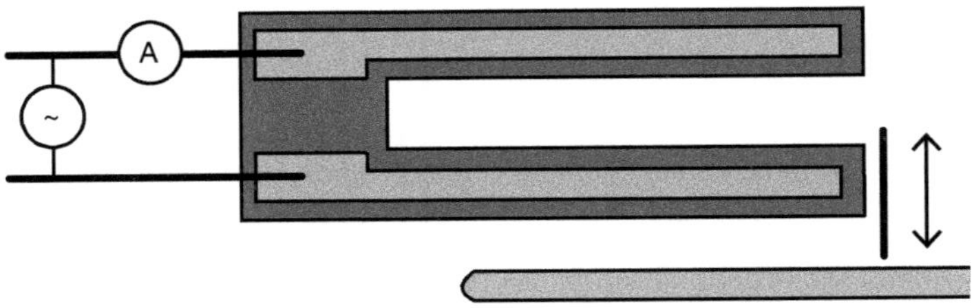

Figure 11.7 Configuration of the QTF, with a tip attached to one of the prongs, as a surface probe in the atomic force microscope. Source: Republished with permission of AIP Publishing, from Edwards et al. [39]. Permission conveyed through Copyright Clearance Center, Inc.

Another use case explored by the authors was the feedback for tip–surface distance control where the QTF was electronically oscillated at the resonant frequency, which was measured and tracked using lock-in amplifiers. In this case, the tip approached the sample surface in a perpendicular direction. As the tip–surface distance reduced, the signal amplitude reduced, based on which evasive steps were taken to prevent the tip from getting too close to the sample surface, while the optical measurements were made. The authors noted that there was a strong correlation between the measurements made optically and those made using the QTF as the force probe [36, 37]. Thus, the optical fiber attached to the QTF acted as a shear force pickup and a near-field optical probe, and the scanning speed was improved a year later in a study published by Walid A. Atia and Christopher C. Davis of University of Maryland [38].

This idea was borrowed by Allan Melmed of Johns Hopkins University and his research team at Texas Instruments. They replaced the cantilever in an atomic force microscope (AFM) with the QTF acting as both an actuator and a sensor, which picked up interactions between the tip and the sample in a study published in 1997. The tip was attached to one of the prongs and oscillated perpendicular to the sample plane as shown in Figure 11.7. This allowed faster modes of imaging using a phase-locked loop detector [39].

A year later, Franz J. Giessibl adopted and improved this technique by increasing the scan speed. This was achieved by fixing the tip etched out of a tungsten wire to one prong and fixing the other prong firmly to a supporting structure, thus effectively converting it into a cantilever as shown in Figure 11.8a. This setup decreased the resonant frequency to around 12–16 kHz, and the QTF was mechanically oscillated by another actuator as shown in Figure 11.8b. This allowed more control of QTF oscillations with lower damping, which allowed increasing the imaging speed and achieving atomic resolution [40, 41]. The success achieved by Giessibl led more researchers to use modified QTFs as force sensors in AFMs, which have become commercial alternatives to cantilevers [42].

11.5.1 Mass-loaded QTFs as Sensors

Due to the level of precision afforded by the tuning forks, in 2001 Sean J. O'Shea and his team at the Institute of Materials Research and Engineering, Singapore, first

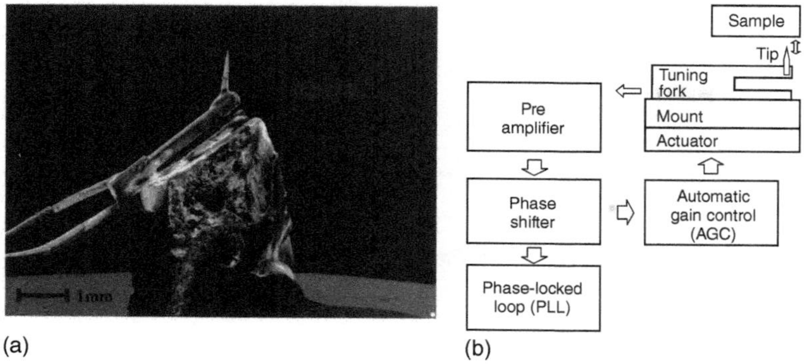

Figure 11.8 QTF as a cantilever in the AFM. Source: Republished with permission of AIP Publishing, from Giessibl [40]. Permission conveyed through Copyright Clearance Center, Inc. (a) Scanning electron microscope image of QTF force sensor. Source: [40] Giessibl et al. / with permission of AIP Publishing LLC. (b) Schematic of AFM tip drive configuration.

published a study [43] investigating the feasibility of QTFs as sensors for biological applications. Specifically, they studied the viability of surface-modified QTFs as mass-sensitive biosensors, which would be coated with specific biomolecules facilitating selective binding of complementary analytes. As documented in the previous studies mentioned earlier, change in oscillation amplitude and change in resonant frequency could be used to quantitatively determine the type and amount of bioanalytes present in the immediate environment of the sensor. They studied two experimental setups: self-excitation of the QTF with the help of oscillatory circuits (Colpitts, Pierce, etc.), of which the QTF would be a part, and mechanical excitation of the QTF with the help of an electrically driven piezoelectric plate, as shown in Figure 11.9.

The QTF was modified by surface coating the prongs with a thin layer of polystyrene (PS) where the fork was dipped in a solution of PS and aniline and the aniline was allowed to evaporate off the fork. For the detection of human IgG, the

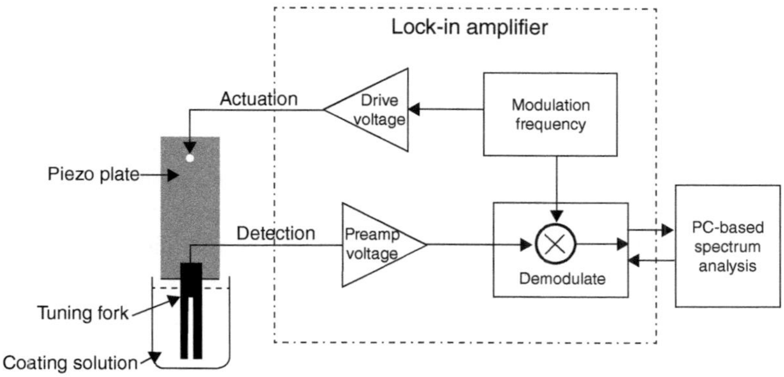

Figure 11.9 Schematic of a mechanically excited QTF biosensor. Source: Su et al. [43] / With permission of Elsevier.

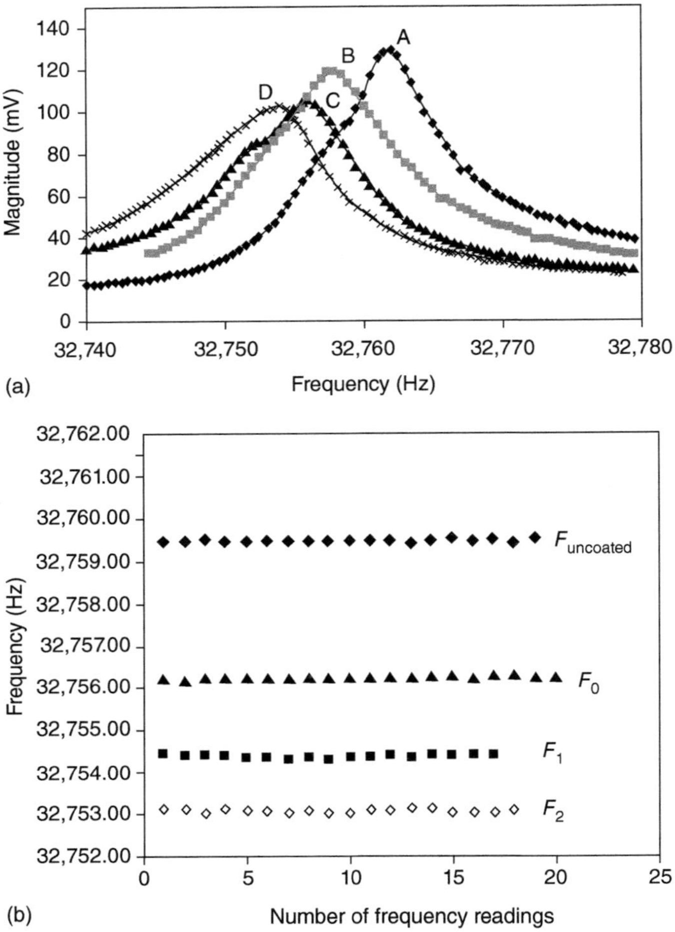

Figure 11.10 (a) Frequency spectrum of (A) bare fork, (B) PS-modified, (C) anti-h IgG immobilized, and (D) h IgG interaction and (b) frequency shift of bare fork, PS-modified (F_0), anti-h IgG immobilized (F_1), and h IgG interaction (F_2). Number of readings refer to the number of times the experiment was repeated. Anti-h IgG immobilization at 0.25 mg ml^{-1} and h-IgG interaction at 0.36 mg ml^{-1}. Source: Su et al. [43] / With permission of Elsevier.

PS-coated QTFs were again dipped in anti-human IgG and allowed to dry. For self-excited QTFs, the prongs were coated with varying concentrations of PS and their responses measured. The amplitude of the mechanically driven QTF, which was a direct function of the amount of mass loaded on the modified QTF, was measured as shown in Figure 11.10a. The shift in resonant frequency of self-excited QTFs, which was also a function of the mass loading, is shown in Figure 11.10b. For human IgG detection, the amount of mass being loaded onto the sensors was proportional to the IgG present in the immediate atmosphere of the QTF, as was apparent from the results. Also, this was verified by the fact that PS-coated QTFs without anti-h IgG showed no changes in amplitude or frequency in the presence of h-IgG.

O'Shea and his team concluded that though both mechanically driven and self-excited QTFs were suitable as mass loading biosensors, inherent issues in the former setup made it a less suitable candidate. Since oscillation amplitude is dependent on the Q factor, and the experimental setup process (especially binding the QTF to the piezo plate) and other external factors are difficult to control and vary with each QTF sensor, there are issues of reproducibility, which limit practical application. The self-excited QTF sensor was found to have a more reproducible frequency shift of 2 ppm with a detection limit of nearly 5 $\mu g\ ml^{-1}$. However, they did flag a few disadvantages of the QTF sensors, which are yet not completely resolved till now, such as long-term frequency drift of the sensors, temperature influence, humidity interference, and the difficulty using the sensors in a liquid medium [43].

Two years later, they used a very similar setup to study how QTF sensors could be used to determine liquid properties, especially density of Newtonian liquids, by immersing the sensor in the liquids. They also extended their study to investigate how mass loading of thin solid films affects the resonant frequency. They concluded that for mass loading of thin solid films, the self-excited mode of operation was the most ideal, while for liquid property measurements the mechanically driven mode of operation was the best. This is because the disturbances in the liquid interfere with the feedback mechanism of the oscillator circuitry. They also studied the theoretical and experimental variation in resonant frequency of the QTF sensor when it was placed in various liquids including acetone, benzene, dichloromethane, toluene, and ethanol. It was concluded that liquid density is the major property affecting the resonant frequency, but with increasing density, other factors such as viscosity and coupling between tines dominate the effect of the liquid on the QTF [44].

11.5.2 Polymer Wire/Film Modified QTF Sensor

The following section discusses the progress of research on QTF-based sensors functionalized with polymer bridges.

11.5.2.1 QTFs Modified with Polymer Wire Bridges

In 2003, S. Boussaad and N. J. Tao published a landmark study entitled "Polymer wire chemical sensor using a microfabricated tuning fork" in *Nano Letters*. This was a pivotal point in the development of the QTF as a nanosensor, as it explored the possibility of using more than simple mass loading to identify analytes. They successfully bridged the two tines of a QTF by gluing on a nitrocellulose/toluene sulfonamide formaldehyde resin wire, formed by pulling the polymer gel dissolved in toluene, under an optical microscope with the help of a stage. The wire with a diameter of a few micrometers was etched off at a certain point using a focused ion beam (FIB), where the diameter was reduced to a few nanometers.

It was observed that the resonant frequency of the QTF–polymer wire system increased by around 2 kHz after modification with the wire. This was attributed to Young's modulus of the wire, which contributed to an increase in the spring constant as the wire acted as a spring trying to pull the two tines together, restricting the oscillation amplitude. At the same time, the change in the effective mass due to the addition of the polymer wire was negligible. The QTF sensor was placed in a small gas chamber and a stream of nitrogen-carrying ethanol vapors was passed through it. The QTF was externally driven close to its resonant frequency, and, as was expected, the oscillation amplitude increased in the presence of ethanol vapors. The quantum of increase was higher for greater concentrations of ethanol vapors in the carrier gas, but the reaction times were slower. Also, when acetone vapors were introduced into the chamber, there was a faster and sharper change in the oscillation amplitude.

This could be attributed to the fact that the wire is soluble in both ethanol and acetone, being more soluble in acetone, and as the analytes are adsorbed onto the wire, vapors dissolve the wire. The authors concluded that the change in mass is negligible and is not significant enough to cause a decrease in the resonant frequency as was observed, and, therefore, the change in the spring constant of the polymer wire is dependent on the interaction between the wire and the analyte. This was also confirmed by the fact that the broken wire did not show any response to the vapors. Also, the sensitivity could be increased by using smaller wires, as demonstrated by the fact that the QTF with FIB cut wire showed better and faster response. Reversibility studies were also done to show that the process was a reversible physical phenomenon that was causing changes in the resonant frequency and oscillation amplitude of the QTF sensors and were thus reusable [45].

The authors also filed a US patent for "Chemical and Biological Sensing using Tuning Forks" in October of 2004, which was granted in July of 2012. During this

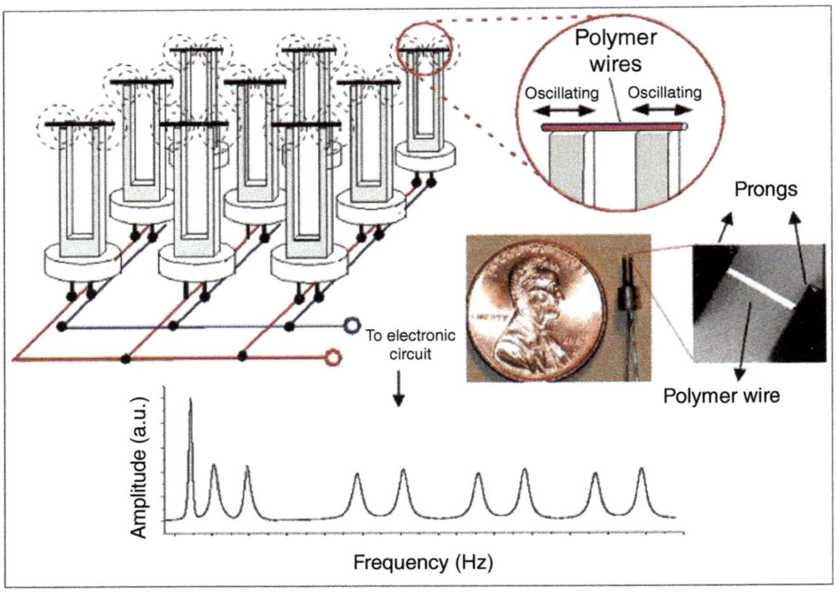

Figure 11.11 Array of QTF sensors functionalized with polymer wires. Source: Republished with permission of American Chemical Society, from Ren et al. [46]. Permission conveyed through Copyright Clearance Center, Inc.

period, a team of researchers at ASU, led by Tao, continued to work on QTF sensors and published nine more papers exploring the use of QTFs functionalized with polymer wires as gas and chemical sensors, where most of the analytes sensed were VOCs. In 2005, they published a study in Analytical Chemistry in which the QTFs were functionalized with wires and coatings of ethyl cellulose (T-10, N-22 and N-7), butylcellulose, poly(vinylphosphonic acid) (PVPA), poly(acrylic acid) (PAA), transparent nail polish, and Super Glue (Duro). An array of such QTF–polymer systems, shown in Figure 11.11, was used to detect the presence of, and quantify the concentration of relative humidity, *p*-nitroethyltoluene vapors, ethanol vapors and analyze beer vapors [46].

The same group also showed how polymer-modified QTFs could be used as temperature and infrared (IR) sensors. They used QTFs functionalized with PS, hyper-branched fluoroalcohol polycarbosilanes (SC-F105) mixture (90–10% by mass), poly(*N*-isopropylacrylamide) (*N*-IPAA), polyvinylchloride (PVC). Of these, N-IPAA is the most sensitive to temperature at 10 Hz °C^{-1}. The experiments were done by using a 635 nm laser, a pulsed resistive heater, and a soldering iron as the sources of heat and IR. They reported that they were able to achieve a noise equivalent temperature difference (NETD) of 0.5 mK at 20 °C, a thermal limited noise level of 5 μ °C, and a time constant of about 50 ms. Some of the results are shown in Figure 11.12 [47].

Similar to other chemical sensors, there were some issues that needed to be ironed out to improve the performance of polymer wire–modified QTF sensors. Issues such

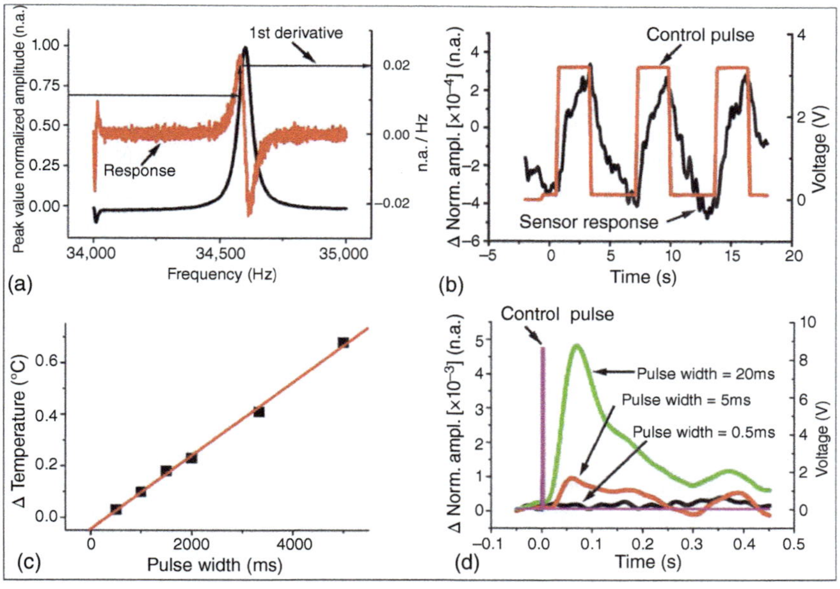

Figure 11.12 (a) QTF with N-IPAA wire spectral graph (black) with the response (red) and the first derivative, (b) PS + SC-F105 wire sensor response to laser (black) and laser pulse (red), (c) relationship between pulse width and temperature control, and (d) sensor response to different pulse widths. Source: Republished with permission of AIP Publishing, from Tsow and Tao [47]. Permission conveyed through Copyright Clearance Center, Inc.

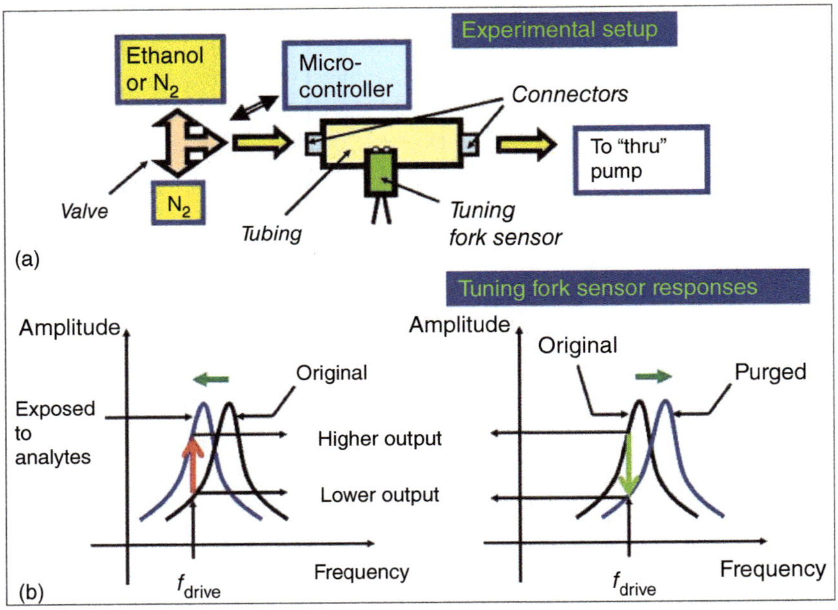

Figure 11.13 (a) Schematic block diagram of frequency-encoded sensing setup. (b) Illustrations of QTF sensor responses. Source: Republished with permission of American Chemical Society, from Tsow et al. [48]. Permission conveyed through Copyright Clearance Center, Inc.

as the dynamic range, saturation, and response time needed to be properly optimized so that significant savings could be attained in terms of response and recovery times and a level of protection could be provided to the sensors while using sensitive sensors built for low-concentration detection for high-concentration detection as well. These issues were explored in a publication by Tao and his team in 2008 in which they demonstrated that a pulsed or sampled exposure of the sensor to the analyte and purging gas cyclically, done automatically with the help of microcontrollers and preset threshold values, would go a long way in mitigating most of these issues. A schematic illustration of the method is shown in Figure 11.13. They also concluded, in the cases of high-concentration exposure, the frequencies of such cycles could itself be used as an indirect method of analyte concentration detection with potential noise reduction [48].

In one of their works published in 2009, they explored how certain sulfur derivatives, namely, dimethyl sulfide (DMS), ethanethiol, and methylsulfide could be selectively detected by QTF sensors in the presence of various kinds of interferents, as shown in Figure 11.14. They devised a sensor array setup that could effectively block off the interferents while allowing the target analytes to interact with the QTF sensor array. The authors concluded that a system-level approach when designing sensor systems goes a long way in enhancing the sensitivity, selectivity, and life of any sensor, and, in this case, particularly QTF–polymer wire sensors [49].

QTFs modified with polybutadiene and silica nanoparticles modified with butadiene silane and SP-IL100 have also been shown to selectively sense ozone in real time.

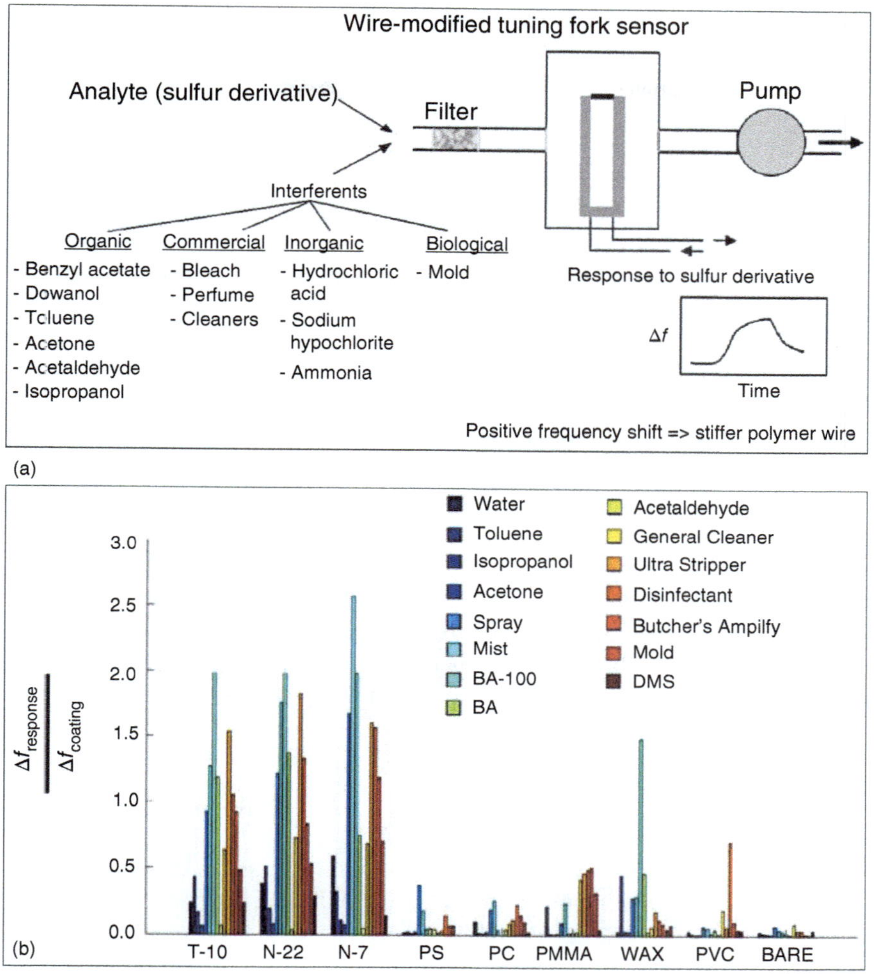

Figure 11.14 Selective sensing of sulfur derivatives using QTF sensors. (a) Schematic of a setup with possible interferents and expected frequency response. (b) Response of various QTF–polymer systems to different analytes. Source: Rai et al. [49] / With permission of Elsevier.

They used a sensor array setup with 10 QTF sensors, including redundancies, which would detect and measure ozone, relative humidity, temperature drifts, and flow rate in the sensing chamber.

QTFs modified with polyacrylamide (PAM) were used as humidity sensors, while sealed and bare QTFs were used for temperature and flow rate measurements. The authors concluded that polybutadiene sensors showed the best response and were able to detect ppb levels of ozone in ambient air, with a mass detection limit of almost 2 pg mm^{-2} and a response time of one second [50].

One of the main advantages of QTF sensors is that they can be easily integrated into peripheral electronics, which makes them accessible to newer electronic features. Therefore, they can be used in real-time applications. Work done by Tao et al. to monitor certain toxic environmental VOCs in real time is one such example. They made two major changes in the sensor system compared to the previous works. The first was the use of molecularly imprinted polymers (MIPs) for QTF modification rather than the conventional coatings or wires. MIPs are essentially custom-built receptors for the target analytes and show extremely high selectivity. In this work, the MIP was made of cross-linked PS with biphenyl as the template and xylenes as progens. The second major change was the development of a new integrated sensing platform composed of two components: the wearable sensing device and a cell phone for data logging. The sensing device housed a sensor array, various filters, valves, pump, and associated electronics with wireless data transmission capabilities, as shown in Figure 11.15. The authors demonstrated that this device had real-world applications by measuring VOC concentrations outdoor and at gas stations. They also successfully detected and discriminated benzene from among the benzene, toluene, ethylbenzene, and xylenes (BTEX) complex [51].

Tao's team published another paper detailing development of a handheld hybrid system that could detect BTEX from complex samples in real time. This work again relied on QTFs modified with MIPs. The sensor system was developed as a two-part system: a separation column system, consisting of two bonded stainless steel capillary columns, feeding into the sensing system composed of the QTF sensor array housed inside the gas chamber and driven by a wireless high-frequency resolution detection circuit, as depicted in Figure 11.15. The authors concluded that

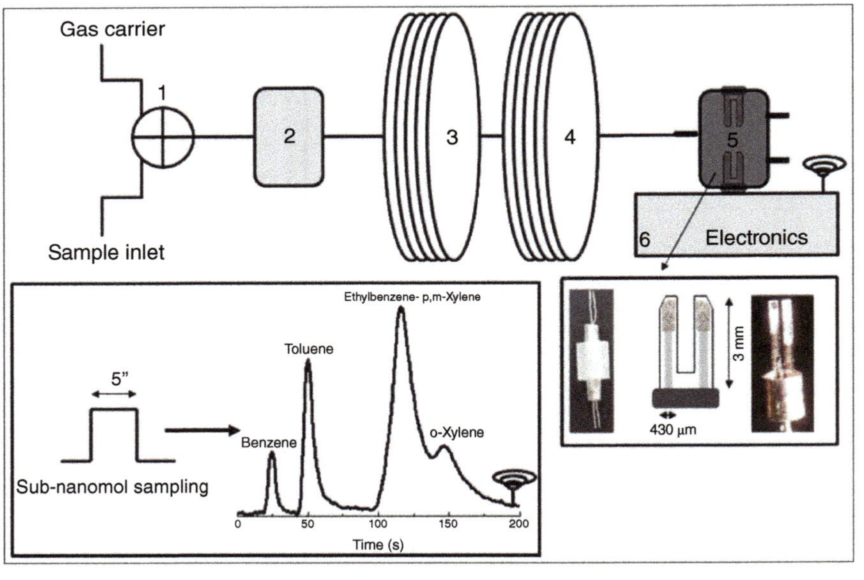

Figure 11.15 Schematic diagram of BTEX sensing system with (1) valve, (2) pump, (3 and 4) bonded stainless steel capillary columns, (5) sensing chamber with an array of QTFs, and (6) supporting electronics with wireless capabilities. The typical chromatogram of BTEX separation is also shown (bottom-left inset). Source: American Chemical Society, from Iglesias [52].

such a hybrid system improved the performance of the sensor without the use of programmed heating elements, preconcentration, or other processes and had potential applications in occupational health and toxicological population studies [52].

The devices and methods perfected in these studies were employed by the team to refine the sensing process and use it for real-world applications. In a study published in 2010, they used the sensor system to detect low ppb level concentrations of VOCs and acids in four different scenarios while being insensitive to environmental factors such as temperature, humidity, or personal care products. The four case studies done were as follows: (i) hazardous waste exposure at the ASU waste management facility, (ii) first- and second-hand cigarette smoke exposure, (iii) exposure of cleaning workers engaged in floor waxing, and (iv) exposure assessment of firefighters and inspectors who were at the site after a fire was extinguished. The environment at these scenarios were also collected in sample bags and tested using conventional methods or other sensing mechanisms. The authors reported that the sensor performed at par with or sometimes superior to the other sensing methods in terms of sensitivity and selectivity [53].

A similar study was published in 2012 reporting the detection of personal exposure to VOCs indoor during paint remodeling, outdoor traffic exposure in parks and by highways, and on shrimp boats traversing the bays near the Deepwater Horizon oil spill of 2010. The same device as described in the previous study was used. In this work also, the results of the sensor data were cross verified to ensure accuracy with the help of gas samples collected as the sources using Tedlar sampling bags. The results were validated using three methods: validation using reference methods such as gas chromatography–mass spectrometry (GC-MS) and selected ion flow tube–mass spectrometry (SIFT-MS), inter-laboratory validation, and validation with other commercial VOC sensors. The authors reported accuracy higher than 81% without false positives from interferents, having an additional benefit of being wearable and offering real-time results in a wide detection range of 4 ppb to 1000 ppm [54]. A very similar research study was published in 2013 in the *IEEE Sensors* journal [55].

A study on the aging properties of the MIP-functionalized QTF sensors was undertaken and reported by Yue Deng et al. at ASU in 2015. The chemical and physical properties of the MIPs were studied with the help of scanning electron microscopy (SEM), Brunauer–Emmett–Teller adsorption isotherm analysis (BET), and Fourier-transform infrared spectroscopy (FTIR), in addition to kinetic binding studies. It was found that the intrinsic adsorption rates of hydrocarbons onto MIP binding sites reduced with time even though there was very little change in the chemical and physical properties of the MIP. The authors concluded that the MIP aging could be slowed down by storing the sensors at lower temperatures [56]. A similar study providing a detailed adsorption thermodynamic analysis was published in 2017 by the same group [57].

The reduction in intrinsic adsorption rates of hydrocarbons on to MIPs over time. Therefore, the same group also provided a possible solution in a paper published in 2016 wherein the sensors can be calibrated on the fly, considering correction factors such as sensitivity, aging, and temperature effects. They proposed that this can be

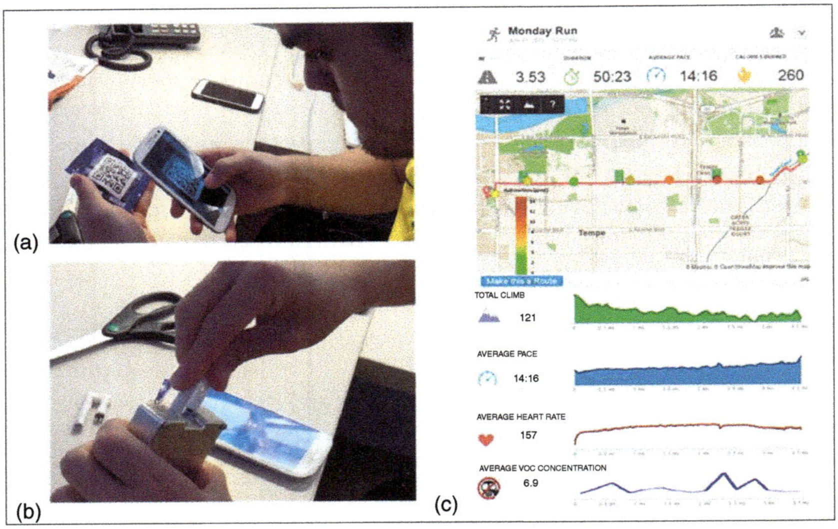

Figure 11.16 (a) Calibration using QR codes, (b) sensor cartridge replacement, and (c) real-time data during personal exercise. Source: [59] Deng et al. MDPI / CC BY 4.0.

achieved by using quick response (QR) codes so that the QTF sensors in the wearable monitoring devices can be used under free living conditions [58]. An extended work was published the same year detailing improvements in the device design, which include miniaturization, better wireless integration, and disposable sensor arrays. They demonstrated how the QR codes with pre-calibrated sensor information can be used to easily change out these plug-and-play sensor cartridges ensuring reusability of the device with no change in sensitivity or selectivity as shown in Figure 11.16 [59].

In 2018, a QTF-based particulate matter ($PM_{2.5}$) sensor was reported by Tao's team, which was to be used as a personal exposure monitor. In this work, they dip coated a QTF with polyvinylpyrrolidone (PVP)-based Elmer's repositionable glue. The resonant frequency of the QTF dropped with increased exposure to $PM_{2.5}$ as this resulted in an increase in the effective mass of the sensor. The $PM_{2.5}$ sensor had a detection range of 2–2000 $\mu g\, m^{-3}$ with a linear response in the typical urban ambient concentrations of 6–100 $\mu g\, m^{-3}$ at a flow rate of $1.7\, l\, min^{-1}$. The QTF sensors were placed in a setup similar to the one described above, with the addition of a special nozzle arrangement to ensure maximum adhesion of the $PM_{2.5}$ particles on the QTF [60].

11.5.2.2 QTFs Modified with Polymer Film Bridges

Meanwhile, around 2012, a research team under Sangmin Jeon, at POSTECH, South Korea, was also using QTFs as a scientific tool to investigate the evaporation dynamics of sessile liquids from solid surfaces. In a study published in 2013, they used both the thickness shear mode and the flexure mode of the QTF to study the evaporation of deionized water and suspensions of PS nanoparticles. This was done by tracking the changes in the resonance frequency and the dissipation factor. The authors concluded that the mass measurements in different modes vary and that while shear mode is better suited to track changes in the physical properties of the nanoparticle suspension, the flexural mode is best suited for determining the nanoparticle concentration [61].

Tao's success with the QTFs bridged with polymer wires inspired them to work on polymer-functionalized QTF sensors. In another study published in 2013, they investigated the feasibility of QTFs functionalized with polymer nanofilms as vapor sensors as shown in Figure 11.17. They used QTFs functionalized with PS membranes

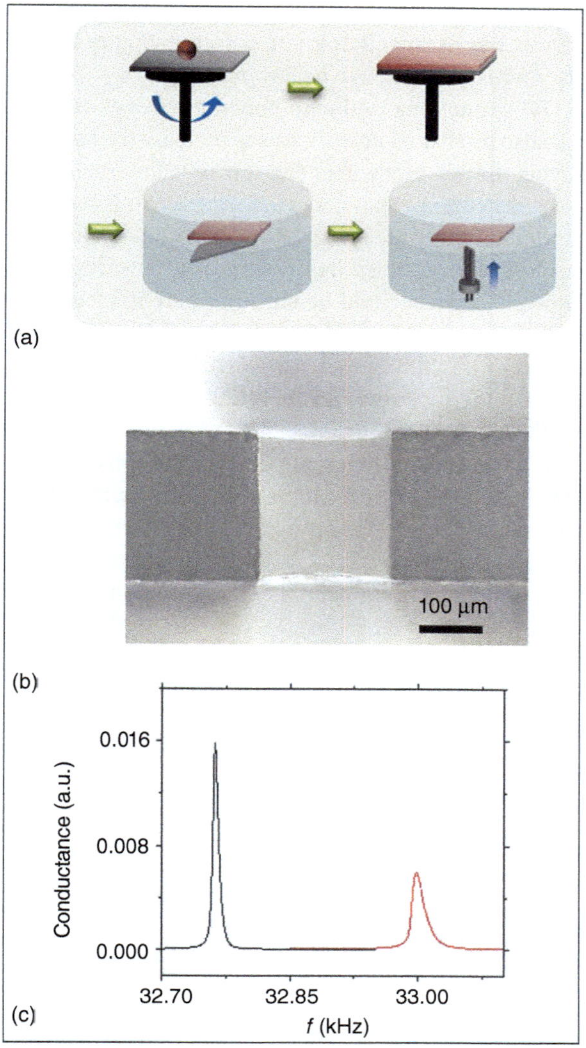

Figure 11.17 (a) Schematic of the functionalization, (b) optical image of the QTF sensor. Source: [25] Yun et al. / with permission of AIP Publishing LLC. (c) Resonance peaks before (black) and after (red) functionalization.

as ethanol sensors and found that there was a reduction in response time by as much as 6.5 times compared to similar QCM sensors. This was attributed to the fact that the analyte now had access to the film from both the top and bottom. They, therefore, concluded that such an approach holds much promise in investigating solvent-induced changes in mechanical properties of polymer nanofilms [25].

The team further studied about PS nanofilms in 2014. They integrated graphene into the PS film by spin-casting PS solution onto a graphene layer synthesized using chemical vapor deposition techniques, thereby creating a bilayer. The authors reported that Young's modulus was twofold due to the presence of graphene, and the frequency response of the graphene–PS bilayer was three times of that observed with only PS films, while there was less degradation in the Q factor (Figure 11.18). The detection limit for the sensor was determined to be 20 ppm [62].

The authors also worked on QTFs functionalized with polymer wires as sensors. The main focus of another work, also published in 2014, was to increase the sensitivity of the QTF sensors by modifying the polymer wires. They proposed two solutions to this. The first was to use photo-cross-linking polymers to form the wires between the two prongs. They demonstrated that by modifying the QTFs with polyvinylcinnamate (PVCN) wires, and then cross-linking the wires by exposing the wires to UV radiation for one to three minutes, the wires could be shrunk, thereby increasing their stiffness. A comparison of the frequency responses of QTFs before and after cross-linking the wires shows that cross-linking increased the sensitivity and the Q factor of the sensor by amplifying the modulus change in the wire due to either the presence or absence of ethanol vapors [63].

In another study, published in 2016, the authors reported that the sensitivity of polymer wire sensors could be enhanced by using multiple electrospun wires. They electrospun wires of polymethyl methacrylate (PMMA) onto the QTF, thus forming polymer bridges. Then the sensors were exposed to water vapor to use them as humidity sensors. The resonance frequency of the sensor reduced with an increase in the humidity level due to adsorption of water molecules onto the PMMA wires, leading to a decrease in the modulus of the polymer bridges. The amount of decrease,

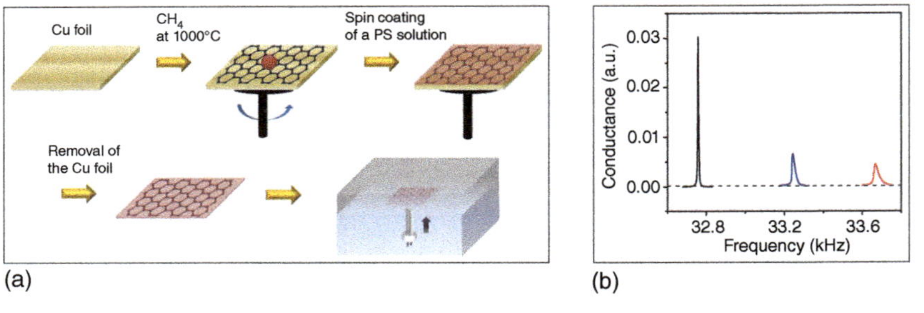

Figure 11.18 QTFs functionalized with bilayer of graphene and PS. (a) Illustration of modified QTF with graphene–PS bilayer membrane. (b) Frequency response of bare QTF (black), QTF with single PS layer (blue), and QTF with the bilayer (red). Source: Republished with permission of AIP Publishing, from Yen et al. [62]. Permission conveyed through Copyright Clearance Center, Inc.

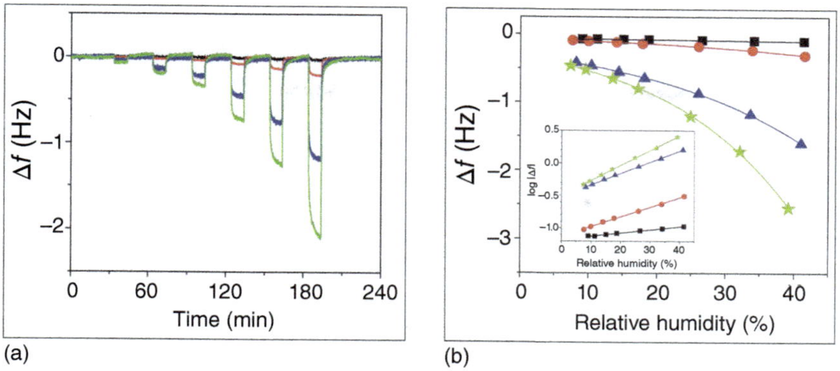

Figure 11.19 Frequency responses of bare QTF (black), QTF with 1 PS wire (red), QTF with 3 PS wires, and QTF with 6 PS wires (green). (a) Frequency variations with time for 8, 11, 15, 18, 26, 33 and 40% water vapor. (b) Frequency variations as a function of relative humidity. Source: Kim et al. [64]/Royal Society of Chemistry.

for a given humidity level, was found to be a function of the number of wires present as shown in Figure 11.19 [64].

An extension of this work was published in 2017, where they modified the electrospinning conditions (higher humidity) to produce nanoporous PMMA wires (Figure 11.20), which were then functionalized onto the QTFs. The sensitivity of such sensors to ethanol vapors was then studied. These sensors showed faster

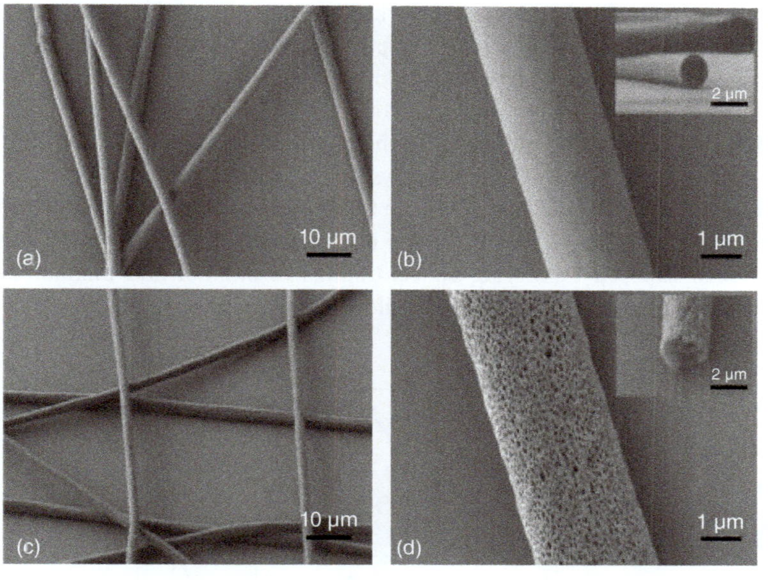

Figure 11.20 SEM images of (a) plain PMMA wires and (b) its magnified image. (c) SEM images of nanoporous PMMA wires and (d) its magnified image. Insets show cross-sectional SEM images of wires. Source: [65] Huang et al. / with permission of American Chemical Society.

response times and increased sensitivity, which were attributed to the higher surface area and better infiltration of ethanol molecules into the wires due to the pore networks. However, when exposed to humidity as in the previous study, there was no significant change, which, the authors concluded, could be due to the fact that ethanol swells PMMA almost 8 times more compared to water molecules [65]. An extension of this work was published in 2020 where they used PMMA wires, which were reinforced with cellulose nanocrystals (CNCs). They also studied how increasing the humidity during the formation of these wires resulted in sensors with increased sensitivity as seen earlier [66].

11.5.2.3 Improving Selectivity of Polymer-modified QTF Sensors and Classification of VOCs

The work done by our group started with an investigation of how sensitivity and selectivity could be improved using QTF sensors functionalized with polyvinyl alcohol (PVA) wires. Different sensors were made by introducing additives in the form of gold (Au) nanoparticles, which were capped with either polyvinylpyrrolidone (PVP) or do-decanethiol (DDT), graphene nanoribbon (GNR) suspensions, and polyaniline (PANI) suspensions. These additives were added to PVA and were drawn with the help of a copper wire as shown in Figure 11.21. The sensors were exposed individually to VOCs such as acetone, chloroform, ethanol, methanol [67].

The studies showed that spring-loaded QTFs could show an increase in sensitivity of around 400% to 3000% compared to mass-loaded sensors because modulus changes of the wire overshadow the mass effects significantly. This work is summarized in Figure 11.22. It was shown in this work that a mass change of 1 ng could correspond to a change of almost 80 N m^{-1} in spring constant. This was supported by a field emission microscopy (FEM) analysis done with the help of COMSOL Multiphysics [67].

An extension of the work was published in the same year in which the authors studied the feasibility of the QTF–polymer system as a sensor that can detect methanol adulteration of ethanol. Two separate QTF sensors were made: one with PVA–PANI composite wires and the other with PVP–aniline composite wires. The sensors were then exposed to varying concentrations of ethanol and methanol vapors cyclically (60 seconds of exposure, followed by 60 seconds of recovery,

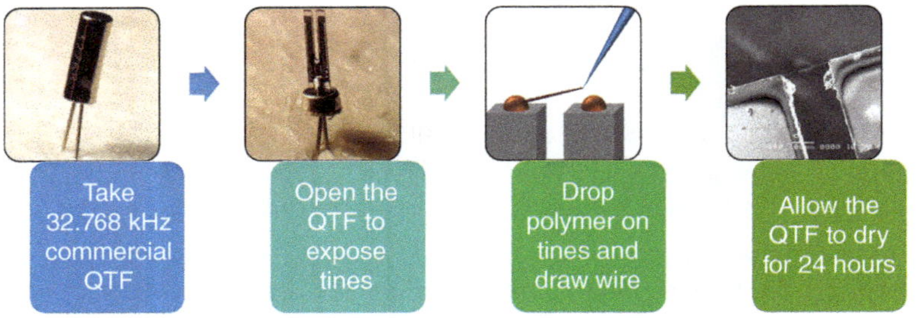

Figure 11.21 Flow diagram of QTF functionalization process.

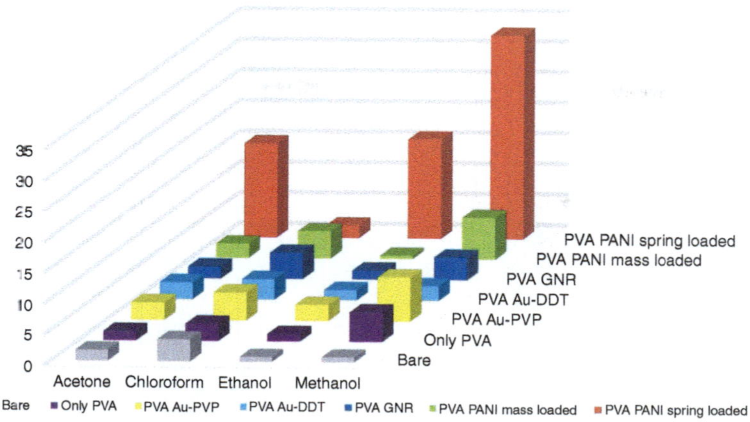

Figure 11.22 Summary of sensor responses of QTFs with varying functionalizations to different analytes.

followed by another 60 seconds of stabilization). With each cycle of analyte exposure, the percentage of ethanol vapors decreased and the percentage of methanol vapors increased, and in the first cycle pure ethanol was injected, while in the last cycle, pure methanol was injected [68].

The experiment was performed for both 1 and 6 ml of VOC vapors. However, spring-loaded PVP–aniline sensors were so sensitive to the vapors – injection of 6 ml of vapors caused a drastic and sudden change in the resonant frequency that the sensor response was not quantifiable. The authors confirmed here again, for both PVA–PANI and PVP–aniline sensors, the responses of spring-loaded QTFs are greater than those of mass-loaded QTFs, as shown in Figure 11.23. The work concluded that QTFs that are spring-loaded with PVP–aniline wires are selective toward methanol vapors in the presence of ethanol. Therefore, the percentage of adulteration of ethanol with methanol could be determined by measuring the amount of shift in resonant frequency of the sensor [68].

The team published a work in 2017, in which they demonstrated a prototype that could detect ethanol adulteration with up to 5% of methanol, with 100% confidence. To achieve this, they started by investigating the suitability of QTFs functionalized with PS wires as methanol vapor sensors in the presence of ethanol vapors. QTF sensors with varying weight percentages of PS in aniline (5, 10, 15, 20, 25, and 30%) were created as described earlier. These sensors were exposed to vapors from the head-spaces of six ethanol solutions of varying adulteration percentages, from 0 (pure ethanol) to 100% (pure methanol). The frequency responses of these sensors were measured, and it was observed that the 10% PS wires showed the best response and the rest of the study focused on this.

The same cyclical exposure to vapors, as done for the previous work, was done. When the sensor was exposed to pure ethanol vapors, the resonant frequency was found to increase. However, with the increasing percentage of methanol in the solution, not only the amount of increase in frequency decreased, there was

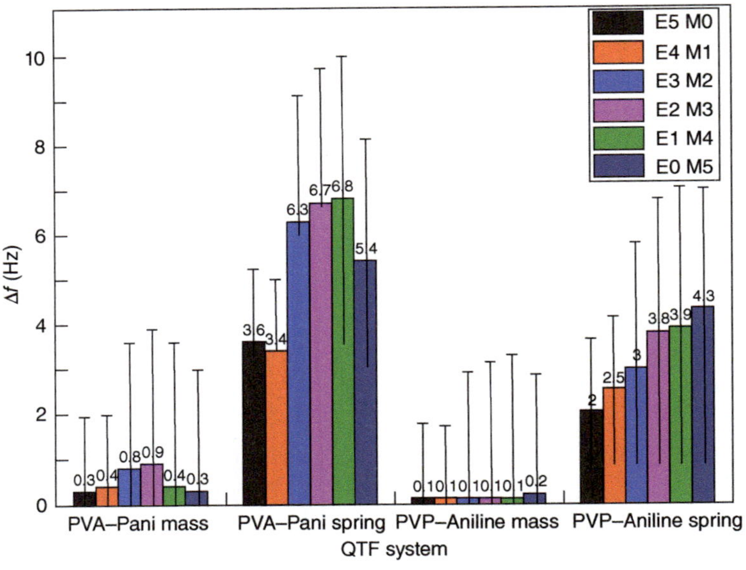

Figure 11.23 Summary of sensor responses of QTF sensors to varying concentrations of ethanol and methanol. Source: Sampson et al. [68]/IEEE.

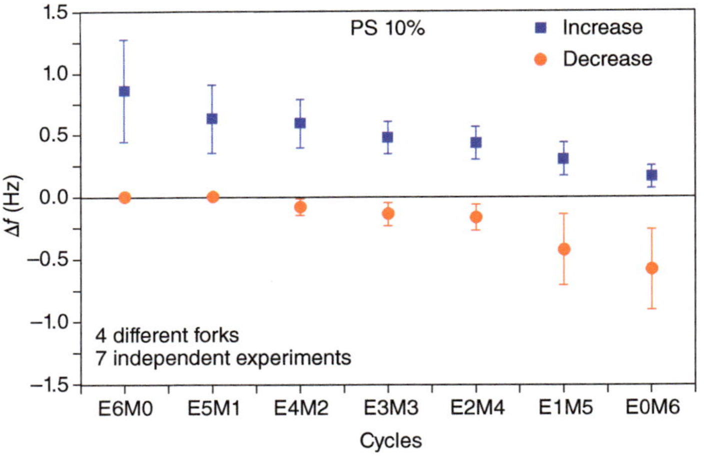

Figure 11.24 Number of shifts in resonant frequency of the QTF sensors on exposure to varying concentrations of ethanol and methanol vapors over 7 cycles per experiment.

also a net decrease in frequency during the exposure cycle. As the concentration of methanol in the solution increased, the amount of decrease in frequency also increased. Multiple experiments were performed to establish the veracity of the results shown in Figure 11.24. An interference study was also done to make sure that the forks were not affected by other common adulterants of ethanol such as pyridine and isopropanol. A handheld prototype was developed that could give

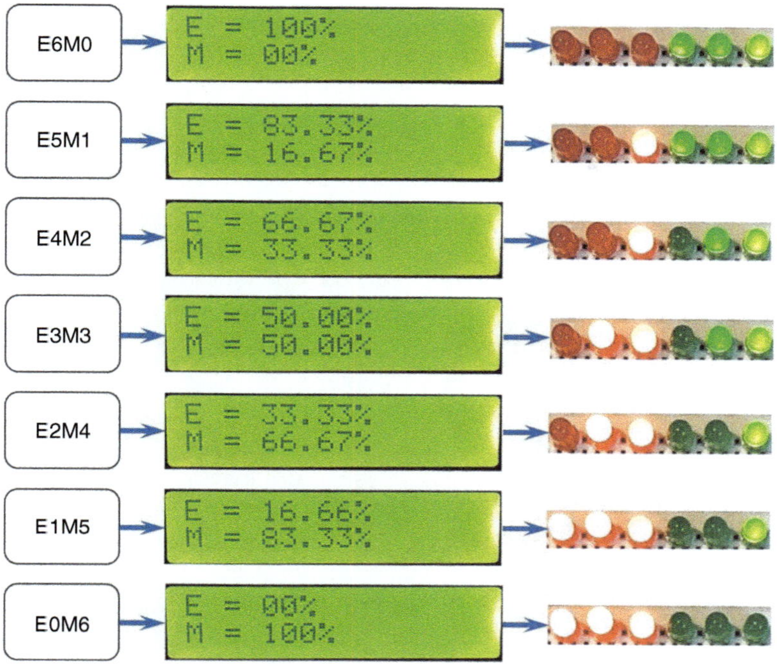

Figure 11.25 The QTF sensor prototype display for corresponding experiment cycles. Source: Republished with permission of Springer Nature, from Abraham Sampson et al. [69]. Permission conveyed through Copyright Clearance Center, Inc.

a readout of the approximate percentage of methanol present in the solution as shown in Figure 11.25 [69].

Drawing from the work on PS nanofilms done by Jeon at POSTECH, the team at DIAT did a preliminary study on another application of QTF sensors, namely, detection of breast cancer biomarkers. For this, two biomarkers, 1,4-dimethoxy-2, 3-butanediol (BD), and cyclohexanone (CH) were chosen, and three possible interferents, acetone, methanol, and ethanol, were identified. The QTFs were functionalized with PS and PMMA films of varying weight percentages, an array of which was created and exposed to these analytes. Here again, it was observed that QTFs with PMMA films showed an increase in resonant frequency when exposed to BD, while in all other cases, resonant frequency decreased. Since there were too many interferents, the authors used linear discriminant analysis (LDA) to classify the VOCs and identify them. The authors concluded that the LDA studies could predict VOCs with an accuracy of 97% in the test data, and the accuracy could be further enhanced by improving the sensor array by increasing individual sensor selectivity and by adding more number of sensors [70].

These changes were made in the further study when the sensor array of three sensors was made and exposed to these analytes simultaneously. Schematic of the setup is shown in Figure 11.26.

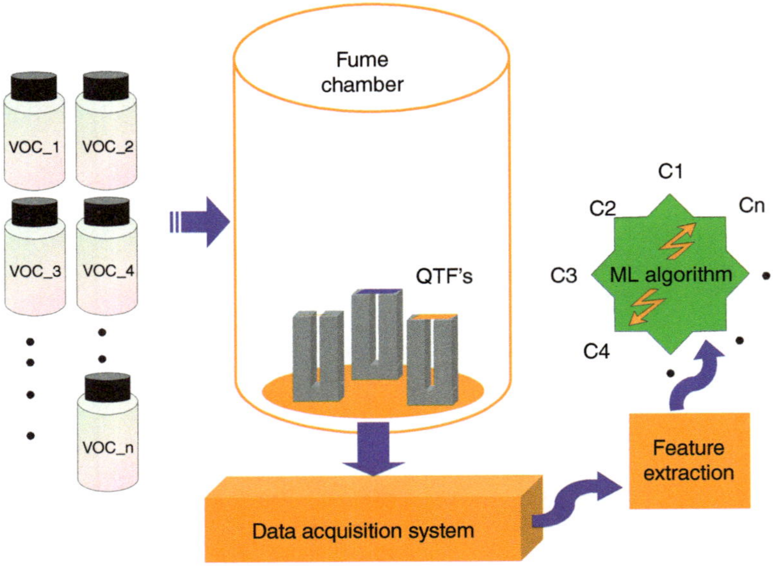

Figure 11.26 Schematic of the QTF sensor array-based VOC classifier.

Table 11.1 Analyte and their concentration used for experiments.

Analyte	Concentration (ppm)
(*R*,*R*)-(+)-1,4-Dimethoxy-2,3-butanediol (BD)	0.5, 1
Cyclohexanone (CH)	10, 20
Ethanol	10, 20
Methanol	10, 20
Acetone	10, 20

For classification of the VOCs, changes in the resonant frequency, response time, and recovery time of the sensor were recorded. Classification of 5 VOCs and measurements of their concentrations were done using various classifiers, and the best suited classifier was identified. Table 11.1 shows the analytes and their concentrations used for classification.

Figure 11.27 shows the response of sensor array toward various concentrations of 5 VOCs in terms of changes in frequency, response time, and recovery time in the form of a box plot. The figure shows that with increasing complexity of the sensor system, adding more number of sensors and detecting more VOCs simultaneously, machine learning tools need to be used for classification. The sensor data was divided into 80% training and 20% testing data. After careful analysis of each classifier, it

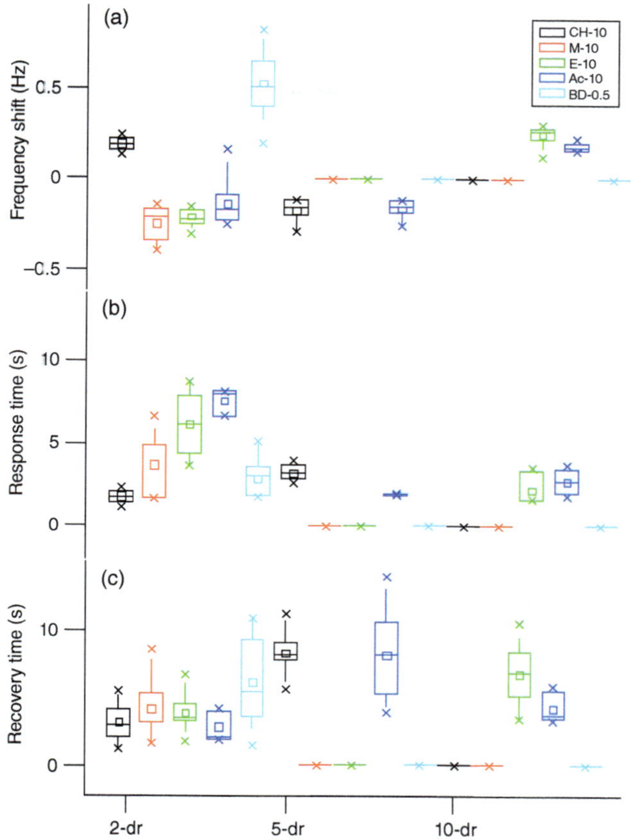

Figure 11.27 Box plot of the response of individual sensors with sensing attributes: (a) shift in frequency, (b) response time, and (c) recovery time.

was observed that the random forest was best suited for the particular problem and gave more than 95% testing accuracy in classification of 5 VOCs and their concentrations [71].

11.6 Conclusions

The QTF has proved itself to be a very useful scientific tool and continues to be used as sensors in various fields of study. A complete review of all the uses of QTFs is too arduous a task and beyond the scope of this book. However, it has firmly been established as a nanosensor capable of measuring physical parameters such as temperature and pressure, chemical analytes such as gases and vapors, and even biological analytes such as proteins and antibodies. Table 11.2 lists such analytes and the corresponding sensing polymers that are used to functionalize the QTFs that are reported

Table 11.2 A list of all materials used to functionalize the QTF sensors and all the analytes detected, as reported in peer reviewed literature.

Analytes/interferents	Sensing polymers
Ethanol	Cellulose nanocrystal–reinforced polymethylmethacrylate Nanoporous polymethylmethacrylate Polymethylmethacrylate Polyvinylcinnamate Nitrocellulose/toluene sulfonamide formaldehyde resin Graphene-decorated polystyrene Ethyl cellulose Polystyrene Apiezon Wax Poly(acrylic acid) Super glue Nail polish
Toluene	Polymethylmethacrylate Molecularly imprinted polymers Apiezon Wax Fluoroalcohol polysiloxane Polystyrene Ethyl cellulose Polycarbonate Poly(vinyl chloride)
Humidity	Polymethylmethacrylate Iron-doped amorphous magnesium silicate Butylcellulose Poly(vinylphosphonic acid) Poly(acrylic acid) Poly(acrylamide) Poly(styrene sulfonate)
Dimethyl disulfide	Polystyrene Poly(3-hexylthiophene-2,5-diyl) Ethyl cellulose Polycarbonate Poly(vinyl chloride) Polymethylmethacrylate Apiezon Wax
Benzyl acetate	Polystyrene Poly(3-hexylthiophene-2,5-diyl) Ethyl cellulose Polycarbonate Poly(vinyl chloride) Polymethylmethacrylate Apiezon Wax
Isopropanol	Polystyrene Ethyl cellulose Polycarbonate Poly(vinyl chloride) Polymethylmethacrylate Apiezon Wax

(Continued)

Table 11.2 (Continued)

Analytes/interferents	Sensing polymers
Acetone	Polystyrene Ethyl cellulose Polycarbonate Poly(vinyl chloride) Polymethylmethacrylate Apiezon Wax
Acetaldehyde	Polystyrene Ethyl cellulose Polycarbonate Poly(vinyl chloride) Polymethylmethacrylate Apiezon Wax
General-purpose cleaner (3M)	Polystyrene Ethyl cellulose Polycarbonate Poly(vinyl chloride) Polymethylmethacrylate Apiezon Wax
Bathroom disinfectant cleaner (3M)	Polystyrene Ethyl cellulose Polycarbonate Poly(vinyl chloride) Polymethylmethacrylate Apiezon Wax
Body spray (Axe)	Polystyrene Ethyl cellulose Body mist (Coty, US)
Wax cleaner (Butchers' Amplify, 3M)	Polycarbonate Poly(vinyl chloride) Polymethylmethacrylate Apiezon Wax
Water	Polystyrene Ethyl cellulose Polycarbonate Poly(vinyl chloride) Polymethylmethacrylate Apiezon Wax
Methanol	Polystyrene Ethyl cellulose Polycarbonate Poly(vinyl chloride) Polymethylmethacrylate Apiezon Wax
Benzene	Polystyrene Ethyl cellulose Polycarbonate Poly(vinyl chloride) Polymethylmethacrylate Apiezon Wax
Xylene	Polyvinyl alcohol Polyvinyl alcohol with nanoadditives Polystyrene Polymethylmethacrylate Polyvinylpyrrolidone
Ethylbenzene	Molecularly imprinted polymers Fluoroalcohol polysiloxane Polymethylmethacrylate Apiezon Wax

(Continued)

Table 11.2 (Continued)

Analytes/interferents	Sensing polymers
p-Nitroethylbenzene	Molecularly imprinted polymers Fluoroalcohol polysiloxane
	Apiezon Wax
Temperature/infrared	Molecularly imprinted polymers Fluoroalcohol polysiloxane
	Apiezon Wax
	Parafilm
	Polystyrene
	Polystyrene–fluoroalcohol polycarbosilanes mixture
	Poly(*N*-isopropylacrylamide)
	Polyvinylchloride
Chloroform	Ethyl cellulose
	Hydroxypropyl cellulose
Ozone	Polybutadiene
	Silica nanoparticles modified with butadiene silane
$PM_{2.5}$	Polyvinylpyrrolidone-based Elmer's repositionable stick
Methylsulfide	Polystyrene
Ethanethiol	Polystyrene
Hydrochloric acid	Polystyrene
Sodium hypochlorite	Polystyrene
Ammonia	Polystyrene

in peer reviewed literature. Undoubtedly, advancements in the field of nanotechnology and material engineering have enabled the development of QTF-based chemical sensors as most of the modifications of the QTF involve a certain nanomaterial that would act as a sensing medium for those gases and vapors.

The aim of this chapter is not to provide an exhaustive list of all the ways in which QTFs have been used as a chemical sensor, as there are other experimental and theoretical studies done on QTF-based systems, which are not mentioned here. Rather, the aim is to give a brief introduction of how QTFs have been developed as sensors in general and chemical sensors in particular. There are endless possibilities of polymer–analyte combinations that can be explored to create highly sensitive and selective sensors. There are also a myriad possibilities in terms of sensing devices, which can augment and support the sensor by increasing lifetime and stability. It is also worth mentioning that all the work represented here are laboratory studies, and even though various prototypes have been developed, none has been commercialized in any grade or form till date. Therefore, QTF-based nanosensors are a very niche field of study, which is in its nascency and has much unbridled potential as highly accurate, sensitive, and selective sensors.

References

1 Wang, Z.L. (2003). Nanobelts, nanowires and nanodiskettes of semiconducting oxides – from materials to nanodevices. *Adv. Mater.* 15 (5): 432–436.

2 Huang, J., Viriji, S., Weiller, B.H., and Kaner, R.B. (2004). Nanostructured polyaniline sensors. *Chem. Eur. J.* 10 (6): 1314–1319.

3 Zhang, Y., Ichihashi, T., Ladn, E. et al. (1999). Heterostructures of single-walled carbon nanotubes and carbide nanorods. *Science* 285: 1719–1722.

4 Ding, B., Wang, M., Yu, J., and Sun, G. (2009). Gas sensors based on electrospun nanofibers. *Sensors* 9: 1609–1624.

5 Liu, Y., Parisi, J., Suna, X., and Lei, Y. (2014). Solid-state gas sensors for high temperature applications – a review. *J. Mater. Chem. A* 2 (26): 9919–9943.

6 Zhang, L., Qi, H., Wang, Y. et al. (2014). Effective visualization assay for alcohol content sensing and methanol differentiation with solvent stimuli-responsive supramolecular ionic materials. *Anal. Chem.* 86 (15): 7280–7285.

7 McAlpine, M.C., Ahmad, H., Wang, D., and Heath, J.R. (2007). Highly ordered nanowire arrays on plastic substrates for ultrasensitive flexible chemical sensors. *Nat. Mater.* 6: 379–384.

8 Grayson, A.C.R., Shawgo, R.S., Johnson, A.M. et al. (2004). A bioMEMS review: MEMS technology for physiologically integrated devices. *Proc. IEEE* 92 (1): 6–21.

9 Polla, D.L., Erdman, A.G., Robbins, W.P. et al. (2000). Microdevices in medicine. *Annu. Rev. Biomed. Eng.* 2: 551–576.

10 Sadek, A.Z., Wlodarski, W., Shin, K. et al. (2006). A layered surface acoustic wave gas sensor based on a polyaniline/In_2O_3 nanofibre composite. *Nanotechnology* 17 (17): 4488–4492.

11 Li, D., Huang, J., and Kaner, R.B. (2009). Polyaniline nanofibers: a unique polymer nanostructure for versatile applications. *Acc. Chem. Res.* 42 (1): 135–145.

12 Pejcic, B., Eadington, P., and Ross, A. (2007). Environmental monitoring of hydrocarbons: a chemical sensor perspective. *Environ. Sci. Technol.* 41: 6333–6342.

13 Wang, F., Gu, H., and Swager, T.M. (2008). Carbon nanotube/polythiophene chemiresistive sensors for chemical warfare agents. *J. Am. Chem. Soc.* 130: 5392–5393.

14 Stetter, J.R., Penrose, W.R., and Yao, S. (2003). Sensors, chemical sensors, electrochemical sensors, and ECS. *J. Electrochem. Soc.* 150 (2): S11–S16.

15 Mansour, M.A., Connick, W.B., Lachicotte, R.J. et al. (1998). Linear chain au(i) dimer compounds as environmental sensors: a luminescent switch for the detection of volatile organic compounds. *J. Am. Chem. Soc.* 120 (6): 1329–1330.

16 Sexton, K., Adgate, J.L., Ramachandran, G. et al. (2004). Comparison of personal, indoor, and outdoor exposures to hazardous air pollutants in three urban communities. *Environ. Sci. Technol.* 38 (2): 423–430.

17 Begerow, J., Jermann, E., Koch, T., and Dunemann, L. (1996). Screening method for the determination of 28 volatile organic compounds in indoor and outdoor

air at environmental concentrations using dual-column capillary gas chromatography with tandem electron- capture-flame ionization detection. *J. Chromatogr. A* 749: 181–191.

18 Srivastava, A.K. (2003). Detection of volatile organic compounds (VOCs) using SNO_2 gas- sensor array and artificial neural network. *Sens. Actuators B Chem.* 96: 24–37.

19 Du, W.Y. (2014). *Resistive, Capacitive, Inductive, and Magnetic Sensor Technologies*, 1e. Boca Raton: CRC Press.

20 Fraden, J. (2004). *Handbook of Modern Sensors: Physics, Designs and Applications*, 3e. Weinheim: Springer.

21 Salvagione, H.J., Martneza, G., and Gmeza, M.A. (2009). Synthesis of poly(vinyl alcohol)/reduced graphite oxide nanocomposites with improved thermal and electrical properties. *J. Mater. Chem.* 19: 5027–5032.

22 Wu, G.H., Datar, R.H., Hansen, K.M. et al. (2001). Bioassay of prostate-specific antigen (PSA) using microcantilevers. *Nat. Biotechnol.* 19: 856–860.

23 Subramanian, A., Oden, P.I., Kennel, S.J. et al. (2002). Glucose biosensing using an enzyme-coated microcantilever. *Appl. Phys. Lett.* 81: 385–387.

24 Arntz, Y., Seelig, J.D., Lang, H.P. et al. (2003). Label-free protein assay based on a nanomechanical cantilever array. *Nanotechnology* 14: 86–90.

25 Yun, M., Lee, S., Yim, C. et al. (2013). Suspended polymer nanobridge on a quartz resonator. *Appl. Phys. Lett.* 103: 053109–1–053109–4.

26 Kosterev, A.A., Tittel, F.K., Serebryakov, D.V. et al. (2005). Applications of quartz tuning forks in spectroscopic gas sensing. *Rev. Sci. Instrum.* 76 (4): 043105–1–043105–9.

27 Valkering, A., Mares, A., Untiedt, C. et al. (2005). A force sensor for atomic point contacts. *Rev. Sci. Instrum.* 76 (10): 103903–1–103903–5.

28 Ma, Y. (2020). Recent advances in QEPAS and QEPTS based trace gas sensing: a review. *Front. Phys.* 8: 268.

29 Forrer, M.P. (1969). A flexure-mode quartz for an electronic wrist-watch. *23rd Annual Symposium on Frequency Control*, Atlantic City, NJ, USA (May 1969). IEEE, pp. 157–162.

30 Yoda, H., Ikeda, H., and Yamaber, Y. (1972). Low power crystal oscillator for electronic wrist watch. *26th Annual Symposium on Frequency Control*, Atlantic City, NJ, USA (June 1972). IEEE, pp. 140–147.

31 Mahgoub, Y., ElGhazaly, A., and El-Badawy, A. (2020). A study on the behavior and characteristics of a quartz tuning fork using finite element method. *12th International Conference on Electrical Engineering (ICEENG)*, Cairo, Egypt (July 2020). IEEE pp. 325–330.

32 Ou, C., Ding, J., Li, X. et al. (2020). Research on liquid characteristic sensing technology of quartz tuning fork based on COMSOL multiphysics. *J. Phys. Conf. Ser.* 1650: 022034.

33 Dinger, R. (1982). The torsional tuning fork as a temperature sensor. *36th Annual Symposium on Frequency Control*, Philadelphia, PA, USA (June 1982). IEEE. pp. 265–274.

34 Christen, M. (1983). Air and gas damping of quartz tuning forks. *Sensors Actuators* 4: 555–564.

35 Günther, P., Fischer, U.C., and Dransfeld, K. (1989). Scanning near-field acoustic microscopy. *Appl. Phys. B* 48 (1): 89–92.

36 Karra, K. and Grober, R.D. (1995). Piezoelectric tip-sample distance control for near field optical microscopes. *Appl. Phys. Lett.* 66 (14): 1842–1844.

37 Karra, K. and Grober, R.D. (1995). Piezo-electric tuning fork tipsample distance control for near field optical microscopes. *Ultramicroscopy* 61 (1): 197–205.

38 Atia, W.A. and Davis, C.C. (1997). A phase-locked shear-force microscope for distance regulation in near-field optical microscopy. *Appl. Phys. Lett.* 70 (4): 405–407.

39 Edwards, H., Taylor, L., Duncan, W., and Melmed, A.J. (1997). Fast, high-resolution atomic force microscopy using a quartz tuning fork as actuator and sensor. *Appl. Phys. Lett.* 82 (3): 980–984.

40 Giessibl, F.J. (1998). High-speed force sensor for force microscopy and profilometry utilizing a quartz tuning fork. *Appl. Phys. Lett.* 73 (6): 3956–3958.

41 Giessibl, F.J. (2000). Atomic resolution on Si (111)–(7 × 7) by noncontact atomic force microscopy with a force sensor based on a quartz tuning fork. *Appl. Phys. Lett.* 76 (11): 1470–1472.

42 Giessibl, F.J. (2003). Advances in atomic force microscopy. *Rev. Mod. Phys.* 75: 949–983.

43 Su, X., Dai, C., Zhang, J., and O'Shea, S.J. (2002). Quartz tuning fork biosensor. *Biosens. Bioelectron.* 17: 111–117.

44 Zhang, J. and O'Shea, S.J. (2003). Tuning forks as micromechanical mass sensitive sensors for bio- or liquid detection. *Sens. Actuators B Chem.* 94 (1): 65–72.

45 Boussaad, S. and Tao, N.J. (2003). Polymer wire chemical sensor using a microfabricated tuning fork. *Nano Lett.* 3 (8): 1173–1176.

46 Ren, M., Forzani, E.S., and Tao, N. (2005). Chemical sensor based on microfabricated wristwatch tuning forks. *Anal. Chem.* 77: 2700–2707.

47 Tsow, F. and Tao, N. (2007). Microfabricated tuning fork temperature and infrared sensor. *Appl. Phys. Lett.* 90: 174102-1–174102-3.

48 Tsow, F., Forzani, E.S., and Tao, N. (2008). Frequency-coded chemical sensors. *Anal. Chem.* 80: 606–611.

49 A. Rai, F. Tsow, S. Nassirpour, J. Bankers, M. Spinatsch, M. P. He, E. Forzani, and N. J. Tao, "Selective detection of sulfur derivatives using microfabricated tuning fork-based sensors," *Sens. Actuators B*, vol. 140, no. 2, pp. 490–499, 2009.

50 Wang, R., Tsow, F., Zhang, X. et al. (2009). Real-time ozone detection based on a microfabricated quartz crystal tuning fork sensor. *Sensors* 9: 5655–5663.

51 Tsow, F., Forzani, E., Rai, A. et al. (2009). A wearable and wireless sensor system for real-time monitoring of toxic environmental volatile organic compounds. *IEEE Sens. J.* 9 (12): 1734–1740.

52 Iglesias, R.A., Tsow, F., Wang, R. et al. (2009). Hybrid separation and detection device for analysis of benzene, toluene, ethylbenzene, and xylenes in complex samples. *Anal. Chem.* 81 (21): 8930–8935.

53 Negi, I., Tsow, F., Tanwar, K. et al. (2011). Novel monitor paradigm for real-time exposure assessment. *J. Expo. Sci. Environ. Epid.* 21 (4): 419–426.

54 Chen, C., Campbell, K.D., Negi, I. et al. (2012). A new sensor for the assessment of personal exposure to volatile organic compounds. *Atmos. Environ.* 54: 679–687.

55 Chen, C., Tsow, F., Campbell, K.D. et al. (2013). A wireless hybrid chemical sensor for detection of environmental volatile organic compounds. *IEEE Sens. J.* 13 (5): 1748–1755.

56 Deng, Y., Chen, C., Qin, X. et al. (2015). Aging effect of a molecularly imprinted polymer on a quartz tuning fork sensor for detection of volatile organic compounds. *Sens. Actuators B Chem.* 211: 25–32.

57 Deng, Y., Liu, N.-Y., Tsow, F. et al. (2017). Adsorption thermodynamic analysis of a quartz tuning fork based sensor for volatile organic compounds detection. *ACS Sens.* 2 (11): 1662–1668.

58 Deng, Y., Chen, C., Tsow, F., Xian, X., and Forzani, E. Unraveling fabrication and calibration of wearable gas monitor for use under free-living conditions. *38th Annual International Conference of the IEEE Engineering in Medicine and Biology Society (EMBC)*. August. IEEE. pp. 4897–4900.

59 Deng, Y., Xian, X., Tsow, F. et al. (2016). A novel wireless wearable volatile organic compound (VOC) monitoring device with disposable sensors. *Sensors* 16 (12): 2060–1–2060–13.

60 Qin, X., Xian, X., Deng, Y. et al. (2018). Micro quartz tuning fork-based $PM_{2.5}$ sensor for personal exposure monitoring. *IEEE Sens. J.* 19 (7): 2482–2489.

61 Jung, N., Yun, M., and Jeon, S. (2013). Evaporation of a sessile liquid droplet from the shear or flexural surfaces of a quartz tuning fork. *Appl. Phys. Lett.* 114: 074511–1–074511–5.

62 Yun, M., Lee, E., Cho, K., and Jeon, S. (2014). Enhanced sensitivity of a microfabricated resonator using a graphene-polystyrene bilayer membrane. *Appl. Phys. Lett.* 105: 073 116– 1–073 116–4.

63 Lee, S., Yun, M., and Jeon, S. (2014). A tunable microresonator sensor based on a photocrosslinking polymer microwire. *Appl. Phys. Lett.* 104: 053506–1–053506–4.

64 Kim, W., Yun, M., Lee, S., and Jeon, S. (2016). Enhanced sensitivity of quartz tuning fork sensors using electrospun polymer wires. *RSC Adv.* 6: 31 131–31 134.

65 Hwang, S., Kim, W., Yoon, H., and Jeon, S. (2017). Performance enhancement of a microfabricated resonator using electrospun nanoporous polymer wire. *ACS Sens.* 2 (9): 1355–1358.

66 Kim, W., Park, E., and Jeon, S. (2020). Performance enhancement of a quartz tuning fork sensor using a cellulose nanocrystal-reinforced nanoporous polymer fiber. *Sensors* 20 (2): 1424–8220.

67 Sampson, S.A., Date, K.S., Panchal, S.V. et al. (2015). Investigation of QTF based gas sensors. *Sens. Actuators B Chem.* 216: 586–594.

68 Sampson, S.A., Panchal, S.V., Date, K.S., and Datar, S.S. (2015). QTF based methanol sensors. *Proceedings of the 2nd International Symposium on Physics and Technology of Sensors (ISPTS)*, Pune, India (8–10 March 2015). IEEE, pp. 303–306.

69 Sampson, S.A., Panchal, S.V., Mishra, A. et al. (2017). Quartz tuning fork based portable sensor for vapor phase detection of methanol adulteration of ethanol by using aniline-doped polystyrene microwires. *Microchim. Acta* 184: 1659–1667.

70 Sampson, A., Panchal, S., Phadke, A. et al. (2018). Quartz tuning fork based sensor for detection of volatile organic compounds: toward breath analysis. *Mater. Res. Exp.* 5 (4): 045407.

71 Panchal, S., Phadke, A., Gopinathan, U., and Datar, S. (2021). Development of a polymer modified quartz tuning fork (QTF) sensor array based volatile organic compound (VOC) classifier. *IEEE Sensors Journal* 1–1. https://doi.org/10.1109/JSEN.2021.3099537.

Index

a
amperometric sensor 32
arch shaped configuration 79
atomic force microscope (AFM) 3

b
bipolar junction transistor (BJT) 99
Brunauer-Emmett- Teller adsorption isotherm analysis (BET) 337
bubble-jet printing model 7, 209

c
calorimetric sensors 265–266
cantilever beams 80–85
carbonization 6
Carboxymethyl Cellulose (CMC) 7, 215
Casimir energy 5
Casimir force 141–142, 149
  carbon nanotubes (CNT)-sheets 143
  carbon nanotubes (CNT) tweezers 144
  circular tweezers 143
  Dirichlet and Neumann modes 148–149
  proximity force approximation 144–148
  rectangular beam plate 142
  rectangular tweezers 143
  wire plate 142
charge transfer band (CTB) 283
chemical bath deposition (CBD) 2
chemical sensors 318–319, 324
chemical vapor deposition (CVD) 29, 225, 340
chemiresistive sensing mechanism 261–262
CNT ink 225–228, 230
Co-precipitation 257
conductive nanomaterials 7, 8
controlled precipitation 67–68
Coulomb scattering 101–102, 114
  limitation 4
current in plane (CIP) 306
current perpendicular to plane (CPP) 306

d
detachment length 149
differential mobility spectrometry (DMS) 261
direct sensors 318
Dirichlet and Neumann modes 148, 149
dodecylamine (DDA) 7
down-conversion photoluminescence 10, 11
down-conversion PL 279
drop-on-demand (DOD) printer 209

e
elastodielectric effects 301
electric arc discharge 225
electrochemical biosensors 35–36
electrochemical methods 251
electrohydrodynamic (EHD) Printing 7, 211

Nanotechnology in Electronics: Materials, Properties, Devices, First Edition.
Edited by Visakh P.M., Artem Semkin, Raneesh Balakrishnan, and Saša Lazović.

electron beam evaporation 264
electron-electron scattering 97
electrophoretic effect 10, 280
electrospinning 6
electrostatic forces
  carbon nanotubes (CNT) tweezers 134–136
  carbon nanotubes (CNTs)-sheet 132–133
  circular tweezers 133, 134
  Mejis-Fokkema model 136, 137
  Palmers model 136
  rectangular beam-plate 131
  rectangular tweezers 132–134
  wire plate 131–132
enzymatic electrochemical biosensors 38, 40
EPDQ 306

f

ferromagnetic (FM) electrodes 307
fiber optic sensors 262–264
field effect transistor (FET) 34–36
flexographic printing 212
fluorescence resonance energy transfer (FRET) 34
fluorine-doped-tin-oxide (FTO) 30
focused ion beam (FIB) 13, 331
Fourier transform infrared spectroscopy (FTIR) 337
freestanding phenomenon
  detachment length 149–150
  surface layer and size-dependent effects 150–151
full scan mode 258

g

Gas Chromatography-Mass Spectrometry (GC-MS) 258–261, 337
gas sensors 318
Gaussian distribution 108
giant magnetoresistance (GMR) 302, 303
glassy carbon electrode (GCE) 35
graphene 25, 26
  -based biosensors 32–33
  based electrochemical biosensors 35–36
  based fluorescent biosensors 46
  -based nano-electronics 30–32
  detection of acute myocardial infarction 36–37
  detection of asthma 37
  detection of cholesterol 41–42
  detection of diabetes 38
  detection of hydrogen peroxide sensing 42
  detection of lung cancer 37
  detection of nucleic acids 47
  detection of pathogens and food toxins 47
  detection of small molecules 46
  detection of toxic heavy metal ions 47
  in electrochemical biosensing platforms 33–34
  enzymes detection 44
  field effect transistors 34
  food toxin sensing 45
  fundamentals of 27–28
  history of 26–27
  heavy metal detection 45
  hydrogen peroxide detection 43
  nano-electronics biosensors 1, 2
  nucleic acid biosensor 43–44
  optical platform for bio-sensing 34
  pesticides detection 45–46
  properties of 27
  synthesis of 28–30
  thermal and electrical properties 25
graphene ink 228, 234
graphene nano-ribbons (GNR) 31
graphene quantum dots (GQDs) 1, 27
gravure printing 211, 212
green chemistry synthesis 2
green synthesis method 3, 68

h

Hänsch reliability model 105
heterogeneous electron transfer (HET) 33
hexamethylenetetramine (HMT) 2

hot carrier degradation (HCD) 97
creation process 104–108
degradation of 108–118
simulation framework of 102–103
static analysis 118–119
transient analysis 119–121
hot carrier injection (HCI) 97
Hummer's method 29
hybrid nanogenerators 75, 76
hydrothermal method 67

i
immunosensor 36
impedometric methods 33
indium-tin-oxide (ITO) 30
Inelastic Electron Tunneling Spectroscopy (IETS) 307
inkjet printing 209
integrated electrode (IDE) 35
inverse direct gravure printing 211

j
JLFET 98
Judd-Ofelt analysis 281
Julliere's model 304

k
Klaassen mobility model 102

l
laser ablation 225
laser beam pulse 67
Lennard-Jones potential 139
li-ion battery 31
line edge roughness (LER) 100
liquid-phase synthesis 67
lossy mode resonance (LMR) 263

m
magnetic dipole transition 284
magnetic tunnel junction (MTJ) 12, 304
magnetoelastic effects 301
magnetoelectric (ME) effects 301
magnetoelectric coupling 301, 305
magnetoresistance 302
mammography 250
mass-selected ion injected 258
mechanochemistry method 66
Mejis-Fokkema model 136, 137
metal oxide semiconductors 8, 10
metallic based inks 214–215
micro-electromechanical systems (MEMS) 4
microcantilever 319
microfabricated tuning fork 13, 331
Miller capacitance 121
molecularly imprinted polymers (MIPs) 336
MOSFET
reliability-based analysis 3–4
junctionless transistor 98–102
creation process 104–108
degradation of 108–118
hot carrier degradation 102–103
multiferroic materials 11–12
magnetoelectric materials 301, 302
spintronics 302
multiferroic tunnel junctions (MFTJs) 307–309
multiple electrospun wires 13, 340
multiple ion monitoring mode 258

n
nano-electromechanical systems (NEMS) 4–5
electrostatic forces 137
Casimir forces 141–149
electrostatic forces 131
freestanding phenomenon 149–151
van der Waals forces 137–141
nano-electronics biosensors 1, 2
nanoelectronics 30
nanogenerator (NG) 3, 71
hybrid 75–76
piezoelectric nanogenerators 71–72
pyroelectric 74–75
triboelectric nanogenerators 73–74
nanomaterials 5–6, 30
based inks 213–215
growth 255–257

needle–wire connections 69
noise margin (NM) 118
non-enzymatic glucose biosensor 40, 41

o

Ohnesorge number 209
organic multiferroic tunnel junctions (OMFTJs) 308
oxide trapping/detrapping process 105

p

Palmer's model 136
piezoelectric nanogenerators (PENGs) 71–72, 76–79
piezoelectric printing model 7, 209
piezomagnetic effects 301
plasma electrolytic oxidation (PEO) 10, 280
polarized neutron reflectometry (PNR) 11, 303
polyacrylamide (PAM) 335
polyacrylic acid 8
polybutadiene 334
polycrystalline graphene 1, 27–28
polymer film bridges 338–342
polymer wire bridges 331–338
polymethyl methacrylate (PMMA) 340
polystyrene (PS) 329
polyvinylcinnamate (PVCN) 13
polyvinylpyrrolidone (PVP) 338
potentiometric methods 33
printable nanomaterials 212
printed electronics (PE) 207
pristine graphene 1, 27
proton transfer reaction mass spectrometry (PTR-MS) 258
proximity force approximation (PFA) 144–148
pulsed laser deposition (PLD) 255
pyroelectric nanogenerators 74, 75

q

QR codes 338
quantum confinement effect 101
quantum mechanical effect 303
quartz crystal microbalance (QCM) 319
quartz enhanced photo-thermal spectroscopy (QEPTS) 321
quartz enhanced photoacoustic spectroscopy (QEPAS) 321
quartz tuning fork (QTF) 12–13
  chemical sensors 318–324
  early development 324–325
  mass loaded sensor 328–331
  modified with polymer film bridges 338–342
  modified with polymer wire bridges 331–338
  sensor 325–329
quick response (QR) codes 338

r

rectangular beam-sheets 138
reduced graphene oxide (RGO) 1, 27, 28
resonant frequency 321
roll-to-roll (R2R) printing 212

s

scanning electron microscope (SEM) 66, 281, 337
Schottky contact 3
screen printed electrode (SPE) 35
screen printing 7, 210
selected ion flow tube-mass spectrometry (SIFT-MS) 257–258, 337
semiconductor metal oxide (SMO) 252
shaft tweezers models 130
silver nanoparticles 215–222
silver nanowire 222–226
SiO_2, graphitization of 27
slot die coating 210–211
sol-gel synthesis 66–67
spin-dependent tunneling (SDT) 304
spin FET 309–311
spin LED 311
spin orbit coupling (SOC) 311
spin valve 306–307
spin-transfer torque (STT) 308

spin–orbit torque field effect transistor (SOTFET) 311
spintronics
  fundamental aspects of 302–303
  giant magnetoresistance 303
  multiferroic tunnel junction 307–309
  spin FET 309–311
  spin LED 311
  spin valve 306–307
  tunneling magnetoresistance 304
spray pyrolysis (SP) 255
Stokes shift 279
substrate carrier injection (SCI) 97
supercapacitors 30
surface plasmon resonance (SPR) sensor 264–265

t
Taylor series expansion 134
Tedlar sampling 337
thermally assisted switching (TAS) 308
thick film deposition 255
thin film deposition 252–254
threshold voltage 118
Tragacanth gum 68
transconductance 97, 110, 111
triboelectric nanogenerators (TENGs) 73–74
tunneling magnetoresistance (TMR) 304

u
ultrasonic irradiation 67
ultrasound 250
up-conversion PL 279

v
Van der Waals force 137
  carbon nanotubes (CNT) tweezers 140–141
  carbon nanotubes (CNT)-sheets 138–139
  circular tweezers 140
  rectangular beam-sheets 138
  rectangular tweezers 139–140
  wire plate 138
vapor transport synthesis 68–69
vapor–liquid solid (VLS) method 3, 70
volatile organic compound (VOC)
  body metabolism for 247–249
  evanescent wave fiber optic sensors 263
  lossy mode resonance fiber optic sensors 263–264
  breast cancer related 250
  components of human breath 249
  composite metal oxides for 252
  sensing, metal oxides for 252
voltage transfer characteristic (VTC) 118
voltammetric methods 33

w
wet-chemical method 2
wet spinning method 6

x
X-ray diffraction (XRD) 66, 281
X-ray energy dispersive spectroscopy (EDS) 281

z
zinc nitrate hexahydrate 2
zinc oxide (ZnO)
  controlled precipitation 67–68
  hydrothermal method 67
  liquid-phase synthesis 67
  mechanochemical method 66
  NPs crystal structure of 63–66
  piezoelectric effect 69–70
  piezoelectric nanogenerators 70–79
  structure of 63
  sol-gel synthesis 66–67
  vapor transport synthesis 68–69
  wurtzite-structure of 72
zinc oxide piezoelectric nanogenerators 3

ZrO_2
chemical and phase composition 281–283
Dy^{3+} 292–293
Er^{3+} 286
Eu^{3+} 283–285
experimental 280–281
films 283
Gd^{3+} 292
Ho^{3+} 288
Ln^{3+} 295
morphology 281–283
Pr^{3+} 290–292
Sm^{3+} 286
Tb^{3+} 287–288
Tm^{3+} 288–290
undoped 283